KU-216-617

PENGUIN BOOKS

SEA OF CORTEZ

Born in Salinas, California, in 1902, JOHN STEINBECK grew up in a fertile agricultural valley about twenty-five miles from the Pacific Coast—and both valley and coast would serve as settings for some of his best fiction. In 1919 he went to Stanford University, where he intermittently enrolled in literature and writing courses until he left in 1925 without taking a degree. During the next five years he supported himself as a laborer and journalist in New York City and then as a caretaker for a Lake Tahoe estate, all the time working on his first novel, *Cup of Gold* (1929). After marriage and a move to Pacific Grove, he published two California fictions, *The Pastures of Heaven* (1932) and *To a God Unknown* (1933), and worked on short stories later collected in *The Long Valley* (1938). Popular success and financial security came only with *Tortilla Flat* (1935), stories about Monterey's paisanos. A ceaseless experimenter throughout his career, Steinbeck changed courses regularly. Three powerful novels of the late 1930s focused on the California laboring class: *In Dubious Battle* (1936), *Of Mice and Men* (1937), and the book considered by many his finest, *The Grapes of Wrath* (1939). Early in the 1940s, Steinbeck became a filmmaker with *The Forgotten Village* (1941) and a serious student of marine biology with *Sea of Cortez*. He devoted his services to the war, writing *Bombs Away* (1942) and the controversial play-novelette *The Moon Is Down* (1942). *Cannery Row* (1945), *The Wayward Bus* (1947), *The Pearl* (1947), *A Russian Journal* (1948), another experimental drama, *Burning Bright* (1950), and *The Log from the* Sea of Cortez (1951) preceded publication of the monumental *East of Eden* (1952), an ambitious saga of the Salinas Valley and his own family's history. The last decades of his life were spent in New York City and Sag Harbor with his third wife, with whom he traveled widely. Later books include *Sweet Thursday* (1954), *The Short Reign of Pippin IV: A Fabrication* (1957), *Once There Was a War* (1958), *The Winter of Our Discontent* (1961), *Travels with Charley in Search of America* (1962), *America and Americans* (1966), and the posthumously published *Journal of a Novel: The* East of Eden *Letters* (1969), *Viva Zapata!* (1975), *The Acts of King Arthur and His Noble Knights* (1976), and *Working Days: The Journals of* The Grapes of Wrath (1989). He died in 1968, having won a Nobel Prize in 1962.

EDWARD F. RICKETTS was director of the Pacific Biological Laboratories, an institution which supplied marine specimens for many of the largest colleges and research laboratories in the country. He also coauthored, with Jack Calvin, *Between Pacific Tides* (1939).

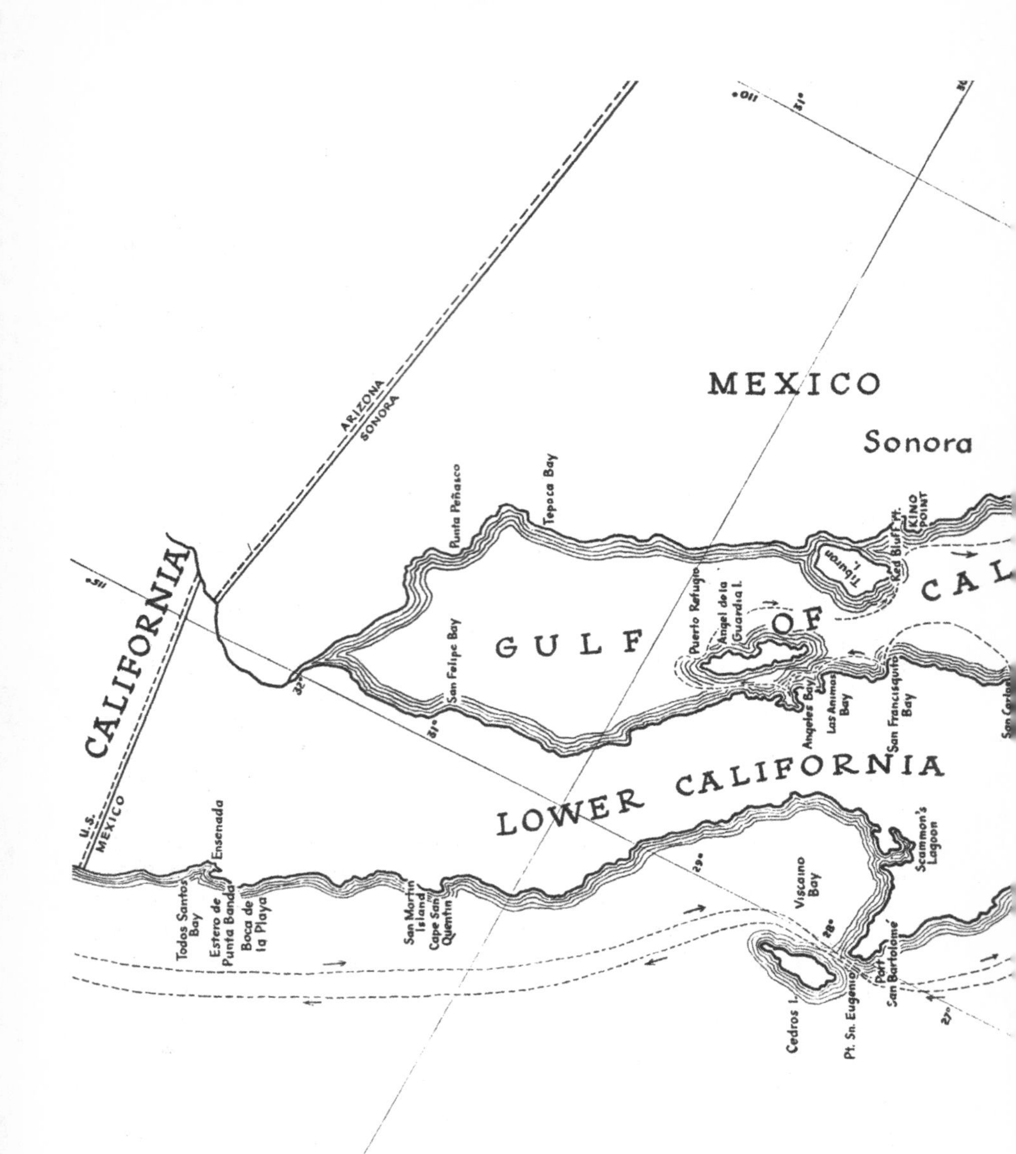
MEXICO
Sonora
GULF OF CAL
LOWER CALIFORNIA
CALIFORNIA
ARIZONA
SONORA
U.S.
MEXICO
Punta Peñasco
Tepoca Bay
San Felipe Bay
Puerto Refugio
Angel de la Guardia I.
Tiburon I.
Red Bluff Pt.
KINO POINT
Angeles Bay
Las Animas Bay
San Francisquito Bay
Ensenada
Todos Santos Bay
Estero de Punta Banda
Boca de la Playa
San Martin Island
Cape San Quentin
Viscaino Bay
Scammon's Lagoon
Cedros I.
Pt. Sn. Eugenio
Port San Bartolomé
32°
31°
29°
28°
27°
110°
115°

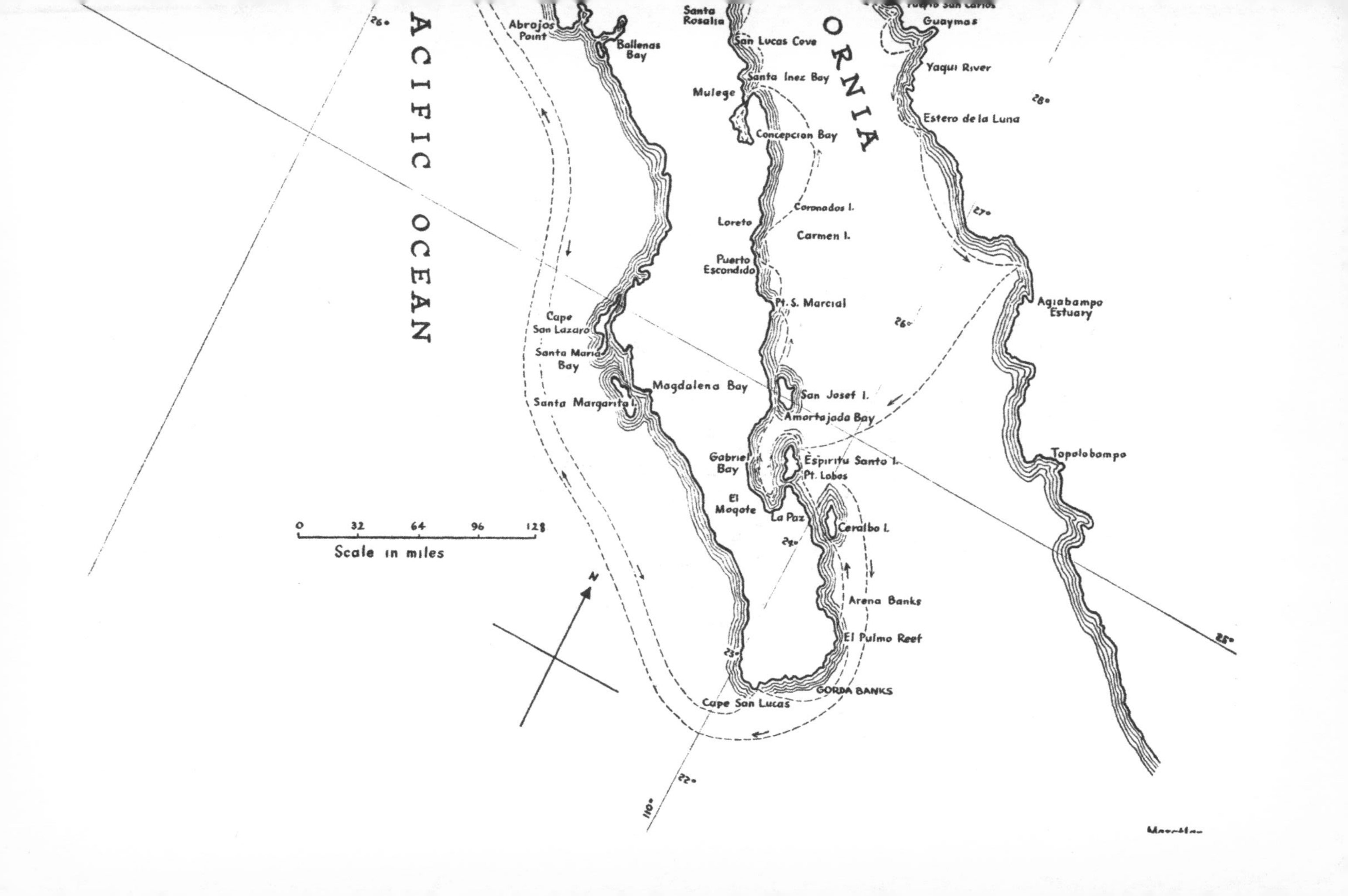

ORNIA
ACIFIC OCEAN
Guaymas
Yaqui River
Estero de la Luna
Agiabampo Estuary
Topolobampo
Santa Rosalia
San Lucas Cove
Santa Inez Bay
Mulege
Concepcion Bay
Coronados I.
Carmen I.
Loreto
Puerto Escondido
Pt. S. Marcial
San Josef I.
Amortajada Bay
Gabriel Bay
Espiritu Santo I.
Pt. Lobos
El Mogote
La Paz
Ceralbo I.
Arena Banks
El Pulmo Reef
GORDA BANKS
Cape San Lucas
Abrajos Point
Ballenas Bay
Cape San Lazaro
Santa Maria Bay
Magdalena Bay
Santa Margarita I.
26°
27°
28°
25°
24°
23°
22°
110°
0
32
64
96
128
Scale in miles
N

BY JOHN STEINBECK

FICTION

Cup of Gold
The Pastures of Heaven
To a God Unknown
Tortilla Flat
In Dubious Battle
Saint Katy the Virgin
Of Mice and Men
The Red Pony
The Long Valley
The Moon Is Down
Cannery Row
The Wayward Bus
The Pearl
Burning Bright
East of Eden
Sweet Thursday
The Winter of Our Discontent
The Short Reign of Pippin IV
The Grapes of Wrath
Las uvas de la ira (Spanish-language edition of *The Grapes of Wrath*)
The Acts of King Arthur and His Noble Knights

NONFICTION

Sea of Cortez: A Leisurely Journal of Travel and Research
(in collaboration with Edward F. Ricketts)
Bombs Away: The Story of a Bomber Team
A Russian Journal *(with pictures by Robert Capa)*
The Log from the *Sea of Cortez*
Once There Was a War
Travels with Charley in Search of America
America and Americans
Journal of a Novel: The *East of Eden* Letters
Working Days: The Journals of *The Grapes of Wrath*

PLAYS

Of Mice and Men
The Moon Is Down

COLLECTIONS

The Portable Steinbeck
The Short Novels of John Steinbeck
Steinbeck: A Life in Letters

OTHER WORKS

The Forgotten Village (documentary)
Zapata (includes the screenplay of *Viva Zapata!*)

CRITICAL LIBRARY EDITION

The Grapes of Wrath (edited by Peter Lisca)

Sea of Cortez

A LEISURELY JOURNAL OF TRAVEL AND RESEARCH

WITH A SCIENTIFIC APPENDIX
COMPRISING MATERIALS FOR A SOURCE BOOK
ON THE MARINE ANIMALS
OF THE PANAMIC FAUNAL PROVINCE

BY

John Steinbeck

AND

Edward F. Ricketts

PENGUIN BOOKS

PENGUIN BOOKS

Published by the Penguin Group
Penguin Group (USA) Inc., 375 Hudson Street,
New York, New York 10014, U.S.A.
Penguin Group (Canada), 90 Eglinton Avenue East, Suite 700, Toronto,
Ontario, Canada M4P 2Y3 (a division of Pearson Penguin Canada Inc.)
Penguin Books Ltd, 80 Strand, London WC2R 0RL, England
Penguin Ireland, 25 St Stephen's Green, Dublin 2, Ireland
(a division of Penguin Books Ltd)
Penguin Group (Australia), 250 Camberwell Road, Camberwell, Victoria 3124,
Australia (a division of Pearson Australia Group Pty Ltd)
Penguin Books India Pvt Ltd, 11 Community Centre, Panchsheel Park,
New Delhi – 110 017, India
Penguin Group (NZ), 67 Apollo Drive, Rosedale, North Shore 0632,
New Zealand (a division of Pearson New Zealand Ltd)
Penguin Books (South Africa) (Pty) Ltd, 24 Sturdee Avenue,
Rosebank, Johannesburg 2196, South Africa

Penguin Books Ltd, Registered Offices:
80 Strand, London WC2R 0RL, England

First published in the United States of America by The Viking Press 1941
Published in Penguin Books 2009

1 3 5 7 9 10 8 6 4 2

ISBN 978-0-14-311721-6

Printed in the United States of America

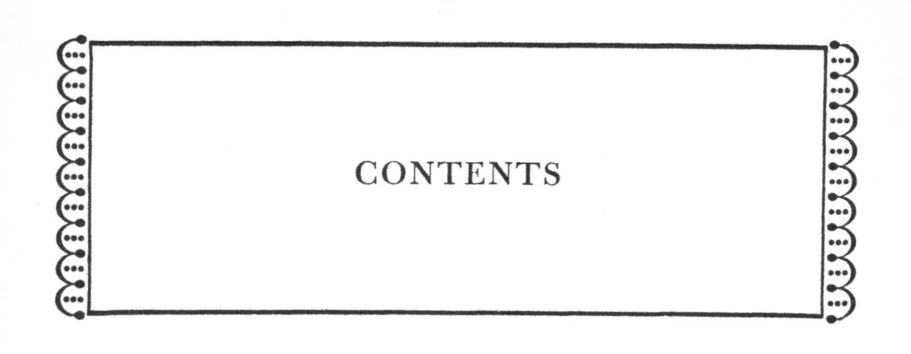
CONTENTS

ILLUSTRATIONS

ALL ILLUSTRATIONS ARE GROUPED FOLLOWING PAGE 278

Color Photographs

Drawings

Drawings

Black and White Photographs

Black and White Photographs

Charts

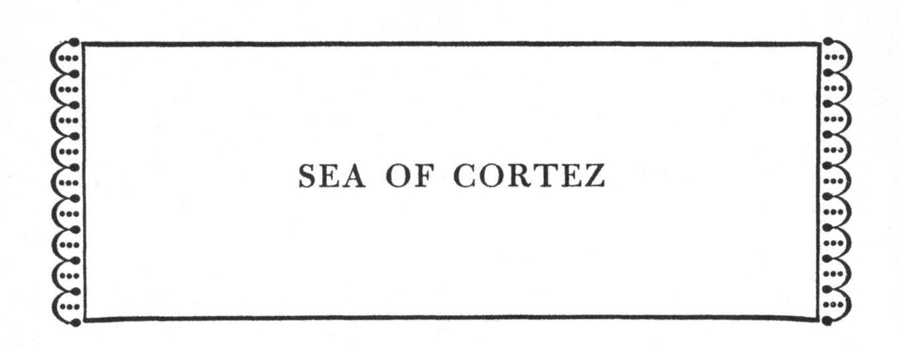

SEA OF CORTEZ

INTRODUCTION

THE design of a book is the pattern of a reality controlled and shaped by the mind of the writer. This is completely understood about poetry or fiction, but it is too seldom realized about books of fact. And yet the impulse which drives a man to poetry will send another man into the tide pools and force him to try to report what he finds there. Why is an expedition to Tibet undertaken, or a sea bottom dredged? Why do men, sitting at the microscope, examine the calcareous plates of a sea-cucumber, and, finding a new arrangement and number, feel an exaltation and give the new species a name, and write about it possessively? It would be good to know the impulse truly, not to be confused by the "services to science" platitudes or the other little mazes into which we entice our minds so that they will not know what we are doing.

We have a book to write about the Gulf of California. We could do one of several things about its design. But we have decided to let it form itself: its boundaries a boat and a sea; its duration a six weeks' charter time; its subject everything we could see and think and even imagine; its limits—our own without reservation.

We made a trip into the Gulf; sometimes we dignified it by calling it an expedition. Once it was called the Sea of Cortez, and that is a better-sounding and a more exciting name. We stopped in many little harbors and near barren coasts to collect and preserve the marine invertebrates of the littoral. One of the reasons we gave ourselves for this trip—and when we used this reason, we called the trip an expedition—was to observe the distribution of invertebrates, to see and to record their kinds and numbers, how they lived together, what they ate, and how they reproduced. That plan was simple, straight-forward, and only a part of the truth. But we did tell the truth to ourselves. We were curious. Our curiosity was not limited, but was as wide and horizonless as that of Darwin or Agassiz or Linnaeus or Pliny. We wanted to see everything our eyes would accommodate, to think what we could, and, out of our seeing and thinking, to build some kind of structure in modeled imitation of the observed reality. We knew that what we would see and record and construct would be warped, as all knowledge patterns are warped, first, by the collective pressure and stream of our time and race, second by the thrust of our individual personalities. But knowing this, we might not fall into too many holes—we might maintain some balance between our warp and the separate thing, the external reality. The oneness of these two might take its contribution from both. For example: the Mexican sierra has "XVII–15–IX" spines in the dorsal fin. These can easily be counted. But if the sierra strikes hard on the line so that our hands are burned, if the fish sounds and nearly escapes and finally comes in over the rail, his colors pulsing and his tail beating the air, a whole new relational externality has come into being—an entity which is more than the sum of the fish plus the fisherman. The only way to count the spines of the sierra unaffected by this second relational reality is to sit in a laboratory, open an evil-smelling jar, remove a stiff colorless fish from formalin solution, count the spines, and write the truth "D. XVII–15–IX." There you have recorded a reality which cannot be assailed—probably the least important reality concerning either the fish or yourself.

It is good to know what you are doing. The man with his

pickled fish has set down one truth and has recorded in his experience many lies. The fish is not that color, that texture, that dead, nor does he smell that way.

Such things we had considered in the months of planning our expedition and we were determined not to let a passion for unassailable little truths draw in the horizons and crowd the sky down on us. We knew that what seemed to us true could be only relatively true anyway. There is no other kind of observation. The man with his pickled fish has sacrificed a great observation about himself, the fish, and the focal point, which is his thought on both the sierra and himself.

We suppose this was the mental provisioning of our expedition. We said, "Let's go wide open. Let's see what we see, record what we find, and not fool ourselves with conventional scientific strictures. We could not observe a completely objective Sea of Cortez anyway, for in that lonely and uninhabited Gulf our boat and ourselves would change it the moment we entered. By going there, we would bring a new factor to the Gulf. Let us consider that factor and not be betrayed by this myth of permanent objective reality. If it exists at all, it is only available in pickled tatters or in distorted flashes. Let us go," we said, "into the Sea of Cortez, realizing that we become forever a part of it; that our rubber boots slogging through a flat of eelgrass, that the rocks we turn over in a tide pool, make us truly and permanently a factor in the ecology of the region. We shall take something away from it, but we shall leave something too." And if we seem a small factor in a huge pattern, nevertheless it is of relative importance. We take a tiny colony of soft corals from a rock in a little water world. And that isn't terribly important to the tide pool. Fifty miles away the Japanese shrimp boats are dredging with overlapping scoops, bringing up tons of shrimps, rapidly destroying the species so that it may never come back, and with the species destroying the ecological balance of the whole region. That isn't very important in the world. And six thousand miles away the great bombs are falling on London and the stars are not moved thereby. None of it is important or all of it is.

We determined to go doubly open so that in the end we could, if we wished, describe the sierra thus: "D. XVII–15–IX; A. II–15–IX," but also we could see the fish alive and swimming, feel it plunge against the lines, drag it threshing over the rail, and even finally eat it. And there is no reason why either approach should be inaccurate. Spine-count description need not suffer because another approach is also used. Perhaps out of the two approaches, we thought, there might emerge a picture more complete and even more accurate than either alone could produce. And so we went.

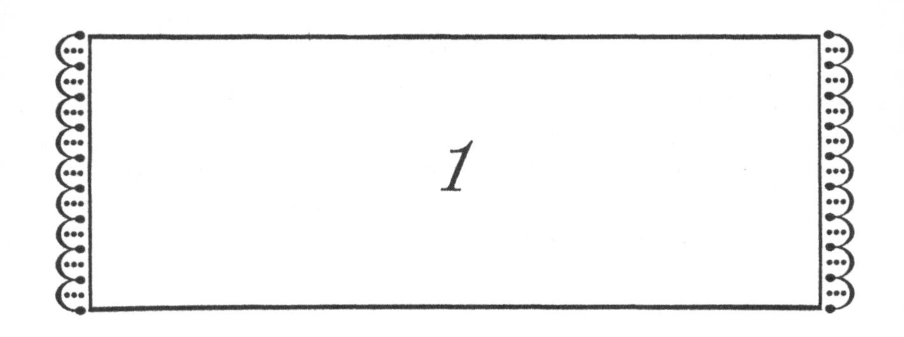

HOW does one organize an expedition: what equipment is taken, what sources read; what are the little dangers and the large ones? No one has ever written this. The information is not available. The design is simple, as simple as the design of a well-written book. Your expedition will be enclosed in the physical framework of start, direction, ports of call, and return. These you can forecast with some accuracy; and in the better-known parts of the world it is possible to a degree to know what the weather will be in a given season, how high and low the tides, and the hours of their occurrence. One can know within reason what kind of boat to take, how much food will be necessary for a given crew for a given time, what medicines are usually needed—all this subject to accident, of course.

We had read what books were available about the Gulf and they were few and in many cases confused. The *Coast Pilot* had not been adequately corrected for some years. A few naturalists with specialties had gone into the Gulf and, in the way of specialists, had seen nothing they hadn't wanted to. Clavigero, a Jesuit of the eighteenth century, had seen more than most and reported what he saw with more accuracy than most. There were some romantic accounts by young people who had gone into the Gulf looking for adventure and, of course, had found it. The same romantic drive aimed at the stockyards would not be disappointed. From the information available, a few facts did emerge. The Sea of Çortez, or the Gulf of California, is a long, narrow, highly dangerous body of water. It is subject to sudden and vicious storms of great intensity. The months of March and April are usually quite calm and dependable and the March–April tides of 1940 were particularly good for collecting in the littoral.

The maps of the region were self-possessed and confident about headlands, coastlines, and depth, but at the edge of the Coast they become apologetic—laid in lagoons with dotted lines, supposed and presumed their boundaries. The *Coast Pilot* spoke as heatedly as it ever does about mirage and treachery of light. Going back from the *Coast Pilot* to Clavigero, we found more visual warnings in his accounts of ships broken up and scattered, of wrecks and wayward currents; of fifty miles of sea more dreaded than any other. The *Coast Pilot,* like an elderly scientist, cautious and restrained, on one side—and the old monk, setting down ships and men lost, and starvation on the inhospitable coasts.

In time of peace in the modern world, if one is thoughtful and careful, it is rather more difficult to be killed or maimed in the outland places of the globe than it is in the streets of our great cities, but the atavistic urge toward danger persists and its satisfaction is called adventure. However, your adventurer feels no gratification in crossing Market Street in San Francisco against the traffic. Instead he will go to a good deal of trouble and expense to get himself killed in the South Seas. In reputedly rough water, he will go in a canoe; he will invade deserts without adequate food and he will expose his tolerant and uninoculated blood to strange viruses. This is adventure. It is possible that his ancestor, wearying of the humdrum attacks of the saber-tooth, longed for the good old days of pterodactyl and triceratops.

We had no urge toward adventure. We planned to collect marine animals in a remote place on certain days and at certain hours indicated on the tide charts. To do this we had, in so far as we were able, to avoid adventure. Our plans, supplies, and equipment had to be more, not less, than adequate; and none of us was possessed of the curious boredom within ourselves which makes adventurers or bridge-players.

Our first problem was to charter a boat. It had to be sturdy and big enough to go to sea, comfortable enough to live on for six weeks, roomy enough to work on, and shallow enough so that little bays could be entered. The purse-seiners of Monterey were ideal for the purpose. They are dependable work boats with comfortable quarters and ample storage room. Furthermore, in March and April the sardine season is over and they are tied up. It would be easy, we thought, to charter such a boat; there must have been nearly a hundred of them anchored in back of the breakwater. We went to the pier and spread the word that we were looking for such a boat for charter. The word spread all right, but we were not overwhelmed with offers. In fact, no boat was offered. Only gradually did we discover the state of mind of the boat owners. They were uneasy about our project. Italians, Slavs, and some Japanese, they were primarily sardine fishers. They didn't even approve of fishermen who fished for other kinds of fish. They frankly didn't believe in the activities of the land—road-building and manufacturing and brick-laying. This was not a matter of ignorance on their part, but of intensity. All the directionalism of thought and emotion that man was capable of went into sardine-fishing; there wasn't room for anything else. An example of this occurred later when we were at sea. Hitler was invading Denmark and moving up towards Norway; there was no telling when the invasion of England might begin; our radio was full of static and the world was going to hell. Finally in all the crackle and noise of the short-wave one of our men made contact with another boat. The conversation went like this:

"This is the *Western Flyer*. Is that you, Johnny?"

"Yeah, that you, Sparky?"

"Yeah, this is Sparky. How much fish you got?"

"Only fifteen tons; we lost a school today. How much fish you got?"

"We're not fishing."

"Why not?"

"Aw, we're going down in the Gulf to collect starfish and bugs and stuff like that."

"Oh, yeah? Well, O.K., Sparky, I'll clear the wave length."

"Wait, Johnny. You say you only got fifteen tons?"

"That's right. If you talk to my cousin, tell him, will you?"

"Yeah, I will, Johnny. *Western Flyer's* all clear now."

Hitler marched into Denmark and into Norway, France had fallen, the Maginot Line was lost—we didn't know it, but we knew the daily catch of every boat within four hundred miles. It was simply a directional thing; a man has only so much. And so it was with the chartering of a boat. The owners were not distrustful of us; they didn't even listen to us because they couldn't quite believe we existed. We were obviously ridiculous.

Now the time was growing short and we began to worry. Finally one boat owner who was in financial difficulty offered his boat at a reasonable price and we were ready to accept when suddenly he raised the price out of question and bolted. He was horrified at what he had done. He raised the price, not to cheat us, but to get out of going.

The boat problem was growing serious when Anthony Berry sailed into Monterey Bay on the *Western Flyer*. The idea was no shock to Tony Berry; he had chartered to the government for salmon tagging in Alaskan waters and was used to nonsense. Besides, he was an intelligent and tolerant man. He knew that he had idiosyncrasies and that some of his friends had. He was willing to let us do any crazy thing that we wanted so long as we (1) paid a fair price, (2) told him where to go, (3) did not insist that he endanger the boat, (4) got back on time, and (5) didn't mix him up in our nonsense. His boat was not busy and he was willing to go. He was a quiet young man, very serious and a good master. He knew some navigation—a rare thing in the fishing fleet—and he had a natural caution which we admired. His boat was new and comfortable and clean, the engines in fine condition. We took the *Western Flyer* on charter.

She was seventy-six feet long with a twenty-five-foot beam; her engine, a hundred and sixty-five horsepower direct reversible Diesel, drove her at ten knots. Her deckhouse had a wheel forward, then combination master's room and radio room, then bunkroom, very comfortable, and behind that the galley. After the galley, a large hatch gave into the fish-hold, and after the hatch were the big turn-table and roller of the purse-seiner. She carried a twenty-foot skiff and a ten-foot skiff. Her engine was a thing of joy, spotlessly clean, the moving surfaces shining and damp with oil and the green paint fresh and new on the housings. The engine-room floor was clean and all the tools polished and hung in their places. One look into the engine-room inspired confidence in the master. We had seen other engines in the fishing fleet and this perfection on the *Western Flyer* was by no means a general thing.

As crew we signed Tex Travis, engineer, and Sparky Enea and Tiny Colletto, seamen. All three were a little reluctant to go, for the whole thing was crazy. None of us had been into the Gulf, although the master had been as far as Cape San Lucas, and the Gulf has a really bad name. It was a thoughtful crew who agreed to go with us.

We could never tell when the change of attitude toward us came, but it came very rapidly. Perhaps it was because Tony Berry was known as a cautious man who would not indulge in nonsense, or perhaps it was pure relief that at last it had been settled. All of a sudden we were overwhelmed with help. We had offers from men to go with us without pay. Sparky was offered a certain price for his job that was more than he would get from us. All he had to do was turn over his job and sit in Monterey and spend the money. But Sparky refused. Our project had become honorable. We had more help than we could use and advice enough to move the navies of the world.

We did not know what our crew thought of the expedition but later, in the field, they became good collectors—a little emotional sometimes, as when Tiny, in outrage at being pinched, declared a war of extermination on the whole Sally Lightfoot species, but on the whole collectors of taste and quickness.

The charter was signed with dignity and reverence. It is impossible to be light-hearted in the face of a ship's charter, for the law has foreseen or remembered the most doleful and arbitrary acts of God and has set them down as possibilities, but in the tone of inevitabilities. Thus, you read what you or the others must do in the case of wreck, or sunken rocks; of death at sea in its most painful and astonishing aspects; of injury to plank and keel; of water shortage and mutiny. Next to marriage settlement or sentence of death, a ship's charter is as portentous a document as has ever been written. Penalties are set down against both parties, and if on some morning the rising sun should find your ship in the middle of the Mojave Desert you have only to look again at the charter to find the blame assigned and the penalty indicated. It took us several hours to get over the solemn feeling the charter put on us. We thought we might live better lives and pay our debts, and one at least of us contemplated for one holy, horrified moment a vow of chastity.

But the charter was signed and food began to move into the *Western Flyer*. It is amazing how much food seven people need to exist for six weeks. Cases of spaghetti, cases and cases of peaches and pineapple, of tomatoes, whole Romano cheeses, canned milk in coveys, flour and cornmeal, gallons of olive oil, tomato paste, crackers, cans of butter and jam, catsup and rice, beans and bacon and canned meats, vegetables and soups in cans; truckloads of food. And all this food was stored eagerly and happily by the crew. It disappeared into cupboards, under little hatches in the galley floor, and many cases went below.

We had done a good deal of collecting, but largely in temperate zones. The equipment for collecting, preserving, and storing specimens was selected on the basis of experience in other waters and of anticipation of difficulties imposed by a hot humid country. In some cases we were right, in others very wrong.

In a small boat, the library should be compact and available. We had constructed a strong, steel-reinforced wooden case, the front of which hinged down to form a desk. This case holds about twenty large volumes and has two filing cases, one for separates (scientific reprints) and one for letters; a small metal box holds pens, pencils, erasers, clips, steel tape, scissors, labels, pins, rubber bands, and so forth. Another compartment contains a three-by-five-inch card file. There are cubby-holes for envelopes, large separates, small separates, typewriter paper, carbon, a box for India ink and glue. The construction of the front makes room for a portable typewriter, drawing board, and T-square. There is a long narrow space for rolled charts and maps. Closed, this compact and complete box is forty-four inches long by eighteen by eighteen; loaded, it weighs between three and four hundred pounds. It was designed to rest on a low table or in an unused bunk. Its main value is compactness, completeness, and accessibility. We took it aboard the *Western Flyer*. There was no table for it to rest on. It did not fit in a bunk. It could not be put on the deck because of moisture. It ended up lashed to the rail on top of the deckhouse, covered with several layers of tarpaulin and roped on. Because of the roll of the boat it had to be tied down at all times. It took about ten minutes to remove the tarpaulin, untie the lashing line, open the cover, squeeze down between two crates of oranges, read the title of the wanted book upside down, remove it, close and lash and cover the box again. But if there had been a low table or a large bunk, it would have been perfect.

For many little errors like this, we have concluded that all collecting trips to fairly unknown regions should be made twice; once to make mistakes and once to correct them. Some of the greatest difficulty lies in the fact that previous collectors have never set down the equipment taken and its success or failure. We propose to rectify this in our account.

The library contained all the separates then available on the Panamic and Gulf fauna. Primary volumes such as Johnson and Snook, Ricketts and Calvin, Russell and Yonge, Flattely and Walton, Keep's *West Coast Shells,* Fisher's three-volume starfish monograph, the Rathbun brachyuran monograph, Schmitt's *Marine Decapod Crustacea of California,* Fraser's *Hydroids,* Barnhart's *Marine Fishes of Southern California, Coast Pilots* for the whole Pacific Coast; charts, both large and small scale, of the whole region to be covered.

The camera equipment was more than adequate, for it was never used. It included a fine German reflex and an 8-mm. movie camera with tripod, light meters, and everything. But we had no camera-man. During low tides we all collected; there was no time to dry hands and photograph at the collecting scene. Later, the anesthetizing, killing, preserving, and labeling of specimens were so important that we still took no pictures. It was an error in personnel. There should be a camera-man who does nothing but take pictures.

Our collecting material at least was good. Shovels, wrecking- and abalone-bars, nets, long-handled dip-nets, wooden fish-kits, and a number of seven-cell flashlights for night collecting were taken. Containers seemed to go endlessly into the hold of the *Western Flyer.* Wooden fish-kits with heads; twenty hard-fir barrels with galvanized hoops in fifteen- and thirty-gallon sizes; cases of gallon jars, quart, pint, eight-ounce, five-ounce, and two-ounce screw-cap jars; several gross of corked vials in four chief sizes, 100x33 mm., six-dram, four-dram, and two-dram sizes. There were eight two-and-a-half-gallon jars with screw caps. And with all these we ran short of containers, and before we were through had to crowd those we had. This was unfortunate, since many delicate animals should be preserved separately to prevent injury.

Of chemicals, we put into the boat a fifteen-gallon barrel of U.S.P. formaldehyde and a fifteen-gallon barrel of denatured alcohol. This was not nearly enough alcohol. The stock had to be replenished at Guaymas, where we bought ten gallons of pure sugar alcohol. We took two gallons of Epsom salts for anesthetization and again ran out and had to buy more in Guaymas. Menthol, chromic acid, and novocain, all for relaxing animals, were included in the chemical kit. Of preparing equipment, there were glass chiton plates and string, lots of rubber gloves, graduates, forceps, and scalpels. Our binocular microscope, Bausch & Lomb A.K.W., was fitted with a twelve-volt light, but on the rolling boat the light was so difficult to handle that we used a spot flashlight instead. We had galvanized iron nested trays of fifteen- to twenty-gallon capacity for gross hardening and preservation. We had enameled and glass trays for the laying out of specimens, and one small examination aquarium.

The medical kit had been given a good deal of thought. There were nembutal, butesin picrate for sunburn, a thousand two-grain quinine capsules, two-percent mercuric oxide salve for barnacle cuts, cathartics, ammonia, mercurochrome, iodine, alcaroid, and, last, some whisky for medicinal purposes. This did not survive our leave-taking, but since no one was ill on the whole trip, it may have done its job very well.

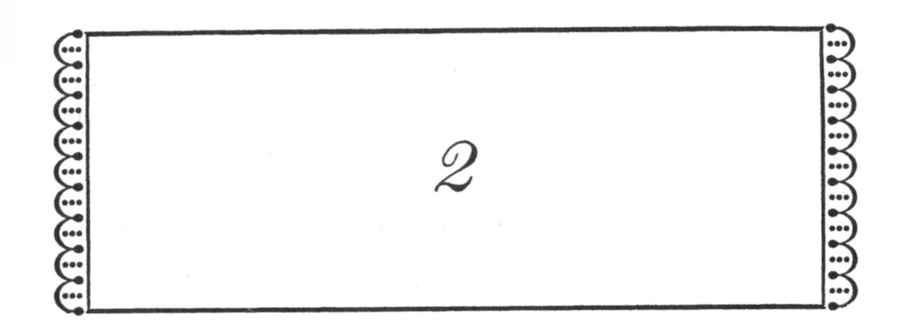

WHAT little time we were not on lists and equipment or in grudging sleep we went to the pier and looked at boats, watched them tied to their buoys behind the breakwater—the dirty boats and the clean painted boats, each one stamped with the personality of its owner. Here, where the discipline was as individual as the owners, every boat was different from every other one. If the stays were rusting and the deck unwashed, paint scraped off and lines piled carelessly, there was no need to see the master; we knew him. And if the lines were coiled and the cables greased and the little luxury of deer horns nailed to the crow's-nest, there was no need to see that owner either. There were deer horns on many of the crow's-nests, and when we asked why, we were told they brought good luck. Out of some ancient time, they brought good luck to these people, most of them out of Sicily, the horns grown sturdily on the structure of their race. If you ask, "Where does the idea come from?" the owner will say, "It brings good luck, we always put them on." And a thousand years ago the horns were on the masts and brought good luck, and probably when the ships of Carthage and Tyre put into the harbors of Sicily, the horns were on the mastheads and brought good luck and no one knew why. Out of some essential race soul the horns come, and not only the horns, but the boats themselves, so that to a man, to nearly all men, a boat more than any other tool he uses is a little representation of an archetype. There is an "idea" boat that is an emotion, and because the emotion is so strong it is probable that no other tool is made with so much honesty as a boat. Bad boats are built, surely, but not many of them. It can be argued that a bad boat cannot survive tide and wave and hence is not worth building, but the same might be said of a bad automobile on a rough road. Apparently the builder of a boat acts under a compulsion greater than himself. Ribs are strong by definition and feeling. Keels are sound, planking truly chosen and set. A man builds the best of himself into a boat—builds many of the unconscious memories of his ancestors. Once, passing the boat department of Macy's in New York, where there are duck-boats and skiffs and little cruisers, one of the authors discovered that as he passed each hull he knocked on it sharply with his knuckles. He wondered why he did it, and as he wondered, he heard a knocking behind him, and another man was rapping the hulls with *his* knuckles, the same tempo—three sharp knocks on each hull. During an hour's observation there no man or boy and few women passed who did not do the same thing. Can this have been an unconscious testing of the hulls? Many who passed could not have been in a boat, perhaps some of the little boys had never seen a boat, and yet everyone tested the hulls, knocked to see if they were sound, and did not even know he was doing it. The observer thought perhaps they and he would knock on any large wooden object that might give forth a resonant sound. He went to the piano department, icebox floor, beds, cedar-chests, and no one knocked on them—only on boats.

How deep this thing must be, the giver and the receiver again; the boat designed through millenniums of trial and error by the human consciousness, the boat which has no counterpart in nature unless it be a dry leaf fallen by accident in a stream. And Man receiving back from Boat a warping of his psyche so that the sight of a boat riding in the water clenches a fist of emotion in his chest. A horse, a beautiful dog, arouses sometimes a quick emotion, but of inanimate things only a boat can do it. And a boat, above all other inanimate things, is personified in man's mind. When we have been steering, the boat has seemed sometimes nervous and irritable, swinging off course before the correction could be made, slapping her nose into the quartering wave. After a storm she has seemed tired and sluggish. Then with the colored streamers set high and snapping, she is very happy, her nose held high and her stern bouncing a little like the buttocks of a proud and confident girl. Some have said they have felt a boat shudder before she struck a rock, or cry when she beached and the surf poured into her. This is not mysticism, but identification; man, building this greatest and most personal of all tools, has in turn received a boat-shaped mind, and the boat, a man-shaped soul. His spirit and the tendrils of his feeling are so deep in a boat that the identification is complete. It is very easy to see why the Viking wished his body to sail away in an unmanned ship, for neither could exist without the other; or, failing that, how it was necessary that the things he loved most, his women and his ship, lie with him and thus keep closed the circle. In the great fire on the shore, all three started at least in the same direction, and in the gathered ashes who could say where man or woman stopped and ship began?

This strange identification of man with boat is so complete that probably no man has even destroyed a boat by bomb or torpedo or shell without murder in his heart; and were it not for the sad trait of self-destruction that is in our species, he could not do it. Only the trait of murder which our species seems to have could allow us the sick, exultant sadness of sinking a ship, for we can murder the things we love best, which are, of course, ourselves.

We have looked into the tide pools and seen the little animals feeding and reproducing and killing for food. We name them and describe them and, out of long watching, arrive at some conclusion about their habits so that we say, "This species typically does thus and so," but we do not objectively observe our own species as a species, although we know the individuals fairly well. When it seems that men may be kinder to men, that wars may not come again, we completely ignore the record of our species. If we used the same smug observation on ourselves that we do on hermit crabs we would be forced to say, with the information at hand, "It is one diagnostic trait of *Homo sapiens* that groups of individuals are periodically infected with a feverish nervousness which causes the individual to turn on and destroy, not only his own kind, but the works of his own kind. It is not known whether this be caused by a virus, some airborne spore, or whether it be a species reaction to some meteorological stimulus as yet undetermined." Hope, which is another species diagnostic trait—the hope that this may not always be—does not in the least change the observable past and present. When two crayfish meet, they usually fight. One would say that perhaps they might not at a future time, but without some mutation it is not likely that they will lose this trait. And perhaps our species is not likely to forgo war without some psychic mutation which at present, at least, does not seem imminent. And if one place the blame for killing and destroying on economic insecurity, on inequality, on injustice, he is simply stating the proposition in another way. We have what we are. Perhaps the crayfish feels the itch of jealousy, or perhaps he is sexually insecure. The effect is that he fights. When in the world there shall come twenty, thirty, fifty years without evidence of our murder trait, under whatever system of justice or economic security, then we may have a contrasting habit pattern to examine. So far there is no such situation. So far the murder trait of our species is as regular and observable as our various sexual habits.

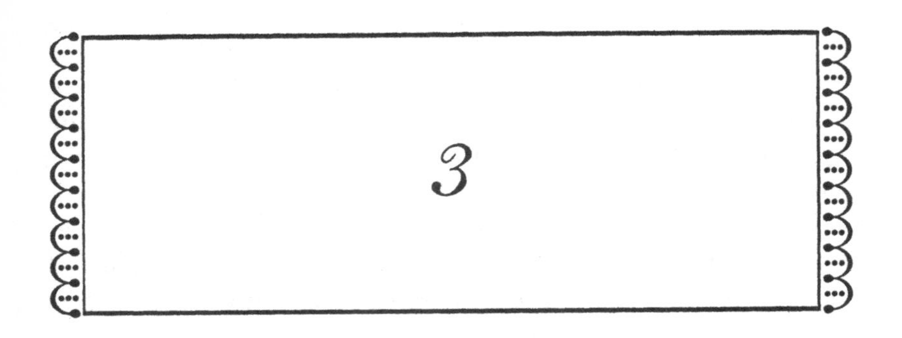

IN THE time before our departure for the Gulf we sat on the pier and watched the sardine purse-seiners riding among the floating grapefruit rinds. A breakwater is usually a dirty place, as though the tampering with the shore line is obscene and impractical to the cleansing action of the sea. And we talked to our prospective crew. Tex, our engineer, was caught in the ways of the harbor. He was born in the Panhandle of Texas and early he grew to love Diesel engines. They are so simple and powerful, blocks of pure logic in shining metal. They appealed to some sense of neat thinking in Tex. He might be sentimental and illogical in some things, but he liked his engines to be true and logical. By an accident, possibly alcoholic, he came to the Coast in an old Ford and sat down beside the Bay, and there he discovered a wonderful thing. Here, combined in one, were the best Diesels to be found anywhere, and boats. He never recovered from his shocked pleasure. He could never leave the sea again, for nowhere else could he find these two perfect things in one. He is a sure man with an engine. When he goes below he is identified with his engine. He moves about, not seeing, not looking, but knowing. No matter how tired or how deeply asleep he may be, one miss of the engine jerks him to his feet and into the engine-room before he is awake, and we truly believe that a burned bearing or a cracked shaft gives him sharp pains in his stomach.

We talked to Tony, the master and part owner of the *Western Flyer,* and our satisfaction with him as master increased constantly. He had the brooding, dark, Slavic eyes and the hawk nose of the Dalmatian. He rarely talked or laughed. He was tall and lean and very strong. He had a great contempt for forms. Under way, he liked to wear a tweed coat and an old felt hat, as though to say, "I keep the sea in my head, not on my back like a Goddamn yachtsman." Tony has one great passion; he loves rightness and he hates wrongness. He thinks speculation a complete waste of time. To our sorrow, and some financial loss, we discovered that Tony never spoke unless he was right. It was useless to bet with him and impossible to argue with him. If he had not been right, he would never have opened his mouth. But once knowing and saying a truth, he became infuriated at the untruth which naturally enough was set against it. Inaccuracy was like an outrageous injustice to him, and when confronted with it, he was likely to shout and to lose his temper. But he did not personally triumph when his point was proven. An ideal judge, hating larceny, feels no triumph when he sentences a thief, and Tony, when he has nailed a true thing down and routed a wrong thing, feels good, but not righteous. He retires grumbling a little sadly at the stupidity of a world which can conceive a wrongness or for one moment defend one. He loves the leadline because it tells a truth on its markers; he loves the Navy charts; and until he went into the Gulf he admired the *Coast Pilot.* The *Coast Pilot* was not wrong, but things had changed since its correction, and Tony is uneasy in the face of variables. The whole relational thinking of modern physics was an obscenity to him and he refused to have anything to do with it. Parallels and compasses and the good Navy maps were things you could trust. A circle is true and a direction is set forever, a shining golden line across the mind. Later, in the mirage of the Gulf where visual distance is a highly variable matter, we wondered whether Tony's certainties were ever tipped. It did not seem so. His qualities made him a good master. He took no chances he could avoid, for his boat and his life and ours were no light things for him to tamper with.

We come now to a piece of equipment which still brings anger to our hearts and, we hope, some venom to our pen. Perhaps in self-defense against suit, we should say, "The outboard motor mentioned in this book is purely fictitious and any resemblance to outboard motors living or dead is coincidental." We shall call this contraption, for the sake of secrecy, a Hansen Sea-Cow—a dazzling little piece of machinery, all aluminum paint and touched here and there with spots of red. The Sea-Cow was built to sell, to dazzle the eyes, to splutter its way into the unwary heart. We took it along for the skiff. It was intended that it should push us ashore and back, should drive our boat into estuaries and along the borders of little coves. But we had not reckoned with one thing. Recently, industrial civilization has reached its peak of reality and has lunged forward into something that approaches mysticism. In the Sea-Cow factory where steel fingers tighten screws, bend and mold, measure and divide, some curious mathematick has occurred. And that secret so long sought has accidentally been found. Life has been created. The machine is at last stirred. A soul and a malignant mind have been born. Our Hansen Sea-Cow was not only a living thing but a mean, irritable, contemptible, vengeful, mischievous, hateful living thing. In the six weeks of our association we observed it, at first mechanically and then, as its living reactions became more and more apparent, psychologically. And we determined one thing to our satisfaction. When and if these ghoulish little motors learn to reproduce themselves the human species is doomed. For their hatred of us is so great that they will wait and plan and organize and one night, in a roar of little exhausts, they will wipe us out. We do not think that Mr. Hansen, inventor of the Sea-Cow, father of the outboard motor, knew what he was doing. We think the monster he created was as accidental and arbitrary as the beginning of any other life. Only one thing differentiates the Sea-Cow from the life that we know. Whereas the forms that are familiar to us are the results of billions of years of mutation and complication, life and intelligence emerged simultaneously in the Sea-Cow. It is more than a species. It is a whole new redefinition of life. We observed the following traits in it and we were able to check them again and again:

1. Incredibly lazy, the Sea-Cow loved to ride on the back of a boat, trailing its propeller daintily in the water while we rowed.

2. It required the same amount of gasoline whether it ran or not, apparently being able to absorb this fluid through its body walls without recourse to explosion. It had always to be filled at the beginning of every trip.

3. It had apparently some clairvoyant powers, and was able to read our minds, particularly when they were inflamed with emotion. Thus, on every occasion when we were driven to the point of destroying it, it started and ran with a great noise and excitement. This served the double purpose of saving its life and of resurrecting in our minds a false confidence in it.

4. It had many cleavage points, and when attacked with a screwdriver, fell apart in simulated death, a trait it had in common with opossums, armadillos, and several members of the sloth family, which also fall apart in simulated death when attacked with a screwdriver.

5. It hated Tex, sensing perhaps that his knowledge of mechanics was capable of diagnosing its shortcomings.

6. It completely refused to run: (a) when the waves were high, (b) when the wind blew, (c) at night, early morning, and evening, (d) in rain, dew, or fog, (e) when the distance to be covered was more than two hundred yards. But on warm, sunny

days when the weather was calm and the white beach close by—in a word, on days when it would have been a pleasure to row—the Sea-Cow started at a touch and would not stop.

7. It loved no one, trusted no one. It had no friends.

Perhaps toward the end, our observations were a little warped by emotion. Time and again as it sat on the stern with its pretty little propeller lying idly in the water, it was very close to death. And in the end, even we were infected with its malignancy and its dishonesty. We should have destroyed it, but we did not. Arriving home, we gave it a new coat of aluminum paint, spotted it at points with new red enamel, and sold it. And we might have rid the world of this mechanical cancer!

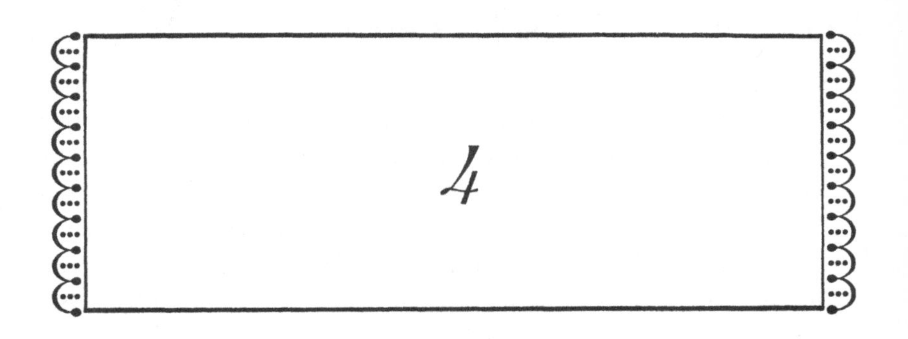

IT WOULD be ridiculous to suggest that ours was anything but a makeshift expedition. The owner of a boat on short charter does not look happily on any re-designing of his ship. In a month or two we could have changed the *Western Flyer* about and made her a collector's dream, but we had neither the time nor the money to do it. The low-tide period was approaching. We had on board no permanent laboratory. There was plenty of room for one in the fish-hold, but the dampness there would have rusted the instruments overnight. We had no dark-room, no permanent aquaria, no tanks for keeping animals alive, no pumps for delivering sea water. We had not even a desk except the galley table. Microscopes and cameras were put away in an empty bunk. The enameled pans for laying out animals were in a large crate lashed to the net-table aft, where it shared the space with the two skiffs. The hatch cover of the fish-hold became laboratory and aquarium, and we carried sea water in buckets to fill the pans. Another empty bunk was filled with flashlights, medicines, and the more precious chemicals. Dip-nets, wooden collecting buckets, and vials and jars in their cases were stowed in the fish-hold. The barrels of alcohol and formaldehyde were lashed firmly to the rail on deck, for all of us had, I think, a horror-thought of fifteen gallons of U.S.P. formaldehyde broken loose and burst. One achieves a respect and a distaste for formaldehyde from working with it. Fortunately, none of us had a developed formalin allergy. Our small refrigerating chamber, powered by a two-cycle gasoline engine and designed to cool sea water for circulation to living animals, began the trip on top of the deckhouse and ended back on the net-table. This unit, by the way, was not very effective, the motor being jerky and not of sufficient power. But on certain days in the Gulf it did manage to cool a little beer or perhaps more than a little, for the crew fell in joyfully with our theory that it is unwise to drink unboiled water, and boiled water isn't any good. In addition, the weather was too hot to boil water, and besides the crew wished to test this perfectly sound scientific observation thoroughly. We tested it by reducing the drinking of water to an absolute minimum.

A big pressure tube of oxygen was lashed to a deck rail, its gauges and valves wrapped in canvas. Gradually, the boat was loaded and the materials put away, some never to be taken out again. It was agreed that we should all stand wheel-watch when we were running night and day; but once in the Gulf, and working at collecting stations, the hired crew should work the boat, since we would anchor at night and run only during the daytime.

Toward the end of the preparation, a small hysteria began to build in ourselves and our friends. There were hundreds of unnecessary trips back and forth. Some materials were stowed on board with such cleverness that we never found them again. Now the whole town of Monterey was becoming fevered and festive—but not because of our going. At the end of the sardine season, canneries and boat owners provide a celebration. There is a huge barbecue on the end of the pier with free beef and beer and salad for all comers. The sardine fleet is decorated with streamers and bunting and serpentine, and the boat with the biggest season catch is queen of a strange nautical parade of boats; and every boat is an open house, receiving friends of owners and of crew. Wine flows beautifully, and the parade of boats that starts with dignity and precision sometimes ends in a turmoil. This fiesta took place on Sunday, and we were to sail on Monday morning. The *Western Flyer* was decorated like the rest with red and blue bunting and serpentine. Master and crew refused to sail before the fiesta was over. We rode in the parade of boats, some of us in the crow's-nest and some on the house. With five thousand other people we crowded on the pier and ate great hunks of meat and drank beer and heard speeches. It was the biggest barbecue the sardine men had ever given, and the potato salad was served out of washtubs. The speeches rose to a crescendo of patriotism and good feeling beyond anything Monterey had ever heard.

There should be here some mention of the permits obtained from the Mexican government. At the time of our preparation, Mexico was getting ready for a presidential election, and the apparent issues were so complex as to cause apprehension that there might be violence. The nation was a little nervous, and it seemed to us that we should be armed with permits which clearly established us as men without politics or business interests. The work we intended to do might well have seemed suspicious to some patriotic customs official or soldier—a small boat that crept to uninhabited points on a barren coast, and a party which spent its time turning over rocks. It was not likely that we could explain our job to the satisfaction of a soldier. It would seem ridiculous to the military mind to travel fifteen hundred miles for the purpose of turning over rocks on the seashore and picking up small animals, very few of which were edible; and doing all this without shooting at anyone. Besides, our equipment might have looked subversive to one who had seen the war sections of *Life* and *Pic* and *Look*. We carried no firearms except a .22-caliber pistol and a very rusty ten-gauge shotgun. But an oxygen cylinder might look too much like a torpedo to an excitable rural soldier, and some of the laboratory equipment could have had a lethal look about it. We were not afraid for ourselves, but we imagined being held in some mud *cuartel* while the good low tides went on and we missed them. In our naïveté, we considered that our State Department, having much business with the Mexican government, might include a paragraph about us in one of its letters, which would convince Mexico of our decent intentions. To this end, we wrote to the State Department explaining our project and giving a list of people who would confirm the purity of our motives. Then we waited with a childlike faith that when a thing is stated simply and evidence of its truth is included there need be no mix-up. Besides, we told ourselves, we were American citizens and the government was our servant. Alas, we did not know diplomatic procedure. In due course, we had an answer from the State Department. In language so diplomatic as to be barely intelligible it gently disabused us. In the first place, the State Department was *not* our servant, however other departments might feel about it. The State Department had little or no interest in the collection of marine invertebrates unless carried on by an institution of learning, preferably with Dr. Butler as its president. The government never made such representations for private citizens. Lastly, the State Department hoped to God we would not get into trouble and appeal to it for aid. All this was concealed in language so beautiful and incomprehensible that we began to understand why diplomats say they are "studying" a message from Japan or England or Italy. We studied this letter for the better part of one night, reduced its sentences to words, built it up again, and came out with the above-mentioned gist. "Gist" is, we imagine, a word which makes the State Department shudder with its vulgarity.

There we were, with no permits and the imaginary soldier still upset by our oxygen tube. In Mexico, certain good friends worked to get us the permits; the consul-general in San Francisco wrote letters about us, and then finally, through a friend, we got in touch with Mr. Castillo Najera, the Mexican ambassador to Washington. To our wonder there came an immediate reply from the ambassador which said there was no reason why we should not go and that he would see the permits were issued immediately. His letter said just that. There was a little sadness in us when we read it. The ambassador seemed such a good man we felt it a pity that he had no diplomatic future, that he could never get anywhere in the world of international politics. We understood his letter the first time we read it. Clearly, Mr. Castillo Najera is a misfit and a rebel. He not only wrote clearly, but he kept his word. The permits came through quickly and in order. And we wish here and now to assure this gentleman that whenever the inevitable punishment for his logic and clarity falls upon him we will gladly help him to get a new start in some other profession.

When the permits arrived, they were beautifully sealed so that even a soldier who could not read would know that if we were not what we said we were, we were at least influential enough spies and saboteurs to be out of his jurisdiction.

And so our boat was loaded, except for the fuel tanks, which we planned to fill at San Diego. Our crew entered the contests at the sardine fiesta—the skiff race, the greased-pole walk, the water-barrel tilt—and they did not win anything, but no one cared. And late in the night when the feast had died out we slept ashore for the last time, and our dreams were cluttered with things we might have forgotten. And the beer cans from the fiesta washed up and down the shore on the little brushing waves behind the breakwater.

We had planned to sail about ten o'clock on March 11, but so many people came to see us off and the leave-taking was so pleasant that it was afternoon before we could think of going. The moment or hour of leave-taking is one of the pleasantest times in human experience, for it has in it a warm sadness without loss. People who don't ordinarily like you very well are overcome with affection at leave-taking. We said good-by again and again and still could not bring ourselves to cast off the lines and start the engines. It would be good to live in a perpetual state of leave-taking, never to go nor to stay, but to remain suspended in that golden emotion of love and longing; to be missed without being gone; to be loved without satiety. How beautiful one is and how desirable; for in a few moments one will have ceased to exist. Wives and fiancées were there, melting and open. How beautiful they were too; and against the hull of the boat the beer cans from the fiesta of yesterday tapped lightly like little bells, and the sea-gulls flew around and around but did not land. There was no room for them—too many people were seeing us off. Even a few strangers were caught in the magic and came aboard and wrung our hands and went into the galley. If our medicine chest had held out we might truly never have sailed. But about twelve-thirty the last dose was prescribed and poured and taken. Only then did we realize that not only were *we* fortified against illness, but that fifty or sixty inhabitants of Monterey could look forward to a long period of good health.

The day of charter had arrived. That instrument said we would leave on the eleventh, and the master was an honest man. We ejected our guests, some forcibly. The lines were cast off. We backed and turned and wove our way out among the boats of the fishing fleet. In our rigging the streamers, the bunting, the serpentine still fluttered, and as the breakwater was cleared and the wind struck us, we seemed, to ourselves at least, a very brave and beautiful sight. The little bell buoy on the reef at Cabrillo Point was excited about it too, for the wind had freshened and the float rolled heavily and the four clappers struck the bell with a quick tempo. We stood on top of the deckhouse and watched the town of Pacific Grove slip by and dark pine-covered hills roll back on themselves as though they moved, not we.

We sat on a crate of oranges and thought what good men most biologists are, the tenors of the scientific world—temperamental, moody, lecherous, loud-laughing, and healthy. Once in a while one comes on the other kind—what used in the university to be called a "dry-ball"—but such men are not really biologists. They are the embalmers of the field, the picklers who see only the preserved form of life without any of its principle. Out of their own crusted minds they create a world wrinkled with formaldehyde. The true biologist deals with life, with teeming boisterous life, and learns something from it, learns that the first rule of life is living. The dry-balls cannot possibly learn a thing every starfish knows in the core of his soul and in the vesicles between his rays. He must, so know the starfish and the student biologist who sits at the feet of living things, proliferate in all directions. Having certain tendencies, he must move along their lines to the limit of their potentialities. And we have known biologists who did proliferate in all directions: one or two have had a little trouble about it. Your true biologist will sing you a song as loud and off-key as will a blacksmith, for he knows that morals are too often diagnostic of prostatitis and stomach ulcers. Sometimes he may proliferate a little too much in all directions, but he is as easy to kill as any other organism, and meanwhile he is very good company, and at least he does not confuse a low hormone productivity with moral ethics.

The *Western Flyer* pushed through the swells toward Point Joe, which is the southern tip of the Bay of Monterey. There was a line of white which marked the open sea, for a strong north wind was blowing, and on that reef the whistling buoy rode, roaring like a perplexed and mournful bull. On the shore road we could see the cars of our recent friends driving along keeping pace with us while they waved handkerchiefs sentimentally. We were all a little sentimental that day. We turned the buoy and cleared the reef, and as we did the boat rolled heavily and then straightened. The north wind drove down on our tail, and we headed south with the big swells growing under us and passing, so that we seemed to be standing still. A squadron of pelicans crossed our bow, flying low to the waves and acting like a train of pelicans tied together, activated by one nervous system. For they flapped their powerful wings in unison, coasted in unison. It seemed that they tipped a wavetop with their wings now and then, and certainly they flew in the troughs of the waves to save themselves from the wind. They did not look around or change direction. Pelicans seem always to know exactly where they are going. A curious sea-lion came out to look us over, a tawny, crusty old fellow with rakish mustaches and the scars of battle on his shoulders. He crossed our bow too and turned and paralleled our course, trod water, and looked at us. Then, satisfied, he snorted and cut for shore and some sea-lion appointment. They always have them, it's just a matter of getting around to keeping them.

And now the wind grew stronger and the windows of houses along the shore flashed in the declining sun. The forward guy-wire of our mast began to sing under the wind, a deep and yet penetrating tone like the lowest string of an incredible bull-fiddle. We rose on each swell and skidded on it until it passed and dropped us in the trough. And from the galley ventilator came the odor of boiling coffee, a smell that never left the boat again while we were on it.

In the evening we came back restlessly to the top of the deckhouse, and we discussed the Old Man of the Sea, who might well be a myth, except that too many people have seen him. There is some quality in man which makes him people the ocean with monsters and one wonders whether they are there or not. In one sense they are, for we continue to see them. One afternoon in the laboratory ashore we sat drinking coffee and talking with Jimmy Costello, who is a reporter on the Monterey *Herald*. The telephone rang and his city editor said that the decomposed body of a sea-serpent was washed up on the beach at Moss Landing, half-way around the Bay. Jimmy was to rush over and get pictures of it. He rushed, approached the evil-smelling monster from which the flesh was dropping. There was a note pinned to its head which said, "Don't worry about it, it's a basking shark. [Signed] Dr. Rolph Bolin of the Hopkins Marine Station." No doubt that Dr. Bolin acted kindly, for he loves true things; but his kindness was a blow to the people of Monterey. They so wanted it to be a sea-serpent. Even we hoped it would be. When sometime a true sea-serpent, complete and undecayed, is found or caught, a shout of triumph will go through the world. "There, you see," men will say, "I knew they were there all the time. I just had a feeling they were there." Men really need sea-monsters in their personal oceans. And the Old Man of the Sea is one of these. In Monterey you can find many people who have seen him. Tiny Colletto has seen him close up and can draw a crabbed sketch of him. He is very large. He stands up in the water, three or four feet emerged above the waves, and watches an approaching boat until it comes too close, and then he sinks slowly out of sight. He looks somewhat like a tremendous diver, with large eyes and fur shaggily hanging from him. So far, he has not been photographed. When he is, probably Dr. Bolin will identify him and another beautiful story will be shattered. For this reason we rather hope he is never photographed, for if the Old Man of the Sea should turn out to be some great malformed sea-lion, a lot of people would feel a sharp personal loss—a Santa Claus loss. And the ocean would be none the better for it. For the ocean, deep and black in the depths, is like the low dark levels of our minds in which the dream symbols incubate and sometimes rise up to sight like the Old Man of the Sea. And even if the symbol vision be horrible, it is there and it is ours. An ocean without its unnamed monsters would be like a completely dreamless sleep. Sparky and Tiny do not question the Old Man of the Sea, for they have looked at him. Nor do we question him because we know he is there. We would accept the testimony of these boys sufficiently to send a man to his death for murder, and we know they saw this monster and that they described him as they saw him.

We have thought often of this mass of sea-memory, or sea-thought, which lives deep in the mind. If one ask for a description of the unconscious, even the answer-symbol will usually be in terms of a dark water into which the light descends only a short distance. And we have thought how the human fetus has, at one stage of its development, vestigial gill-slits. If the gills are a component of the developing human, it is not unreasonable to suppose a parallel or concurrent mind or psyche development. If there be a life-memory strong enough to leave its symbol in vestigial gills, the preponderantly aquatic symbols in the individual unconscious might well be indications of a group psyche-memory which is the foundation of the whole unconscious. And what things must be there, what monsters, what enemies, what fear of dark and pressure, and of prey! There are numbers of examples wherein even invertebrates seem to remember and to react to stimuli no longer violent enough to cause the reaction. Perhaps, next to that of the sea, the strongest memory in us is that of the moon. But moon and sea and tide are one. Even now, the tide establishes a measurable, although minute, weight differential. For example, the steamship *Majestic* loses about fifteen pounds of its weight under a full moon.[1] According to a theory of George Darwin (son of Charles Darwin), in pre-Cambrian times, more than a thousand million years ago, the tides were tremendous; and the weight differential would have been correspondingly large. The moon-pull must have been the most important single environmental factor of littoral animals. Displacement and body weight then must certainly have decreased and increased tremendously with the rotation and phases of the moon, particularly if the orbit was at that time elliptic. The sun's reinforcement was probably slighter, relatively.

---

[1] Marmer, *The Tide,* 1926, p. 26.

Consider, then, the effect of a decrease in pressure on gonads turgid with eggs or sperm, already almost bursting and awaiting the slight extra pull to discharge. (Note also the dehiscence of ova through the body walls of the polychaete worms. These ancient worms have their ancestry rooted in the Cambrian and they are little changed.) Now if we admit for the moment the potency of this tidal effect, we have only to add the concept of inherited psychic pattern we call "instinct" to get an inkling of the force of the lunar rhythm so deeply rooted in marine animals and even in higher animals and in man.

When the fishermen find the Old Man rising in the pathways of their boats, they may be experiencing a reality of past and present. This may not be a hallucination; in fact, it is little likely that it is. The interrelations are too delicate and too complicated. Tidal effects are mysterious and dark in the soul, and it may well be noted that even today the effect of the tides is more valid and strong and widespread than is generally supposed. For instance, it has been reported that radio reception is related to the rise and fall of Labrador tides,[2] and that there may be a relation between tidal rhythms and the recently observed fluctuations in the speed of light.[3] One could safely predict that all physiological processes correspondingly might be shown to be influenced by the tides, could we but read the indices with sufficient delicacy.

It appears that the physical evidence for this theory of George Darwin is more or less hypothetical, not in fact, but by interpretation, and that critical reasoning could conceivably throw out the whole process and with it the biologic connotations, because of unknown links and factors. Perhaps it should read the other way around. The animals themselves would seem to offer a striking confirmation to the tidal theory of cosmogony. One is almost forced to postulate some such

---

[2] *Science Supplement*, Vol. 80, No. 2069, p. 7, Aug. 24, 1934.
[3] *Science*, Vol. 81, No. 2091, p. 101, Jan. 25, 1935.

theory if he would account causally for this primitive impress. It would seem far-fetched to attribute the strong lunar effects actually observable in breeding animals to the present fairly weak tidal forces only, or to coincidence. There is tied up to the most primitive and powerful racial or collective instinct a rhythm sense or "memory" which affects everything and which in the past was probably more potent than it is now. It would at least be more plausible to attribute these profound effects to devastating and instinct-searing tidal influences active during the formative times of the early race history of organisms; and whether or not any mechanism has been discovered or is discoverable to carry on this imprint through the germ plasms, the fact remains that the imprint is there. The imprint is in us and in Sparky and in the ship's master, in the palolo worm, in mussel worms, in chitons, and in the menstrual cycle of women. The imprint lies heavily on our dreams and on the delicate threads of our nerves, and if this seems to come a long way from sea-serpents and the Old Man of the Sea, actually it has not come far at all. The harvest of symbols in our minds seems to have been planted in the soft rich soil of our pre-humanity. Symbol, the serpent, the sea, and the moon might well be only the signal light that the psycho-physiologic warp exists.

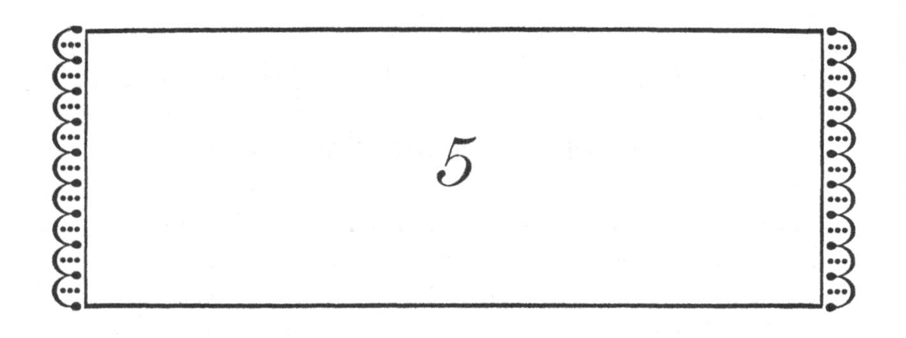

THE evening came down on us and as it did the wind dropped but the tall waves remained, not topped with whitecaps any more. A few porpoises swam near and looked at us and swam away. The watches changed and we ate our first meal aboard, the cold wreckage of farewell snacks, and when our watch was done we were reluctant to go down to the bunks. We put on heavier coats and hung about the long bench where the helmsman sat. The little light on the compass card and the port and starboard lights were our outmost boundaries. Then we passed Point Sur and the waves flattened out into a ground-swell and increased in speed. Tony the master said, "Of course, it's always that way. The point draws the waves." Another might say, "The waves come greatly to the point," and in both statements there would be a good primitive exposition of the relation between giver and receiver. This relation would be through waves; wave to wave to wave, each of which is connected by torsion to its inshore fellow and touches it enough, although it has gone before, to be affected by its torsion. And so on and on to the shore, and to the point where the last wave, if you think from the sea, and the first if you think from the shore, touches and breaks. And it is important where you are thinking from.

The sharp, painful stars were out and bright enough to make the few whitecaps gleam against the dark surrounding water. From the wheel the little flag-jack on the peak stood against the course and swung back and forth over the horizon stars, blotting out each one as it passed. We tried to cover a star with the flag-jack and keep it covered, but this was impossible; no one could do that, not even Tony. But Tony, who knew his boat so well, could feel the yaw before it happened, could correct an error before it occurred. This is no longer reason or thought. One achieves the same feeling on a horse he knows well; one almost feels the horse's impulse in one's knees, and knows, but does not know, not only when the horse will shy, but the direction of his jump. The landsman, or the man who has been long ashore, is clumsy with the wheel, and his steering in a heavy sea is difficult. One grows tense on the wheel, particularly if someone like Tony is watching sardonically. Then keeping the compass card steady becomes impossible and the swing, a variable arc from two to ten degrees. And as weariness creeps up it is not uncommon to forget which way to turn the wheel to make the compass card swing back where you want it. The wheel turns only two ways, left or right. The fact of the lag, and the boat swinging rapidly so that a slow correcting allows it to pass the course and err on the other side, becomes a maddening thing when Tony the magnificent sits beside you. He does not correct you, he doesn't even speak. But Tony loves the truth, and the course is the truth. If the helmsman is off course he is telling a lie to Tony. And as the course projects, hypothetically, straight off the bow and around the world, so the wake drags out behind, a tattler on the conduct of the steersman. If one should steer mathematically perfectly, which is of course impossible, the wake will be a straight line; but even if, when drawn, it may have been straight, it bends to currents and to waves, and your true effort is wiped out. There is probably a unified-field hypothesis available in navigation as in all things. The internal factors would be the boat, the controls, the engine, and the crew, but chiefly the will and intent of the master, sub-headed with his conditioning experience, his sadness and ambitions and pleasures. The external factors would be the ocean with its bordering land, the waves and currents and the winds with their constant and varying effect in modifying the influence of the rudder against the changing tensions exerted on it.

If you steer *toward an object,* you cannot perfectly and indefinitely steer directly at it. You must steer to one side, or run it down; but you can steer exactly at a compass point, indefinitely. That does not change. Objects achieved are merely its fulfillment. In going toward a headland, for example, you can steer directly for it while you are at a distance, only changing course as you approach. Or you may set your compass course for the point and correct it by vision when you approach. The working out of the ideal into the real is here—the relationship between inward and outward, microcosm to macrocosm. The compass simply represents the ideal, present but unachievable, and sight-steering a compromise with perfection which allows your boat to exist at all.

In the development of navigation as thought and emotion—and it must have been a slow, stumbling process frightening to its innovators and horrible to the fearful—how often must the questing mind have wished for a constant and unvarying point on the horizon to steer by. How simple if a star floated unchangeably to measure by. On clear nights such a star is there, but it is not trustworthy and the course of it is an arc. And the happy discovery of Stella Polaris—which, although it too shifts very minutely in an arc, is constant relatively—was encouraging. Stella Polaris will get you there. And so to the crawling minds Stella Polaris must have been like a very goddess of constancy, a star to love and trust.

What we have wanted always is an *unchangeable,* and we have found that only a compass point, a thought, an individual ideal, does not change—Schiller's and Goethe's *Ideal* to be worked out in terms of reality. And from such a thing as this, Beethoven writes a Ninth Symphony to Schiller's *Ode to Joy.*

A tide pool has been called a world under a rock, and so it might be said of navigation, "It is the world within the horizon."

Of steering, the external influences to be overcome are in the nature of oscillations; they are of short or long periods or both. The mean levels of the extreme ups and downs of the oscillations symbolize opposites in a Hegelian sense. No wonder, then, that in physics the symbol of oscillation, $\sqrt{-1}$, is fundamental and primitive and ubiquitous, turning up in every equation.

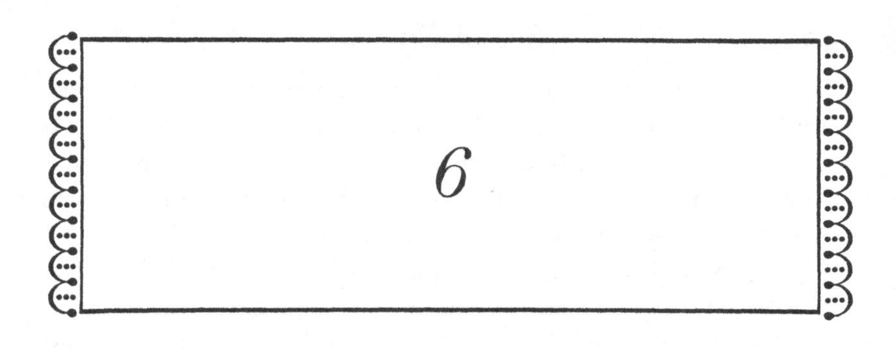

6

March 12

IN THE morning we had come to the Santa Barbara Channel and the water was slick and gray, flowing in long smooth swells, and over it, close down, there hung a little mist so that the sea-birds flew in and out of sight. Then, breaking the water as though they swam in an obscure mirror, the porpoises surrounded us. They really came to us. We have seen them change course to join us, these curious animals. The Japanese will eat them, but rarely will Occidentals touch them. Of our crew, Tiny and Sparky, who loved to catch every manner of fish, to harpoon any swimming thing, would have nothing to do with porpoises. "They cry so," Sparky said, "when they are hurt, they cry to break your heart." This is rather a difficult thing to understand; a dying cow cries too, and a stuck pig raises his protesting voice piercingly and few hearts are broken by those cries. But a porpoise cries like a child in sorrow and pain. And we wonder whether the general seaman's real affection for porpoises might not be more complicated than the simple fear of hearing them cry. The nature of the animal might parallel certain traits in ourselves—the outrageous boastfulness of porpoises, their love of play, their joy in speed. We have watched them for many hours, making designs in the water, diving and rising and then seeming to turn over to see if they are watched. In bursts of speed they hump their backs and the beating tails take power from the whole body. Then they slow down and only the muscles near the tails are strained. They break the surface, and the blow-holes, like eyes, open and gasp in air and then close like eyes before they submerge. Suddenly they seem to grow tired of playing; the bodies hump up, the incredible tails beat, and instantly they are gone.

The mist lifted from the water but the oily slickness remained, and it was like new snow for keeping the impressions of what had happened there. Near to us was the greasy mess where a school of sardines had been milling, and on it the feathers of gulls which had come to join the sardines and, having fed hugely, had sat on the water and combed themselves in comfort. A Japanese liner passed us, slipping quickly through the smooth water, and for a long time we rocked in her wake. It was a long lazy day, and when the night came we passed the lights of Los Angeles with its many little dangling towns. The searchlights of the fleet at San Pedro combed the sea constantly, and one powerful glaring beam crept several miles and lay on us so brightly that it threw our shadows on the exhaust stack.

In the early morning before daylight we came into the harbor at San Diego, in through the narrow passage, and we followed the lights on a changing course to the pier. All about us war bustled, although we had no war; steel and thunder, powder and men—the men preparing thoughtlessly, like dead men, to destroy things. The planes roared over in formation and the submarines were quiet and ominous. There is no playfulness in a submarine. The military mind must limit its thinking to be able to perform its function at all. Thus, in talking with a naval officer who had won a target competition with big naval guns, we asked, "Have you thought what happens in a little street when one of your shells explodes, of the families torn to pieces, a thousand generations influenced when you signaled *Fire?*" "Of course not," he said. "Those shells travel so far that you couldn't possibly see where they land." And he was quite correct. If he could really see where they land and what they do, if he could really feel the power in his dropped hand and the waves radiating out from his gun, he would not be able to perform his function. He himself would be the weak point of his gun. But by not seeing, by insisting that it be a problem of ballistics and trajectory, he is a good gunnery officer. And he is too humble to take the responsibility for thinking. The whole structure of his world would be endangered if he permitted himself to think. The pieces must stick within their pattern or the whole thing collapses and the design is gone. We wonder whether in the present pattern the pieces are not straining to fall out of line; whether the paradoxes of our times are not finally mounting to a conclusion of ridiculousness that will make the whole structure collapse. For the paradoxes are becoming so great that leaders of people must be less and less intelligent to stand their own leadership.

The port of San Diego in that year was loaded with explosives and the means of transporting and depositing them on some enemy as yet undetermined. The men who directed this mechanism were true realists. They knew an enemy would emerge, and when one did, they had explosives to deposit on him.

In San Diego we filled the fuel tanks and the water tanks. We filled the icebox and took on the last perishable foods, bread and eggs and fresh meat. These would not last long, for when the ice was gone only the canned goods and the foods we could take from the sea would be available. We tied up to the pier all day and a night; got our last haircuts and ate broiled steaks.

This little expedition had become tremendously important to us; we felt a little as though we were dying. Strangers came to the pier and stared at us and small boys dropped on our deck like monkeys. Those quiet men who always stand on piers asked where we were going and when we said, "To the Gulf of California," their eyes melted with longing, they wanted to go so badly. They were like the men and women who stand about airports and railroad stations; they want to go away, and most of all they want to go away from themselves. For they do not know that they would carry their globes of boredom with them wherever they went. One man on the pier who wanted to participate made sure he would be allowed to cast us off, and he waited at the bow line for a long time. Finally he got the call and he cast off the bow line and ran back and cast off the stern line; then he stood and watched us pull away and he wanted very badly to go.

* * *

Below the Mexican border the water changes color; it takes on a deep ultramarine blue—a washtub bluing blue, intense and seeming to penetrate deep into the water; the fishermen call it "tuna water." By Friday we were off Point Baja. This is the region of the sea-turtle and the flying fish. Tiny and Sparky put out the fishing lines, and they stayed out during the whole trip.

Sparky Enea and Tiny Colletto grew up together in Monterey and they were bad little boys and very happy about it. It is said lightly that the police department had a special detail to supervise the growth and development of Tiny and Sparky. They are short and strong and nearly inseparable. An impulse seems to strike both of them at once. Let Tiny make a date with a girl and Sparky make a date with another girl—it then becomes necessary for Tiny, by connivance and trickery, to get Sparky's girl. But it is all right, since Sparky has been moving mountains to get Tiny's girl.

These two shared a watch, and on their watches we often went strangely off course and no one ever knew why. The compass had a way of getting out of hand so that the course invariably arced inshore. These two rigged the fishing lines with feathered artificial squid. Where the tackle was tied to the stays on either side, they looped the line and inset automobile inner tubes. For the tuna strikes so hard that something must give, and if the line does not break, the jaws tear off, so great is the combination of boat speed and tuna speed. The inner tube solves this problem by taking up the strain of the first great strike until direction and speed are equalized.

When Sparky and Tiny had the watch they took care of the fishing, and when the rubber tubes snapped and shook, one of them climbed down to take in the fish. If it were a large one, or a sharp-fighting fish, hysterical shrieks came from the fisherman. Whereupon the one left at the wheel came down to help and the wheel swung free. We wondered if this habit might not have caused the wonderful course we sailed sometimes. It is not beyond reason that coming back to the wheel, arguing and talking, they might have forgotten the set course and made one up almost as good. "Surely," they might think, "that is kinder and better than waking up the master to ask the course again, and five or ten degrees isn't so important when you aren't going far." If Tony loved the truth for itself, he was more than counterbalanced by Sparky and Tiny. They have little faith in truth, or, for that matter, in untruth. The police who had overseen their growing up had given them a nice appreciation of variables; they tested everything to find out whether it were true or not. In a like manner they tested the compass for a weakness they suspected was in it. And if Tony should say, "You are way off course," they could answer, "Well, we didn't hit anything, did we?"

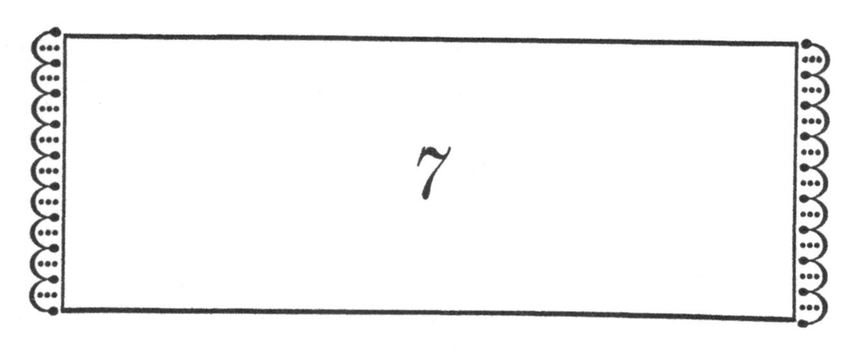

7

March 16

BY TWO P.M. we were in the region of Magdalena Bay. The sea was still oily and smooth, and a light lacy fog lay on the water. The flying fish leaped from the forcing bow and flew off to right and left. It seemed, although this has not been verified, that they could fly farther at night than during the day. If, as is supposed, the flight is terminated when the flying fins dry in the air, this observation would seem to be justified, for at night they would not dry so quickly. Again, the whole thing might be a trick of our eyes. Often we played the searchlight on a fish in flight. The strangeness of light may have made the flight seem longer.

Tiny is a natural harpooner; often he had stood poised on the bow, holding the lance, but thus far nothing had appeared except porpoises, and these he would not strike. But now the sea-turtles began to appear in numbers. He stood for a long time waiting, and finally he drove his lance into one of them. Sparky promptly left the wheel, and the two of them pulled in a small turtle, about two and a half feet long. It was a tortoiseshell turtle.[1] Now we were able to observe the tender hearts of our crew. The small arrow-harpoon had penetrated the fairly soft shell, then turned sideways in the body. They hung the turtle to a stay where it waved its flippers helplessly and stretched its old wrinkled neck and gnashed its parrot beak.

[1] *Eretmochelys imbricata* (Linn.). Nelson, 1921, § Y-29, p. 114, but usually known as *Chelone imbricata.*

The small dark eyes had a quizzical pained look and a quantity of blood emerged from the pierced shell. Suddenly remorse seized Tiny; he wanted to put the animal out of its pain. He lowered the turtle to the deck and brought out an ax. With his first stroke he missed the animal entirely and sank the blade into the deck, but on his second stroke he severed the head from the body. And now a strange and terrible bit of knowledge came to Tiny; turtles are very hard to kill. Cutting off the head seems to have little immediate effect. This turtle was as lively as it had been, and a large quantity of very red blood poured from the trunk of the neck. The flippers waved frantically and there was none of the constricting motion of a decapitated animal. We were eager to examine this turtle and we put Tiny's emotion aside for the moment. There were two barnacle bases on the shell and many hydroids which we preserved immediately. In the hollow beside the small tail were two pelagic crabs [2] of the square-fronted group, a male and a female; and from the way in which they hid themselves in the fold of turtle skin they seemed to be at home there. We were eager to examine the turtle's intestinal tract, both to find the food it had been eating and to look for possible tapeworms. To this end we sawed the shell open at the sides and opened the body cavity. From gullet to anus the digestive tract was crammed with small bright-red rock-lobsters [3]; a few of those nearest the gullet were whole enough to preserve. The gullet itself was lined with hard, sharp-pointed spikes, not of bone, but of a specialized tissue hard enough to macerate the small crustacea the turtle fed on. A curious peristalsis of the gullet (still observable, since even during dissection the reflexes were quite active) brought these points near together in a grinding motion and at the same time passed the increasingly macerated material downward toward the stomach. A good adaptation to food supply by structure, or

---

[2] *Planes minutus* (Linn.).
[3] *Pleuroncodes planipes* Stimpson.

perhaps vice versa. The heart continued to beat regularly. We removed it and placed it in a jar of salt water, where it continued to pulse for several hours; and twenty-four hours later, when it had apparently stopped, a touch with a glass rod caused it to pulse several times before it relaxed again. Tiny did not like this process of dissection. He wants his animals to die and be dead when he chops them; and when we cut up the muscular tissue, intending to cook it, and even the little cubes of white meat responded to touch, Tiny swore that he would give up sea-turtles and he never again tried to harpoon one. In his mind they joined the porpoises as protected animals. Probably he identified himself with the writhing tissue of the turtle and was unable to see it objectively.

The cooking was a failure. We boiled the meat, and later threw out the evil-smelling mess. (Subsequently, we discovered that one has to know how to cook a turtle.) But the turtle shell we wished to preserve. We scraped it as well as we could and salted it. Later we hung it deep in the water, hoping the isopods would clean it for us, but they never did. Finally we impregnated it with formaldehyde, then let it dry in the sun, and after all that we threw it away. It was never pretty and we never loved it.

During the night we crossed a school of bonito,[4] fast, clean-cut, beautiful fish of the mackerel family. The boys on watch caught five of them on the lines and during the process we got quite badly off course. We tried to take moving pictures of the color and of the color-pattern change which takes place in these fish during their death struggles. In the flurry when they beat the deck with their tails, the colors pulse and fade and brighten and fade again, until, when they are dead, a new pattern is visible. We wished to take color photographs of many of the animals because of the impossibility of retaining color in preserved specimens, and also because many animals, in fact most

[4] *Sarda chiliensis* (Girard).

animals, have one color when they are alive and another when they are dead. However, none of us was expert in photography and we had a very mediocre success. The bonitos were good to eat, and Sparky fried big thick fillets for us.

That night we netted two small specimens of the northern flying fish.[5] Sparky, when we were looking at Barnhart's *Marine Fishes of Southern California,* saw a drawing of a lantern-fish entitled *"Monoceratias acanthias* after Gilbert" and he asked, "What's he after Gilbert for?"

This smooth blue water runs out of time very quickly, and a kind of dream sets in. Then a floating box cast overboard from some steamship becomes a fascinating thing, and it is nearly impossible not to bring the wheel over and go to pick it up. A new kind of porpoise began to appear, gray, where the northern porpoise had been dark brown. They were slim and very fast, the noses long and paddle-shaped. They move about in large schools, jumping out of the water and seeming to have a very good time. The abundance of life here gives one an exuberance, a feeling of fullness and richness. The playing porpoises, the turtles, the great schools of fish which ruffle the water surface like a quick breeze, make for excitement. Sometimes in the distance we have seen a school of jumping tuna, and as they threw themselves clear of the water, the sun glittered on them for a moment. The sea here swarms with life, and probably the ocean bed is equally rich. Microscopically, the water is crowded with plankton. This is the tuna water—life water. It is complete from plankton to gray porpoises. The turtle was complete with the little almost-commensal crab living under his tail and with barnacles and hydroids riding on his back. The pelagic rock-lobsters[6] littered the ocean with red spots. There was food everywhere. Everything ate everything else with a furious exuberance.

---

[5] *Cypselurus californicus.*

[6] *Pleuroncodes.*

About five P.M. on the sixteenth, seventy miles north of Point Lazaro, we came upon hosts of the red rock-lobsters on the surface, brilliant red and beautiful against the ultramarine of the water. There was no protective coloration here—a greater contrast could not have been chosen. The water seemed almost solid with the little red crustacea, called "*langustina*" by the Mexicans. According to Stimpson, on March 8, 1859, a number of them were thrown ashore at Monterey in California, many hundreds of miles from their usual range. It was probably during one of those queer cycles when the currents do amazing things. We idled our engine and crept slowly along catching up the *langustina* in dip-nets. We put them in white porcelain pans and took some color moving pictures of them—some of the few good moving pictures, incidentally, made during the whole trip. In the pans we saw that these animals do not swim rapidly, but rather wriggle and crawl through the water. Finally, we immersed them in fresh water and when they were dead, preserved them in alcohol, which promptly removed their brilliant color.

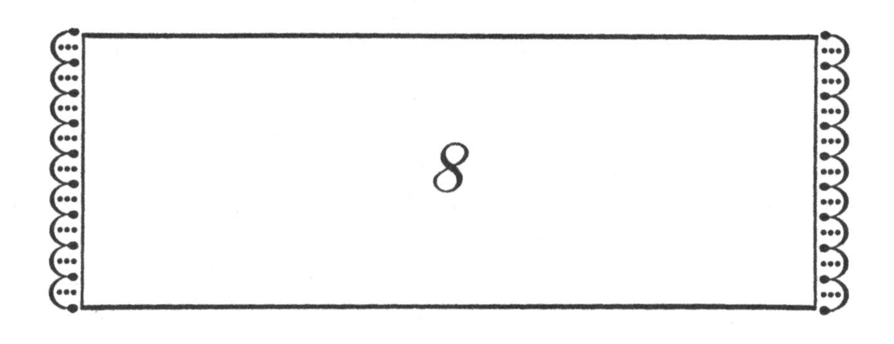

8

March 17

AT TWO A.M. we passed Point Lazaro, one of the reputedly dangerous places of the world, like Cedros Passage, or like Cape Horn, where the weather is always bad even when it is good elsewhere. There is a sense of relief when one is safely past these half-mythical places, for they are not only stormy but treacherous, and again the atavistic fear arises—the Scylla-Charybdis fear that made our ancestors people such places with monsters and enter them only after prayer and propitiation. It was only reasonably rough when we passed, and immediately south the water was very calm. About five in the morning we came upon an even denser concentration of the little red *Pleuroncodes,* and we stopped again and took a great many of them. While we netted the *langustina,* a skipjack struck the line and we brought him in and had him for breakfast. During the meal we said the fish was *Katsuwonus pelamis,* and Sparky said it was a skipjack because he was eating it and he was quite sure he would not eat *Katsuwonus pelamis* ever. A few hours later we caught two small dolphins,[1] startlingly beautiful fish of pure gold, pulsing and fading and changing colors. These fish are very widely distributed.

We were coming now toward the end of our day-and-night running; the engine had never paused since we left San Diego except for idling the little time while we took the *langustina.*

[1] *Coryphaena equisetis* Linn.

The coastline of the Peninsula slid along, brown and desolate and dry with strange flat mountains and rocks torn by dryness, and the heat shimmer hung over the land even in March. Tony had kept us well offshore, and only now we approached closer to land, for we would arrive at Cape San Lucas in the night, and from then on we planned to run only in the daytime. Some collecting stations we had projected, like Pulmo Reef and La Paz and Angeles Bay, but except for those, we planned to stop wherever the shore looked interesting. Even this little trip of ninety hours, though, had grown long, and we were glad to be getting to the end of it. The dry hills were red gold that afternoon and in the night no one left the top of the deckhouse. The Southern Cross was well above the horizon, and the air was warm and pleasant. Tony spent a long time in the galley going over the charts. He had been to Cape San Lucas once before. Around ten o'clock we saw the lighthouse on the false cape. The night was extremely dark when we rounded the end; the great tall rocks called "The Friars" were blackly visible. The *Coast Pilot* spoke of a light on the end of the San Lucas pier, but we could see no light. Tony edged the boat slowly into the dark harbor. Once a flashlight showed for a moment on the shore and then went out. It was after midnight, and of course there would be no light in a Mexican house at such a time. The searchlight on our deckhouse seemed to be sucked up by the darkness. Sparky on the bow with the leadline found deep water, and we moved slowly in, stopping and drifting and sounding. And then suddenly there was the beach, thirty feet away, with little waves breaking on it, and still we had eight fathoms on the lead. We backed away a little and dropped the anchor and waited until it took a firm grip. Then the engine stopped, and we sat for a long time on the deckhouse. The sweet smell of the land blew out to us on a warm wind, a smell of sand verbena and grass and mangrove. It is so quickly forgotten, this land smell. We know it so well on shore that the nose forgets it, but after a few days at sea the odor memory pattern is lost so that the first land smell strikes a powerful emotional nostalgia, very sharp and strangely dear.

In the morning the black mystery of the night was gone and the little harbor was shining and warm. The tuna cannery against the gathering rocks of the point and a few houses along the edge of the beach were the only habitations visible. And with the day came the answer to the lightlessness of the night before. The *Coast Pilot* had not been wrong. There is indeed a light on the end of the cannery pier, but since the electricity is generated by the cannery engine, and since the cannery engine runs only in the daytime, so the light burns only in the daytime. With the arrived day, this light came on and burned bravely until dusk, when it went off again. But the *Coast Pilot* was absolved, it had not lied. Even Tony, who had been a little bitter the night before, was forced to revise his first fierceness. And perhaps it was a lesson to Tony in exact thinking, like those carefully worded puzzles in joke books; the *Pilot* said a light burned—it only neglected to say when, and we ourselves supplied the fallacy.

The great rocks on the end of the Peninsula are almost literary. They are a fitting Land's End, standing against the sea, the end of a thousand miles of peninsula and mountain. Good Hope is this way too, and perhaps we take some of our deep feelings of termination from these things, and they make our symbols. The Friars stood high and protective against an interminable sea.

Clavigero, a Jesuit monk, came to the Point and the Peninsula over two hundred years ago. We quote from the Lake and Gray translation of his history of Lower California,[2] page fifteen: "This Cape is its southern terminus, the Red River [Colorado] is the eastern limit, and the harbor of San Diego, situated at 33 degrees north latitude and about 156 degrees longitude,

---

[2] Stanford University Press, 1937.

can be called its western limit. To the north and the northeast it borders on the countries of barbarous nations little known on the coasts and not at all in the interior. To the west it has the Pacific Sea and on the east the Gulf of California, already called the Red Sea because of its similarity to the Red Sea, and the Sea of Cortés, named in honor of the famous conqueror of Mexico who had it discovered and who navigated it. The length of the Peninsula is about 10 degrees, but its width varies from 30 to 70 miles and more.

"The name, California," Clavigero goes on, "was applied to a single port in the beginning, but later it was extended to mean all the Peninsula. Some geographers have even taken the liberty of comprising under this denomination New Mexico, the country of the Apaches, and other regions very remote from the true California and which have nothing to do with it."

Clavigero says of its naming, "The origin of this name is not known, but it is believed that the conqueror, Cortés, who pretended to have some knowledge of Latin, named the harbor, where he put in, '*Callida fornax*' because of the great heat which he felt there; and that either he himself or some one of the many persons who accompanied him formed the name California from these two words. If this conjecture be not true, it is at least credible."

We like Clavigero for these last words. He was a careful man. The observations set down in his history of Baja California are surprisingly correct, and if not all true, they are at least all credible. He always gives one his choice. Perhaps his Jesuit training is never more evident than in this. "If you believe this," he says in effect, "perhaps you are not right, but at least you are not a fool."

Lake and Gray include an interesting footnote in their translation. "The famous corsair, Drake, called California 'New Albion' in honor of his native land. Father Scherer, a German Jesuit, and M. de Fer, a French geographer, used the name

'Carolina Island' to designate California, which name began to be used in the time of Charles II, King of Spain, when that Peninsula was considered an island, but these and other names were soon forgotten and that given it by the conqueror, Cortés, prevailed."

And in a second footnote, Lake and Gray continue, "We shall add the opinion of the learned ex-Jesuit, Don José Campoi, on the etymology of the name, 'California,' or 'Californias' as others say. This Father believes that the said name is composed of the Spanish word *'Cala'* which means a small cove of the sea, and the latin word *'fornix'* which means an arch; because there is a small cove at the cape of San Lucas on the western side of which there overhangs a rock pierced in such a way that in the upper part of that great opening is seen an arch formed so perfectly that it appears made by human skill. Therefore Cortés, noticing the cove and arch, and understanding Latin, probably gave to that port the name 'California' or *Cala-y-fornix,* speaking half Spanish and half Latin.

"To these conjectures we could add a third one, composed of both, by saying that the name is derived from *Cala,* as Campoi thinks, and *fornax,* as the author believes, because of the cove, and the heat which Cortés felt there, and that the latter might have called that place *Cala, y fornax.*" This ends the footnote.

Our feeling about this, and all the erudite discussion of the origin of this and other names, is that none of these is true. Names attach themselves to places and stick or fall away. When men finally go to live in Antarctica it is unlikely that they will ever speak of the Rockefeller Mountains or use the names designated by breakfast food companies. More likely a name emerges almost automatically from a place as well as from a man and the relationship between name and thing is very close. In the naming of places in the West this has seemed apparent. In this connection there are two examples: in the Sierras there are two little mountains which were called by the early settlers "Maggie's Bubs." This name was satisfactory and descriptive, but it seemed vulgar to later and more delicate lovers of nature, who tried to change the name a number of times and failing, in usage at least, finally surrendered and called them "The Maggies," explaining that it was an Indian name. In the same way Dog ---- Point (and I am delicate only for those same nature lovers) has had finally to be called in print "The Dog." It does not look like a dog, but it does look like that part of a dog which first suggested its name. However, anyone seeing this point immediately reverts to the designation which was anatomically accurate and strangely satisfying to the name-giving faculty. And this name-giving faculty is very highly developed and deeply rooted in our atavistic magics. To name a thing has always been to make it familiar and therefore a little less dangerous to us. "Tree" the abstract may harbor some evil until it has a name, but once having a name one can cope with it. A tree is not dangerous, but the forest is. Among primitives sometimes evil is escaped by never mentioning the name, as in Malaysia, where one never mentions a tiger by name for fear of calling him. Among others, as even among ourselves, the giving of a name establishes a familiarity which renders the thing impotent. It is interesting to see how some scientists and philosophers, who are an emotional and fearful group, are able to protect themselves against fear. In a modern scene, when the horizons stretch out and your philosopher is likely to fall off the world like a Dark Ages mariner, he can save himself by establishing a taboo-box which he may call "mysticism" or "supernaturalism" or "radicalism." Into this box he can throw all those thoughts which frighten him and thus be safe from them. But in geographic naming it seems almost as though the place contributed something to its own name. As Tony says, "The point draws the waves"—we say, "The place draws the name." It doesn't matter what California means; what does matter is that with all the names bestowed upon this place, "California" has seemed right to those who have seen it. And the meaningless word "California" has completely routed all the "New Albions" and "Carolinas" from the scene.

The strangest case of nicknaming we know concerns a man whose first name is Copeland. In three different parts of the country where he has gone, not knowing anyone, he has been called first "Copenhagen" and then "Hagen." This has happened automatically. He is Hagen. We don't know what quality of Hagen-ness he has, but there must be some. Why not "Copen" or "Cope"? It is never that. He is invariably Hagen. This, we realize, has become mystical, and anyone who wishes may now toss the whole thing into his taboo-box and slam the lid down on it.

The tip of the Cape at San Lucas, with the huge gray Friars standing up on the end, has behind the rocks a little beach which is a small boy's dream of pirates. It seems the perfect place to hide and from which to dart out in a pinnace on the shipping of the world; a place to which to bring the gold bars and jewels and beautiful ladies, all of which are invariably carried by the shipping of the world. And this little beach must so have appealed to earlier men, for the names of pirates are still in the rock, and the pirate ships did dart out of here and did come back. But now in back of the Friars on the beach there is a great pile of decaying hammer-head sharks, the livers torn out and the fish left to rot. Some day, and that soon, the more mature piracy which has abandoned the pinnace for the coast gun will stud this point with gray monsters and will send against the shipping of the Gulf, not little bands of ragged men, but projectiles filled with TNT. And from that piracy no jewels or beautiful ladies will come back to the beach behind the rocks.

On that first morning we cleaned ourselves well and shaved while we waited for the Mexican officials to come out and give us the right to land. They were late in coming, for they had to find their official uniforms, and they too had to shave. Few boats put in here. It would not be well to waste the occasion of the visit of even a fishing boat like ours. It was noon before the well-dressed men in their sun helmets came down to the beach and were rowed out to us. They were armed with the .45-caliber automatics which everywhere in Mexico designate officials. And they were armed also with the courtesy which is unique in official Mexico. No matter what they do to you, they are nice about it. We soon learned the routine in other ports as well as here. Everyone who has or can borrow a uniform comes aboard—the collector of customs in a washed and shiny uniform; the business agent in a business suit having about him what Tiny calls "a double-breasted look"; then soldiers if there are any; and finally the Indians, who row the boat and rarely have uniforms. They come over the side like ambassadors. We shake hands all around. The galley has been prepared: coffee is ready and perhaps a drop of rum. Cigarettes are presented and then comes the ceremonial of the match. In Mexico cigarettes are cheap, but matches are not. If a man wishes to honor you, he lights your cigarette, and if you have given him a cigarette, he must so honor you. But having lighted your cigarette and his, the match is still burning and not being used. Anyone may now make use of this match. On a street, strangers who have been wishing for a light come up quickly and light from your match, bow, and pass on.

We were impatient for the officials, and this time we did not have to wait long. It developed that the Governor of the southern district had very recently been to Cape San Lucas and just before that a yacht had put in. This simplified matters, for, having recently used them, the officials knew exactly where to find their uniforms, and, having found them, they did not, as sometimes happens, have to send them to be laundered before they could come aboard. About noon they trooped to the beach, scattering the pigs and Mexican vultures which browsed happily there. They filled the rowboat until the gunwales just missed dipping, and majestically they came alongside. We conducted the ceremony of clearing with some dignity, for if we spoke to them in very bad Spanish, they in turn honored us with very bad English. They cleared us, drank coffee, smoked, and finally left, promising to come back. Much as we had enjoyed them, we were impatient, for the tide was dropping and the exposed rocks looked very rich with animal life.

All the time we were indulging in courtliness there had been light gunfire on the cliffs, where several men were shooting at black cormorants; and it developed that everyone in Cape San Lucas hates cormorants. They are the flies in a perfect ecological ointment. The cannery cans tuna; the entrails and cuttings of the tuna are thrown into the water from the end of the pier. This refuse brings in schools of small fish which are netted and used for bait to catch tuna. This closed and tight circle is interfered with by the cormorants, who try to get at the bait-fish. They dive and catch fish, but also they drive the schools away from the pier out of easy reach of the baitmen. Thus they are considered interlopers, radicals, subversive forces against the perfect and God-set balance on Cape San Lucas. And they are rightly slaughtered, as all radicals should be. As one of our number remarked, "Why, pretty soon they'll want to vote."

Finally we could go. We unpacked the Hansen Sea-Cow and fastened it on the back of the skiff. This was our first use of the Sea-Cow. The shore was very close and we were able just by pulling on the starter rope to spin the propeller enough to get us to shore. The Sea-Cow did not run that day but it seemed to enjoy having its flywheel spun.

The shore-collecting equipment usually consisted of a number of small wrecking bars; wooden fish-kits with handles; quart jars with screw caps; and many glass tubes. These tubes are invaluable for small and delicate animals: the chance of bringing them back uninjured is greatly increased if each individual, or at least only a few of like species, are kept in separate containers. We filled our pockets with these tubes. The soft animals must never be put in the same container with any of the livelier crabs, for these, when restrained or inhibited in any way, go into paroxysms of rage and pinch everything at random, even each other; sometimes even themselves.

The exposed rocks had looked rich with life under the lowering tide, but they were more than that: they were ferocious with life. There was an exuberant fierceness in the littoral here, a vital competition for existence. Everything seemed speeded-up; starfish and urchins were more strongly attached than in other places, and many of the univalves were so tightly fixed that the shells broke before the animals would let go their hold. Perhaps the force of the great surf which beats on this shore has much to do with the tenacity of the animals here. It is noteworthy that the animals, rather than deserting such beaten shores for the safe cove and protected pools, simply increase their toughness and fight back at the sea with a kind of joyful survival. This ferocious survival quotient excites us and makes us feel good, and from the crawling, fighting, resisting qualities of the animals, it almost seems that they are excited too.

We collected down the littoral as the water went down. We didn't seem to have time enough. We took samples of everything that came to hand. The uppermost rocks swarmed with Sally Lightfoots, those beautiful and fast and sensitive crabs. With them were white periwinkle snails. Below that, barnacles and Purpura snails; more crabs and many limpets. Below that many serpulids—attached worms in calcareous tubes with beautiful purple floriate heads. Below that, the multi-rayed starfish, *Heliaster kubiniji* of Xanthus. With *Heliaster* were a few urchins, but not many, and they were so placed in crevices as to be hard to dislodge. Several resisted the steel bar to the extent of breaking—the mouth remaining tight to the rock while the shell fell away. Lower still there were to be seen swaying in the water under the reefs the dark gorgonians, or sea-fans. In the lowest surf-levels there was a brilliant gathering of the moss animals known as bryozoa; flatworms; flat crabs; the large sea-cucumber[3]; some anemones; many sponges of two types, a smooth, encrusting purple one, the other erect, white, and calcareous. There were great colonies of tunicates, clusters of tiny individuals joined by a common tunic and looking so like the sponges that even a trained worker must await the specialist's determination to know whether his find is sponge or tunicate. This is annoying, for the sponge being one step above the protozoa, at the bottom of the evolutionary ladder, and the tunicate near the top, bordering the vertebrates, your trained worker is likely to feel that a dirty trick has been played upon him by an entirely too democratic Providence.

We took many snails, including cones and murexes; a small red tectibranch (of a group to which the sea-hares belong) ; hydroids; many annelid worms; and a red pentagonal starfish.[4] There were the usual hordes of hermit crabs, but oddly enough we saw no chitons (sea-cradles), although the region seemed ideally suited to them.

We collected in haste. As the tide went down we kept a little ahead of it, wading in rubber boots, and as it came up again it drove us back. The time seemed very short. The incredible beauty of the tide pools, the brilliant colors, the swarming species ate up the time. And when at last the afternoon surf began to beat on the littoral and covered it over again, we seemed barely to have started. But the buckets and jars and tubes were full, and when we stopped we discovered that we were very tired.

Our collecting ends were different from those ordinarily en-

---

[3] *Holothuria lubrica.*

[4] *Oreaster.*

tertained. In most cases at the present time, collecting is done by men who specialize in one or more groups. Thus, one man interested in hydroids will move out on a reef, and if his interest is sharp enough, he will not even see other life forms about him. For him, the sponge is something in the way of his hydroids. Collecting large numbers of animals presents an entirely different aspect and makes one see an entirely different picture. Being more interested in distribution than in individuals, we saw dominant species and changing sizes, groups which thrive and those which recede under varying conditions. In a way, ours is the older method, somewhat like that of Darwin on the *Beagle*. He was called a "naturalist." He wanted to see everything, rocks and flora and fauna; marine and terrestrial. We came to envy this Darwin on his sailing ship. He had so much room and so much time. He could capture his animals and keep them alive and watch them. He had years instead of weeks, and he saw so many things. Often we envied the inadequate transportation of his time—the *Beagle* couldn't get about rapidly. She moved slowly along under sail. And we can imagine that young Darwin, probably in a bos'n's chair hung over the side, with a dip-net in his hands, scooping up jellyfish. When he went inland, he rode a horse or walked. This is the proper pace for a naturalist. Faced with all things he cannot hurry. We must have time to think and to look and to consider. And the modern process—that of looking quickly at the whole field and then diving down to a particular—was reversed by Darwin. Out of long long consideration of the parts he emerged with a sense of the whole. Where we wished for a month at a collecting station and took two days, Darwin stayed three months. Of course he could see and tabulate. It was the pace that made the difference. And in the writing of Darwin, as in his thinking, there is the slow heave of a sailing ship, and the patience of waiting for a tide. The results are bound up with the pace. We *could* not do this even if we could. We have thought in this connection that the speed and tempo and tone of modern writing might be built on the nervous clacking of a typewriter; that the brittle jerky thinking of the present might rest on the brittle jerky curricula of our schools with their urge to "turn them out." To turn them out. They use the phrase in speeches; turn them out to what? And the young biologists tearing off pieces of their subject, tatters of the life forms, like sharks tearing out hunks of a dead horse, looking at them, tossing them away. This is neither a good nor a bad method; it is simply the one of our time. We can look with longing back to Charles Darwin, staring into the water over the side of the sailing ship, but for us to attempt to imitate that procedure would be romantic and silly. To take a sailing boat, to fight tide and wind, to move four hundred miles on a horse when we could take a plane, would be not only ridiculous but ineffective. For we first, before our work, are products of our time. We might produce a philosophical costume piece, but it would be completely artificial. However, we can and do look on the measured, slow-paced accumulation of sight and thought of the Darwins with a nostalgic longing.

Even our boat hurried us, and while the Sea-Cow would not run, it had nevertheless infected us with the idea of its running. Six weeks we had, and no more. Was it a wonder that we collected furiously; spent every low-tide moment on the rocks, even at night? And in the times between low tides we kept the bottom nets down and the lines and dip-nets working. When the charter was up, we would be through. How different it had been when John Xanthus was stationed in this very place, Cape San Lucas, in the sixties. Sent down by the United States Government as a tidal observer, but having lots of time, he collected animals for our National Museum. The first fine collections of Gulf forms came from Xanthus. And we do not feel that we are injuring his reputation, but rather broadening it, by repeating a story about him. Speaking to the manager of the cannery at the Cape, we remarked on what a great man Xanthus had been. Where another would have kept his tide charts and brooded and wished for the Willard Hotel, Xanthus had collected animals widely and carefully. The manager said, "Oh, he was even better than that." Pointing to three little Indian children he said, "Those are Xanthus's great-grandchildren," and he continued, "In the town there is a large family of Xanthuses, and a few miles back in the hills you'll find a whole tribe of them." There were giants in the earth in those days.

We wonder what modern biologist, worried about titles and preferment and the gossip of the Faculty Club, would have the warmth and breadth, or even the fecundity for that matter, to leave a "whole tribe of Xanthuses." We honor this man for all his activities. He at least was one who literally did proliferate in all directions.

Many people have spoken at length of the Sally Lightfoots. In fact, everyone who has seen them has been delighted with them. The very name they are called by reflects the delight of the name. These little crabs, with brilliant cloisonné carapaces, walk on their tiptoes. They have remarkable eyes and an extremely fast reaction time. In spite of the fact that they swarm on the rocks at the Cape, and to a less degree inside the Gulf, they are exceedingly hard to catch. They seem to be able to run in any one of four directions; but more than this, perhaps because of their rapid reaction time, they appear to read the mind of their hunter. They escape the long-handled net, anticipating from what direction it is coming. If you walk slowly, they move slowly ahead of you in droves. If you hurry, they hurry. When you plunge at them, they seem to disappear in little puffs of blue smoke—at any rate, they disappear. It is impossible to creep up on them. They are very beautiful, with clear brilliant colors, reds and blues and warm browns. We tried for a long time to catch them. Finally, seeing fifty or sixty in a big canyon of rock, we thought to outwit them. Surely we were more intelligent, if slower, than they. Accordingly, we pitted our obviously superior intelligence against the equally obvious physical superiority of Sally Lightfoot. Near the top of the crevice a boulder protruded. One of our party, taking a secret and circuitous route, hid himself behind this boulder, net in hand. He was completely concealed even from the stalk eyes of the crabs. Certainly they had not seen him go there. The herd of Sallys drowsed on the rocks in the lower end of the crevice. Two more of us strolled in from the seaward side, nonchalance in our postures and ingenuousness on our faces. One might have thought that we merely strolled along in a contemplation which severely excluded Sally Lightfoots. In time the herd moved ahead of us, matching our nonchalance. We did not hurry, they did not hurry. When they passed the boulder, helpless and unsuspecting, a large net was to fall over them and imprison them. But they did not know that. They moved along until they were four feet from the boulder, and then as one crab they turned to the right, climbed up over the edge of the crevice and down to the sea again.

Man reacts peculiarly but consistently in his relationship with Sally Lightfoot. His tendency eventually is to scream curses, to hurl himself at them, and to come up foaming with rage and bruised all over his chest. Thus, Tiny, leaping forward, slipped and fell and hurt his arm. He never forgot nor forgave his enemy. From then on he attacked Lightfoots by every foul means he could contrive (and a training in Monterey street fighting had equipped him well for this kind of battle). He hurled rocks at them; he smashed at them with boards; and he even considered poisoning them. Eventually we did catch a few Sallys, but we think they were the halt and the blind, the simpletons of their species. With reasonably well-balanced and non-neurotic Lightfoots we stood no chance.

We came back to the boat loaded with specimens, and immediately prepared to preserve them. The square, enameled pans were laid out on the hatch, the trays and bowls and watch-glasses (so called because at one time actual watch-crystals were used). The pans and glasses were filled with fresh sea water, and into them we distributed the animals by families—all the crabs in one, anemones in another, snails in another, and delicate things like flatworms and hydroids in others. From this distribution it was easier to separate them finally by species.[5]

[5] Because the information is not readily and compactly available and because of the many and bitter complaints by specialists against the condition of labeling of specimens brought in by many expeditions, we have devoted a separate note, page 272, to a description of some methods of killing and preserving marine animals. These are the results of many trials and errors by ourselves and others. Some of the methods given are far from perfect, but they are the best we know.

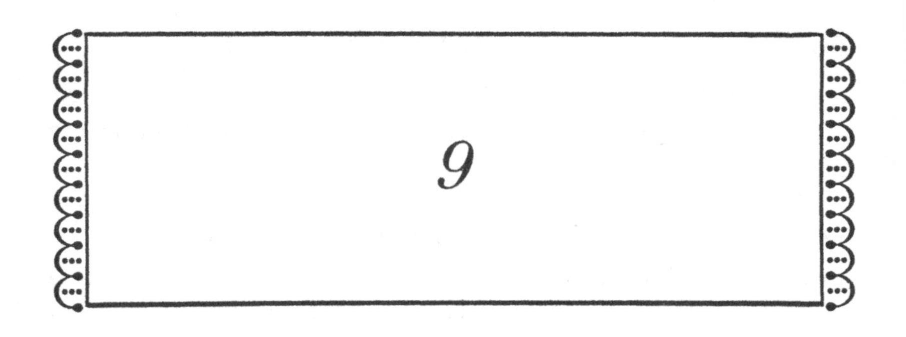

WHEN the catch was sorted and labeled, we went ashore to the cannery and later drove with Chris, the manager, and Señor Luis, the port captain, to the little town of San Lucas. It was a sad little town, for a winter storm and a great surf had wrecked it in a single night. Water had driven past the houses, and the streets of the village had been a raging river. "Then there were no roofs over the heads of the people," Señor Luis said excitedly. "Then the babies cried and there was no food. Then the people suffered."

The road to the little town, two wheel-ruts in the dust, tossed us about in the cannery truck. The cactus and thorny shrubs ripped at the car as we went by. At last we stopped in front of a mournful *cantina* where morose young men hung about waiting for something to happen. They had waited a long time—several generations—for something to happen, these good-looking young men. In their eyes there was a hopelessness. The storm of the winter had been discussed so often that it was sucked dry. And besides, they all knew the same things about it. Then we happened to them. The truck pulled up to the *cantina* door and we—strangers, foreigners—stepped out, as disorderly-looking a group as had ever come to their *cantina*. Tiny wore a Navy cap of white he had traded for, he said, in a washroom in San Diego. Tony still had his snap-brim felt. There were yachting caps and sweaters, and jeans stiff with fish blood. The young men stirred to life for a little while, but we were not enough. The flood had been much better. They relapsed again into their gloom.

There is nothing more doleful than a little *cantina.* In the first place it is inhabited by people who haven't any money to buy a drink. They stand about waiting for a miracle that never happens: the angel with golden wings who settles on the bar and orders drinks for everyone. This never happens, but how are the sad handsome young men to know it never will happen? And suppose it did happen and they were somewhere else? And so they lean against the wall; and when the sun is high they sit down against the wall. Now and then they go away into the brush for a while, and they go to their little homes for meals. But that is an impatient time, for the golden angel might arrive. Their faith is not strong, but it is permanent.

We could see that we did not greatly arouse them. The *cantina* owner promptly put his loudest records on the phonograph to force a gaiety into this sad place. But he had Carta Blanca beer and (at the risk of a charge that we have sold our souls to this brewery) we love Carta Blanca beer. There was no ice, no electric lights, and the gasoline lanterns hissed and drew the bugs from miles away. The cockroaches in their hordes rushed in to see what was up. Big, handsome cockroaches, with almost human faces. The loud music only made us sadder, and the young men watched us. When we lifted a split of beer to our lips the eyes of the young men rose with our hands, and even the cockroaches lifted their heads. We couldn't stand it. We ordered beer all around, but it was too late. The young men were too far gone in sorrow. They drank their warm beer sadly. Then we bought straw hats, for the sun is deadly here. There should be a kind of ridiculous joy in buying a floppy hat, but those young men, so near to tears, drained even that joy. Their golden angel had come, and they did not find him good. We felt rather as God would feel when, after all the preparation of Paradise, all the plannings for eternities of joy, all the making and tuning of harps, the street-paving with gold, and the writing of hosannas, at last He let in the bleacher customers and they looked at the heavenly city and wished to be again in Brooklyn. We told funny stories, knowing they wouldn't be enjoyed, tiring of them ourselves before the point was reached. Nothing was fun in that little *cantina.* We started back for the boat. I think those young men were glad to see us go; because once we were gone, they could begin to build us up, but present, we inhibited their imaginations.

At the bar Chris told us of a native liquor called *damiana,* made from an infusion of a native herb, and not much known outside of Baja California. Chris said it was an aphrodisiac, and told some interesting stories to prove it. We felt a scientific interest in his stories, and bought a bottle of *damiana,* intending to subject it to certain tests under laboratory conditions. But the customs officials of San Diego took it away from us, not because of its romantic aspect, but because it had alcohol in it. Thus we were never able to give it a truly scientific testing. We think we were going to use it on a white rat. Tiny said he didn't want any such stuff getting in his way when he felt lustful.

There doesn't seem to be a true aphrodisiac; there are excitants like cantharides, and physical aids to the difficulties of psychic traumas, like yohimbine sulphate; there are strong protein foods like *bêche-de-mer* and the gonads of sea-urchins, and the much over-rated oyster; even chiles, with their irritating qualities, have some effect, but there seems to be no true aphrodisiac, no sweet essence of that goddess to be taken in a capsule. A certain young person said once that she found sexual intercourse an aphrodisiac; certainly it is the only good one.

So many people are interested in this subject but most of them are forced to pretend they are not. A man, for his own ego's sake, must, publicly at least, be over-supplied with libido.

But every doctor knows so well the "friend of the client" who needs help. He is the same "friend" who has gonorrhea, the same "friend" who needs the address of an abortionist. This elusive friend—what will we not do to help him out of his difficulties; the nights we spend sleepless, worrying about him! He is interested in an aphrodisiac; we must try to find him one. But the *damiana* we brought back for our "friend" possibly just now is in the hands of the customs officials in San Diego. Perhaps they too have a friend. Since we suggested the qualities of *damiana* to them, it is barely possible that this fascinating liquor has already been either devoted to a friend or even perhaps subjected to a stern course of investigation under laboratory conditions.

We have wondered about the bawdiness this book must have if it is to be true. Bawdiness, vulgarity—call it what you will—is such a relative matter, so much a matter of attitude. A man we know once long ago worked for a wealthy family in a country place. One morning one of the cows had a calf. The children of the house went down with him to watch her. It was a good normal birth, a perfect presentation, and the cow needed no help. The children asked questions and he answered them. And when the emerged head cleared through the sac, the little black muzzle appeared, and the first breath was drawn, the children were fascinated and awed. And this was the time for their mother to come screaming down on the vulgarity of letting the children see the birth. This "vulgarity" had given them a sense of wonder at the structure of life, while the mother's propriety and gentility supplanted that feeling with dirtiness. If the reader of this book is "genteel," then this is a very vulgar book, because the animals in a tide pool have two major preoccupations: first, survival, and second, reproduction. They reproduce all over the place. We could retire into obscure phrases or into Greek or Latin. This, for some reason, protects the delicate. In an earlier time biologists made their little jokes that way, as in the naming of the animals. But some later men found their methods vulgar. Verrill, in *The Actinaria of the Canadian Arctic Expeditions,* broke out in protest. He cries, "Prof. McMurrich has endeavored to restore for this species a name (*senilis*) used by Linnaeus for a small indeterminable species very imperfectly described in 1761. . . . The description does not in the least apply to this species. He described the thing as the size of the last joint of a finger, sordid, rough, with a sub-coriaceous tunic. Such a description could not possibly apply to this soft and smooth species . . . but it would be mere guesswork to say what species he had in view. . . . Moreover, aside from this uncertainty, most modern writers have rejected most of the Linnaean names of actinians on account of their obscenity or indecency. All this confusion shows the impossibility of fixing the name, even if it were not otherwise objectionable. It should be forgotten or ignored, like the generic names used by Linnaeus in 1761, and by some others of that period, for species of Actinia. Their indecent names were usually the Latinized forms of vulgar names used by fishermen, some of which are still in use among the fishermen of our own coasts, for similar things."

This strange attempt to "clean up" biology will have, we hope, no effect whatever. We at least have kept our vulgar sense of wonder. We are no better than the animals; in fact in a lot of ways we aren't as good. And so we'll let the book fall as it may.

* * *

We left the truck and walked through the sandy hills in the night, and in this latitude the sky seemed very black and the stars very white. Already the smell of the land was gone from our noses, for we were used to the smell of vegetation again. The beer was warm in us and pleasant, and the air had a liquid warmth that was really there without the beer, for we tested it later. In the brush beside the track there was a little heap of light, and as we came closer to it we saw a rough wooden cross lighted indirectly. The cross-arm was bound to the staff with a thong, and the whole cross seemed to glow, alone in the darkness. When we came close we saw that a kerosene can stood on the ground and that in it was a candle which threw its feeble light upward on the cross. And our companion told us how a man had come from a fishing boat, sick and weak and tired. He tried to get home, but at this spot he fell down and died. And his family put the little cross and the candle there to mark the place. And eventually they would put up a stronger cross. It seems good to mark and to remember for a little while the place where a man died. This is his one whole lonely act in all his life. In every other thing, even in his birth, he is bound close to others, but the moment of his dying is his own. And in nearly all of Mexico such places are marked. A grave is quite a different thing. Here one's family boasts, or lies, or excuses, in material of elegance and extravagance. But that is a family or a social matter, not the dead man's own at all. The unmarked cross and the secret light are his; almost a reflection of the last piercing loneliness that comes into a dying man's eyes.

From a few feet away the cross seemed to flicker unsubstantially with a small yellow light, seemed to be almost a memory while we saw it. And the man who tried to get home and crawled this far—we never knew his name but he stays in our memory too, for some reason—a supra-personal being, a slow, painful symbol and a pattern of his whole species which tries always from generation to generation, man and woman, which struggles always to get home but never quite makes it.

We came back to the pier and got into our little boat. The Sea-Cow of course would not start, it being night time, so we rowed out to the *Western Flyer*. Before we started, by some magic, there on the end of the pier stood the sad beautiful young men watching us. They had not moved; some jinni had picked them up and transported them and set them down. They watched us put out into the darkness toward our riding lights, and then we suppose they were whisked back again to the *cantina,* where the proprietor was putting the records away and feeling with delicate thumbs the dollar bills we had left. On the pier no light burned, for the engine had stopped at sundown. We went to bed; there was a tide to be got to in the morning.

On the beach at San Lucas there is a war between the pigs and the vultures. Sometimes one side dominates and sometimes the other. On occasion the swine feel a dynamism and demand *Lebensraum,* and in the pride of their species drive the vultures from the decaying offal. And again, when their thousand years of history is over, the vultures spring to arms, tear up treaties, and flap the pigs from the garbage. And on the beach there are certain skinny dogs, without any dynamisms whatever and without racial pride, who nevertheless manage to get the best snacks. They don't thrive on it—always they are meager and skinny and cowardly—but when the *Gauleiter* swine has just captured a fish belly, and before he can shout his second *"Sieg Heil!"* the dog has it.

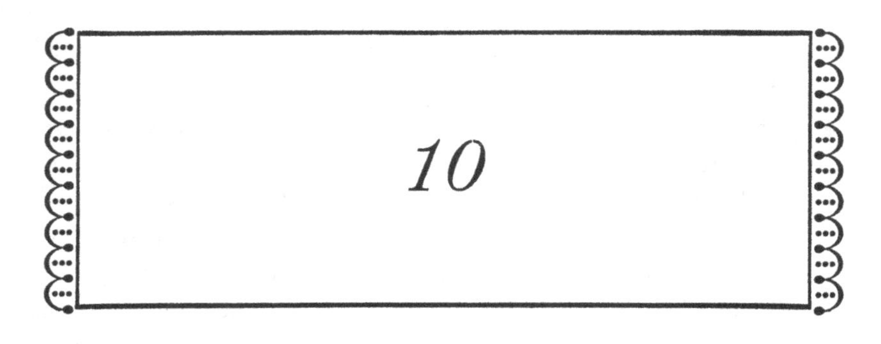

10

March 18

THE tidal series was short. We wished to cover as much ground as possible, to establish as many collecting stations as we could, for we wanted a picture as nearly whole of the Gulf as possible. The next morning we got under way to run the short distance to Pulmo Reef, around the tip and on the eastern shore of the Peninsula. It was a brilliant day, the water riffled and very blue, the sandy beaches of the shore shining with yellow intensity. Above the beaches the low hills were dark with brush. Many people had come to Cape San Lucas, and many had described it. We had read a number of the accounts, and of course agreed with none of them. To a man straight off a yacht, it is a miserable little flea-bitten place, poor and smelly. But to one who puts in hungry, in a storm-beaten boat, it must be a place of great comfort and warmth. These are extremes, but the area in between them also has its multiform conditioning, and what we saw had our conditioning. Once we read a diary, written by a man who came through Panama in 1839. He had read about the place before he got there, but the account he read was about the old city, and in his diary, written after he had gone through, he set down a description of the city he had read about. He didn't know that the town in the book had been destroyed, and that the new one was not even in the same place, but he was not disturbed by these discrepancies. He knew what he would find there and he found it.

There is a curious idea among unscientific men that in scientific writing there is a common plateau of perfectionism. Nothing could be more untrue. The reports of biologists are the measure, not of the science, but of the men themselves. There are as few scientific giants as any other kind. In some reports it is impossible, because of inept expression, to relate the descriptions to the living animals. In some papers collecting places are so mixed or ignored that the animals mentioned cannot be found at all. The same conditioning forces itself into specification as it does into any other kind of observation, and the same faults of carelessness will be found in scientific reports as in the witness chair of a criminal court. It has seemed sometimes that the little men in scientific work assumed the awe-fullness of a priesthood to hide their deficiencies, as the witch-doctor does with his stilts and high masks, as the priesthoods of all cults have, with secret or unfamiliar languages and symbols. It is usually found that only the little stuffy men object to what is called "popularization," by which they mean writing with a clarity understandable to one not familiar with the tricks and codes of the cult. We have not known a single great scientist who could not discourse freely and interestingly with a child. Can it be that the haters of clarity have nothing to say, have observed nothing, have no clear picture of even their own fields? A dull man seems to be a dull man no matter what his field, and of course it is the right of a dull scientist to protect himself with feathers and robes, emblems and degrees, as do other dull men who are potentates and grand imperial rulers of lodges of dull men.

* * *

As we neared Pulmo Reef, Tony sent a man up the mast to the crow's-nest to watch for concealed rocks. It is possible to see deep into the water from that high place; the rocks seem to float suddenly up from the bottom like dark shadows. The water in this shallow area was green rather than blue, and the sandy bottom was clearly visible. We pulled in as close as was safe and dropped our anchor. About a mile away we could see the proper reef with the tide beginning to go down on it. On the shore behind the white beach was one of those lonely little *rancherias* we came to know later. Usually a palm or two are planted near by, and by these trees sticking up out of the brush one can locate the houses. There is usually a small corral, a burro or two, a few pigs, and some scrawny chickens. The cattle range wide for food. A dugout canoe lies on the beach, for a good part of the food comes from the sea. Rarely do you see a light from the sea, for the people go to sleep at dusk and awaken with the first light. They must be very lonely people, for they appear on shore the moment a boat anchors, and paddle out in their canoes. At Pulmo Reef the little canoe put off and came alongside. In it were two men and a woman, very ragged, their old clothes patched with the tatters of older clothes. The *serapes* of the men were so thin and threadbare that the light shone through them, and the woman's *rebozo* had long lost its color. They sat in the canoe holding to the side of the *Western Flyer*, and they held their greasy blankets carefully over their noses and mouths to protect themselves from us. So much evil the white man had brought to their ancestors: his breath was poisonous with the lung disease; to sleep with him was to poison the generations. Where he set down his colonies the indigenous people withered and died. He brought industry and trade but no prosperity, riches but no ease. After four hundred years of him these people have ragged clothes and the shame that forces the wearing of them; iron harpoons for their hands, syphilis and tuberculosis; a few of the white man's less complex neuroses, and a curious devotion to a God who was sacrificed long ago in the white man's country. They know the white man is poisonous and they cover their noses against him. They do find us fascinating. However, they sit on the rail for many hours watching us and waiting. When we feed them they eat and are courteous about it, but they did not come for food, they are not beggars. We give the men some shirts and they fold them and put them into the bow of the canoe, but they did not come for clothing. One of the men at last offers us a match-box in which are a few misshapen little pearls like small pale cancers. Five pesos he wants for the pearls, and he knows they aren't worth it. We give him a carton of cigarettes and take his pearls, although we do not want them, for they are ugly little things. Now these three should go, but they do not. They would stay for weeks, not moving nor talking except now and then to one another in soft little voices as gentle as whispers. Their dark eyes never leave us. They ask no questions. They seem actually to be dreaming. Sometimes we asked of the Indians the local names of animals we had taken, and then they consulted together. They seemed to live on remembered things, to be so related to the seashore and the rocky hills and the loneliness that they are these things. To ask about the country is like asking about themselves. "How many toes have you?" "What, toes? Let's see—of course, ten. I have known them all my life, I never thought to count them. Of course it will rain tonight, I don't know why. Something in me tells me I will rain tonight. Of course, I am the whole thing, now that I think about it. I ought to know when I will rain." The dark eyes, whites brown and stained, have curious red lights in the pupils. They seem to be a dreaming people. If finally you must escape their eyes, their timeless dreaming eyes, you have only to say, *"Adiós, señor,"* and they seem to start awake. *"Adiós,"* they say softly. *"Que vaya con Dios."* And they paddle away. They bring a hush with them, and when they go away one's own voice sounds loud and raw.

We loaded the smaller skiff with collecting materials: the containers and bars, tubes and buckets. We put the Sea-Cow on the stern and it made one of its few mistakes. It thought we were going directly to the beach instead of to the reef a mile away. It started up with a great roar and ran for a quarter of a mile before it became aware of its mistake. It was rarely fooled again. We rowed on to the reef.

Collecting in this region, we always wore rubber boots. There are many animals which sting, some severely, and at least one urchin which is highly poisonous. Some of the worms, such as *Eurythoë,* leave spines in the skin which burn unmercifully. And even a barnacle cut infects readily. It is impossible to wear gloves; one must simply be as careful as possible and look where the finger is going before it is put down. Some of the little beasts are incredibly gallant and ferocious. On one occasion, a moray eel not more than eight inches long lashed out from under a rock, bit one of us on the finger, and retired. If one is not naturally cautious, painful and bandaged hands very soon teach caution. The boots protect one's feet from nearly everything, but there is an urchin which has spines so sharp that they pierce the rubber and break off in the flesh, and they sting badly and usually cause infection.

Pulmo is a coral reef. It has often been remarked that reef-building corals seem to live only on the eastern sides of large land bodies, not on the western sides. This has been noticed many times, and even here at Pulmo the reef-building coral [1] occurs only on the eastern side of the Peninsula. This can have nothing to do with wave-shock or current, but must be governed by another of those unknown factors so ever-present and so haunting to the ecologist.

The complexity of the life-pattern on Pulmo Reef was even greater than at Cape San Lucas. Clinging to the coral, growing on it, burrowing into it, was a teeming fauna. Every piece of the soft material broken off skittered and pulsed with life—little crabs and worms and snails. One small piece of coral might conceal thirty or forty species, and the colors on the reef were

[1] *Pocillopora capitata* Verrill.

electric. The sharp-spined urchins [2] gave us trouble immediately, for several of us, on putting our feet down injudiciously, drove the spines into our toes.

The reef was gradually exposed as the tide went down, and on its flat top the tide pools were beautiful. We collected as widely and rapidly as possible, trying to take a cross-section of the animals we saw. There were purple pendent gorgonians like lacy fans; a number of small spine-covered puffer fish which bloat themselves when they are attacked, erecting the spines; and many starfish, including some purple and gold cushion stars. The club-spined sea-urchins [3] were numerous in their rock niches. They seemed to move about very little, for their niches always just fit them, and have the marks of constant occupation. We took a number of the slim green and brown starfish [4] and the large slim five-rayed starfish with plates bordering the ambulacral grooves.[5] There were numbers of barnacles and several types of brittle-stars. We took one huge, magnificent murex snail. One large hemispherical snail was so camouflaged with little plants, corallines, and other algae that it could not be told from the reef itself until it was turned over. Rock oysters there were, and oysters; limpets and sponges; corals of two types; peanut worms; sea-cucumbers; and many crabs, particularly some disguised in dresses of growing algae which made them look like knobs on the reef until they moved. There were many worms, including our enemy *Eurythoë,* which stings so badly. This worm makes one timid about reaching without looking. The coral clusters were violently inhabited by snapping shrimps, red smooth crabs,[6] and little fuzzy black and white spider crabs.[7] Autotomy in these crabs, shrimps,

---

[2] *Arbacia incisa.*
[3] *Eucidaris thouarsii.*
[4] *Phataria unifascialis* Gray.
[5] *Pharia pyramidata.*
[6] *Trapezia* spp.
[7] *Mithrax areolatus.*

and brittle-stars is very highly developed. At last, under the reef, we saw a large fleshy gorgonian, or sea-fan, waving gently in the clear water, but it was deep and we could not reach it. One of us took off his clothes and dived for it, expecting at any moment to be attacked by one of those monsters we do not believe in. It was murky under the reef, and the colors of the sponges were more brilliant than in those exposed to greater light. The diver did not stay long; he pulled the large sea-fan free and came up again. And although he went down a number of times, this was the only one of this type of gorgonian he could find. Indeed, it was the only one taken on the entire trip.

The collecting buckets and tubes and jars were very full of specimens—so full that we had constantly to change the water to keep the animals alive. Several large pieces of coral were taken and kept submerged in buckets and later were allowed to lie in stale sea water in one of the pans. This is an interesting thing, for as the water goes stale, the thousands of little roomers which live in the tubes and caves and interstices of the coral come out of hiding and scramble for a new home. Worms and tiny crabs appear from nowhere and are then easily picked up.

The sea bottom inside the reef was of white sand studded with purple and gold cushion stars, of which we collected many. And lying on the sandy bottom were heads and knobs of another coral,[8] much harder and more regularly formed than the reef-building coral. The rush of collecting as much as possible before the tide re-covered the reef made us indiscriminate in our collecting, but in the long run this did not matter. For once on board the boat again we could re-collect, going over the pieces of coral and rubble carefully and very often finding animals we had not known were there.

El Pulmo was the only coral reef we found on the entire expedition, and the fauna and even the algae were rather specialized to it. No very great surf could have beaten it, for extremely

---

[8] *Porites porosa* Verrill.

delicate animals lived on its exposed top where they would have been crushed or washed away had strong seas struck them. And the competition for existence was as great as it had been at San Lucas, but it seemed to us that different methods were employed for frustrating enemies. Whereas at San Lucas speed and ferocity were the attributes of most animals, at Pulmo concealment and camouflage were largely employed. The little crabs wore masks of algae and bryozoa and even hydroids, and most animals had little tunnels or some protected place to run to. The softness of the coral made this possible, where the hard smooth granite of San Lucas had forbidden it. On several occasions we wished for diving equipment, but never more than here at Pulmo, for the under-cut shoreward side of the reef concealed hazy wonders which we could not get at. It is not satisfactory to hold one's breath and to look with unglassed eyes through the dim waters.

The water behind the reef was very warm. We abandoned our boots and, putting on tennis shoes to protect our feet from various stingers, we dived again and again for perfect knobs of coral.

Again we tried to start the Sea-Cow—and then rowed back to the *Western Flyer*. There we complained so bitterly to Tex, the engineer, that he took the evil little thing to pieces. Piece by piece he examined it, with a look of incredulity in his eyes. He admired, I think, the ingenuity which could build such a perfect little engine, and he was astonished at the concept of building a whole motor for the purpose of not running. Having put it together again, he made a discovery. The Sea-Cow would run perfectly out of water—that is, in a barrel of water with the propeller and cooling inlet submerged. Placed thus, the Sea-Cow functioned perfectly and got good mileage.

Immediately on arriving back at the *Western Flyer* we pulled up the anchor and got under way again. It was efficient that we preserve and label while we sailed as long as the sea was calm, and now it was very calm. The great collection from the reef required every enameled pan and glass dish we had. The killing and relaxing and preserving took us until dark, and even after dark we sat and made the labels to go into the tubes. As the jars filled and were labeled, we put them back in their corrugated-paper cartons and stowed them in the hold. The corked tubes were tested for leaks, then wrapped in paper toweling and stacked in boxes. Thus there was very small loss from breakage or leakage, and by labeling the same day as collecting, there had thus far been virtually no confusion in the tabulation of animals. But we knew already that we had made one error in planning: we had not brought nearly enough small containers. It is best to place an animal alone in a jar or a tube which accommodates him, but not too freely. The enormous numbers of animals we took strained our resources and containers long before we were through.

As we moved up the Gulf, the mirage we had heard about began to distort the land. While it is worse on the Sonora coast, it is sufficiently interesting on the Peninsula to produce a heady, crazy feeling in the observer. As you pass a headland it suddenly splits off and becomes an island and then the water seems to stretch inward and pinch it to a mushroom-shaped cliff, and finally to liberate it from the earth entirely so that it hangs in the air over the water. Even a short distance offshore one cannot tell what the land really looks like. Islands too far off, according to the map, are visible; while others which should be near by cannot be seen at all until suddenly they come bursting out of the mirage. The whole surrounding land is unsubstantial and changing. One remembers the old stories of invisible kingdoms where princes lived with ladies and dragons for company; and the more modern fairy-tales in which heroes drift in and out of dimensions more complex than the original three. We are open enough to miracles of course, but what must have been the feeling of the discovering Spaniards? Miracles were daily happenings to them. Perhaps to that extent their feet were more firmly planted on the ground. Subject as they were to the constant apparitions of saints, to the trooping of holy virgins into their dreams and reveries, perhaps mirages were commonplaces. We have seen many miraculous figures in Mexico. They are usually Christs which have supernaturally appeared on mountains or in caves and usually at times of crisis. But it does seem odd that the heavenly authorities, when they wished a miraculous image to appear, invariably chose bad Spanish wood-carving of the seventeenth century. But perhaps art criticism in heaven was very closely related to the sensibilities of the time. Certainly it would have been a little shocking to find an Epstein Christ under a tree on a mountain in Mexico, or a Brancusi bird, or a Dali *Descent from the Cross*.

It must have been a difficult task for those first sturdy Jesuit fathers to impress the Indians of the Gulf. The very air here is miraculous, and outlines of reality change with the moment. The sky sucks up the land and disgorges it. A dream hangs over the whole region, a brooding kind of hallucination. Perhaps only the shock of seventeenth-century wood-carving could do the trick; surely the miracle must have been very virile to be effective.

Tony grew restive when the mirage was working, for here right and wrong fought before his very eyes, and how could one tell which was error? It is very well to say, "The land is here and what blots it out is a curious illusion caused by light and air and moisture," but if one is steering a boat, he must sail by what he sees, and if air and light and moisture—three realities —plot together and perpetrate a lie, what is a realistic man to believe? Tony did not like the mirage at all.

While we worked at the specimens, the trolling lines were out and we caught another skipjack, large and fat and fast. As it came in on the line, one of us ran for the moving-picture camera, for we wanted to record on color film the changing tints and patterns of the fish's dying. But the exposure was wrong as usual, and we did not get it.

Near the moving boat swordfishes played about. They seemed to play in pure joy or exhibitionism. It is thought that they leap to clear themselves of parasites; they jump clear of the water and come crashing down, and sometimes they turn over in the air and flash in the sunshine. This afternoon, too, we saw the first specimens of the great manta ray (a giant skate), and we rigged the harpoons and coiled the line ready. One light harpoon just pierced a swordfish's tail, but he swished away, for the barb had not penetrated. And we did not turn and pursue the great rays, for we wished to anchor that night near Point Lobos on Espíritu Santo Island.

In the evening we came near to it, but as we prepared to anchor, the wind sprang up full on us, and Tony decided to run for the shelter of Pescadero Point on the mainland. The wind seemed to grow instantly out of the evening, and the sea with it. The jars and collecting pans were in danger of flying overboard. For half an hour we were very busy tying the equipment down and removing the flapping canvas we had stretched to keep the sun off our specimen pans. Under the powerful wind we crossed the channel which leads to La Paz, and saw the channel light—the first one we had seen since the big one on the false cape. This one seemed very strange in the Gulf. The waves were not high, but the wind blew with great intensity, making whitecaps rather than rollers, and only when we ran in under Pescadero Point did we drop the wind. We eased in slowly, sounding as we went. When the anchor was finally down we cooked and ate the skipjack, a most delicious fish. And after dinner a group action took place.

We carried no cook and dishwasher; it had been understood that we would all help. But for some time Tex had been secretly mutinous about washing dishes. At the proper times he had things to do in the engine-room. He might have succeeded in this crime, if he had ever varied his routine, but gradually a suspicion grew on us that Tex did not like to wash dishes. He denied this vigorously. He said he liked very much to wash dishes. He appealed to our reason. How would we like it, he argued, if we were forever in the engine-room, getting our hands dirty? There was danger down there too, he said. Men had been killed by engines. He was not willing to see us take the risk. We met his arguments with a silence that made him nervous. He protested then that he had once washed dishes from west Texas to San Diego without stopping, and that he had learned to love it so much that he didn't want to be selfish about it now. A circle of cold eyes surrounded him. He began to sweat. He said that later (he didn't say how much later) he was going to ask us for the privilege of washing all the dishes, but right now he had a little job to do in the engine-room. It was for the safety of the ship, he said. No one answered him. Then he cried, "My God, are you going to hang me?" At last Sparky spoke up, not unkindly, but inexorably. "Tex," he said, "you're going to wash 'em or you're going to sleep with 'em." Tex said, "Now just as soon as I do one little job there's nothing I'd rather do than wash four or five thousand dishes." Each of us picked up a load of dishes, carried them in, and laid them gently in Tex's bunk. He got up resignedly then and carried them back and washed them. He didn't grumble, but he was broken. Some joyous light had gone out of him, and he never did get the catsup out of his blankets.

That night Sparky worked at the radio and made contact with the fishing fleet that was operating in the region from Cedros Island and around the tip into the Gulf, fishing for tuna. Fishermen are no happier than farmers. It is difficult to see why anyone becomes a farmer or a fisherman. Dreadful things happen to them constantly: they lose their nets; the fish are wild; sea-lions get into the nets and tear their way out; snags are caught; there are no fish, and the price high; there are too many fish, and the price is low; and if some means could be devised so that the fish swam up to a boat, wriggled up a trough, squirmed their way into the fish-hold, and pulled ice over themselves with their own fins, the imprecations would be terrible because they had not removed their own entrails and brought their own ice. There is no happiness for fishermen anywhere. Cries of anguish at the injustice of the elements inundated the short-wave receiver as we lay at anchor.

The pattern of a book, or a day, of a trip, becomes a characteristic design. The factors in a trip by boat, the many-formed personality phases all shuffled together, changing a little to fit into the box and yet bringing their own lumps and corners, make the trip. And from all these factors your expedition has a character of its own, so that one may say of it, "That was a good, kind trip." Or, "That was a mean one." The character of the whole becomes defined and definite. We ran from collecting station to new collecting station, and when the night came and the anchor was dropped, a quiet came over the boat and the trip slept. And then we talked and speculated, talked and drank beer. And our discussions ranged from the loveliness of remembered women to the complexities of relationships in every other field. It is very easy to grow tired at collecting; the period of a low tide is about all men can endure. At first the rocks are bright and every moving animal makes his mark on the attention. The picture is wide and colored and beautiful. But after an hour and a half the attention centers weary, the colors fade, and the field is likely to narrow to an individual animal. Here one may observe his own world narrowed down until interest and, with it, observation, flicker and go out. And what if with age this weariness become permanent and observation dim out and not recover? Can this be what happens to so many men of science? Enthusiasm, interest, sharpness, dulled with a weariness until finally they retire into easy didacticism? With this weariness, this stultification of the attention centers, perhaps there comes the pained and sad memory of what the old excitement was like, and regret might turn to envy of the men who still have it. Then out of the shell of didacticism, such a used-up man might attack the unwearied, and he would have in his hands proper weapons of attack. It does seem certain that to a wearied man an error in a mass of correct data wipes out all the correctness and is a focus for attack; whereas the unwearied man, in his energy and receptivity, might consider the little dross of error a by-product of his effort. These two may balance and produce a purer thing than either in the end. These two may be the stresses which hold up the structure, but it is a sad thing to see the interest in interested men thin out and weaken and die. We have known so many professors who once carried their listeners high on their single enthusiasm, and have seen these same men finally settle back comfortably into lectures prepared years before and never vary them again. Perhaps this is the same narrowing we observe in relation to ourselves and the tide pool—a man looking at reality brings his own limitations to the world. If he has strength and energy of mind the tide pool stretches both ways, digs back to electrons and leaps space into the universe and fights out of the moment into non-conceptual time. Then ecology has a synonym which is ALL.

It is strange how the time sense changes with different peoples. The Indians who sat on the rail of the *Western Flyer* had a different time sense—"time-world" would be the better term—from ours. And we think we can never get into them unless we can invade that time-world, for this expanding time seems to trail an expanding universe, or perhaps to lead it. One considers the durations indicated in geology, in paleontology, and, thinking out of our time-world with its duration between time-stone and time-stone, says, "What an incredible interval!" Then, when one struggles to build some picture of astro-physical time, he is faced with a light-year, a thought-deranging duration unless the relativity of all things intervenes and time expands and contracts, matching itself relatively to the pulsings of a relative universe.

It is amazing how the strictures of the old teleologies infect our observation, causal thinking warped by hope. It was said earlier that hope is a diagnostic human trait, and this simple cortex symptom seems to be a prime factor in our inspection of our universe. For hope implies a change from a present bad condition to a future better one. The slave hopes for freedom, the weary man for rest, the hungry for food. And the feeders of hope, economic and religious, have from these simple strivings of dissatisfaction managed to create a world picture which is very hard to escape. Man grows toward perfection; animals grow toward man; bad grows toward good; and down toward up, until our little mechanism, hope, achieved in ourselves probably to cushion the shock of thought, manages to warp our whole world. Probably when our species developed the trick of memory and with it the counterbalancing projection called "the future," this shock-absorber, hope, had to be included in the series, else the species would have destroyed itself in despair. For if ever any man were deeply and unconsciously sure that his future would be no better than his past, he might deeply wish to cease to live. And out of this therapeutic poultice we build our iron teleologies and twist the tide pools and the stars into the pattern. To most men the most hateful statement possible is, *"A thing is because it is."* Even those who have managed to drop the leading-strings of a Sunday-school deity are still led by the unconscious teleology of their developed trick. And in saying that hope cushions the shock of experience, that one trait balances the directionalism of another, a teleology is implied, unless one know or feel or think that we *are* here, and that without this balance, hope, our species in its blind mutation might have joined many, many others in extinction. Dr. Torsten Gislén, in his fine paper on fossil echinoderms called "Evolutional Series toward Death and Renewal," [9] has shown that as often as not, in his studied group at least, mutations have had destructive, rather than survival value. Extending this thesis, it is interesting to think of the mutations of our own species. It is said and thought there has been none in historical times. We wonder, though, where in man a mutation might take place. Man is the only animal whose interest and whose drive are outside himself. Other animals may dig holes to live in; may weave nests or take possession of hollow trees. Some species, like bees or spiders, even create complicated homes, but they do it with the fluids and processes of their own bodies. They make little impression on the world. But the world is furrowed and cut, torn and blasted by man. Its flora has been swept away and changed; its mountains torn down by man; its flat lands littered by the debris of his living. And these changes have been wrought, not because any inherent technical ability has demanded them, but because his desire has created that technical ability. Physiological man does not require this paraphernalia to exist, but the whole man does. He is the only animal who lives outside of himself, whose drive is in external things—property, houses, money, concepts of power. He lives in his cities and his factories, in his business and job and art. But having projected himself into these external complexities, he *is* them. His house, his automobile are a part of him and a large part of him. This is beautifully demonstrated by a thing doctors know—that when a man loses his possessions a very common result is sexual impotence. If then the projection, the preoccupation of man, lies in external things so that even his subjectivity is a mirror of houses and cars and grain elevators, the place to look for his mutation would be in the direction of his drive, or in other words in the external things he deals with. And here we can indeed readily find evidence of mutation. The

---

[9] *Ark. f. zool. K. Svenska Vetens.*, Vol. 26 A, No. 16, Stockholm, Jan. 1934.

industrial revolution would then be indeed a true mutation, and the present tendency toward collectivism, whether attributed to Marx or Hitler or Henry Ford, might be as definite a mutation of the species as the lengthening neck of the evolving giraffe. For it must be that mutations take place in the direction of a species drive or preoccupation. If then this tendency toward collectivization is mutation there is no reason to suppose it is for the better. It is a rule in paleontology that ornamentation and complication precede extinction. And our mutation, of which the assembly line, the collective farm, the mechanized army, and the mass production of food are evidences or even symptoms, might well correspond to the thickening armor of the great reptiles—a tendency that can end only in extinction. If this should happen to be true, nothing stemming from thought can interfere with it or bend it. Conscious thought seems to have little effect on the action or direction of our species. There is a war now which no one wants to fight, in which no one can see a gain—a zombie war of sleep-walkers which nevertheless goes on out of all control of intelligence. Some time ago a Congress of honest men refused an appropriation of several hundreds of millions of dollars to feed our people. They said, and meant it, that the economic structure of the country would collapse under the pressure of such expenditure. And now the same men, just as honestly, are devoting many billions to the manufacture, transportation, and detonation of explosives to protect the people they would not feed. And it must go on. Perhaps it is all a part of the process of mutation and perhaps the mutation will see us done for. We have made our mark on the world, but we have really done nothing that the trees and creeping plants, ice and erosion, cannot remove in a fairly short time. And it is strange and sad and again symptomatic that most people, reading this speculation which is *only* speculation, will feel that it is a treason to our species so to speculate. For in spite of overwhelming evidence to the contrary, the trait of hope still controls the future, and man, not a species, but a triumphant race, will approach perfection, and, finally, tearing himself free, will march up the stars and take his place where, because of his power and virtue, he belongs: on the right hand of the $\sqrt[\pi]{-1}$. From which majestic seat he will direct with pure intelligence the ordering of the universe. And perhaps when that occurs—when our species progresses toward extinction or marches into the forehead of God—there will be certain degenerate groups left behind, say, the Indians of Lower California, in the shadows of the rocks or sitting motionless in the dugout canoes. They may remain to sun themselves, to eat and starve and sleep and reproduce. Now they have many legends as hazy and magical as the mirage. Perhaps then they will have another concerning a great and godlike race that flew away in four-motored bombers to the accompaniment of exploding bombs, the voice of God calling them home.

* * *

Nights at anchor in the Gulf are quiet and strange. The water is smooth, almost solid, and the dew is so heavy that the decks are soaked. The little waves rasp on the shell beaches with a hissing sound, and all about in the darkness the fishes jump and splash. Sometimes a great ray leaps clear and falls back on the water with a sharp report. And again, a school of tiny fishes whisper along the surface, each one, as it breaks clear, making the tiniest whisking sound. And there is no feeling, no smell, no vibration of people in the Gulf. Whatever it is that makes one aware that men are about is not there. Thus, in spite of the noises of waves and fishes, one has a feeling of deadness and of quietness. At anchor, with the motor stopped, it is not easy to sleep, and every little sound starts one awake. The crew is restless and a little nervous. If a dog barks on shore or a cow bellows, we are reassured. But in many places of anchorage there were utterly no sounds associated with man. The crew read books they have not known about—Tony reads *Studs Lonigan* and says he does not like to see such words in print. And we are reminded that we once did not like to hear them spoken because we were not used to them. When we became used to hearing them, they took their place with the simple speech-sounds of the race of man. Tony read on in *Studs Lonigan,* and the shock of the new words he had not seen printed left him and he grew into the experience of Studs. Tiny read the book too. He said, "It's like something that happened to me."

Sometimes in the night a little breeze springs up and the boat tugs experimentally at the anchor and swings slowly around. There is nothing so quiet as a boat when the motor has stopped; it seems to lie with held breath. One gets to longing for the deep beat of the cylinders.

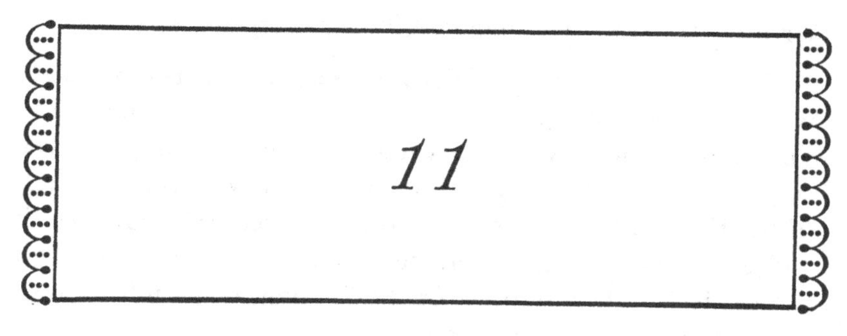

March 20

WE HAD marked the southern end of Espíritu Santo Island as our next collecting stop. This is a long narrow island which makes the northern side of the San Lorenzo Channel. It is mountainous and stands high and sheer from the blue water. We wanted particularly to collect there so that we could contrast the fauna of the eastern tip of this island with that of the secluded and protected bay of La Paz. Throughout we attempted to work in stations in the same area which nevertheless contrasted conditions for living, such as wave-shock, bottom, rock formation, exposure, depth, and so forth. The most radical differences in life forms are discovered in this way.

Early in the morning we sailed from our shelter under Pescadero Point and crossed the channel again. It was a very short run. There were many manta rays cruising slowly near the surface, with only the tips of their "wings" protruding above the water. They seemed to hover, and when we approached too near, they disappeared into the blue depths. Their effortless speed is astonishing. On the lines we caught two yellowfin tunas,[1] speedy and efficient fish. They struck the line so hard that it is impossible to see why they did not tear their heads off.

We anchored near a bouldery shore. This would be the first station in the Gulf where we would be able to turn over rocks, and a new ecological set-up was indicated by the fact that the small boulders rested in sand.

---

[1] *Neothunnus macropterus.*

This time everyone but Tony went ashore. Sparky and Tiny were already developing into good collectors, and now Tex joined us and quickly became excited in the collecting. We welcomed this help, for in general work, what with the shortness of the time and the large areas to be covered, the more hands and eyes involved, the better. Besides, these men who lived by the sea had a great respect for the sea and all its inhabitants. Association with the sea does not breed contempt.

The boulders on this beach were almost a perfect turning-over size—heavy enough to protect the animals under them from grinding by the waves, and light enough to be lifted. They were well coated with short algae and bedded in very coarse sand. The dominant species on this beach was a sulphury cucumber,[2] a dark, almost black-green holothurian which looks as though it were dusted with sulphur. As the tide dropped on the shallow beach we saw literally millions of these cucumbers. They lay in clusters and piles between the rocks and under the rocks, and as the tide went down and the tropical sun beat on the beach, many of them became quite dry without apparent injury. Most of these holothurians were from five to eight inches long, but there were great numbers of babies, some not more than an inch in length. We took a great many of them.

Easily the second most important animal of this shore in point of quantity was the brittle-star. We had read of their numbers in the Gulf and here they were, mats and clusters of them, giants under the rocks. It was simple to pick up a hundred at a time in black, twisting, squirming knots. There were five species of them, and these we took in large numbers also, for in preservation they sometimes cast off their legs or curl up into knots, and we wished to have a number of perfect specimens. Starfish were abundant here and we took six varieties. The difference between the brittle-star and the starfish is interestingly reflected in the scientific names—"Ophio" is a Greek root sig-

---

[2] *Holothuria lubrica.*

nifying "serpent"—the round compact body and long serpent-like arms of the brittle-star are suggested in the generic name "ophiuran," while the more truly star-like form of the starfish is recognizable in the Latin root "Astra," which occurs in so many of its proper names, "Heliaster," "Astrometis," etc. We found three species of urchins, among them the very sharp-spined and poisonous *Centrechinus mexicanus;* approximately ten different kinds of crabs, four of shrimps, a number of anemones of various types, a great number of worms, including our enemy *Eurythoë,* which seems to occur everywhere in the Gulf, several species of naked mollusks, and a good number of peanut worms. The rocks and the sand underneath them were heavily populated. There were chitons and keyhole limpets, a number of species of clams, flatworms, sponges, bryozoa, and numerous snails.

Again the collecting buckets were very full, but already we had begun the elimination of animals to be taken. On this day we took enough of the sulphury cucumbers and brittle-stars for our needs. These were carefully preserved, but when found again at a new station they would simply be noted in the collecting record, unless some other circumstance such as color change or size variation prevailed. Thus, as we proceeded, we gradually stopped collecting certain species and only noted them as occurring.

On board the *Western Flyer,* again we laid out the animals in pans and prepared them for anesthetization. In one of the sea-cucumbers we found a small commensal fish [3] which lived well inside the anus. It moved in and out with great ease and speed, resting invariably head inward. In the pan we ejected this fish by a light pressure on the body of the cucumber, but it quickly returned and entered the anus again. The pale, colorless appearance of this fish seemed to indicate that it habitually lived there.

---

[3] See § W-21, Appendix.

It is interesting to see how areas are sometimes dominated by one or two species. On this beach the yellow-green cucumber was everywhere, with giant brittle-stars a close second. Neither of these animals has any effective offensive property as far as we know, although neither of them seems to be a delicacy enjoyed by other animals. There does seem to be a balance which, when passed by a certain species, allows that animal numerically to dominate a given area. When this threshold of successful reproduction and survival is crossed, the area becomes the special residence of this form. Then it seems other animals which might be either hostile or perhaps the prey of the dominating animal would be wiped out or would desert the given area. In many cases the arrival and success of a species seem to be by chance entirely. In some northern areas, where the ice of winter yearly scours and cleans the rocks, it has been noted that summer brings sometimes one dominant species and sometimes another, the success factor seeming to be prior arrival and an early start.[4] With marine fauna, as with humans, priority and possession appear to be vastly important to survival and dominance. But sometimes it is found that the very success of an animal is its downfall. There are examples where the available food supply is so exhausted by the rapid and successful reproduction that the animal must migrate or die. Sometimes, also, the very by-products of the animals' own bodies prove poisonous to a too great concentration of their own species.

It is difficult, when watching the little beasts, not to trace human parallels. The greatest danger to a speculative biologist is analogy. It is a pitfall to be avoided—the industry of the bee, the economics of the ant, the villainy of the snake, all in human terms have given us profound misconceptions of the animals. But parallels are amusing if they are not taken too seriously as

---

[4] Gislén, T., "Epibioses of the Gullmar Fjord II." 1930, p. 157. Kristinebergs Zool. Sta. 1877–1927, *Skrift. ut. av K. Svenska Vetens.* N:r 4.

regards the animal in question, and are downright valuable as regards humans. The routine of changing domination is a case in point. One can think of the attached and dominant human who has captured the place, the property, and the security. He dominates his area. To protect it, he has police who know him and who are dependent on him for a living. He is protected by good clothing, good houses, and good food. He is protected even against illness. One would say that he is safe, that he would have many children, and that his seed would in a short time litter the world. But in his fight for dominance he has pushed out others of his species who were not so fit to dominate, and perhaps these have become wanderers, improperly clothed, ill fed, having no security and no fixed base. These should really perish, but the reverse seems true. The dominant human, in his security, grows soft and fearful. He spends a great part of his time in protecting himself. Far from reproducing rapidly, he has fewer children, and the ones he does have are ill protected inside themselves because so thoroughly protected from without. The lean and hungry grow strong, and the strongest of them are selected out. Having nothing to lose and all to gain, these selected hungry and rapacious ones develop attack rather than defense techniques, and become strong in them, so that one day the dominant man is eliminated and the strong and hungry wanderer takes his place.

And the routine is repeated. The new dominant entrenches himself and then softens. The turnover of dominant human families is very rapid, a few generations usually sufficing for their rise and flowering and decay. Sometimes, as in the case of Hearst, the rise and glory and decay take place in one generation and nothing is left. One dominant thing sometimes does survive and that is not even well defined; some quality of the spirit of an individual continues to dominate. Whereas the great force which was Hearst has died before the death of the man and will soon be forgotten except perhaps as a ridiculous and vulgar fable, the spirit and thought of Socrates not only survive, but continue as living entities.

There is a strange duality in the human which makes for an ethical paradox. We have definitions of good qualities and of bad; not changing things, but generally considered good and bad throughout the ages and throughout the species. Of the good, we think always of wisdom, tolerance, kindliness, generosity, humility; and the qualities of cruelty, greed, self-interest, graspingness, and rapacity are universally considered undesirable. And yet in our structure of society, the so-called and considered good qualities are invariable concomitants of failure, while the bad ones are the cornerstones of success. A man—a viewing-point man—while he will love the abstract good qualities and detest the abstract bad, will nevertheless envy and admire the person who through possessing the bad qualities has succeeded economically and socially, and will hold in contempt that person whose good qualities have caused failure. When such a viewing-point man thinks of Jesus or St. Augustine or Socrates he regards them with love because they are the symbols of the good he admires, and he hates the symbols of the bad. But actually he would rather be successful than good. In an animal other than man we would replace the term "good" with "weak survival quotient" and the term "bad" with "strong survival quotient." Thus, man in his thinking or reverie status admires the progression toward extinction, but in the unthinking stimulus which really activates him he tends toward survival. Perhaps no other animal is so torn between alternatives. Man might be described fairly adequately, if simply, as a two-legged paradox. He has never become accustomed to the tragic miracle of consciousness. Perhaps, as has been suggested, his species is not set, has not jelled, but is still in a state of becoming, bound by his physical memories to a past of struggle and survival, limited in his futures by the uneasiness of thought and consciousness.

Back on the *Western Flyer*, Sparky cooked the tuna in a sauce of tomatoes and onions and spices and we ate magnificently. Each rock turned over had not been heavy, but we had turned over many tons of rocks in all. And now the work with the animals had to go on, the preservation and labeling. But we rested and drank a little beer, which in this condition of weariness is rest itself.

While we were eating, a boat came alongside and two Indians climbed aboard. Their clothing was better than that of the poor people of the day before. They were, after all, within a day's canoe trip of La Paz, and some of the veneer of that city had stuck to them. Their clothing was patched and ragged, but at least not falling apart from decay. We asked Sparky and Tiny to bring them a little wine, and after two glasses they became very affable, making us think of the intolerance of the Indian for alcohol. Later it developed that Sparky and Tiny had generously laced the wine with whisky, which proved just the opposite about the Indians' tolerance for alcohol. None of us could have drunk two tumblers, half whisky and half wine, but these men did and became gay and companionable. They were barefoot and carried the iron harpoons of the region, and in the bottom of their canoe lay a huge fish. Their canoe was typical of the region and was interesting. There are no large trees in the southern part of the Peninsula, hence all the canoes come from the mainland, most of them being made near Mazatlán. They are double-ended canoes carved from a single log of light wood, braced inside with struts. Sometimes a small sail is set, but ordinarily they are paddled swiftly by two men, one at either end. They are seaworthy and fast. The wood inside and out is covered with a thin layer of white or blue plaster, waterproof and very hard. This is made by the people themselves and applied regularly. It is not a paint, but a hard, shell-like plaster, and we could not learn how it is made although this is probably well known to many people. Equipped with one of these canoes, an iron harpoon, a pair of trousers, shirt, and hat, a young man is fairly well set up in life. In fact, the acquiring of a Nayarit canoe will probably give a young man so much security in his own eyes and make him so desirable in the eyes of others that he will promptly get married.

It is said so often and in such ignorance that Mexicans are contented, happy people. "They don't want anything." This, of course, is not a description of the happiness of Mexicans, but of the unhappiness of the person who says it. For Americans, and probably all northern peoples, are all masses of wants growing out of inner insecurity. The great drive of our people stems from insecurity. It is often considered that the violent interest in little games, the mental rat-mazes of contract bridge, and the purposeful striking of little white balls with sticks, comes from an inner sterility. But more likely it comes from an inner complication. Boredom arises not so often from too little to think about, as from too much, and none of it clear nor clean nor simple. Bridge is a means of forgetting the thousands of little irritations of a mind over-crowded with anarchy. For bridge has a purpose, that of taking as many tricks as possible. The end is clear and very simple. But nothing in the lives of bridge-players is clean-cut, and no ends are defined. And so they retire into some orderly process, even in a game, from the messy complication of their lives. It is possible, although we do not know this, that the poor Mexican Indian is a little less messy in his living, having a baby, spearing a fish, getting drunk, backing a political candidate; each one of these is a clear, free process, ending in a result. We have thought of this in regard to the bribes one sometimes gives to Mexican officials. This is universally condemned by Americans, and yet it is a simple, easy process. A bargain is struck, a price named, the money paid, a graceful compliment exchanged, the service performed, and it is over. He is not your man nor you his. A little process has been terminated. It is rather like the old-fashioned buying and selling for cash or produce.

We find we like this cash-and-carry bribery as contrasted with our own system of credits. With us, no bargain is struck, no price named, nothing is clear. We go to a friend who knows a judge. The friend goes to the judge. The judge knows a senator who has the ear of the awarder of contracts. And eventually we sell five carloads of lumber. But the process has only begun. Every member of the chain is tied to every other. Ten years later the son of the awarder of contracts must be appointed to Annapolis. The senator must have traffic tickets fixed for many years to come. The judge has a political lien on your friend, and your friend taxes you indefinitely with friends who need jobs. It would be simpler and cheaper to go to the awarder of contracts, give him one-quarter of the price of the lumber, and get it over with. But that is dishonest, that is a bribe. Everyone in the credit chain eventually hates and fears everyone else. But the bribe-bargain, having no enforcing mechanism, promotes mutual respect and a genuine liking. If the accepter of a bribe cheats you, you will not go to him again and he will soon have to leave the public service. But if he fulfills his contract, you have a new friend whom you can trust.

We do not know whether Mexicans are happier than we; it is probable that they are exactly as happy. However, we do know that the channels of their happiness or unhappiness are different from ours, just as their time sense is different. We can invade neither, but it is some gain simply to know that it is so.

As the men on our deck continued with what we thought was wine and they probably considered some expensive foreign beverage (it must have tasted bad enough to be very foreign and very expensive), they uncovered a talent for speech we have often noticed in these people. They are natural orators, filling their sentences with graceful forms, with similes and elegant parallels. Our oldest man delivered for us a beautiful political speech. He was an ardent admirer of General Almazán, who was then a candidate for the Mexican presidency. Our Indian likened the General militarily to the god of war, but whether Mars, or Huitzilopochtli, he did not say. In physical beauty the General stemmed from Apollo, not he of the Belvedere, but an earlier, sturdier Apollo. In kindness and forethought and wisdom Almazán was rather above the lesser deities. Our man even touched on the General's abilities in bed, although how he knew he did not say. We gathered, though, that the General was known and well thought of in this respect by his total feminine constituency. "He is a strong man," said our orator, holding himself firmly upright by the port stay. One of us interposed, "When he is elected there will be more fish in the sea for the poor people of Mexico." And the orator nodded wisely. "That is so my friend," he said. It was later that we learned that General Camacho, the other candidate, had many of the same beautiful qualities as General Almazán. And since he won, perhaps he had them more highly developed. For political virtues always triumph, and when two such colossi as these oppose each other, one can judge their relative excellences only by counting the vote.

We had known that sooner or later we must develop an explanation for what we were doing which would be short and convincing. It couldn't be the truth because that wouldn't be convincing at all. How can you say to a people who are preoccupied with getting enough food and enough children that you have come to pick up useless little animals so that perhaps your world picture will be enlarged? That didn't even convince us. But there had to be a story, for everyone asked us. One of us had once taken a long walking trip through the southern United States. At first he had tried to explain that he did it because he liked to walk and because he saw and felt the country better that way. When he gave this explanation there was unbelief and dislike for him. It sounded like a lie. Finally a man said to him, "You can't fool me, you're doing it on a bet." And after that, he used this explanation, and everyone liked and understood him from then on. So with these men we developed our story and stuck to it thereafter. We were collecting curios, we said. These beautiful little animals and shells, while they abounded so greatly here as to be valueless, had, because of their scarcity in the United States, a certain value. They would not make us rich but it was at least profitable to take them. And besides, we liked taking them. Once we had developed this story we never had any more trouble. They all understood us then, and brought us what they thought were rare articles for the collection. They considered that we might get very rich. Thank heaven they do not know that when at last we came back to San Diego the customs fixed a value on our thousands of pickled animals of five dollars. We hope these Indians never find it out; we would go down steeply in their estimations.

Our men went away finally a trifle intoxicated, but not forgetting to take an armload of empty tomato cans. They value tin cans very highly.

It would not have done to sail for La Paz harbor that night, for the pilot has short hours and any boat calling for him out of his regular hours must pay double. But we wanted very much to get to La Paz; we were out of beer and already the water in our tanks was stale-tasting. It had seemed to us that it was stale when we put it in and time did not improve it. It isn't likely that we would have died of thirst. The second or third day would undoubtedly have seen us drinking the unpleasant stuff. But there were other reasons why we longed for La Paz. Cape San Lucas had not really been a town, and our crew had convinced itself that it had been a very long time out of touch with civilization. In civilization we think they included some items which, if anything, are attenuated in highly civilized groups. In addition, there is the genuine fascination of the city of La Paz. Everyone in the area knows the greatness of La Paz.

You can get anything in the world there, they say. It is a huge place—not of course so monstrous as Guaymas or Mazatlán, but beautiful out of all comparison. The Indians paddle hundreds of miles to be at La Paz on a feast day. It is a proud thing to have been born in La Paz, and a cloud of delight hangs over the distant city from the time when it was the great pearl center of the world. The robes of the Spanish kings and the stoles of bishops in Rome were stiff with the pearls from La Paz. There's a magic-carpet sound to the name, anyway. And it is an old city, as cities in the West are old, and very venerable in the eyes of Indians of the Gulf. Guaymas is busier, they say, and Mazatlán gayer, perhaps, but La Paz is *antigua*.

The Gulf and Gulf ports have always been unfriendly to colonization. Again and again attempts were made before a settlement would stick. Humans are not much wanted on the Peninsula. But at La Paz the pearl oysters drew men from all over the world. And, as in all concentrations of natural wealth, the terrors of greed were let loose on the city again and again. An event which happened at La Paz in recent years is typical of such places. An Indian boy by accident found a pearl of great size, an unbelievable pearl. He knew its value was so great that he need never work again. In his one pearl he had the ability to be drunk as long as he wished, to marry any one of a number of girls, and to make many more a little happy too. In his great pearl lay salvation, for he could in advance purchase masses sufficient to pop him out of Purgatory like a squeezed watermelon seed. In addition he could shift a number of dead relatives a little nearer to Paradise. He went to La Paz with his pearl in his hand and his future clear into eternity in his heart. He took his pearl to a broker and was offered so little that he grew angry, for he knew he was cheated. Then he carried his pearl to another broker and was offered the same amount. After a few more visits he came to know that the brokers were only the many hands of one head and that he could not sell his pearl for more. He took it to the beach and hid it under a stone, and that night he was clubbed into unconsciousness and his clothing was searched. The next night he slept at the house of a friend and his friend and he were injured and bound and the whole house searched. Then he went inland to lose his pursuers and he was waylaid and tortured. But he was very angry now and he knew what he must do. Hurt as he was he crept back to La Paz in the night and he skulked like a hunted fox to the beach and took out his pearl from under the stone. Then he cursed it and threw it as far as he could into the channel. He was a free man again with his soul in danger and his food and shelter insecure. And he laughed a great deal about it.

This seems to be a true story, but it is so much like a parable that it almost can't be. This Indian boy is too heroic, too wise. He knows too much and acts on his knowledge. In every way, he goes contrary to human direction. The story is probably true, but we don't believe it; it is far too reasonable to be true.

La Paz, the great city, was only a little way from us now, we could almost see its towers and smell its perfume. And it was right that it should be so hidden here out of the world, inaccessible except to the galleons of a small boy's imagination.

While we were anchored at Espíritu Santo Island a black yacht went by swiftly, and on her awninged after-deck ladies and gentlemen in white clothing sat comfortably. We saw they had tall cool drinks beside them and we hated them a little, for we were out of beer. And Tiny said fiercely, "Nobody but a pansy'd sail on a thing like that." And then more gently, "But I've never been sure I ain't queer." The yacht went down over the horizon, and up over the horizon climbed an old horror of a cargo ship, dirty and staggering. And she stumbled on toward the channel of La Paz; her pumps must have been going wide open. Later, at La Paz, we saw her very low in the water in the channel. We said to a man on the beach, "She is sinking." And he replied calmly, "She always sinks."

On the *Western Flyer,* vanity had set in. Clothing was washed unmercifully. The white tops of caps were laundered, and jeans washed and patted smooth while wet and hung from the stays to dry. Shoes were even polished and the shaving and bathing were deafening. The sweet smell of unguents and hair oils, of deodorants and lotions, filled the air. Hair was cut and combed; the mirror over the washstand behind the deckhouse was in constant use. We regarded ourselves in the mirror with the long contemplative coy looks of chorus girls about to go on stage. What we found was not good, but it was the best we had. Heaven knows what we expected to find in La Paz, but we wanted to be beautiful for it.

And in the morning, when we got under way, we washed the fish blood off the decks and put away the equipment. We coiled the lines in lovely spirals and washed all the dishes. It seemed to us we made a rather gallant show, and we hoped that no beautiful yacht was anchored in La Paz. If there were a yacht, we would be tough and seafaring, but if no such contrast was available some of us at least proposed to be not a little jaunty. Even the least naïve of us expected Spanish ladies in high combs and mantillas to be promenading along the beach. It would be rather like the opening scene of a Hollywood production of *Life in Latin America,* with dancers in the foreground and cabaret tables upstage from which would rise a male chorus to sing "I met my love in La Paz—satin and Latin she was."

We assembled on top of the deckhouse, the *Coast Pilot* open in front of us. Even Tony had succumbed; he wore a gaudy white seaman's cap with a gold ornament on the front of it which seemed to be a combination of field artillery and submarine service, except that it had an arrow-pierced heart superimposed on it.

We have so often admired the literary style and quality of the *Coast Pilot* that it might be well here to quote from it. In the first place, the compilers of this book are cynical men. They know that they are writing for morons, that if by any effort their descriptions can be misinterpreted or misunderstood by the reader, that effort will be made. These writers have a contempt for almost everything. They would like an ocean and a coastline unchanging and unchangeable; lights and buoys that do not rust and wash away; winds and storms that come at specified times; and, finally, reasonably intelligent men to read their instructions. They are gratified in none of these desires. They try to write calmly and objectively, but now and then a little bitterness creeps in, particularly when they deal with Mexican lights, buoys, and port facilities. The following quotation is from H. O. No. 84, "Sailing Directions for the West Coasts of Mexico and Central America, 1937, Corrections to January 1940," page 125, under "La Paz Harbor."

La Paz Harbor is that portion of La Paz Channel between the eastern end of El Mogote and the shore in the vicinity of La Paz. El Mogote is a low, sandy, bush-covered peninsula, about 6 miles long, east and west, and 1½ miles wide at its widest part, that forms the northern side of Ensenada de Anpe, a large lagoon. This lagoon lies in a low plain that is covered with a thick growth of trees, bushes, and cactus. The water is shoal over the greater part of the lagoon, but a channel in which there are depths of 2 to 4 fathoms leads from La Paz Harbor to its northwestern part.

La Paz Harbor is ½ to ¾ mile wide, but it is nearly filled with shoals through which there is a winding channel with depths of 3 to 4 fathoms. A shoal with depths of only 1 to 8 feet over it extends northward from the eastern end of El Mogote to within 400 yards of Prieta Point and thus protects La Paz Harbor from the seas caused by northwesterly winds.

La Paz Channel, leading between the shoal just mentioned and the mainland, and extending from Prieta Point to abreast the town of La Paz, has a length of about 3½ miles and a least charted depth of 3¼ fathoms, but this depth can not be depended upon. Vessels of 13-foot draft may pass through the channel at any stage of the tide. The channel is narrow, with steep banks on either side, the water in some places shoaling from 3 fathoms to 3 or 4 feet within a distance of 20 yards. The deep water of the channel and the projecting points of the shoals on either side can readily be distinguished from aloft. In 1934 the controlling depth in the channel was reported to be 16 feet.

A 9-foot channel, frequently used by coasters, leads across the shoal bank and into La Paz Channel at a position nearly 1 mile south-southeastward of Prieta Point. Caymancito Rock, on the eastern side of La Paz Channel, bearing 129°, leads through this side channel.

Beacons—Off Prieta Point, at the entrance to the channel leading to La Paz, there are three beacons consisting of lengths of 3-inch pipe driven into the bottom and extending only a few feet above the surface of the water. They are difficult to make out at high tide in the daytime, and are not lighted at night [here the hatred creeps in subtly].

Light Beacons—Three pairs of concrete range beacons, from each of which a light is shown, mark La Paz Channel. The outer range is situated on the shore near the entrance to the channel, about 1 mile southeastward of Prieta Point; the middle range is on a hillside about ¼ mile south-southeastward of Caymancito Rock; and the inner range is situated about ¾ mile northeastward of the municipal pier at La Paz. . . .

Harbor Lights—A light is shown from a wooden post 18 feet high and another from a post 20 feet high on the north and south ends, respectively, of the T-head of the municipal pier at La Paz. . . .

Anchorage—Vessels waiting for a pilot can anchor southward of Prieta Point in depths of 7 to 10 fathoms. Anchorage can also be taken northward of El Mogote, but it is exposed to wind and sea. . . .

The best berth off the town is 200 to 300 yards westward of the pier in a depth of about 3½ fathoms, sand. . . .

Pilotage is compulsory for all foreign merchant vessels. Pilots come out in a small motor launch carrying a white flag on which is the letter P, and board incoming vessels in the vicinity of Prieta Point. Although pilots will take vessels in at night, it is not advisable to attempt to enter the harbor after dark.

This is a good careful description by men whose main drive is toward accuracy, and they must be driven frantic as man and tide and wave undermine their work. The shifting sands of the channel; the three-inch pipe driven into the bottom; the T-head municipal pier with its lights on wooden posts, none of which has been there for some time; and, last, their conviction that the pilots cannot find the channel at night, make for their curious, cold, tactful statement. We trust these men. They are controlled, and only now and then do their nerves break and a cry of pain escape them thus, in the "Supplement" dated 1940:

Page 109, Line 1, for *"LIGHTS"* read *"LIGHT"* and for *"TWO LIGHTS ARE"* read *"WHEN THE CANNERY IS IN OPERATION, A LIGHT IS."*

Or again:

Page 149, Line 2, after *"line"* add: *"two piers project inward from this mole, affording berths for vessels and, except alongside these two piers, the mole is foul with debris and wrecked cranes."*

These coast pilots are constantly exasperated; they are not happy men. When anything happens they are blamed, and their writing takes on an austere tone because of it. No matter how hard they work, the restlessness of nature and the carelessness of man are always two jumps ahead of them.

We ran happily up under Prieta Point as suggested, and dropped anchor and put up the American flag and under it the yellow quarantine flag. We would have liked to fire a gun, but we had only the ten-gauge shotgun, and its hammer was rusted down. It was only for a show of force anyway; we had never intended it for warlike purposes. And then we sat and waited. The site was beautiful—the highland of Prieta Point and a tower on the hillside. In the distance we could see the beach of La Paz, and it really looked like a Hollywood production, the fine, low buildings close down to the water and trees flanking them and a colored bandstand on the water's edge. The little canoes of Nayarit sailed by, and the sea was ruffled with a fair breeze. We took some color motion pictures of the scene, but they didn't come out either.

After what seemed a very long time, the little launch mentioned in the *Coast Pilot* started for us. But it had no white flag with the letter "P." Like the municipal pier, that was gone. The pilot, an elderly man in a business suit and a dark hat, came stiffly aboard. He had great dignity. He refused a drink, accepted cigarettes, took his position at the wheel, and ordered us on grandly. He looked like an admiral in civilian clothes. He governed Tex with a sensitive hand—a gentle push forward against the air meant "ahead." A flattened hand patting downward signified "slow." A quick thumb over the shoulder, "reverse." He was not a talkative man, and he ran us through the channel with ease, hardly scraping us at all, and signaled our anchor down 250 yards westward of the municipal pier—if there had been one—the choicest place in the harbor.

La Paz grew in fascination as we approached. The square, iron-shuttered colonial houses stood up right in back of the beach with rows of beautiful trees in front of them. It is a lovely place. There is a broad promenade along the water lined with benches, named for dead residents of the city, where one may rest oneself.

Soon after we had anchored, the port captain, customs man, and agent came aboard. The captain read our papers, which complimented us rather highly, and was so impressed that he immediately assigned us an armed guard—or, rather, three shifts of armed guards—to protect us from theft. At first we did not like this, since we had to pay these men, but we soon found the wisdom of it. For we swarmed with visitors from morning to night; little boys clustered on us like flies, in the rigging and on the deck. And although we were infested and crawling with very poor people and children, we lost nothing; and this in spite of the fact that there were little gadgets lying about that any one of us would have stolen if we had had the chance. The guards simply kept our visitors out of the galley and out of the cabin. But we do not think they prevented theft, for in other ports where we had no guard nothing was stolen.

The guards, big pleasant men armed with heavy automatics, wore uniforms that were starched and clean, and they were helpful and sociable. They ate with us and drank coffee with us and told us many valuable things about the town. And in the end we gave each of them a carton of cigarettes, which seemed valuable to them. But they were the reverse of what is usually thought and written of Mexican soldiers—they were clean, efficient, and friendly.

With the port captain came the agent, probably the finest invention of all. He did everything for us, provisioned us, escorted us, took us to dinner, argued prices for us in local stores, warned us about some places and recommended others. His fee was so small that we doubled it out of pure gratitude.

As soon as we were cleared, Sparky and Tiny and Tex went ashore and disappeared, and we did not see them until late that night, when they came back with the usual presents: shawls and carved cow-horns and colored handkerchiefs. They were so delighted with the exchange (which was then six pesos for a dollar) that we were very soon deeply laden with curios. There were five huge stuffed sea-turtles in one bunk alone, and Japanese toys, combs from New England, Spanish shawls from New Jersey, machetes from Sheffield and New York; but all of them, from having merely lived a while in La Paz, had taken on a definite Mexican flavor. Tony, who does not trust foreigners, stayed aboard, but later even he went ashore for a while.

The tide was running out and the low shore east of the town was beginning to show through the shallow water. We gathered our paraphernalia and started for the beach, expecting and finding a fauna new to us. Here on the flats the water is warm, very warm, and there is no wave-shock. It would be strange indeed if, with few exceptions of ubiquitous animals, there should not be a definite change. The base of this flat was of rubble in which many knobs and limbs of old coral were imbedded, making an easy hiding place for burrowing animals. In rubber boots we moved over the flat uncovered by the dropping tide; a silty sand made the water obscure when a rock or a piece of coral was turned over. And as always when one is collecting, we were soon joined by a number of small boys. The very posture of search, the slow movement with the head down, seems to draw people. "What did you lose?" they ask.

"Nothing."

"Then what do you search for?" And this is an embarrassing question. We search for something that will seem like truth to us; we search for understanding; we search for that principle which keys us deeply into the pattern of all life; we search for the relations of things, one to another, as this young man searches for a warm light in his wife's eyes and that one for the hot warmth of fighting. These little boys and young men on the tide flat do not even know that they search for such things too. We say to them, "We are looking for curios, for certain small animals."

Then the little boys help us to search. They are ragged and dark and each one carries a small iron harpoon. It is the toy of La Paz, owned and treasured as tops or marbles are in America. They poke about the rocks with their little harpoons, and now and then a lazing fish which blunders too close feels the bite of the iron.

There is a small ghost shrimp which lives on these flats, an efficient little fellow who lives in a burrow. He moves very rapidly, and is armed with claws which can pinch painfully. He retires backward into his hole, so that to come at him from above is to invite his weapons. The little boys solved the problem for us. We offered ten centavos for each one they took. They dug into the rubble and old coral until they got behind the ghost shrimp in his burrow, then, prodding, they drove him outraged from his hole. Then they banged him good to reduce his pinching power. We refused to buy the banged-up ones—they had to get us lively ones. Small boys are the best collectors in the world. Soon they worked out a technique for catching the shrimps with only an occasionally pinched finger, and then the ten-centavo pieces began running out, and an increasing cloud of little boys brought us specimens. Small boys have such sharp eyes, and they are quick to notice deviation. Once they know you are generally curious, they bring amazing things. Perhaps we only practice an extension of their urge. It is easy to remember when we were small and lay on our stomachs beside a tide pool and our minds and eyes went so deeply into it that size and identity were lost, and the creeping hermit crab was our size and the tiny octopus a monster. Then the waving algae covered us and we hid under a rock at the bottom and leaped out at fish. It is very possible that we, and even those who probe space with equations, simply extend this wonder.

Among small-boy groups there is usually a stupid one who understands nothing, who brings dull things, rocks and pieces of weed, and pretends that he knows what he does. When we think of La Paz, it is always of the small boys that we think first, for we had many dealings with them on many levels.

The profile of this flat was easy to get. The ghost shrimps, called *"langusta,"* were quite common; our enemy the stinging worm was about, to make us careful of our fingers; the big brittle-stars were there under the old coral, but not in such great masses as at Espíritu Santo. A number of sponges clung to the stones, and small decorated crabs skulked in the interstices. Beautiful purple polyclad worms crawled over lawns of purple tunicates; the giant oyster-like hacha [5] was not often found, but we took a few specimens. There were several growth forms of the common corals [6]; the larger and handsomer of the two slim asteroids [7]; anemones of at least three types; some club urchins and snails and many hydroids.

Some of the exposed snails were so masked with forests of algae and hydroids that they were invisible to us. We found

---

[5] *Pinna* sp.
[6] *Porites.*
[7] *Phataria.*

a worm-like fixed gastropod,[8] many bivalves, including the long peanut-shaped boring clam [9]; large brilliant-orange nudibranchs; hermit crabs; mantids; flatworms which seemed to flow over the rocks like living gelatin; sipunculids; and many limpets. There were a few sun-stars, but not so many or so large as they had been at Cape San Lucas.

The little boys ran to and fro with full hands, and our buckets and tubes were soon filled. The ten-centavo pieces had long run out, and ten little boys often had to join a club whose center and interest was a silver peso, to be changed and divided later. They seemed to trust one another for the division. And certainly they felt there was no chance of their being robbed. Perhaps they are not civilized and do not know how valuable money is. The poor little savages seem not to have learned the great principle of cheating one another.

The population of small boys at La Paz is tremendous, and we had business dealings with a good part of it. Hardly had we returned to the *Western Flyer* and begun to lay out our specimens when we were invaded. Word had spread that there were crazy people in port who gave money for things a boy could pick up on the rocks. We were more than invaded—we were deluged with small boys bearing specimens. They came out in canoes, in flatboats, some even swam out, and all of them carried specimens. Some of the things they brought we wanted and some we did not want. There were hurt feelings about this, but no bitterness. Battalions of boys swarmed back to the flats and returned again. The second day little boys came even from the hills, and they brought every conceivable living thing. If we had not sailed the second night they would have swamped the boat. Meanwhile, in our dealings on shore, more small boys were involved. They carried packages, ran errands, directed us (mostly wrongly), tried to anticipate our wishes; but one boy

---

[8] *Aletes,* or similar.
[9] *Lithophaga plumula,* or similar.

soon emerged. He was not like the others. His shoulders were not slender, but broad, and there was a hint about his face and expression that seemed Germanic or perhaps Anglo-Saxon. Whereas the other little boys lived for the job and the payment, this boy created jobs and looked ahead. He did errands that were not necessary, he made himself indispensable. Late at night he waited, and the first dawn saw him on our deck. Further, the other small boys seemed a little afraid of him, and gradually they faded into the background and left him in charge.

Some day this boy will be very rich and La Paz will be proud of him, for he will own the things other people must buy or rent. He has the look and the method of success. Even the first day success went to his head, and he began to cheat us a little. We did not mind, for it is a good thing to be cheated a little; it causes a geniality and can be limited fairly easily. His method was simple. He performed a task, and then, getting each of us alone, he collected for the job so that he was paid several times. We decided we would not use him any more, but the other little boys decided even better than we. He disappeared, and later we saw him in the town, his nose and lips heavily bandaged. We had the story from another little boy. Our financial wizard told the others that he was our sole servant and that we had said that they weren't to come around any more. But they discovered the lie and waylaid him and beat him very badly. He wasn't a very brave little boy, but he will be a rich one because he wants to. The others wanted only sweets or a new handkerchief, but the aggressive little boy wishes to be rich, and they will not be able to compete with him.

On the evening of our sailing we had rather a sad experience with another small boy. We had come ashore for a stroll, leaving our boat tied to a log on the beach. We walked up the curiously familiar streets and ended, oddly enough, in a bar to have a glass of beer. It was a large bar with high ceilings, and nearly deserted. As we sat sipping our beer we saw a ferocious face scowling at us. It was a very small, very black Indian boy, and the look in his eyes was one of hatred. He stared at us so long and so fiercely that we finished our glasses and got up to go. But outside he fell into step with us, saying nothing. We walked back through the softly lighted streets, and he kept pace. But near the beach he began to pant deeply. Finally we got to the beach and as we were about to untie the skiff he shouted in panic, *"Cinco centavos!"* and stepped back as from a blow. And then it seemed that we could see almost how it was. We have been the same way trying to get a job. Perhaps the father of this little boy said, "Stupid one, there are strangers in the town and they are throwing money away. Here sits your father with a sore leg and you do nothing. Other boys are becoming rich, but you, because of your sloth, are not taking advantage of this miracle. Señor Ruiz had a cigar this afternoon and a glass of beer at the *cantina* because his fine son is not like you. When have you known me, your father, to have a cigar? Never. Now go and bring back some little piece of money."

Then that little boy, hating to do it, was burdened with it nevertheless. He hated us, just as we have hated the men we have had to ask for jobs. And he was afraid, too, for we were foreigners. He put it off as long as he could, but when we were about to go he had to ask and he made it very humble. Five centavos. It did seem that we knew how hard it had been. We gave him a peso, and then he smiled broadly and he looked about for something he could do for us. The boat was tied up, and he attacked the water-soaked knot like a terrier, even working at it with his teeth. But he was too little and he could not do it. He nearly cried then. We cast off and pushed the boat away, and he waded out to guide us as far as he could. We felt both good and bad about it; we hope his father bought a cigar and an *aguardiente,* and became mellow and said to a group of men in that little boy's hearing, "Now you take Juanito. You have rarely seen such a good son. This very cigar is a gift to his father who has hurt his leg. It is a matter of pride, my friends, to have a son like Juanito." And we hope he gave Juanito, if that was his name, five centavos to buy an ice and a paper bull with a firecracker inside.

No doubt we were badly cheated in La Paz. Perhaps the boatmen cheated us and maybe we paid too much for supplies—it is very hard to know. And besides, we were so incredibly rich that we couldn't tell, and we had no instinct for knowing when we were cheated. Here we were rich, but in our own country it was not so. The very rich develop an instinct which tells them when they are cheated. We knew a rich man who owned several large office buildings. Once in reading his reports he found that two electric-light bulbs had been stolen from one of the toilets in one of his office buildings. It hurt him; he brooded for weeks about it. "Civilization is dying," he said. "Whom can you trust any more? This little theft is an indication that the whole people is morally rotten."

But we were so newly rich that we didn't know, and besides we were a little flattered. The boatmen raised their price as soon as they saw the Sea-Cow wouldn't work, but as they said, times are very hard and there is no money.

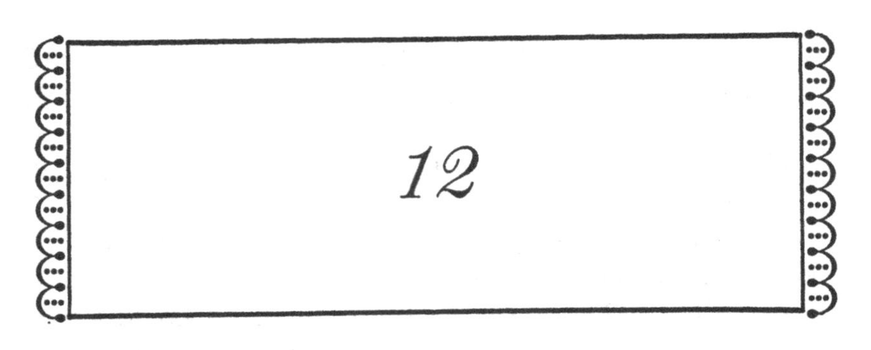

12

March 22

THIS was Good Friday, and we scrubbed ourselves and put on our best clothes and went to church, all of us. We were a kind of parade on the way to church, feeling foreign and out of place. In the dark church it was cool, and there were a great many people, old women in their black shawls and Indians kneeling motionless on the floor. It was not a very rich church, and it was old and out of repair. But a choir of small black children made the Stations of the Cross. They sang music that sounded like old Spanish madrigals, and their voices were shrill and sharp. Sometimes they faltered a little bit on the melody, but they hit the end of each line shrieking. When they had finished, a fine-looking young priest with a thin ascetic face and the hot eyes of fervency preached from over their heads. He filled the whole church with his faith, and the people were breathlessly still. The ugly bloody Christs and the simpering Virgins and the over-dressed saints were suddenly out of it. The priest was purer and cleaner and stronger than they. Out of his own purity he seemed to plead for them. After a long time we got up and went out of the dark cool church into the blinding white sunlight.

The streets were very quiet on Good Friday, and no wind blew in the trees, the air was full of the day—a kind of hush, as though the world awaited a little breathlessly the dreadful experiment of Christ with death and Hell; the testing in a furnace of an idea. And the trees and the hills and the people seemed to wait as a man waits when his wife is having a baby, expectant and frightened and horrified and half unbelieving.

There is no certainty that the Easter of the Resurrection will really come. We were probably literarily affected by the service and the people and their feeling about it, the crippled and the pained who were in the church, the little half-hungry children, the ancient women with eyes of patient tragedy who stared up at the plaster saints with eyes of such pleading. We liked them and we felt at peace with them. And strolling slowly through the streets we thought a long time of these people in the church. We thought of the spirits of kindness which periodically cause them to be fed, a little before they are dropped back to hunger. And we thought of the good men who labored to cure them of disease and poverty.

And then we thought of what they are, and we are—products of disease and sorrow and hunger and alcoholism. And suppose some all-powerful mind and will should cure our species so that for a number of generations we would be healthy and happy? We are the products of our disease and suffering. These are factors as powerful as other genetic factors. To cure and feed would be to change the species, and the result would be another animal entirely. We wonder if we would be able to tolerate our own species without a history of syphilis and tuberculosis. We don't know.

Certain communicants of the neurological conditioning religions practiced by cowardly people who, by narrowing their emotional experience, hope to broaden their lives, lead us to think we would not like this new species. These religionists, being afraid not only of pain and sorrow but even of joy, can so protect themselves that they seem dead to us. The new animal resulting from purification of the species might be one we wouldn't like at all. For it is through struggle and sorrow that people are able to participate in one another—the heartlessness of the healthy, well-fed, and unsorrowful person has in it an infinite smugness.

On the water's edge of La Paz a new hotel was going up, and it looked very expensive. Probably the airplanes will bring week-enders from Los Angeles before long, and the beautiful poor bedraggled old town will bloom with a Floridian ugliness.

Hearing a burst of chicken voices, we looked over a mud wall and saw that there were indeed chickens in the yard behind it. We asked then of a woman if we might buy several. They could be sold, she said, but they were not what one calls "for sale." We entered her yard. One of the proofs that they were not for sale was that we had to catch them ourselves. We picked out two which looked a little less muscular than the others, and went for them. Whatever has been said, true or not, of the indolence of the Lower Californian is entirely untrue of his chickens. They were athletes, highly trained both in speed and in methods of escape. They could run, fly, and, when cornered, disappear entirely and re-materialize in another part of the yard. If the owner did not want to catch them, that hesitancy was not shared by the rest of La Paz. People and children came from everywhere; a mob collected, first to give excited advice and then to help. A pillar of dust arose out of that yard. Small boys hurled themselves at the chickens like football-players. We were bound to catch them sooner or later, for as one group became exhausted, another took up the chase. If we had played fair and given those chickens rest periods, we would never have caught them. But by keeping at them, we finally wore them down and they were caught, completely exhausted and almost shorn of their feathers. Everyone in the mob felt good and happy then and we paid for the chickens and left.

On board it was Sparky's job to kill them, and he hated it. But finally he cut their heads off and was sick. He hung them over the side to bleed and a boat came along and mashed them flat against our side. But even then they were tough. They had the most highly developed muscles we have ever seen. Their legs were like those of ballet dancers and there was no softness in their breasts. We stewed them for many hours and it did no good whatever. We were sorry to kill them, for they were gallant, fast chickens. In our country they could easily have got scholarships in one of our great universities and had collegiate careers, for they had spirit and fight and, for all we know, loyalty.

On the afternoon tide we were to collect on El Mogote, a low sandy peninsula with a great expanse of shallows which would be exposed at low tide. The high-tide level was defined by a heavy growth of mangrove. The area was easily visible from our anchorage, and the sand was smooth and not filled with rubble or stones or coral. A tall handsome boy of about nineteen had been idling about the *Western Flyer*. He had his own canoe, and he offered to paddle us to the tide flats. This boy's name was Raúl Velez; he spoke some English and was of great service to us, for his understanding was quick and he helped valuably at the collecting. He told us the local names of many of the animals we had taken; "cornuda" was the hammer-head shark; "barco," the red snapper; "caracol," and also "burral," all snails in general, but particularly the large conch. Urchins were called "erizo" and sea-fans, "abanico." "Bromas" were barnacles and "hacha" the pinna, or large clam.

The sand flats were very interesting. We dug up a number of Dentaliums of two species, the first we had found. These animals, which look like slender curved teeth, belong to a small class of mollusks, little known popularly, called "tooth shells."

On the shallow bottom, attached to very small stones, we found little anemones of three types. There were also sand anemones,[1] in long filthy-looking gray cases when they were dug out. But when they were imbedded in the bottom and ex-

[1] *Cerianthus.*

panded, they looked like lovely red and purple flowers. A great number of small black cucumbers of a type we had not taken crawled on the bottom, as well as one large pepper-and-salt cucumber. We found many heart-urchins, two species of the ordinary ophiurans (brittle-stars), and one burrowing ophiuran. Sponges and tunicates were fastened to the insecure footing of very small stones, but since there is probably very little churning of water on the tide flats, they were safe enough. There were flat worms of several species; stinging worms, peanut worms, echiuroid worms, and what in the collecting notes are listed rather tiredly as "worms." We took one specimen of the sea-whip, a rather spectacular colony of animals looking exactly like a long white whip. The lower portion is a horny stalk and the upper part consists of zooids carrying on their own life processes but connected by a series of canals which unite their body cavities with the main stalk.

As the tide came up we moved upward in the intertidal toward the mangrove trees, and the foul smell of them reached us. They were in bloom, and the sharp sweet smell of their flowers, combined with the filthy odor of the mud about their roots, was sickening. But they are fascinating to look into. Huge hermit crabs seem to live among their stilted roots; the black mud, product of the root masses, swarms as a meeting place for land and sea animals. Flies and insects in great numbers crawl and buzz about the mud, and the scavenging hermit crabs steal secretly in and out and even climb into the high roots.

We suppose it is the combination of foul odor and the impenetrable quality of the mangrove roots which gives one a feeling of dislike for these salt-water-eating bushes. We sat quietly and watched the moving life in the forests of the roots, and it seemed to us that there was stealthy murder everywhere. On the surf-swept rocks it was a fierce and hungry and joyous killing, committed with energy and ferocity. But here it was like stalking, quiet murder. The roots gave off clicking sounds, and the odor was disgusting. We felt that we were watching something horrible. No one likes the mangroves. Raúl said that in La Paz no one loved them at all.

On the level flats the tide covers the area very quickly. We waded out to a wrecked boat lying turned over on the sand, and took a number of barnacles from the rotten wood and even from the rusted engine. It was a good rich collecting day, and it had been a curiously emotional day beginning with the church. Sometimes one has a feeling of fullness, of warm wholeness, wherein every sight and object and odor and experience seems to key into a gigantic whole. That day even the mangrove was part of it. Perhaps among primitive peoples the human sacrifice has the same effect of creating a wholeness of sense and emotion—the good and bad, beautiful, ugly, and cruel all welded into one thing. Perhaps a whole man needs this balance. And we had been as excited at finding the Dentaliums as though they had been nuggets of gold.

Raúl had a La Paz harpoon in his canoe, and we bought it from him, hoping to bring it home. It was a shaft of iron with a ring on one end for the line and a point and hinged barb at the other. A little circle of cord holds the barb against the shaft until the friction of the flesh of the victim pushes the cord free and allows the barb to open out inside the flesh. We wanted to keep this harpoon, but we lost it in a manta ray later. At this reading, there are many manta rays in the Gulf cruising about with our harpoons in their hides.

We also wanted one of the Nayarit canoes to take back, for they are light and of shallow draft, ideal for collecting in the lagoons and seaworthy even in rough water. But no one would sell a canoe. They came from too far away and were too well loved. Some very old ones were solid with braces and patches.

It was dusk when we came back to the *Western Flyer,* and the deck was filled with waiting little boys holding mashed and mangled specimens of all kinds. We bought what we needed and then we bought a lot of things we didn't need. The boys had waited a long time for us, under the stern eye of our military man. And it was interesting to see how our soldier loved the ragged little boys of La Paz. When they got out of hand or ran too fast over the deck, he cautioned them, but there was none of the bluster of the policeman. And had we not been just to the little boys, he would have joined them; for they were his people and our great wealth would not have deflected him from them. He wore his automatic, but it was only a badge with no show of force about it, and when he entered the galley or sat down with us he removed the gun belt and hung it up. We liked the tone of voice he used on the boys. It had dignity and authority but no bullying quality, and the boys of the town seemed to respect him without fearing him.

Once when a little boy practiced the most ancient trick in the list of boy skulduggery—that of removing a specimen and selling it again—the soldier spoke to him shortly with contempt, and that boy lost his standing and even his friends.

One boy had, on a light harpoon, a fish which looked something like the puffers—a gray and black fish with a large flat head. When we wished to buy it he refused, saying that a man had commissioned him to get this fish and he was to receive ten centavos for it because the man wanted to poison a cat. This was the *botete,* and our first experience with it. It is thought in La Paz that the poison concentrates in the liver and this part is used for poisoning small animals and even flies. We did not make this test, but we found *botete* everywhere in the warm shallow waters of the Gulf. Probably he is the most prevalent fish of all in lagoons and eel-grass flats. He lies on the bottom, and his marking makes him nearly invisible. Sometimes he lies in a small cleared place in eel-grass or in a slight depression on the silt bottom which indicates, but does not prove, that he has a fairly permanent resting-place to which he returns. When one is wading in the shallows, *botete* lies quiet until he is nearly stepped on before he streaks away, drawing a cloud of disturbed mud after him.

In the press of collecting and preserving, we neglected to dissect the stomach of this fish, so we do not know what he eats.

The literature on *botete* is scattered and hard to come by. Members of his genus, having his poisonous qualities, are distributed all over the world where there are shallows of warm water. Since this fish is very dangerous to eat and is so widely found, it is curious that so little has been written about it. Eating him almost invariably causes death in agony. If he were rare, it would be understandable why he has been so little discussed. But more has been written about some of the seldom-seen fishes of the great depths than of this deadly little *botete*. We were fascinated with him and took a number of specimens. Following are some of the few reports available on his nature and misdemeanors. We still do not know whether he kills flies.

From Herre [2] we learn that "In at least two or three of the sub-orders the flesh nearly always is not only thin, hard, often bitter and usually unpalatable, but also contains poisonous alkaloids. These produce the disease known as ciguatera, in which the nervous system is attacked and violent gastric disturbances, paralysis, and death may follow."

On page 423 he discusses the Balistidae, or trigger-fish such as the Gulf puerco: "Although seen in fish markets throughout the Orient, none of the Balistidae are much used as human food. In some localities of the Philippines, those of moderate size are eaten, but their sale here should be forbidden as their flesh is always more or less poisonous. In such places as Cuba and Mauritius they are not allowed in the markets as they are known to cause ciguatera.

"Francis Day says (*Fishes of India,* 1878, p. 686): 'Dr.

---

[2] "Poisonous and Worthless Fishes: An Account of the Philippine Plectognaths," 1924 (§ W-8), p. 415.

Meunier, at Mauritius, considers that the poisonous flesh acts primarily on the nervous tissue of the stomach, causing violent spasms of that organ and, shortly afterwards, of all the muscles of the body. The frame becomes wracked with spasms, the tongue thickened, the eye fixed, the breathing laborious, and the patient expires in a paroxysm of extreme suffering. The first remedy to be given is a strong emetic, and subsequently oils and demulcents to allay irritability.'

"In his account of the backboned animals of Abyssinia Rüppel states that *Balistes flavomarginatus* is very common in the Red Sea at Djetta, where it is often brought to market, although only pilgrims who do not know the quality of the flesh will buy it. He goes on to say that as a whole the Balistidae not only have a bad taste, but also are unwholesome as food."

Referring to the Tetraodontidae, page 479, Herre uses the name *batete,* or *botete,* as used in most Philippine languages. "This dangerous group of fishes," he says, "is widely distributed in warm seas all over the world and is common throughout the Philippines. Although most people are more or less aware of the poisonous properties of the flesh, it is eaten in practically every Philippine fishing village and not a year goes by without several deaths from this cause.

"A Japanese investigator (I have been unable to obtain a copy of his paper, which appeared in *Archiv für Pathologie und Pharmacologie*) has studied carefully the alkaloid present in the flesh of the Tetraodontidae and finds it to be very near to muscarine, the active poisonous principle of *Amanita muscaria* and other fungi. It is a tasteless, odorless, and very poisonous crystalline alkaloid."

He goes on to state that the natives consider the gall-bladder, the milt, and the eggs to be particularly poisonous. But in La Paz it was the liver which was thought to be the most poisonous part. Only the liver was used to poison animals and flies, although this might be because the liver was more attractive as bait than other portions.

Herre continues on page 488 concerning *Tetraodon:* "The *Medical Journal of Australia* under the date of December 1, 1923, tells of two Malays who ate of a species of Tetraodon although warned of the danger. They ate at noon with no serious effects, but on eating some for supper they were taken violently ill, one dying in an hour, the other about three hours later." Of the Diodontidae, page 503 (the group to which the puffer fish belong): "The fishes of this family have a well-deserved reputation for being poisonous and their flesh should never be eaten."

Botete is sluggish, fairly slow, unarmored, and not very clever at either concealment, escape, or attack. It is amusing but valueless to speculate anthropomorphically in the chicken-egg manner regarding the relationship between his habits and his poison. Did he develop poison in his flesh as a protection in lieu of speed and cleverness, or being poisonous and quite unattractive, was he able to "let himself go," to abandon speed and cleverness? The protected human soon loses his power of defense and attack. Perhaps *botete,* needing neither brains nor tricks nor techniques to protect himself except from a man who wants to poison a cat, has become a frump.

In the evening Tiny returned to the *Western Flyer,* having collected some specimens of *Phthirius pubis,* but since he made no notes in the field, he was unable or unwilling to designate the exact collecting station. His items seemed to have no unusual qualities but to be members of the common species so widely distributed throughout the world.

We were to sail in the early morning, and that night we walked a little in the dim-lighted streets of La Paz. And we wondered why so much of the Gulf was familiar to us, why this town had a "home" feeling. We had never seen a town which even looked like La Paz, and yet coming to it was like returning rather than visiting. Some quality there is in the whole Gulf that trips a trigger of recognition so that in fantastic and exotic scenery one finds oneself nodding and saying inwardly, "Yes, I know." And on the shore the wild doves mourn in the evening and then there comes a pang, some kind of emotional jar, and a longing. And if one followed his whispering impulse he would walk away slowly into the thorny brush following the call of the doves. Trying to remember the Gulf is like trying to re-create a dream. This is by no means a sentimental thing, it has little to do with beauty or even conscious liking. But the Gulf does draw one, and we have talked to rich men who own boats, who can go where they will. Regularly they find themselves sucked into the Gulf. And since we have returned, there is always in the backs of our minds the positive drive to go back again. If it were lush and rich, one could understand the pull, but it is fierce and hostile and sullen. The stone mountains pile up to the sky and there is little fresh water. But we know we must go back if we live, and we don't know why.

Late at night we sat on the deck. They were pumping water out of the hold of the trading boat, preparing her to float and flounder away to Guaymas for more merchandise. But La Paz was asleep; not a soul moved in the streets. The tide turned and swung us around, and in the channel the ebbing tide whispered against our hull and we heard the dogs of La Paz barking in the night.

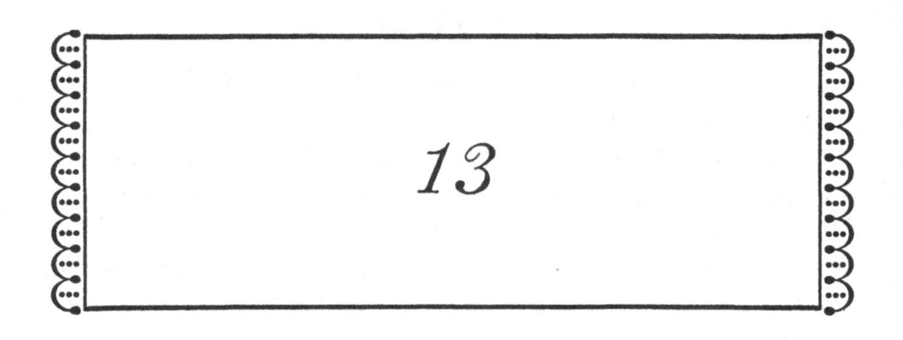

March 23

WE SAILED in the morning. The mustached old pilot came aboard and steered us out, then bowed deeply and stepped into the launch which had followed us. The sea was calm and very blue, almost black-blue, as we turned northward along the coast. We wished to stop near San José Island as our next collecting station. It was good to be under way again and good to be out from under the steady eyes of those ubiquitous little boys who waited interminably for us to do something amusing.

In mid-afternoon we came to anchorage at Amortajada Bay on the southwest tip of San José Island. A small dark islet had caught our attention as we came in. For although the day was bright this islet, called Cayo on the map, looked black and mysterious. We had a feeling that something strange and dark had happened there or that it was the ruined work of men's hands. Cayo is only a quarter of a mile long and a hundred yards wide. Its northern end is a spur and its southern end a flat plateau about forty feet high. Even in the distance it had a quality which we call "burned." One knows there will be few animals on a "burned" coast; that animals will not like it, will not be successful there. Even the algae will be like lost colonists. Whether or not this is the result of a deadly chemistry we cannot say. But we can say that it is possible, after long collecting, to recognize a shore which is "burned" even if it is so far away that details cannot be seen.

Cayo lay about a mile and a half from our anchorage and seemed to blacken even the air around it. This was the first time that the Sea-Cow could have been of great service to us. It was for just such occasions that we had bought it. We were kind to it that day—selfishly of course. We said nice things about it and put it tenderly on the stern of the skiff, pretending to ourselves that we expected it to run, that we didn't dream it would not run. But it would not. We rowed the boat—and the Sea-Cow—to Cayo Islet. There is so much that is strange about this islet that we will set much of it down. It is nearly all questions, but perhaps someone reading this may know the answers and tell us. There is no landing place; all approaches are strewn with large sea-rounded boulders which even in fairly still water would beat the bottom out of a boat. On its easterly side, the one we approached, a cliff rises in back of a rocky beach and there are a number of shallow caves in the cliffside. Set in the great boulders in the intertidal zone there are large iron rings and lengths of big chain, but so rusted and disintegrated that they came off in our hands. Also, set in the cliff six to eight feet above the beach, are other iron rings with loops eight inches in diameter. They look very old, but the damp air of the Gulf and the rapid oxidation caused by it make it impossible to say exactly how old they are. In the shallow caves in the cliff there were evidences of many fires' having been built, and piled about the fireplaces, some old and some fresh, were not only thousands of clam-shells but turtle-shells also, as though these animals had been brought here to be smoked. A heap of fairly fresh diced turtle-meat lay beside one of the fireplaces. The mysterious quality of all this lies here. There are no clams in this immediate vicinity and turtles do not greatly abound. There is no wood whatever on the island with which to build fires; it would have to be brought here. There is no water whatever. And once arrived, there is no anchorage. Why people would bring clams and turtles and wood and water to an islet where there was no protection we do not know. A mile and a half away they could have beached easily and have found both wood and water. It is a riddle we cannot answer, just as we can think of no reason for the big iron rings. They could not have been for fastening a big boat to, since there is no safe water for a boat to lie in and no cove protection from wind and storm. We are very curious about this. We climbed the cliff by a trail that was well beaten in a crevice and on the flat top found a sparse growth of brown grass and some cactus. Nothing more. On the southernmost end of the cliff sat one large black crow who shrieked at us with dislike, and when we approached flew off and disappeared in the direction of San José Island.

The cliffs were light buff in color, and the grass light brown. It is impossible to say why distance made Cayo look black. Boulders and fixed stones of the reef were of a reddish igneous rock and the island, like the whole region, was volcanic in origin.

Collecting on the rocks we found, as we knew we would, a sparse and unhappy fauna. The animals were very small. *Heliaster,* the sun-star of which there were a few, was small and pale in color. There were anemones, a few sea-cucumbers, and a few sea-rabbits. The one animal which seemed to like Cayo was Sally Lightfoot. These beautiful crabs crawled on the rocks and dominated the life of the region. We took a few *Aletes* (worm-like snails) and some serpulid worms, two or three types of snails, and a few isopods and beach-hoppers.

The tide came up and endangered our boat, which we had balanced on top of a boulder, and we rowed back toward the *Western Flyer,* one of us in the stern pulling with a quiet fury on the starting rope of the Sea-Cow. We wished we had left it dangling by its propeller on one of the cliff rings, and its evil and mysterious magneto would have liked that too.

As soon as we pulled away, Cayo looked black again, and we hope someone can tell us something about this island.

Back on the *Western Flyer* we asked Tex to take the Sea-Cow apart down to the tiniest screw and to find out in truth, once for all, whether its failure were metaphysical or something which could be fixed. This he did, under a deck-light. When he put it together and attached it to the boat, it ran perfectly and he went for a cruise with it. Now at last we felt we had an outboard motor we could depend on.

We were anchored quite near San José Island and that night we were visited by little black beetle-like flies which bit and left a stinging, itching burn. Covering ourselves did not help, for they crawled down inside our bedding and bit us unmercifully. Being unable to sleep, we talked and Tiny told us a little of his career, which, if even part of it is true, is one of the most decoratively disreputable sagas we have ever heard. It is with sadness that we do not include some of it, but certain members of the general public are able to keep from all a treatise on biology unsurpassed in our experience. The great literature of this kind is kept vocal by the combined efforts of Puritans and postal regulations, and so the saga of Tiny must remain unwritten.

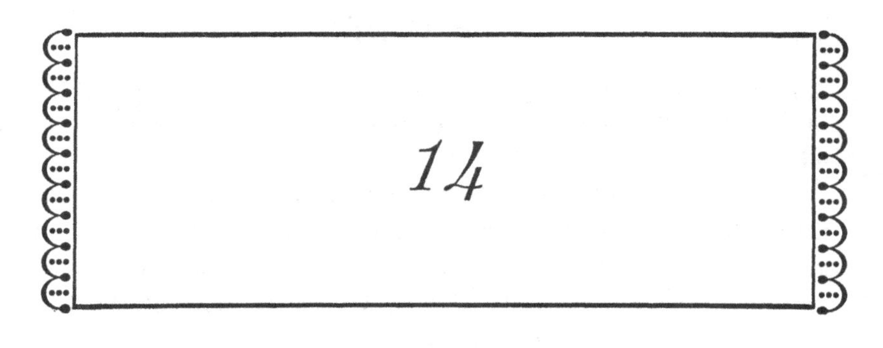

March 24, Easter Sunday

THE beach was hot and yellow. We swam, and then walked along on the sand and went inland along the ridge between the beach and a large mangrove-edged lagoon beyond. On the lagoon side of the ridge there were thousands of burrows, presumably of large land-crabs, but it was hopeless to dig them out. The shores of the lagoon teemed with the little clicking bubbling fiddler crabs and estuarian snails. Here we could smell the mangrove flowers without the foul root smell, and the odor was fresh and sweet, like that of new-cut grass. From where we waded there was a fine picture, still reflecting water and the fringing green mangroves against the burnt red-brown of the distant mountains, all like some fantastic Doré drawing of a pressed and embattled heaven. The air was hot and still and the lagoon rippleless. Now and then the surface was ringed as some lagoon fish came to the air. It was a curious quiet resting-place and perhaps because of the quiet we heard in our heads the children singing in the church at La Paz. We did not collect strongly or very efficiently, but rather we half dozed through the day, thinking of old things, each one in himself. And later we discussed manners of thinking and methods of thinking, speculation which is not stylish any more. On a day like this the mind goes outward and touches in all directions. We discussed intellectual methods and approaches, and we thought that through inspection of thinking technique a kind of purity of approach might be consciously achieved—that non-teleological or "is" thinking might be substituted in part for the usual cause-effect methods.

The hazy Gulf, with its changes of light and shape, was rather like us, trying to apply our thoughts, but finding them always pushed and swayed by our bodies and our needs and our satieties. It might be well here to set down some of the discussions of non-teleological thinking.

During the depression there were, and still are, not only destitute but thriftless and uncareful families, and we have often heard it said that the county had to support them because they were shiftless and negligent. If they would only perk up and be somebody everything would be all right. Even Henry Ford in the depth of the depression gave as his solution to that problem, "Everybody ought to roll up his sleeves and get to work."

This view may be correct as far as it goes, but we wonder what would happen to those with whom the shiftless would exchange places in the large pattern—those whose jobs would be usurped, since at that time there was work for only about seventy percent of the total employable population, leaving the remainder as government wards.

This attitude has no bearing on what might be or could be if so-and-so happened. It merely considers conditions "as is." No matter what the ability or aggressiveness of the separate units of society, at that time there were, and still there are, great numbers necessarily out of work, and the fact that those numbers comprised the incompetent or maladjusted or unlucky units is in one sense beside the point. No causality is involved in that; collectively it's just "so"; collectively it's related to the fact that animals produce more offspring than the world can support. The units may be blamed as individuals, but as members of society they cannot be blamed. Any given individual very possibly may transfer from the underprivileged into the more fortunate group by better luck or by improved aggressiveness or competence, but all cannot be so benefited whatever their strivings, and the large population will be unaffected. The seventy-thirty ratio will remain, with merely a reassortment of the units. And no blame, at least no social fault, imputes to these people; they are where they are "because" natural conditions are what they are. And so far as we selfishly are concerned we can rejoice that they, rather than we, represent the low extreme, since there must be one.

So if one is very aggressive he will be able to obtain work even under the most sub-normal economic conditions, but only because there are others, less aggressive than he, who serve in his stead as potential government wards. In the same way, the sight of a half-wit should never depress us, since his extreme, and the extreme of his kind, so affects the mean standard that we, hatless, coatless, often bewhiskered, thereby will be regarded only as a little odd. And similarly, we cannot justly approve the success manuals that tell our high school graduates how to get a job—there being jobs for only half of them!

This type of thinking unfortunately annoys many people. It may especially arouse the anger of women, who regard it as cold, even brutal, although actually it would seem to be more tender and understanding, certainly more real and less illusionary and even less blaming, than the more conventional methods of consideration. And the value of it as a tool in increased understanding cannot be denied.

As a more extreme example, consider the sea-hare *Tethys,* a shell-less, flabby sea-slug, actually a marine snail, which may be seen crawling about in tidal estuaries, somewhat resembling a rabbit crouched over. A California biologist estimated the number of eggs produced by a single animal during a single breeding season to be more than 478 million. And the adults sometimes occur by the hundred! Obviously all these eggs cannot mature, all this potential cannot, *must not,* become reality, else the ocean would soon be occupied exclusively by sea-hares. There would be no kindness in that, even for the sea-hares themselves, for in a few generations they would overflow the earth; there would be nothing for the rest of us to eat, and nothing for them unless they turned cannibal. On the average, probably no more than the biblical one or two attain full maturity. Somewhere along the way all the rest will have been eaten by predators whose life cycle is postulated upon the presence of abundant larvae of sea-hares and other forms as food—as all life itself is based on such a postulate. Now picture the combination mother-father sea-hare (the animals are hermaphroditic, with the usual cross-fertilization) parentally blessing its offspring with these words: "Work hard and be aggressive, so you can grow into a nice husky *Tethys* like your ten-pound parent." Imagine it, the hypocrite, the illusionist, the Pollyanna, the genial liar, saying that to its millions of eggs *en masse,* with the dice loaded at such a ratio! Inevitably, 99.999 percent are destined to fall by the wayside. No prophet could foresee which specific individuals are to survive, but the most casual student could state confidently that no more than a few are likely to do so; any given individual has *almost* no chance at all—but still there is the "almost," since the race persists. And there is even a semblance of truth in the parent sea-hare's admonition, since even here, with this almost infinitesimal differential, the race is still to the swift and/or to the lucky.

What we personally conceive by the term "teleological thinking," as exemplified by the notion about the shiftless unemployed, is most frequently associated with the evaluating of causes and effects, the purposiveness of events. This kind of thinking considers changes and cures—what "should be" in the terms of an end pattern (which is often a subjective or an anthropomorphic projection); it presumes the bettering of conditions, often, unfortunately, without achieving more than a most superficial understanding of those conditions. In their sometimes intolerant refusal to face facts as they are, teleological notions may substitute a fierce but ineffectual attempt to change conditions which are assumed to be undesirable, in place of the understanding-acceptance which would pave the way for a more sensible attempt at any change which might still be indicated.

Non-teleological ideas derive through "is" thinking, associated with natural selection as Darwin seems to have understood it. They imply depth, fundamentalism, and clarity—seeing beyond traditional or personal projections. They consider events as outgrowths and expressions rather than as results; conscious acceptance as a desideratum, and certainly as an all-important prerequisite. Non-teleological thinking concerns itself primarily not with what should be, or could be, or might be, but rather with what actually "is"—attempting at most to answer the already sufficiently difficult questions *what* or *how,* instead of *why.*

An interesting parallel to these two types of thinking is afforded by the microcosm with its freedom or indeterminacy, as contrasted with the morphologically inviolable pattern of the macrocosm. Statistically, the electron is free to go where it will. But the destiny pattern of any aggregate, comprising uncountable billions of these same units, is fixed and certain, however much that inevitability may be slowed down. The eventual disintegration of a stick of wood or a piece of iron through the departure of the presumably immortal electrons is assured, even though it may be delayed by such protection against the operation of the second law of thermodynamics as is afforded by painting and rustproofing.

Examples sometimes clarify an issue better than explanations or definitions. Here are three situations considered by the two methods.

A. *Why are some men taller than others?*

Teleological "answer": because of the underfunctioning of the growth-regulating ductless glands. This seems simple enough. But the simplicity is merely a function of inadequacy and incompleteness. The finality is only apparent. A child, being wise and direct, would ask immediately if given this answer: "Well, why do the glands underfunction?" hinting instantly towards non-teleological methods, or indicating the rapidity with which teleological thinking gets over into the stalemate of first causes.

In the non-teleological sense there can be no "answer." There can be only pictures which become larger and more significant as one's horizon increases. In this given situation, the steps might be something like this:

(1) Variation is a universal and truly primitive trait. It occurs in any group of entities—razor blades, measuring-rods, rocks, trees, horses, matches, or men.

(2) In this case, the apropos variations will be towards shortness or tallness from a mean standard—the height of adult men as determined by the statistics of measurements, or by common-sense observation.

(3) In men varying towards tallness there seems to be a constant relation with an underfunctioning of the growth-regulating ductless glands, of the sort that one can be regarded as an index of the other.

(4) There are other known relations consistent with tallness, such as compensatory adjustments along the whole chain of endocrine organs. There may even be other factors, separately not important or not yet discovered, which in the aggregate may be significant, or the integration of which may be found to wash over some critical threshold.

(5) The men in question are taller "because" they fall in a

group within which there are the above-mentioned relations. In other words, "they're tall because they're tall."

This is the statistical, or "is," picture to date, more complex than the teleological "answer"—which is really no answer at all—but complex only in the sense that reality is complex; actually simple, inasmuch as the simplicity of the word "is" can be comprehended.

Understandings of this sort can be reduced to this deep and significant summary: "It's so because it's so." But exactly the same words can also express the hasty or superficial attitude. There seems to be no explicit method for differentiating the deep and participating understanding, the "all-truth" which admits infinite change or expansion as added relations become apparent, from the shallow dismissal and implied lack of further interest which may be couched in the very same words.

B. *Why are some matches larger than others?*

Examine similarly a group of matches. At first they seem all to be of the same size. But to turn up differences, one needs only to measure them carefully with calipers or to weigh them with an analytical balance. Suppose the extreme comprises only a .001 percent departure from the mean (it will be actually much more); even so slight a differential we know can be highly significant, as with the sea-hares. The differences will group into plus-minus variations from a hypothetical mean to which not one single example will be found exactly to conform. Now the ridiculousness of the question becomes apparent. There is no *particular* reason. It's just so. There may be in the situation some factor or factors more important than the others: owing to the universality of variation (even in those very factors which "cause" variation), there surely *will* be, some even predominantly so. But the question as put is seen to be beside the point. The good answer is: "It's just in the nature of the beast." And this needn't imply belittlement; to have understood the "nature" of a thing is in itself a considerable achievement.

But if the size variations should be quite obvious—and especially if uniformity were to be a desideratum—then there might be a particularly dominant "causative" factor which could be searched out. Or if a person must have a stated "cause"—and many people must, in order to get an emotional understanding, a sense of relation to the situation and to give a name to the thing in order to "settle" it so it may not bother them any more—he can examine the automatic machinery which fabricates the products, and discover in it the variability which results in variation in the matches. But in doing so, he will become involved with a larger principle or pattern, the universality of variation, which has little to do with causality as we think of it.

C. *Leadership.*

The teleological notion would be that those in the forefront are leaders in a given movement and actually direct and consciously lead the masses in the sense that an army corporal orders "Forward march" and the squad marches ahead. One speaks in such a way of church leaders, of political leaders, and of leaders in scientific thought, and of course there is some limited justification for such a notion.

Non-teleological notion: that the people we call leaders are simply those who, at the given moment, are moving in the direction behind which will be found the greatest weight, and which represents a future mass movement.

For a more vivid picture of this state of affairs, consider the movements of an ameba under the microscope. Finger-like processes, the pseudopodia, extend at various places beyond the confines of the chief mass. Locomotion takes place by means of the animal's flowing into one or into several adjacent pseudopodia. Suppose that the molecules which "happened" to be situated in the forefront of the pseudopodium through which the animal is progressing, or into which it will have flowed subsequently, should be endowed with consciousness and should say to themselves and to their fellows: "We are directly leading this great procession, our leadership 'causes' all the rest of the population to move this way, the mass follows the path we blaze." This would be equivalent to the attitude with which we commonly regard leadership.

As a matter of fact there are three distinct types of thinking, two of them teleological. Physical teleology, the type we have been considering, is by far the commonest today. Spiritual teleology is rare. Formerly predominant, it now occurs metaphysically and in most religions, especially as they are popularly understood (but not, we suspect, as they were originally enunciated or as they may still be known to the truly adept). Occasionally the three types may be contrasted in a single problem. Here are a couple of examples:

(1) Van Gogh's feverish hurrying in the Arles epoch, culminating in epilepsy and suicide.

Teleological "answer": Improper care of his health during times of tremendous activity and exposure to the sun and weather brought on his epilepsy out of which discouragement and suicide resulted.

Spiritual teleology: He hurried because he innately foresaw his imminent death, and wanted first to express as much of his essentiality as possible.

Non-teleological picture: Both the above, along with a good many other symptoms and expressions (some of which could probably be inferred from his letters), were parts of his essentiality, possibly glimpsable as his "lust for life."

(2) The thyroid-neurosis syndrome.

Teleological "answer": Over-activity of the thyroid gland irritates and over-stimulates the patient to the point of nervous breakdown.

Spiritual teleology: The neurosis is causative. Something psychically wrong drives the patient on to excess mental irritation which harries and upsets the glandular balance, especially the thyroid, through shock-resonance in the autonomic system, in the sense that a purely psychic shock may spoil one's appetite, or may even result in violent illness. In this connection, note the army's acceptance of extreme homesickness as a reason for disability discharge.

Non-teleological picture: Both are discrete segments of a vicious circle, which may also include other factors as additional more or less discrete segments, symbols or maybe parts of an underlying but non-teleological pattern which comprises them and many others, the ramifications of which are *n*, and which has to do with causality only reflectedly.

Teleological thinking may even be highly fallacious, especially where it approaches the very superficial but quite common *post hoc, ergo propter hoc* pattern. Consider the situation with reference to dynamiting in a quarry. Before a charge is set off, the foreman toots warningly on a characteristic whistle. People living in the neighborhood come to associate the one with the other, since the whistle is almost invariably followed within a few seconds by the shock and sound of an explosion for which one automatically prepares oneself. Having experienced this many times without closer contact, a very naïve and unthinking person might justly conclude not only that there was a cause-effect relation, but that the whistle actually caused the explosion. A slightly wiser person would insist that the explosion caused the whistle, but would be hard put to explain the transposed time element. The normal adult would recognize that the whistle no more caused the explosion than the explosion caused the whistle, but that both were parts of a larger pattern out of which a "why" could be postulated for both, but more immediately and particularly for the whistle. Determined to chase the thing down in a cause-effect sense, an observer would have to be very wise indeed who could follow the intricacies of cause through more fundamental cause to primary cause, even in this largely man-made series about which we presumably know most of the motives, causes, and ramifications. He would eventually find himself in a welter of thoughts on production, and ownership of the means of production, and economic whys and wherefores about which there is little agreement.

The example quoted is obvious and simple. Most things are far more subtle than that, and have many of their relations and most of their origins far back in things more difficult of access than the tooting of a whistle calculated to warn bystanders away from an explosion. We know little enough even of a man-made series like this—how much less of purely natural phenomena about which also there is apt to be teleological pontificating!

Usually it seems to be true that when even the most definitely apparent cause-effect situations are examined in the light of wider knowledge, the cause-effect aspect comes to be seen as less rather than more significant, and the statistical or relational aspects acquire larger importance. It seems safe to assume that non-teleological is more "ultimate" than teleological reasoning. Hence the latter would be expected to prove to be limited and constricting except when used provisionally. But while it is true that the former is more open, for that very reason its employment necessitates greater discipline and care in order to allow for the dangers of looseness and inadequate control.

Frequently, however, a truly definitive answer seems to arise through teleological methods. Part of this is due to wish-fulfillment delusion. When a person asks "Why?" in a given situation, he usually deeply expects, and in any case receives, only a relational answer in place of the definitive "because" which he thinks he wants. But he customarily accepts the actually relational answer (it couldn't be anything else unless it comprised the whole, which is unknowable except by "living into") as a definitive "because." Wishful thinking probably fosters that error, since everyone continually searches for absolutisms (hence the value placed on diamonds, the most permanent physical things in the world) and imagines continually that he finds them. More justly, the relational picture should be regarded only as a glimpse—a challenge to consider also the rest of the relations as they are available—to envision the whole picture as well as can be done with given abilities and data. But one accepts it instead of a real "because," considers it settled, and, having named it, loses interest and goes on to something else.

Chiefly, however, we seem to arrive occasionally at definitive answers through the workings of another primitive principle: the universality of quanta. No one thing ever merges gradually into anything else; the steps are discontinuous, but often so very minute as to seem truly continuous. If the investigation is carried deep enough, the factor in question, instead of being graphable as a continuous process, will be seen to function by discrete quanta with gaps or synapses between, as do quanta of energy, undulations of light. The apparently definitive answer occurs when causes and effects both arise on the same large plateau which is bounded a great way off by the steep rise which announces the next plateau. If the investigation is extended sufficiently, that distant rise will, however, inevitably be encountered; the answer which formerly seemed definitive now will be seen to be at least slightly inadequate and the picture will have to be enlarged so as to include the plateau next further out. Everything impinges on everything else, often into radically different systems, although in such cases faintly. We doubt very much if there are any truly "closed systems." Those so called represent kingdoms of a great continuity bounded by the sudden discontinuity of great synapses which eventually must be bridged in any unified-field hypothesis. For instance, the ocean, with reference to waves of water, might be considered as a closed system. But anyone who has lived in Pacific

Grove or Carmel during the winter storms will have felt the house tremble at the impact of waves half a mile or more away impinging on a totally different "closed" system.

But the greatest fallacy in, or rather the greatest objection to, teleological thinking is in connection with the emotional content, the belief. People get to believing and even to professing the apparent answers thus arrived at, suffering mental constrictions by emotionally closing their minds to any of the further and possibly opposite "answers" which might otherwise be unearthed by honest effort—answers which, if faced realistically, would give rise to a struggle and to a possible rebirth which might place the whole problem in a new and more significant light. Grant for a moment that among students of endocrinology a school of thought might arise, centering upon some belief as to etiology—upon the belief, for instance, that all abnormal growth is caused by glandular imbalance. Such a clique, becoming formalized and powerful, would tend, by scorn and opposition, to wither any contrary view which, if untrammeled, might discover a clue to some opposing "causative" factor of equal medical importance. That situation is most unlikely to arise in a field so lusty as endocrinology, with its relational insistence, but the principle illustrated by a poor example is thought nevertheless to be sound.

Significant in this connection is the fact that conflicts may arise between any two or more of the "answers" brought forth by either of the teleologies, or between the two teleologies themselves. But there can be no conflict between any of these and the non-teleological picture. For instance, in the condition called hyperthyroidism, the treatments advised by believers in the psychic or neurosis etiology very possibly may conflict with those arising out of a belief in the purely physical cause. Or even within the physical teleology group there may be conflicts between those who believe the condition due to a strictly thyroid upset and those who consider causation derived through a general imbalance of the ductless glands. But there can be no conflict between any or all of these factors and the non-teleological picture, because the latter includes them—evaluates them relationally or at least attempts to do so, or maybe only accepts them as time-place truths. Teleological "answers" necessarily must be included in the non-teleological method—since they are part of the picture even if only restrictedly true—and as soon as their qualities of relatedness are recognized. Even erroneous beliefs are real things, and have to be considered proportional to their spread or intensity. "All-truth" must embrace all extant apropos errors also, and know them as such by relation to the whole, and allow for their effects.

The criterion of validity in the handling of data seems to be this: that the summary shall say in substance, significantly and understandingly, "It's so because it's so." Unfortunately the very same words might equally derive through a most superficial glance, as any child could learn to repeat from memory the most abstruse of Dirac's equations. But to know a thing emergently and significantly is something else again, even though the understanding may be expressed in the self-same words that were used superficially. In the following example [1] note the deep significance of the emergent as contrasted with the presumably satisfactory but actually incorrect original naïve understanding. At one time an important game bird in Norway, the willow grouse, was so clearly threatened with extinction that it was thought wise to establish protective regulations and to place a bounty on its chief enemy, a hawk which was known to feed heavily on it. Quantities of the hawks were exterminated, but despite such drastic measures the grouse disappeared actually more rapidly than before. The naïvely applied customary remedies had obviously failed. But instead of becoming discouraged and quietistically letting this bird go the way of the

[1] Abstracted from the article on ecology by Elton, *Encyclopaedia Britannica,* 14th Edition, Vol. VII, p. 916.

great auk and the passenger pigeon, the authorities enlarged the scope of their investigations until the anomaly was explained. An ecological analysis into the relational aspects of the situation disclosed that a parasitic disease, coccidiosis, was endemic among the grouse. In its incipient stages, this disease so reduced the flying speed of the grouse that the mildly ill individuals became easy prey for the hawks. In living largely off the slightly ill birds, the hawks prevented them from developing the disease in its full intensity and so spreading it more widely and quickly to otherwise healthy fowl. Thus the presumed enemies of the grouse, by controlling the epidemic aspects of the disease, proved to be friends in disguise.

In summarizing the above situation, the measure of validity wouldn't be to assume that, even in the well-understood factor of coccidiosis, we have the real "cause" of any beneficial or untoward condition, but to say, rather, that in this phase we have a highly significant and possibly preponderantly important relational aspect of the picture.

However, many people are unwilling to chance the sometimes ruthless-appearing notions which may arise through nonteleological treatments. They fear even to use them in that they may be left dangling out in space, deprived of such emotional support as had been afforded them by an unthinking belief in the proved value of pest control in the conservation of game birds; in the institutions of tradition; religion; science; in the security of the home or the family; or in a comfortable bank account. But for that matter emancipations in general are likely to be held in terror by those who may not yet have achieved them, but whose thresholds in those respects are becoming significantly low. Think of the fascinated horror, or at best tolerance, with which little girls regard their brothers who have dispensed with the Santa Claus belief; or the fear of the devout young churchman for his university senior who has grown away from depending on the security of religion.

As a matter of fact, whoever employs this type of thinking with other than a few close friends will be referred to as detached, hard-hearted, or even cruel. Quite the opposite seems to be true. Non-teleological methods more than any other seem capable of great tenderness, of an all-embracingness which is rare otherwise. Consider, for instance, the fact that, once a given situation is deeply understood, no apologies are required. There are ample difficulties even to understanding conditions "as is." Once that has been accomplished, the "why" of it (known now to be simply a relation, though probably a near and important one) seems no longer to be preponderantly important. It needn't be condoned or extenuated, it just "is." It is seen merely as part of a more or less dim whole picture. As an example: A woman near us in the Carmel woods was upset when her dog was poisoned—frightened at the thought of passing the night alone after years of companionship with the animal. She phoned to ask if, with our windows on that side of the house closed as they were normally, we could hear her ringing a dinner bell as a signal during the night that marauders had cut her phone wires preparatory to robbing her. Of course that was, in fact, an improbable contingency to be provided against; a man would call it a foolish fear, neurotic. And so it was. But one could say kindly, "We can hear the bell quite plainly, but if desirable we can adjust our sleeping arrangements so as to be able to come over there instantly in case you need us," without even stopping to consider whether or not the fear was foolish, or to be concerned about it if it were, correctly regarding all that as secondary. And if the woman had said apologetically, "Oh, you must forgive me; I know my fears are foolish, but I am so upset!" the wise reply would have been, "Dear person, nothing to forgive. If you have fears, they *are;* they are real things and to be considered. Whether or not they're foolish is beside the point. *What* they are is unimportant alongside the fact *that* they are." In other words, the badness or goodness, the teleology of the fears, was decidedly secondary. The whole notion could be conveyed by a smile or by a pleasant intonation more readily than by the words themselves. Teleological treatment which one might have been tempted to employ under the circumstances would first have stressed the fact that the fear was foolish—would say with a great show of objective justice, "Well, there's no use in *our* doing anything; the fault is that *your* fear is foolish and improbable. Get over that" (as a judge would say, "Come into court with clean hands") ; "then if there's anything *sensible* we can do, we'll see," with smug blame implied in every word. Or, more kindly, it would try to reason with the woman in an attempt to help her get over it—the business of propaganda directed towards change even before the situation is fully understood (maybe as a lazy substitute for understanding). Or, still more kindly, the teleological method would try to understand the fear causally. But with the non-teleological treatment there is only the love and understanding of instant acceptance; after that fundamental has been achieved, the next step, if any should be necessary, can be considered more sensibly.

Strictly, the term non-teleological thinking ought not to be applied to what we have in mind. Because it involves more than thinking, that term is inadequate. *Modus operandi* might be better—a method of handling data of any sort. The example cited just above concerns feeling more than thinking. The method extends beyond thinking even to living itself; in fact, by inferred definition it transcends the realm of thinking possibilities, it postulates "living into."

In the destitute-unemployed illustration, thinking, as being the evaluatory function chiefly concerned, was the point of departure, "the crust to break through." There the "blame approach" considered the situation in the limited and inadequate teleological manner. The non-teleological method included that viewpoint as correct but limited. But when it came to the feeling aspects of a human relation situation, the non-teleological method would probably ameliorate the woman's fears in a loving, truly mellow, and adequate fashion, whereas the teleological would have tended to bungle things by employing the limited and sophisticated approach.

Incidentally, there is in this connection a remarkable etiological similarity to be noted between cause in thinking and blame in feeling. One feels that one's neighbors are to be blamed for their hate or anger or fear. One thinks that poor pavements are "caused" by politics. The non-teleological picture in either case is the larger one that goes beyond blame or cause. And the non-causal or non-blaming viewpoint seems to us very often relatively to represent the "new thing," the Hegelian "Christ-child" which arises emergently from the union of two opposing viewpoints, such as those of physical and spiritual teleologies, especially if there is conflict as to causation between the two or within either. The new viewpoint very frequently sheds light over a larger picture, providing a key which may unlock levels not accessible to either of the teleological viewpoints. There are interesting parallels here: to the triangle, to the Christian ideas of trinity, to Hegel's dialectic, and to Swedenborg's metaphysic of divine love (feeling) and divine wisdom (thinking).

The factors we have been considering as "answers" seem to be merely symbols or indices, relational aspects of things—of which they are integral parts—not to be considered in terms of causes and effects. The truest reason for anything's being so is that it *is*. This is actually and truly a reason, more valid and clearer than all the other separate reasons, or than any group of them short of the whole. Anything less than the whole forms part of the picture only, and the infinite whole is unknowable except by *being* it, by living into it.

A thing may be *so* "because" of a thousand and one reasons of greater or lesser importance, such as the man oversized because of glandular insufficiency. The integration of these many reasons which are in the nature of relations rather than reasons is that he *is*. The separate reasons, no matter how valid, are only fragmentary parts of the picture. And the whole necessarily includes all that it impinges on as object and subject, in ripples fading with distance or depending upon the original intensity of the vortex.

The frequent allusions to an underlying pattern have no implication of mysticism—except inasmuch as a pattern which comprises infinity in factors and symbols might be called mystic. But infinity as here used occurs also in the mathematical aspects of physiology and physics, both far away from mysticism as the term is ordinarily employed. Actually, the underlying pattern is probably nothing more than an integration of just such symbols and indices and mutual reference points as are already known, except that its power is *n*. Such an integration might include nothing more spectacular than we already know. But, equally, it *could* include anything, even events and entities as different from those already known as the vectors, tensors, scalars, and ideas of electrical charges in mathematical physics are different from the mechanical-model world of the Victorian scientists.

In such a pattern, causality would be merely a name for something that exists only in our partial and biased mental reconstructings. The pattern which it indexes, however, would be real, but not intellectually apperceivable because the pattern goes everywhere and is everything and cannot be encompassed by finite mind or by anything short of life—which it is.

The psychic or spiritual residua remaining after the most careful physical analyses, or the physical remnants obvious, particularly to us of the twentieth century, in the most honest and disciplined spiritual speculations of medieval philosophers, all bespeak such a pattern. Those residua, those most minute differentials, the 0.001 percentages which suffice to maintain the races of sea animals, are seen finally to be the most important things in the world, not because of their sizes, but because they are everywhere. The differential is the true universal, the true catalyst, the cosmic solvent. Any investigation carried far enough will bring to light these residua, or rather will leave them still unassailable as Emerson remarked a hundred years ago in "The Oversoul"—will run into the brick wall of the *impossibility* of perfection while at the same time insisting on the *validity* of perfection. Anomalies especially testify to that framework; they are the commonest intellectual vehicles for breaking through; all are solvable in the sense that any *one* is understandable, but that one leads with the power n to still more and deeper anomalies.

This deep underlying pattern inferred by non-teleological thinking crops up everywhere—a relational thing, surely, relating opposing factors on different levels, as reality and potential are related. But it must not be considered as causative, it simply exists, it *is*, things are merely expressions of it as it is expressions of them. And they *are* it, also. As Swinburne, extolling Hertha, the earth goddess, makes her say: "Man, equal and one with me, man that is made of me, man that is I," so all things which are *that*—which is all—equally may be extolled. That pattern materializes everywhere in the sense that Eddington finds the non-integer q "number" appearing everywhere, in the background of all fundamental equations,[2] in the sense that the speed of light, constant despite compoundings or subtractions, seemed at one time almost to be conspiring against investigation.

The whole is necessarily everything, the whole world of fact and fancy, body and psyche, physical fact and spiritual truth, individual and collective, life and death, macrocosm and microcosm (the greatest quanta here, the greatest synapse between these two), conscious and unconscious, subject and object. The

[2] *The Nature of the Physical World,* pp. 208–10.

whole picture is portrayed by *is,* the deepest word of deep ultimate reality, not shallow or partial as reasons are, but deeper and participating, possibly encompassing the Oriental concept of *being.*

And all this against the hot beach on an Easter Sunday, with the passing day and the passing time. This little trip of ours was becoming a thing and a dual thing, with collecting and eating and sleeping merging with the thinking-speculating activity. Quality of sunlight, blueness and smoothness of water, boat engines, and ourselves were all parts of a larger whole and we could begin to feel its nature but not its size.

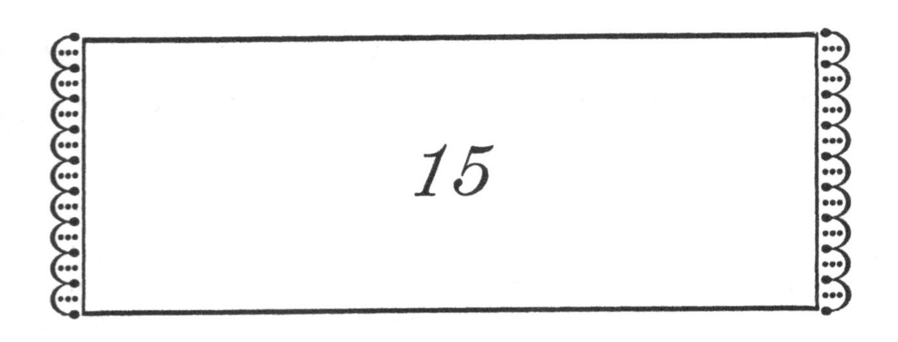

ABOUT noon we sailed and moved out of the shrouded and quiet Amortajada Bay and up the coast toward Marcial Reef, which was marked as our next collecting station. We arrived in mid-afternoon and collected on the late tide, on a northerly pile of boulders, part of the central reef. This was just south of Marcial Point, which marks the southern limit of Agua Verde Bay.

It was not a good collecting tide, although it should have been according to the tide chart. The water did not go low enough for exhaustive collecting. There were a few polyclads which here were high on the rocks. We found two large and many small chitons—the first time we had discovered them in numbers. There were many urchins visible but too deep below the surface to get to. Swarms of larval shrimps were in the water swimming about in small circles. The collecting was not successful in point of view of numbers of forms taken.

That night we rigged a lamp over the side, shaded it with a paper cone, and hung it close down to the water so that the light was reflected downward. Pelagic isopods and mysids immediately swarmed to the illuminated circle until the water seemed to heave and whirl with them. The small fish came to this horde of food, and on the outer edges of the light ring large fishes flashed in and out after the small fishes. Occasionally we interrupted this mad dance with dip-nets, dropping the catch into porcelain pans for closer study, and out of the nets came animals small or transparent that we had not noticed in the sea at all.

Having had no good tide at Marcial Reef, we arose at four o'clock the following morning and went in the darkness to collect again. We carried big seven-cell focusing flashlights. In some ways they make collecting in the dark, in a small area at least, more interesting than daytime collecting, for they limit the range of observation so that in the narrowed field one is likely to notice more detail. There is a second reason for our preference for night collecting—a number of animals are more active at night than in the daytime and they seem to be not much disturbed or frightened by artificial light. This time we had a very fair tide. The light fell on a monster highly colored spiny lobster in a crevice of the reef. He was blue and orange and spotted with brown. The taking of him required caution, for these big lobsters are very strong and are so armed with spikes and points that in struggling with one the hands can be badly cut. We approached with care, bent slowly down, and then with two hands grabbed him about the middle of the body. And there was no struggle whatever. He was either sick or lazy or hurt by the surf, and did not fight at all.

The cavities in Marcial Reef held a great many club-spined urchins and a number of the sharp-spined purple ones which had hurt us before. There were numbers of sea-fans, two of the usual starfish and a new species [1] which later we were to find common farther north in the Gulf. We took a good quantity of the many-rayed sun-stars, and a flat kind of cucumber which was new to us.[2] This was the first time we had collected at night, and under our lights we saw the puffer fish lazily feeding near the surface in the clear water. On the bottom, the brittle-stars, which we had always found under rocks, were crawling about like thousands of little snakes. They rarely move

---

1 *Othilia tenuispinus.*

2 Probably *Stichopus fuscus*—the specimen has since been lost sight of.

about in the daylight. Wherever the sharp, powerful rays of the flashlight cut into the water we could see the moving beautiful fish and the bottoms alive with busy feeding invertebrates. But collecting with a flashlight is difficult unless it is arranged that two people work together—one to hold the light and the other to take the animals. Also, from constant wetting in salt water the life of a flashlight is very short.

The one huge and beautiful lobster was the prize of this trip. We tried to photograph him on color film and as usual something went wrong but we got a very good likeness of one end of him, which was an improvement on our previous pictures. In most of our other photographs we didn't get either end.

We took several species of chitons and a great number of tunicates. There were several turbellarian flatworms, but these are so likely to dissolve before they preserve that we had great difficulties with them. There were in the collecting pans several species of brittle-stars, numbers of small crabs and snapping shrimps, plumularian hydroids, bivalves of a number of species, snails, and some small sea-urchins. There were worms, hermit crabs, sipunculids, and sponges. The pools too had been thick with pelagic larval shrimps, pelagic isopods—tiny crustacea similar to sow-bugs—and tiny shrimps (mysids). In this area the water seemed particularly peopled with small pelagic animals—"bugs," so the boys said. Everywhere there were bugs, flying, crawling, and swimming. The shallow and warm waters of the area promoted a competitive life that was astonishing.

After breakfast we pulled up the anchor and set out again northward. The pattern of the technique of the trip had by now established itself almost as a habit with us; collecting, running to a new station, collecting again. The water was intensely blue on this run, and the fish were very many. We could see the splashing of great schools of tuna in the distance where they beat the water to spray in their millions. The swordfish leaped all about us, and someone was on the bow the whole time trying to drive a light harpoon into one, but we never could get close enough. Cast after cast fell short.

We preserved and labeled as we went, and the water was so smooth that we had no difficulty with delicate animals. If the boat rolls, retractile animals such as anemones and sipunculids are more than likely to draw into themselves and refuse to relax under the Epsom-salts treatment, but this sea was as smooth as a lawn, and our wake fanned out for miles behind us.

The fish-lines on the stays snapped and jerked and we brought in skipjack, Sparky's friend of the curious name, and the Mexican sierra. This golden fish with brilliant blue spots is shaped like a trout. In size it ranges from fifteen inches to two feet, is slender and a very rapid swimmer. The sierra does not seem to travel in dense, surface-beating schools as the tuna does. Although it belongs with the mackerel-like forms, its meat is white and delicate and sweet. Simply fried in big hunks, it is the most delicious fish of all.

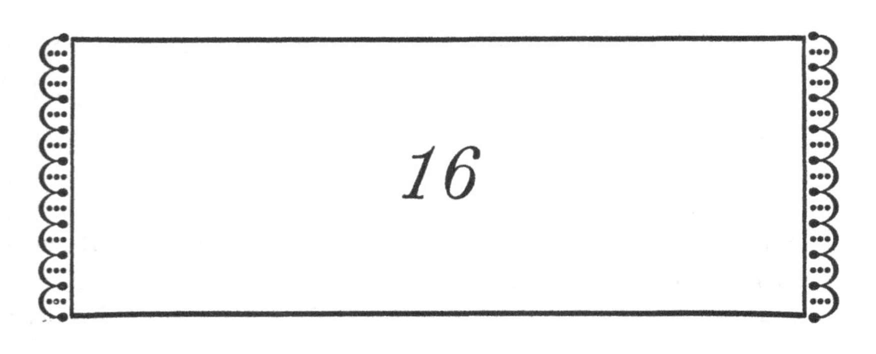

March 25

ABOUT noon we arrived at Puerto Escondido, the Hidden Harbor, a place of magic. If one wished to design a secret personal bay, one would probably build something very like this little harbor. A point swings about, making a small semicircular bay fringed with bright-green mangroves, and only when one has turned inside this outer bay can one see that there is a second, secret bay beyond—a long narrow bay with an entrance not more than fifty feet wide at flood. The charts gave three fathoms at the center of the entrance, but the tide run was so furious that we did not attempt to take the *Western Flyer* in, but anchored in back of the first point, called Piedra de la Marina. Here we had more than ten fathoms, and Tony felt better about it.

In the distance, and from the south, a canoe came up the coast with a small sail set. The Indians move great distances in their tiny boats. As soon as the anchor was out, we dropped the fishing lines and immediately hooked several hammer-head sharks and a large red snapper. The air here was hot and filled with the smell of mangrove flowers. The little outer bay was our first collecting station, a shallow warm cove with a mud bottom and edged with small boulders, smooth and unencrusted with algae. On the bottom we could see long snake-like animals, gray with black markings, with purplish-orange floriate heads like chrysanthemums. They were about three feet long and new to us. Wading in rubber boots, we captured some of them and they proved to be giant synaptids.[1] They were strange and frightening to handle, for they stuck to anything they touched, not with slime but as though they were coated with innumerable suction-cells. On being taken from the water, they collapsed to skin, for their bodily shape is maintained by the current of water which they draw through themselves. When lifted out, this water escapes and they hang as limp as unfilled sausage skins. Since they were new and fascinating to us, we took many specimens, maneuvering them gently to the surface and then sliding them into submerged wooden collecting buckets to prevent them from dropping their water. On the bottom they crawled about, their flower-heads moving gently, while the current of water passing through their bodies drew food into their stomachs. When we took them on board, we found they had to a high degree the habit of a number of holothurians: eviscerating. These *Euapta* were a nervous lot. We tried to relax them with Epsom salts so that we might kill them with their floriate heads extended, but the salts, no matter how carefully administered, caused the heads to retract, and soon afterwards they threw their stomachs out into the water. The word "stomach" is used here inadvisedly, for what they actually disgorge is the intestinal tract and respiratory tree.

We intoxicated them with pure oxygen and then tried the salts, but with the same result. Finally, by administering the salts in minute quantities and very slowly, we were able to preserve some uneviscerated specimens, but none with the head extended. The color motion pictures of the living animals, while not very good, at least showed the color and shape and movement of the extended heads. Again we got photographs of only one end, but this time the more important end, the floriate head.

In the little shallow bay there were many bright-green gars,

---

[1] A worm-like sea-cucumber, *Euapta godeffroyi*.

or needle-fish, but they were too fast for our dip-nets and we were unable to take them. *Botete,* the poison fish, was here also in great numbers, and the boys took some of them with a light seine. We found here two new starfishes and many *Cerianthus* anemones.

While we were collecting on the shore, Tiny rowed about in the little skiff in slightly deeper water. He carried a light three-pronged spear with which he picked up an occasional cushion star from the bottom. We heard him shout, and looked up to see a giant manta ray headed for him, the tips of the wings more than ten feet apart. It was rare to see them in such shallow water. As it passed directly under his boat we yelled at him to spear it, since he wanted to so badly, but he simply sat in the bottom of the boat, gazing after the retreating ray, weakly swearing at us. For a long time he sat there quietly, not quite believing what he had seen. This great fish could have flicked Tiny and boat and all into the air with one flap of its wing. Tiny wanted to sit still and think for a long time and he did. For an hour afterward he could only repeat, "Did you see that Goddamned thing!" And from that moment it became Tiny's ambition to catch and kill one of the giant rays.

The canoe which had been sailing up the coast came alongside and a man and a little boy boarded us. They had with them what they called "abalon"—not true abalones, but gigantic fixed scallops, very good for food. They had also some of the hacha, the huge fan-shaped clam; pearl oysters, which are growing rare; and several huge conchs. We bought from the man what he had and asked him to get us more of the large shellfish. We might look for weeks for animals he could go to directly. Everywhere it is the same: if an animal is good to eat or poisonous or dangerous the natives of the place will know about it and where it lives. But if it have none of these qualities, no matter how highly colored or beautiful, he may never in his life have seen it.

On the stone-bordered sandspit which is the southern block to the true inner Puerto Escondido there was a new stone building not quite finished, with no one about it. Around the point there now came a large rowboat pushed by a fast outboard motor of a species distinct from the Sea-Cow, for it seemed controlled and dominated by its master. In this boat there were several Indians and three men dressed in riding breeches and hiking boots. They came aboard and introduced themselves as Leopoldo Pérpuly, who owned a ranch on the edge of Puerto Escondido, Gilbert Baldibia, a school-teacher from Loreto, and Manuel Madinabeitia C., of the customs service, also of Loreto. These last two were on a vacation and hunting trip. They were strong, fine-looking men wearing the ever-present .45-caliber automatics of the government service. We served them canned fruit salad and discussed with them the country we had covered, and they asked us to go hunting the *borrego,* or big-horn sheep, with them, starting that afternoon and getting back the next day. We were to go into the tremendous and desolate stone mountains to camp and hunt. We accepted immediately, and went with them to the little ranch set back half a mile from Puerto Escondido. We didn't want to kill a big-horn sheep, but we wanted to see the country. As it turned out, none of them—the rancher, the teacher, or the customs man—had any intention of killing a big-horn sheep.

The little ranch was set deep in the brush. It was watered by deep wells of brackish brown water out of which endless chains of buckets emerged at the insistence of mules which turned the windlass. This rancher in the desert has dug sixty-foot wells, and he is raising tomatoes and he has planted many grapevines. But so dry is the earth that a few weeks without the rising buckets would destroy all his work. The houses of the ranch were simply roofs and low walls of woven palm, enough to keep out the wind but no obstruction to the air. The floors were of swept hard-packed earth, and there was an air of comfort about the place. The Indian workmen worked very slowly, and the babies peeked out of the woven houses at us. We were to ride to the mountains on mules and one small horse while two Indian men walked ahead. We were sorry for them until we discovered that their main irritation lay in the fact that horses and mules are so slow. Often they disappeared ahead of us, and we found them later sitting beside the trail waiting for us. The line of us started out on a clear but unfinished road that was eventually to go to Loreto. The thick and thorny brush and cactus had been grubbed, but no scraping had been done yet. It was a fantastic country; heavy xerophytic plants: cacti, mimosa, and thorned bushes and trees crackled with the heat. There were the lichens which bleed bright red when they are broken and were once a source of dye before the anilines were developed. There were poison bushes which we were warned about, for if one touches them and then rubs one's eyes, blindness ensues. We learned some of the uses of plants of this country; maidenhair fern, we were told, is boiled to an infusion and given to women after childbirth. It is said that no hemorrhage can follow this treatment. We rode over a rolling, rocky, desolate country, then left the cleared, some-day road and turned up a trail toward the stone mountains, steep and slippery with shale. And here our Indians were even more impatient, for the mules went more slowly while the Indians did not change gait for the steep places.

"My mule was a complainer. For a while I thought he simply didn't like me, but I believe now that he had a sour eye for the world. With every step he groaned with pain so convincingly that once I removed the saddle to see whether he might not be saddle-burned. He did not grunt, but drew from deep in his belly great groans of an agonized soul left to molder in Purgatory. It is impossible to see why he did this, for certainly no Mexican would believe him and he had never carried one of the more sentimental northern race before. I was heart-broken for him, but not sufficiently to get off and walk. We both suffered up the trail, he with pain and I with sorrow for him." (Extract from the personal journal of one of us.)

The trail cut back on itself again and again, and the bare mountains towered high and brooding over us. Far below we could see the brilliant blue water of the Gulf with a fantastic mirage cast over it.

There was in our party one horse, a spindle-legged, small-buttocked little animal with eyes haunted by social inadequacy; one horse in a society of mules, and a gelding at that. We thought how often one mule is surrounded by socially dominant horses, all grace and prance, conscious of their power and loveliness. In this pattern the mule has developed his anti-social self-sufficiency. He knows he can out-think a horse and he is pretty sure he can out-think a human. In both respects he is correct. And so your socially outcast mule dwells inward in sneering intellectuality; his mental pattern, conditioned by centuries of this cynical intellectualism, is set, and he is complete, sullen, treacherous, loving no one, selfish and self-centered. But this horse, having no such background, was unable to make the change in one generation. Surrounded by mules, he sorrowed and his spirit broke and his eyes were sad. The stiffening was gone from his ears and his mouth hung open. He slunk ashamedly along behind the mules. Stripped of his regalia and his titles, he was a pitiful thing. Refugee princes usually become waiters, but this poor horse was not even able to be a waiter, let alone a horse. And just as one is irritated by a grand duke if he has no robes and garters and large metal-and-enamel decorations, so we found ourselves disliking this poor horse; and he knew it and it didn't help him.

We came at last to a trail of broken stone and rubble so steep that the mules could not carry us any more. We dismounted and crawled on all fours, and we don't know how the mules got up. After a short climb we emerged on a level place in a deep cleft in the granite mountains. In this cleft a tiny stream of water fell hundreds of feet from pool to pool. There were palm trees and wild grapevines and large ferns, and the water was cool and sweet. This little stream, coming from so high up in the mountains and falling so far, never had the final dignity of reaching the ocean. The desert sucked it down and the heat dried it up and on the level it disappeared in a light mist of frustration. We sat beside a pool of the waterfall and our Indians made coffee for us and unpacked a lunch, and one item of this lunch was so delicious that we have wanted it again. It is made in this way: a warm tortilla is laid down and spread with well-cooked beans, and another tortilla laid on top and spread, and another, until it is ten or twelve layers thick. Then it is wrapped in cloth. Before eating it one slices downward through the layers as with a cake. It is a fine dish and very filling. While we ate, the Indians made our beds on the ground, and we fired a few shots at a rock across the canyon. Then it was dark and we lay in our blankets and talked, and here we suffered greatly. For the funny stories began. We suppose they weren't clean stories, but we couldn't be sure. Nearly every one began, "Once there was a school-teacher with large black eyes —very sympathetic—" *"Muy simpática"* has a slightly different connotation from that of "sympathetic," for sympathy is a passive state of receptivity, but to be *"simpática"* is to be more active or co-operative, even sometimes a little forward. At any rate, this *"simpática"* school-teacher invariably had as one of her students "a tall strong boy, *con cojones, pero cojones*"—this last with a gesture easily seen in the firelight. The stories progressed until they came to the snappers; we leaned forward studiously intent, but the snappers were either so colloquial that we could not understand them or so filled with the laughter of the teller that we couldn't make out the words. Story after story was told, and we didn't get a single snapper, not one. Our suspicions were aroused of course. We knew something was bound to happen when a school-teacher *"muy simpática"* asks a large boy *"con cojones"* to stay after school, but whether it ever did or not we do not know.

It grew cold in the night, and the mosquitoes were unmerciful. In this sparsely populated country human blood must be a rarity. We were a seldom-found dessert to them, and they whooped and screamed and attacked, power-diving and wheeling up and diving again. The visibility was good, and we made excellent targets. Only when it became bitterly cold did they go away.

We have noticed many times how lightly Mexican Indians sleep. Often in the night they awaken to smoke a cigarette and talk softly together for a while, and then go to sleep again rather like restless birds, which sing a little in the dark, dreaming that it is already day. Half a dozen times a night they may awaken thus, and it is pleasant to hear them, for they talk very quietly as though they were dreaming.

When the dawn came, our Indians made coffee for us and we ate more of the lunch. Then, with some ceremony, the ranch-owner presented a Winchester .30-30 carbine with a broken stock to those Indians, and they set off straight up the mountainside. This, our first hunt for the *borrego,* or big-horn sheep, was the nicest hunting we have ever had. We did not raise a hand in our own service during the entire trip. Besides, we do not like to kill things—we do it when it is necessary, but we take no pleasure in it; and those fine Indians did it for us—the hunting, that is—while we sat beside the little waterfall and discussed many things with our hosts—how all Americans are rich and own new Fords; how there is no poverty in the United States and everyone sees a moving picture every night and is drunk as often as he wishes; how there are no political animosities; no need; no fear; no failure; no unemployment or hunger. It was a wonderful country we came from and our hosts knew all about it and told us. We could not spoil such a dream. After each one of his assurances we said, "*Cómo no?*" which is the most cautious understatement in the world, for "*Cómo no?*" means nothing at all. It is a polite filler between two statements from your companion. And we sat in that cool place and looked out over the hot desert country to the blue Gulf. In a couple of hours our Indians came back; they had no *borrego,* but one of them had a pocketful of droppings. It was time by now to start back to the boat. We intend to do all our future hunting in exactly this way. The ranch-owner said a little sadly, "If they had killed one we could have had our pictures taken with it," but except for that loss, there was no loss, for none of us likes to have the horns of dead animals around.

We had sat beside the little pool and watched the tree-frogs and the horsehair worms and the water-skaters, and had wondered how they got there, so far from other water. It seemed to us that life in every form is incipiently everywhere waiting for a chance to take root and start reproducing; eggs, spores, seeds, bacilli—everywhere. Let a raindrop fall and it is crowded with the waiting life. Everything is everywhere; and we, seeing the desert country, the hot waterless expanse, and knowing how far away the nearest water must be, say with a kind of disbelief, "How did they get clear here, these little animals?" And until we can attack with our poor blunt weapon of reason that causal process and reduce it, we do not quite believe in the horsehair worms and the tree-frogs. The great fact is that they are there. Seeing a school of fish lying quietly in still water, all the heads pointing in one direction, one says, "It is unusual that this is so" —but it isn't unusual at all. We begin at the wrong end. They simply lie that way, and it is remarkable only because with our blunt tool we cannot carve out a human reason. Everything is potentially everywhere—the body is potentially cancerous, phthisic, strong to resist or weak to receive. In one swing of the balance the waiting life pounces in and takes possession and grows strong while our own individual chemistry is distorted past the point where it can maintain its balance. This we call dying, and by the process we do not give nor offer but are taken by a multiform life and used for its proliferation. These things are balanced. A man is potentially all things too, greedy and cruel, capable of great love or great hatred, of balanced or unbalanced so-called emotions. This is the way he is—one factor in a surge of striving. And he continues to ask "why" without first admitting to himself his cosmic identity. There are colonies of pelagic tunicates [2] which have taken a shape like the finger of a glove. Each member of the colony is an individual animal, but the colony is another individual animal, not at all like the sum of its individuals. Some of the colonists, girdling the open end, have developed the ability, one against the other, of making a pulsing movement very like muscular action. Others of the colonists collect the food and distribute it, and the outside of the glove is hardened and protected against contact. Here are two animals, and yet the same thing—something the early Church would have been forced to call a mystery. When the early Church called some matter "a mystery" it accepted that thing fully and deeply as *so,* but simply not accessible to reason because reason had no business with it. So a man of individualistic reason, if he must ask, "Which is the animal, the colony or the individual?" must abandon his particular kind of reason and say, "Why, it's two animals and they aren't alike any more than the cells of my body are like me. I am much more than the sum of my cells and, for all I know, they are much more than the division of me." There is no quietism in such acceptance, but rather the basis for a far deeper understanding of us and our world. And now this is ready for the taboo-box.

It is not enough to say that we cannot know or judge because all the information is not in. The process of gathering knowledge does not lead to knowing. A child's world spreads only a little beyond his understanding while that of a great

[2] *Pyrosoma giganteum.*

scientist thrusts outward immeasurably. An answer is invariably the parent of a great family of new questions. So we draw worlds and fit them like tracings against the world about us, and crumple them when they do not fit and draw new ones. The tree-frog in the high pool in the mountain cleft, had he been endowed with human reason, on finding a cigarette butt in the water might have said, "Here is an impossibility. There is no tobacco hereabouts nor any paper. Here is evidence of fire and there has been no fire. This thing cannot fly nor crawl nor blow in the wind. In fact, this thing cannot be and I will deny it, for if I admit that this thing is here the whole world of frogs is in danger, and from there it is only one step to anti-frogicentricism." And so that frog will for the rest of his life try to forget that something that is, is.

On the way back from the mountain one of the Indians offered us his pocketful of sheep droppings, and we accepted only a few because he did not have many and he probably had relatives who wanted them. We came back through heat and dryness to Puerto Escondido, and it seemed ridiculous to us that the *Western Flyer* had been there all the time. Our hosts had been kind to us and considerate as only Mexicans can be. Furthermore, they had taught us the best of all ways to go hunting, and we shall never use any other. We have, however, made one slight improvement on their method: we shall not take a gun, thereby obviating the last remote possibility of having the hunt cluttered up with game. We have never understood why men mount the heads of animals and hang them up to look down on their conquerors. Possibly it feels good to these men to be superior to animals, but it does seem that if they were sure of it they would not have to prove it. Often a man who is afraid must constantly demonstrate his courage and, in the case of the hunter, must keep a tangible record of his courage. For ourselves, we have had mounted in a small hardwood plaque one perfect *borrego* dropping. And where another man can say, "There was an animal, but because I am greater than he, he is dead and I am alive, and there is his head to prove it," we can say, "There was an animal, and for all we know there still is and here is the proof of it. He was very healthy when we last heard of him."

After the dryness of the mountain it was good to come back to the sea again. One who was born by the ocean or has associated with it cannot ever be quite content away from it for very long.

Sparky made us a great dish of his spaghetti, the veritable Enea spaghetti, and we ate until we were bloated with it.

Now our equipment began to show its weaknesses. The valve of the oxygen cylinder gave trouble owing to the humidity. The little ice-plant was not powerful enough, and where it should have cooled sea water for us, it was all it could do to keep the beer chilled. Besides, it broke down very often.

By now, some animals began to emerge as ubiquitous. *Heliaster kubiniji,* the sun-star, was virtually everywhere, but we did observe that the farther up the Gulf we went, the smaller he became. *Eurythoë,* the stinging worm, occurred wherever there were loosely imbedded rocks or coral under which he could hide. In this connection it is interesting that in the description of this worm in Chamberlin,[3] the one descriptive item completely ignored is the one most important to the collector—that he stings like the devil, his hair-like fringe breaking off in the hands and leaving a burn which does not disappear for a long time. Tiny, who is able to translate experience readily into emotion, found that anger did not overcome *Eurythoë,* and he grew to have the greatest respect for the worm, even to the point of adopting the usual collector's caution of never putting the hands where one hasn't looked first.

The purple sharp-spined urchin [4] occurred wherever there

---

[3] 1919 (§ J-5), p. 28.
[4] *Arbacia incisa.*

was rock or reef exposed to wave-shock or fast-scouring currents. There were the usual barnacles and limpets on the rocks high up in the littoral wherever their pattern of alternating water and air was available. Anemones, the small bunodid forms, were everywhere too. And, of course, the porcelain crabs, hermit crabs, and sea-cucumbers.

We had taken a great many animals and, as compared with the work of some expensive, well-equipped, well-manned expeditions, our results began to cause us to wonder what methods were used by those collectors. For instance, the best reports to date (with the possible exception of the Hancock Expedition reports—and these are so expensive and rare that an amateur cannot afford them, and even university libraries do not always have them) are those of a well-known scientific expedition into the Gulf, about thirty years ago. There were eight naturalists aboard a specially built and equipped steamboat, with a complete and well-trained crew. In two months out of San Francisco they occupied thirty-five stations and took a total of 2351 individuals of 118 species of echinoderms both from deep water (including dredge hauls down to 1760 fathoms) and from along shore, and in two great faunal provinces. Only 39 species were from shallow water; 31 of these, in about 387 individuals, were from the Gulf. Already, in only nine days of Gulf collecting, in the one zoogeographical province and entirely along shore, we had taken almost double their 31 Gulf echinoderm species—the only group we had so far tabulated—and had begun to restrain our enthusiasm owing to the lack of containers. We worked hard, but not beyond reason, and our wonder is caused not by the numbers we took, but by the small numbers they did. We had time to play and to talk, and even to drink a little beer. (We took 2160 individuals of two species of beer.)

The shores of the Gulf, so rich for the collector, must still be fairly untouched (again except for the largely unreported Hancock collections). We had not the time for the long careful collecting which is necessary before the true picture of the background of life can be established. We rushed through because it was all we could afford, but our results seem to indicate that energy and enthusiasm can offset lack of equipment and personnel.

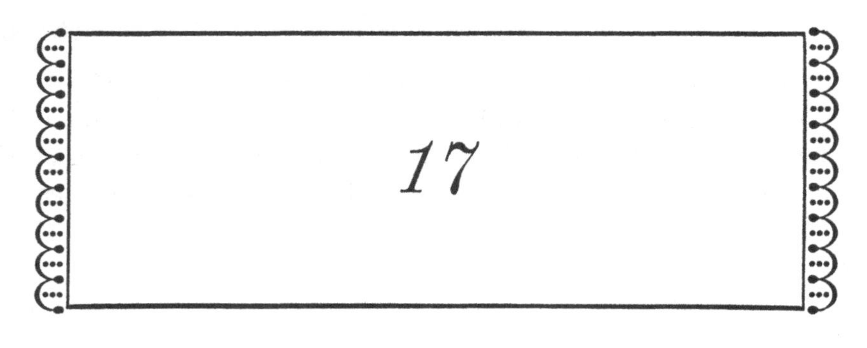

March 27

WE HAD collected extensively on the outer parts of Puerto Escondido, but not in the inner bay itself. At five-thirty A.M. Mexican time, we set out to circle this inner bay in the little skiff. It was dark when we started, and we used the big flashlights for collecting. There was a good low tide, and we moved slowly along the shore, one rowing while the other inspected the bottom with the light. There was no ripple to distort the surface. The eastern shore was dominated by the big, flat, chocolate-brown holothurian.[1] They moved slowly along, feeding on the bottom, many hundreds of them. They far overshadowed in number any other animals in this area. There were many of the ruffled clams[2] with hard, thick, wavy shells. The under-rock fauna was not very rich. The eastern and northern shores were littered with shattered rock, recently enough splintered so that the edges were still sharp, and in this quiet bay no waves would have ground the edges smooth. Mangroves bordered a great part of the bay, and the spicy smell of their flowers was strong and pleasant. A few of the giant, snake-like synaptids that we had taken in the outer bay waved and moved on the bottom. As we rounded toward the westerly side of the bay, we came to sand flats and a change of fauna, for the big brown cucumbers did not live here. The dawn

---

1 *Stichopus fuscus.*
2 *Carditamera affinis.*

came as we moved along the sand flats. Two animals were at the waterside, about as large as small collies, dark brown, with a cat-like walk. In the half-light we could not see them clearly, and as we came nearer they melted away through the mangroves. Possibly they were something like giant civet-cats. They had undoubtedly been fishing at the water's edge. On the smooth sand bottom of this area there were clusters of knobbed, green coral (probably *Porites porosa*—no samples were gathered), but except for *Cerianthus* and a few bivalves this bottom was comparatively sterile.

Rounding the southern end of the bay, we came again to the single narrow entrance where the water was rushing in on the returning tide, and here, suddenly, the area was incredibly rich in fauna. Here, where the water rushes in and out, bringing with it food and freshness, there was a remarkable gathering. Beautiful red and green cushion stars littered the rocky bottom. We found clusters of a solitary soft coral-like form [3] in great knobs and heads in one restricted location on the rocks. Caught against the rocks by the current was a very large pelagic coelenterate, in appearance like an anemone with long orange-pink tentacles, apparently not retractable. On picking him up we were badly stung. His nettle-cells were vicious, stinging even through the calluses of the palms, and hurting like a great many bee-stings. At this entrance also we took several giant sea-hares,[4] a number of clams, and one small specimen of the clam-like hacha. For hours afterwards the sting of the anemone remained. So very many things are poisonous and hurtful in these Gulf waters: urchins, sting-rays, morays, heart-urchins, this beastly anemone, and many more. One becomes very timid after a while. Barnacle-cuts, which are impossible to avoid, cause irritating sores. The fingers and palms become cross-hatched with

---

[3] In superficial appearance it was identical with the figures of the West Indies *Zoanthus pulchellus* illustrated in Duerden's "Actinians of Porto Rico," 1902, *U. S. Fish Comm. Bulletin* for 1900, Vol. 2, pp. 321–74.

[4] *Dolabella californica.*

cuts, and then very quickly, possibly owing to the constant soaking in salt water and the regular lifting of rocks, the hands become covered with a hard, almost horny, callus.

The Puerto Escondido station was one of the richest we visited, for it combined many kinds of environment in a very small area; sand bottom, stone shore, boulders, broken rock, coral, still, warm, shallow places, and racing tide. It is highly probable that careful and extended collecting would show that individuals of species of a very respectable proportion of the total Panamic fauna could be found in this tiny world. Barring surf-battered reef, every probable environment occurs within these few acres—a textbook exhibit for ecologists.

We took rock isopods, sponges, tunicates, turbellarians, chitons, bivalves, snails, hermit crabs, and many other crabs, Heteronereids and mysids pelagic at night, small ophiurans, limpets, and worms and even listed in our collecting notes for the day the horsehair worms from the little waterfall in the mountains.[5] We took six to eight species of cucumbers and eleven of starfish at this one station.

When we came back from the early morning collecting we sailed immediately for the port of Loreto. We were eager to see this town, for it was the first successful settlement on the Peninsula, and its church is the oldest mission of all. Here the inhospitability of Lower California had finally been conquered and a colony had taken root in the face of hunger and mishap. From the sea, the town was buried in a grove of palms and greenery. We dropped anchor and searched the shore with our glasses. A line of canoes lay on the beach and a group of men sat on the sand by the canoes and watched us; comfortable, lazy-looking men in white clothes. When our anchor dropped they got up and made for the town. Of course, they had to find their uniforms, and since Loreto was not very often visited and since

---

[5] *Chorodes* sp., probably *C. occidentalis* Montgomery, according to J. T. Lucker of the U. S. National Museum, their No. 159124.

the Governor had *not* recently been there, this may not have been so easy. There may have been some scurrying of errand-bound children from house to house, looking for tunics or belts or borrowing clean shirts. Señor the official had to shave and scent himself and dress. It all takes time, and the boat in the harbor will wait. It didn't look like much of a boat anyway, but at least it was a boat.

One fine thing about Mexican officials is that they greet a fishing boat with the same serious ceremony they would afford the *Queen Mary,* and the *Queen Mary* would have to wait just as long. This made us feel very good and not rebellious about the port fees—absent in this case! We came to them and they made us feel, not like stodgy people in a purse-seiner but like ambassadors from Ultra-Marina bringing letters of greeting out of the distances. It is no wonder that we too scurried for clean shirts, that Tony put on his master's cap, and Tiny polished the naval insignia on his, which he had come by no doubt honorably in a washroom in San Diego. We were not smart, not very alert, but we were clean and we smelled rather delicious. Sparky sprinkled us with shaving lotion and we filled the air with an odor of flowers. If the *brazo,* the double embrace, should be indicated by any feeling of uncontrollable good-will, we were ready.

The men came back to the beach in their uniforms, paddled out, and we passed the ceremony of induction. Loreto was asleep in the sunshine, a lovely town, with gardens in every yard and only the streets white and hot. The young males watched us from the safe shade of the *cantina* and passed greetings as we went by, and a covey of young girls grew tight-faced and rushed around a corner and giggled. How strange we were in Loreto! Our trousers were dark, not white; the silly caps we wore were so outlandish that no store in Loreto would think of stocking them. We were neither soldiers nor sailors—the little girls just couldn't take it. We could hear their strangled giggling from around the corner. Now and then they peeked back around the corner to verify for themselves our ridiculousness, and then giggled again while their elders hissed in disapproval. And one woman standing in a lovely garden shaded with purple bougainvillaea explained, "Everyone knows what silly things girls are. You must forgive their ill manners; they will be ashamed later on." But we felt that the silly girls had something worthwhile in their attitude. They were definitely amused. It is often so, particularly in our country, that the first reaction to strangeness is fear and hatred; we much preferred the laughter. We don't think it was even unkind—they'd simply never seen anything so funny in their lives.

As usual, a good serious small boy attached himself to us. It would be interesting to see whether a nation governed by the small boys of Mexico would not be a better, happier nation than those ruled by old men whose prejudices may or may not be conditioned by ulcerous stomachs and perhaps a little drying up of the stream of love.

This small boy could have been an ambassador to almost any country in the world. His straight-seeing dark eyes were courteous, yet firm. He was kind and dignified. He told us something of Loreto; of its poverty, and how its church was tumbled down now; and he walked with us to the destroyed mission. The roof had fallen in and the main body of the church was a mass of rubble. From the walls hung the shreds of old paintings. But the bell-tower was intact, and we wormed our way deviously up to look at the old bells and to strike them softly with the palms of our hands so that they glowed a little with tone. From here we could look down on the low roofs and into the enclosed gardens of the town. The white sunlight could not get into the gardens and a sleepy shade lay in them.

One small chapel was intact in the church, but the door to it was barred by a wooden grille, and we had to peer through into the small, dark, cool room. There were paintings on the walls, one of which we wanted badly to see more closely, for it looked very much like an El Greco, and probably was *not* painted by El Greco. Still, strange things have found their way here. The bells on the tower were the special present of the Spanish throne to this very loyal city. But it would be good to see this picture more closely. The Virgin Herself, Our Lady of Loreto, was in a glass case and surrounded by the lilies of the recently past Easter. In the dim light of the chapel she seemed very lovely. Perhaps she is gaudy; she has not the look of smug virginity so many have—the "I-am-the-Mother-of-Christ" look —but rather there was a look of terror in her face, of the Virgin Mother of the world and the prayers of so very many people heavy on her.

To the people of Loreto, and particularly to the Indians of the outland, she must be the loveliest thing in the world. It doesn't matter that our eyes, critical and thin with *good taste,* should find her gaudy. And actually we did not. We too found her lovely in her dim chapel with the lilies of Easter around her. This is a very holy place, and to question it is to question a fact as established as the tide. How easily and quickly we slide into our race-pattern unless we keep intact the stiff-necked and blinded pattern of the recent intellectual training.

We threw it over, and there wasn't much to throw over, and we felt good about it. This Lady, of plaster and wood and paint, is one of the strong ecological factors of the town of Loreto, and not to know her and her strength is to fail to know Loreto. One could not ignore a granite monolith in the path of the waves. Such a rock, breaking the rushing waters, would have an effect on animal distribution radiating in circles like a dropped stone in a pool. So has this plaster Lady a powerful effect on the deep black water of the human spirit. She may disappear and her name be lost, as the Magna Mater, as Isis, have disappeared. But something very like her will take her place, and the longings which created her will find somewhere in the world a similar altar on which to pour their force. No matter what her name is, Artemis, or Venus, or a girl behind a Woolworth counter vaguely remembered, she is as eternal as our species, and we will continue to manufacture her as long as we survive.

We came back slowly through the deserted streets of Loreto, and we walked quietly laden with submergence in a dim chapel.

* * *

A few supplies went aboard, and we pulled up the anchor and moved northward again. On the way we caught a Mexican sierra and another fish, apparently a cross between a yellow-fin tuna [6] and an albacore.[7] Tiny and Sparky, who have fished in tuna water a good deal, say that this cross is often found and taken, although never in numbers.

We sailed north and found anchorage on the northern end of Coronado Island, and went immediately to collect on a long, westerly-extending point. This reef of water-covered stones was not very rich. In high boots we moved slowly about, turning over the flattened algae-covered boulders. We found here many solitary corals,[8] and with great difficulty took some of them. They are very hard, and shatter easily when they are removed. If one could saw out the small section of rock to which they are fastened, it would be easy to take them. The next best method is to use a thin, very sharp knife and, by treating them as delicately as jewels, to remove them from their hard anchorage. Even with care, only about one in five is unbroken. Here also we found clustered heads of hard zoanthidean anemones of two types, one much larger than the other. We found a great number of large hemispherical yellow sponges which were noted in the collecting reports as "strikingly similar to the

[6] *Neothunnus macropterus.*
[7] *Germo alalunga.*
[8] *Astrangia pedersenі.*

Monterey Bay *Tethya aurantia* or *Geodia*"—a similarity partly explainable by the fact that they turned out to be *Tethya aurantia* and a species of *Geodia!* Our collecting included the usual assortment of creatures, ranging from the crabs which plant algae on their backs for protection to the bryozoa which look more like moss than animals. With all these, the region was still not rich, but "burned," and again we felt the thing which had been at the strange Cayo Islet, a resentment of the shore toward animal life, an inhospitable quality in the stones that would make an animal think twice about living there.

It is so strange, this burned quality. We have seen places which seemed hostile to human life, too. There are parts of the coast of California which do not like humans. It is as though they were already inhabited by another and invisible species which resented humans. Perhaps such places are "burned" for us; perhaps a petrologist could say why. Might there not be a mild radio-activity which made one nervous in such a place so that he would say, trying to put words to his feeling, "This place is unfriendly. There is something here that will not tolerate my kind"? While some radio-activities have been shown to encourage not only life but mutation (note experimentation with fruit-flies), there might well be some other combinations which have an opposite effect.

Little fragments of seemingly unrelated information will sometimes accumulate in a process of speculation until a tenable hypothesis emerges. We had come on a riddle in our reading about the Gulf and now we were able to see this riddle in terms of the animals. There is an observable geographic differential in the fauna of the Gulf of California. The Cape San Lucas–La Paz area is strongly Panamic. Many warm-water mollusks and crustaceans are not known to occur in numbers north of La Paz, and some not even north of Cape San Lucas. But the region north of Santa Rosalia, and even of Puerto Escondido, is known to be inhabited by many colder-water animals, including *Pachygrapsus crassipes,* the commonest California shore crab, which ranges north as far as Oregon. These animals are apparently trapped in a blind alley with no members of their kind to the south of them.

The problem is: "How did they get there?" In 1895 Cooper [9] noticed the situation and advanced an explanation. He remarks, referring to the northern part of the Gulf: "It appears that the species found there are more largely of the temperate fauna, many of them being identical with those of the same latitude on the west [outer] coast of the Peninsula. This seems to indicate that the dividing ridge, now three thousand feet or more in altitude, was crossed by one or more channels within geologically recent times."

This differential, which we ourselves saw, has been remarked a number of times in the literature of the region, especially by conchologists. Eric Knight Jordan, son of David Starr Jordan, an extremely promising young paleontologist who was killed some years ago, studied the geological and present distribution of mollusks along the west coast of Lower California. He says [10]: "Two distinct faunas exist on the west coast of Lower California. The southern Californian *now* ranges southward from Point Conception to Cedros Island . . . probably extends a little farther. . . . The fauna of the Gulf of California ranges to the north on the west coast of the Peninsula approximately to Scammon's Lagoon, which is a little farther up than Cedros Island." Present geographical ranges are given for 124 species, collected in lower Quaternary beds at Magdalena Bay, all of which occur living today, but farther to the north. Two pages later he remarks: "It . . . appears that when these Quaternary beds were laid down there was a southward displacement of the isotherms sufficient to carry the conditions

---

[9] § S-9, p. 37.
[10] § S-18, p. 146.

today prevailing at Cedros down as far as the latitude of Magdalena Bay."

Having reviewed the literature, we can confirm the significance of the Cedros Island complex as a present critical horizon (as Carpenter did eighty years ago) where the north and south fauna to some extent intermingle. Apparently this is the very condition that obtained at Magdalena Bay or southward when the lower Quaternary beds were being laid down. The present Magdalena Plain, extending to La Paz on the Gulf side, was at that time submerged. Then it was cold enough to permit a commingling of cold-water and warm-water species at that point. The hypothesis is tenable that when the isotherms retreated northward, the cold-water forms were no longer able to inhabit southern Lower California shores, which included the then Gulf entrance. In these increasingly warm waters they would have perished or would have been pushed northward, both along the outside coast, where they could retreat indefinitely, and into the Gulf. In the latter case the migrating waves of competing animals from the south, which were invading the Gulf and spilling upward, would have pocketed the northern species in the upper reaches, where they have remained to this day. These animals, hemmed in by tropical waters and fortunate competitors, have maintained themselves for thousands of years, though in the struggle they have been modified toward pauperization.

This hypothesis would seem to offset Cooper's assumption of a channel through ridges some 350 miles to the north which show no signs of Quaternary submergence.

It is interesting that a paleontologist, working in one area, should lay the groundwork for a very reasonable hypothesis concerning the distribution of animals in another. It is, however, only one example among many of the obliqueness of investigation and the accident quotient involved in much investigation. The literature of science is filled with answers found when the question propounded had an entirely different direction and end.

There is one great difficulty with a good hypothesis. When it is completed and rounded, the corners smooth and the content cohesive and coherent, it is likely to become a thing in itself, a work of art. It is then like a finished sonnet or a painting completed. One hates to disturb it. Even if subsequent information should shoot a hole in it, one hates to tear it down because it once was beautiful and whole. One of our leading scientists, having reasoned a reef in the Pacific, was unable for a long time to reconcile the lack of a reef, indicated by soundings, with the reef his mind told him was there. A parallel occurred some years ago. A learned institution sent an expedition southward, one of whose many projects was to establish whether or not the sea-otter was extinct. In due time it returned with the information that the sea-otter was indeed extinct. One of us, some time later, talking with a woman on the coast below Monterey, was astonished to hear her describe animals living in the surf which could only be sea-otters, since she described accurately animals she couldn't have known about except by observation. A report of this to the institution in question elicited no response. It had extincted sea-otters and that was that. It was only when a reporter on one of our more disreputable newspapers photographed the animals that the public was informed. It is not yet known whether the institution of learning has been won over.

This is not set down in criticism; it is no light matter to make up one's mind about anything, even about sea-otters, and once made up, it is even harder to abandon the position. When a hypothesis is deeply accepted it becomes a growth which only a kind of surgery can amputate. Thus, beliefs persist long after their factual bases have been removed, and practices based on beliefs are often carried on even when the beliefs which stimulated them have been forgotten. The practice must follow the belief. It is often considered, particularly by reformers and legislators, that law is a stimulant to action or an inhibitor of action, when actually the reverse is true. Successful law is simply the publication of the practice of the majority of units of a society, and by it the inevitable variable units are either driven to conform or are eliminated. We have had many examples of law trying to be the well-spring of action; our prohibition law showed how completely fallacious that theory is.

The things of our minds have for us a greater toughness than external reality. One of us has a beard, and one night when this one was standing wheel-watch, the others sat in the galley drinking coffee. We were discussing werewolves and their almost universal occurrence in regional literature. From this beginning, we played with a macabre thought, "The moon will soon be full," we said, "and he of the beard will begin to feel the pull of the moon. Last night," we said, "we heard the scratch of claws on the deck. When you see him go down on all fours, when you see the red light come into his eyes, then look out, for he will slash your throat." We were delighted with the game. We developed the bearded one's tendencies, how his teeth, the canines at least, had been noticeably longer of late, how for the past week he had torn his dinner apart with his teeth. It was night as we talked thus, and the deck was dark and the wind was blowing. Suddenly he appeared in the doorway, his beard and hair blown, his eyes red from the wind. Climbing the two steps up from the galley, he seemed to arise from all fours, and everyone of us started, and felt the prickle of erecting hairs. We had actually talked and thought ourselves into this pattern, and it took a while for it to wear off.

These mind things are very strong; in some, so strong as to blot out the external things completely.

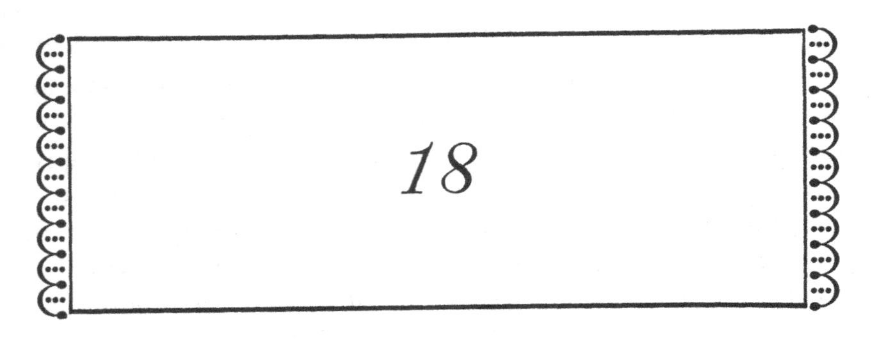

18

March 28

AFTER the collecting on Coronado Island, on the twenty-seventh, and the preservation and labeling, we found that we were very tired. We had worked constantly. On the morning of the twenty-eighth we slept. It was a good thing, we told ourselves; the eyes grow weary with looking at new things; sleeping late, we said, has its genuine therapeutic value; we would be better for it, would be able to work more effectively. We have little doubt that all this was true, but we wish we could build as good a rationalization every time we are lazy. For in some beastly way this fine laziness has got itself a bad name. It is easy to see how it might have come into disrepute, if the result of laziness were hunger. But it rarely is. Hunger makes laziness impossible. It has even become sinful to be lazy. We wonder why. One could argue, particularly if one had a gift for laziness, that it is a relaxation pregnant of activity, a sense of rest from which directed effort may arise, whereas most busyness is merely a kind of nervous tic. We know a lady who is obsessed with the idea of ashes in an ashtray. She is not lazy. She spends a good half of her waking time making sure that no ashes remain in any ashtray, and to make sure of keeping busy she has a great many ashtrays. Another acquaintance, a man, straightens rugs and pictures and arranges books and magazines in neat piles. He is not lazy, either; he is very busy. To what end? If he should relax, perhaps with his feet up on a chair and a glass of cool beer beside him—not cold, but cool—if he should examine from this position a rumpled rug or a crooked picture, saying to himself between sips of beer (preferably Carta Blanca beer), "This rug irritates me for some reason. If it were straight, I should be comfortable; but there is only one straight position (and this is of course, only my own personal discipline of straightness) among all possible positions. I am, in effect, trying to impose my will, my insular sense of rightness, on a rug, which of itself can have no such sense, since it seems equally contented straight or crooked. Suppose I should try to straighten people," and here he sips deeply. "Helen C., for instance, is not neat, and Helen C."—here he goes into a reverie—"how beautiful she is with her hair messy, how lovely when she is excited and breathing through her mouth." Again he raises his glass, and in a few minutes he picks up the telephone. He is happy; Helen C. may be happy; and the rug is not disturbed at all.

How can such a process have become a shame and a sin? Only in laziness can one achieve a state of contemplation which is a balancing of values, a weighing of oneself against the world and the world against itself. A busy man cannot find time for such balancing. We do not think a lazy man can commit murders, nor great thefts, nor lead a mob. He would be more likely to think about it and laugh. And a nation of lazy contemplative men would be incapable of fighting a war unless their very laziness were attacked. Wars are the activities of busy-ness.

With such a background of reasoning, we slept until nine A.M. And then the engines started and we moved toward Concepción Bay. The sea, with the exception of one blow outside of La Paz, had been very calm. This day, a little wind blew over the ultramarine water. The swordfish in great numbers jumped and played about us. We set up our lightest harpoon on the bow with a coil of cotton line beside it, and for hours we stood watch. The helmsman changed course again and again to try to bring the bow over a resting fish, but they seemed to wait until we were barely within throwing range and then they sounded so quickly that they seemed to snap from view. We made many wild casts and once we got the iron in, near the tail of a monster. But he flicked his tail and tore it out and was gone. We could see schools of leaping tuna all about us, and whenever we crossed the path of a school, our lines jumped and snapped under the strikes, and we brought the beautiful fish in.

We had set up a salt barrel near the stern, and we cut the fish into pieces and put them into brine to take home. It developed after we got home that several of us had added salt to the brine and the whole barrel was hopelessly salty and inedible.

As we turned Aguja Point and headed southward into the deep pocket of Concepción Bay, we could see Mulege on the northern shore—a small town in a blistering country. We had no plan for stopping there, for the story is that the port charges are mischievous and ruinous. We do not know that this is so, but it is repeated about Mulege very often. Also, there may be malaria there. We had been following the trail of malaria for a long time. At the Cape they said there was no malaria there but at La Paz it was very serious. At La Paz, they said it was only at Loreto. At Loreto they declared that Mulege was full of it. And there it must remain, for we didn't stop at Mulege; so we do not know what the Mulegeños say about it. Later, we picked up the malaria on the other side, ran it down to Topolobambo, and left it there. We would say offhand, never having been to either place, that the malaria is very bad at Mulege and Topolobambo.

A strong, north-pointing peninsula is the outer boundary of Concepción Bay. At the mouth it is three and a quarter miles wide and it extends twenty-two miles southward, varying in width from two to five miles. The eastern shore, along which we collected, is regular in outline, with steep beaches of sand and pebbles and billions of bleaching shells and many clams and great snails. From the shore, the ascent is gradual toward mountains which ridge the little peninsula and protect this small gulf from the Gulf of California. Along the shore are many pools of very salty water, where thousands of fiddler crabs sit by their moist burrows and bubble as one approaches. The beach was beautiful with the pink and white shells of the murex.[1] Sparky found them so beautiful that he collected a washtubful of them and stored them in the hold. And even then, back in Monterey, he found he did not have enough for his friends.

Behind the beach there was a little level land, sandy and dry and covered with cactus and thick brush. And behind that, the rising dry hills. Now again the wild doves were calling among the hills with their song of homesickness. The quality of longing in this sound, the memory response it sets up, is curious and strong. And it has also the quality of a dying day. One wishes to walk toward the sound—to walk on and on toward it, forgetting everything else. Undoubtedly there are sound symbols in the unconscious just as there are visual symbols—sounds that trigger off a response, a little spasm of fear, or a quick lustfulness, or, as with the doves, a nostalgic sadness. Perhaps in our pre-humanity this sound of doves was a signal that the day was over and a night of terror due—a night which perhaps this time was permanent. Keyed to the visual symbol of the sinking sun and to the odor symbol of the cooling earth, these might all cause the little spasm of sorrow; and with the long response-history, one alone of these symbols might suffice for all three. The smell of a musking goat is not in our experience, but it is in some experience, for smelled faintly, or in perfume, it is not without its effect even on those who have not smelled the passionate gland nor seen the play which follows its discharge. But some great group of shepherd peoples must have known

[1] *Phyllonotus bicolor.*

this odor and its result, and must, from the goat's excitement, have taken a very strong suggestion. Even now, a city man is stirred deeply when he smells it in the perfume on a girl's hair. It may be thought that we produce no musk nor anything like it, but this we do not believe. One has the experience again and again of suddenly turning and following with one's eyes some particular girl among many girls, even trotting after her. She may not be beautiful, indeed, often is not. But what are the stimuli if not odors, perhaps above or below the conscious olfactory range? If one follows such an impulse to its conclusion, one is not often wrong. If there be visual symbols, strong and virile in the unconscious, there must be others planted by the other senses. The sensitive places, ball of thumb, ear-lobe, skin just below the ribs, thigh, and lip, must have their memories too. And smell of some spring flowers when the senses thaw, and smell of a ready woman, and smell of reptiles and smell of death, are deep in our unconscious. Sometimes we can say truly, "That man is going to die." Do we smell the disintegrating cells? Do we see the hair losing its luster and uneasy against the scalp, and the skin dropping its tone? We do not know these reactions one by one, but we say, that man or cat or dog or cow is going to die. If the fleas on a dog know it and leave their host in advance, why do not we also know it? Approaching death, the pre-death of the cells, has informed the fleas and us too.

* * *

The shallow water along the shore at Concepción Bay was littered with sand dollars, two common species [2] and one [3] very rare. And in the same association, brilliant-red sponge arborescences [4] grew in loose stones in the sand or on the knobs of old coral. These are the important horizon markers. On other rocks,

---

[2] *Encope californica* and *E. grandis.*
[3] *Clypeaster rotundus.*
[4] *Tedania ignis.*

imbedded in the sand, there were giant hachas, clustered over with tunicates and bearing on their shells the usual small ophiurans and crabs. One of the masked rock-clams had on it a group of solitary corals. Close inshore were many brilliant large snails, the living animals the shells of which had so moved Sparky. In this area we collected from the skiff, leaning over the edge, bringing up animals in a dip-net or spearing them with a small trident, sometimes jumping overboard and diving for a heavier rock with a fine sponge growing on it.

The ice we had taken aboard at La Paz was all gone now. We started our little motor and ran it for hours to cool the ice-chest, but the heat on deck would not permit it to drop the temperature below about thirty-eight degrees F., and the little motor struggled and died often, apparently hating to run in such heat. It sounded tired and sweaty and disgusted. When the evening came, we had fried fish, caught that day, and after dark we lighted the deck and put our reflecting lamp over the side. We netted a serpent-like eel, thinking from its slow, writhing movement through the water that it might be one of the true viperine sea-snakes which are common farther south. Also we captured some flying fish.

We used long-handled dip-nets in the lighted water, and set up the enameled pans so that the small pelagic animals could be dropped directly into them. The groups in the pans grew rapidly. There were *heteronereis* (the free stages of otherwise crawling worms who develop paddle-like tails upon sexual maturity). There were swimming crabs, other free-swimming annelids, and ribbon-fish which could not be seen at all, so perfectly transparent were they. We should not have known they were there, if they had not thrown faint shadows on the bottom of the pans. Placed in alcohol, they lost their transparency and could easily be seen. The pans became crowded with little skittering animals, for each net brought in many species. When the hooded light was put down very near the water, the smallest animals came to it and scurried about in a dizzying dance so rapidly that they seemed to draw crazy lines in the water. Then the small fishes began to dart in and out, snapping up this concentration, and farther out in the shadows the large wise fishes cruised, occasionally swooping and gobbling the small fishes. Several more of the cream-colored spotted snake-eels wriggled near and were netted. They were very snake-like and they had small bright-blue eyes. They did not swim with a beating tail as fishes do, but rather squirmed through the water.

While we worked on the deck, we put down crab-nets on the bottom, baiting them with heads and entrails of the fish we had had for dinner. When we pulled them up they were loaded with large stalk-eyed snails [5] and with sea-urchins having long vicious spines.[6] The colder-water relatives of both these animals are very slow-moving, but these moved quickly and were completely voracious. A net left down five minutes was brought up with at least twenty urchins in it, and all attacking the bait. In addition to the speed with which they move, these urchins are clever and sensitive with their spines. When approached, the long sharp little spears all move and aim their points at the approaching body until the animal is armed like a Macedonian phalanx. The main shafts of the spines were cream-yellowish-white, but a half-inch from the needle-points they were blue-black. The prick of one of the points burned like a bee-sting. They seemed to live in great numbers at four fathoms; we do not know their depth range, but their physical abilities and their voraciousness would indicate a rather wide one. In the same nets we took several dromiaceous crabs, reminiscent of hermits, which had adjusted themselves to life in half the shell of a bivalve, and had changed their body shapes accordingly.

It is probable that no animal tissue ever decays in this water.

---

[5] *Strombus* spp.
[6] *Astropyga pulvinata.*

The furious appetites which abound would make it unlikely that a dead animal, or even a hurt animal, should last more than a few moments. There would be quick death for the quick animal which became slow, for the shelled animal which opened at the wrong time, for the fierce animal which grew timid. It would seem that the penalty for a mistake or an error would be instant death and there would be no second chance.

It would have been good to keep some of the sensitive urchins alive and watch their method of getting about and their method of attack. Indeed, we will never go again without a full-sized observation aquarium into which we can put interesting animals and keep them for some time. The aquaria taken were made with polarized glass. Thus, the fish could look out but we could not look in. This, it turned out, was an error on our part.

There are three ways of seeing animals: dead and preserved; in their own habitats for the short time of a low tide; and for long periods in an aquarium. The ideal is all three. It is only after long observation that one comes to know the animal at all. In his natural place one can see the normal life, but in an aquarium it is possible to create abnormal conditions and to note the animal's adaptability or lack of it. As an example of this third method of observation, we can use a few notes made during observation of a small colony of anemones in an aquarium. We had them for a number of months.

In their natural place in the tide pool they are thick and close to the rock. When the tide covers them they extend their beautiful tentacles and with their nettle-cells capture and eat many micro-organisms. When a powerful animal, a small crab for example, touches them, they paralyze it and fold it into the stomach, beginning the digestive process before the animal is dead, and in time ejecting the shell and other indigestible matter. On being touched by an enemy, they fold in upon themselves for protection. We brought a group of these on their own stone into the laboratory and placed them in an aquarium. Cooled and oxygenated sea water was sprayed into the aquarium to keep them alive. Then we gave them various kinds of food, and found that they do not respond to simple touch-stimulus on the tentacles, but have something which is at least a vague parallel to taste-buds, whatever may be the chemical or mechanical method. Thus, protein food was seized by the tentacles, taken and eaten without hesitation; fat was touched gingerly, taken without enthusiasm to the stomach, and immediately rejected; starches were not taken at all—the tentacles touched starchy food and then ignored it. Sugars, if concentrated, seemed actually to burn them so that the tentacles moved away from contact. There did really appear to be a chemical method of differentiation and choice. We circulated the same sea water again and again, only cooling and freshening it. Pure oxygen, introduced into the stomach in bubbles, caused something like drunkenness; the animal relaxed and its reaction to touch was greatly slowed, and sometimes completely stopped for a while. But the reaction to chemical stimulus remained active, although slower. In time, all the microscopic food was removed from the water through constant circulation past the anemones, and then the animals began to change their shapes. Their bodies, which had been thick and fat, grew long and neck-like; from a normal inch in length, they changed to three inches long and very slender. We suspected this was due to starvation. Then one day, after three months, we dropped a small crab into the aquarium. The anemones, moving on their new long necks, bent over and attacked the crab, striking downward like slow snakes. Their normal reaction would have been to close up and draw in their tentacles, but these animals had changed their pattern in hunger, and now we found that when touched on the body, even down near the base, they moved downward, curving on their stalks, while their tentacles hungrily searched for food. There seemed even to be competition among the individuals, a thing we have never seen in a tide pool among anemones. This versatility had never been observed by us and is not mentioned in any of the literature we have seen.

The aquarium is a very valuable extension of shore observation. Quick-eyed, timid animals soon become used to having humans about, and quite soon conduct their business under lights. If we could have put our sensitive urchins in an aquarium, we could have seen how it is that they move so rapidly and how they are stimulated to aim their points at an approaching body. But we preserved them, and of course they lost color and dropped many of their beautiful sharp spines. Also, we could have seen how the great snails are able to consume animal tissue so quickly. As it is, we do not know these things.

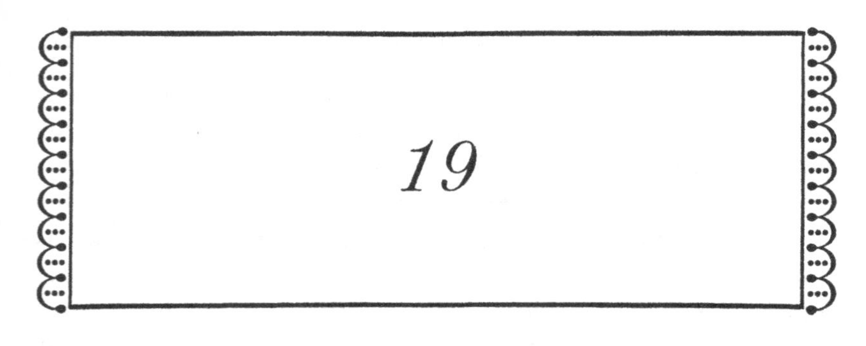

March 29

TIDES had been giving us trouble, for we were now far enough up the Gulf so that the tidal run had to be taken into time consideration. In the evening we had set up a flagged stake at the waterline, so that with glasses we could see from the deck the rise and fall of the tide in relation to the stick. At seven-thirty in the morning the tide was going down from our marker. We had abandoned our tide charts as useless by now, and since we stayed such a short time at each station we could not make new ones. The irregular length of our jumps made it impossible for us to forecast with accuracy from a preceding station. Besides all this, a good, leisurely state of mind had come over us which had nothing to do with the speed and duration of our work. It is very possible to work hard and fast in a leisurely manner, or to work slowly and clumsily with great nervousness.

On this day, the sun glowing on the morning beach made us feel good. It reminded us of Charles Darwin, who arrived late at night on the *Beagle* in the Bay of Valparaiso. In the morning he awakened and looked ashore and he felt so well that he wrote, "When morning came everything appeared delightful. After Tierra del Fuego, the climate felt quite delicious, the atmosphere so dry and the heavens so clear and blue with the sun shining brightly, that all nature seemed sparkling with life." [1]

[1] *Voyage of the Beagle*, Chap. 12, July 23.

Darwin was not saying how it was with Valparaiso, but rather how it was with him. Being a naturalist, he said, "All nature seemed sparkling with life," but actually it was he who was sparkling. He felt so very fine that he can, in these charged though general adjectives, translate his ecstasy over a hundred years to us. And we can feel how he stretched his muscles in the morning air and perhaps took off his hat—we hope a bowler—and tossed it and caught it.

On this morning, we felt the same way at Concepción Bay. "Everything appeared delightful." The tiny waves slid up and down the beach, hardly breaking at all; out in the Bay the pelicans were fishing, flying along and then folding their wings and falling in their clumsy-appearing dives, which nevertheless must be effective, else there would be no more pelicans.

By nine A.M. the water was well down, and by ten seemed to have passed low and to be flowing again. We went ashore and followed the tide down. The beach is steep for a short distance, and then levels out to a gradual slope. We took two species of cake urchins which commingled at one-half to one and one-half feet of water at low tide. The ordinary cake urchin here, with holes, is *Encope californica* Verrill. The grotesquely beautiful keyhole sand dollar [2] was very common here. Finally, there was a rare member of the same group,[3] which we collected unknowingly, and turned out only three individuals of the species when the animals were separated on deck. A little deeper, about two feet submerged, at low tide, a species of cucumber new to us was taken, a flat, sand-encrusted fellow.[4] Giant heart-urchins [5] in some places were available in the thousands. They ranged between two feet and three feet below the surface at low water, and very few were deeper. The greatest number occurred at three feet.

---

[2] *Encope grandis* L. Agassiz.
[3] *Clypeaster rotundus* (A. Agassiz).
[4] *Holothuria inhabilis.*
[5] *Meoma grandis.*

The shore line here is much like that at Puget Sound: in the high littoral is a foreshore of gravel to pebbles to small rocks; in the low littoral, gravelly sand and fine sand with occasional stones below the low tide level. In this zone, with a maximum at four feet, were heavy groves of algae, presumably *Sargassum*, lush and tall, extending to the surface. Except for the lack of eel-grass, it might have been Puget Sound. We took giant stalk-eyed conchs,[6] several species of holothurians and *Cerianthus*, the sand anemone whose head is beautiful but whose encased body is very ugly, like rotting gray cloth. Tiny christened *Cerianthus* "sloppy-guts," and the name stuck. By diving, we took a number of hachas, the huge mussel-like clams. Their shells were encrusted with sponges and tunicates under which small crabs and snapping shrimps hid themselves. Large scalloped limpets also were attached to the shells of the hachas. This creature closes itself so tightly with its big adductor muscle that a knife cannot penetrate it and the shell will break before the muscle will relax. The best method for opening them is to place them in a bucket of water and, when they open a little, to introduce a sharp, thin-bladed knife and sever the muscle quickly. A finger caught between the closing shells would probably be injured. In many of the hachas we found large, pale, commensal shrimps[7] living in the folds of the body. They are soft-bodied and apparently live there always.

About noon we got under way for San Lucas Cove, and as usual did our preserving and labeling while the boat was moving. Some of the sand dollars we killed in formalin and then set in the sun to dry, and many more we preserved in formaldehyde solution in a small barrel. We had taken a great many of them. Sparky had, by now, filled several sacks with the fine white rose-lined murex shells, explaining, as though he were asked for

---

[6] *Strombus galeatus.*
[7] *Pontonia pinnae.*

an explanation, that they would be nice for lining a garden path. In reality, he simply loved them and wanted to have them about.

We passed Mulege, that malaria-ridden town, that town of high port fees—so far as we know—and it looked gay against the mountains, red-roofed and white-walled. We wished we were going ashore there, but the wall of our own resolve kept us out, for we had said, "We will *not* stop at Mulege," and having said it, we could not overcome our own decision. Sparky and Tiny looked longingly at it as we passed; they had come to like the quick excursions into little towns: they found that their Italian was understood for any purposes they had in mind. It was their practice to wander through the streets, carrying their cameras, and in a very short time they had friends. Tony and Tex were foreigners, but Tiny and Sparky were very much at home in the little towns—and they never inquired whose home. This was not reticence, but rather a native tactfulness.

Now that we were engaged in headland navigating, Tiny's and Sparky's work at the wheel had improved, and except when they chased a swordfish (which was fairly often) we were not off course more than two or three times during their watch. They had abandoned the compass with relief and blue water was no longer thrust upon them.

At about this time it was discovered that Tex was getting fat, and inasmuch as he was to be married soon after his return, we decided to diet him and put him in a marrying condition. He protested feebly when we cut off his food, and for three days he sneaked food and stole food and cozened us out of food. During the three days of his diet, he probably ate twice as much as he did before, but the idea that he was starving made him so hungry that at the end of the third day he said he couldn't stand it any longer, and he ate a dinner that nearly killed him. Actually, with his thefts of food he had picked up a few pounds during his diet, but always afterwards he shuddered at the memory of those three days. He said, "A man doesn't feel his best when he is starving" and he asked what good it would do him to be married if he were weak and sick.

At five P.M. on March 29 we arrived at San Lucas Cove and anchored outside. The cove, a deep salt-water lagoon, guarded by a large sandspit, has an entrance that might mave been deep enough for us to enter, but the current is strong and there were no previous soundings available. Besides, Tony was nervous about taking his boat into such places. There was another reason for anchoring outside; in the open Gulf where the breeze moves there are no bugs, while if one anchors in still water near the mangroves little visitors come and spend the night. There is one small, beetle-like black fly which crawls down into bed with you and has a liking for very tender places. We had suffered from this fellow when the wind blew over the mangroves to us. This bug hates light, but finds security and happiness under the bedding, nestling over one's kidneys, munching contentedly. His bite leaves a fiery itch; his collective soul is roasting in Hell, if we have any influence in the court of Heaven. After one experience with him, we anchored always a little farther out.

When we came to San Lucas, the tide was flowing and the little channel was a mill-race. It would be necessary to wait for the morning tide. We were eager to see whether on this sand-bar, so perfectly situated, we could not find amphioxus, that most primitive of vertebrates. As we dropped anchor a large shark cruised about us, his fin high above the water. We shot at him with a pistol and one shot went through his fin. He cut away like a razor blade and we could hear the hiss of the water. What incredible speed sharks can make when they hurry! We wonder how their greatest speed compares with that of a porpoise. The variations in speed among individuals of these fast-swimming species must be very great too. There must be incredible sharks, like Man o' War or Charlie Paddock, which make other sharks seem slow.

That night we hung the light over the side again and captured some small squid, the usual *heteronereis,* a number of free-swimming crustacea, quantities of crab larvae and the transparent ribbon-fish again. The boys developed a technique for catching flying fish: one jabbed at it with a net, making it fly into the net of another. But even in the nets they were not caught, for they struggled and fluttered away with ease. That night we had a mild celebration of some minor event which did not seem important enough to remember. The pans of animals were still lying on the deck and one of our members, confusing Epsom salts with cracker-crumbs, tried to anesthetize a large pan of holothurians with cracker-meal. The resulting thick paste seemed to have no narcotic qualities whatever.

Late, late in the night we recalled that Horace says fried shrimps and African snails will cure a hangover. Neither was available. And we wonder whether this classical remedy for a time-bridging ailment has been prescribed and tried since classical times. We do not know what snail he refers to, or whether it is a marine snail or an escargot. It is too bad that such imaginative remedies have been abandoned for the banalities of anti-acids, heart stimulants, and analgesics. The Bacchic mystery qualified and nullified by a biochemistry which is almost but not quite yet a mystic science. Horace suggests that wine of Cos taken with these shrimps and snails guarantees the remedy. Perhaps it would. In that case, his remedy is in one respect like those unguents used in witchcraft which combine such items as dried babies' brains, frog-eyes, lizards' tongues, and mold from a hanged man's skull with a quantity of good raw opium, and thus serve to stimulate the imagination and the central nervous system at the same time. In our pained discussion at San Lucas Cove we found we had no snails nor shrimps nor wine of Cos. We tore the remedy down to its fundamentals, and decided that it was a good strong dose of proteins and alcohol, so we substituted a new compound—fried fish and a dash of medicinal whisky—and it did the job.

The use of euphemism in national advertising is giving the hangover a bad name. "Over-indulgence" it is called. There is a curious nastiness about over-indulgence. We would not consider over-indulging. The name is unpleasant, and the word "over" indicates that one shouldn't have done it. Our celebration had no such implication. We did *not* drink too much. We drank just enough, and we refuse to profane a good little time of mild inebriety with that slurring phrase "over-indulgence."

There was a reference immediately above to the medicine chest. On leaving Monterey, it may be remembered, we had exhausted the medicine, but no sooner had we put to sea when it was discovered that each one of us, with the health of the whole party in mind, had laid in auxiliary medicine for emergencies. We had indeed, when the good-will of all was assembled, a medicine chest which would not have profaned a fair-sized bar. And the emergencies did occur. Who is to say that an emergency of the soul is not worse than a bad cold? What was good enough for Li-Po was good enough for us. There have been few enough immortals who did not love wine; offhand we cannot think of any and we do not intend to try very hard. The American Indians and the Australian Bushmen are about the only great and intellectual peoples who have not developed an alcoholic liquor and a cult to take care of it. There are, indeed, groups among our own people who have abandoned the use of alcohol, due no doubt to Indian or Bushman blood, but we do not wish to claim affiliation with them. One can imagine such a specimen of Bushman reading this journal and saying, "Why, it was all drinking—beer—and at San Lucas Cove, whisky." So might a night-watchman cry out, "People sleep all the time!" So might a blind man complain, "Among some people there is a pernicious and wicked practice called 'seeing.' This eventually causes death and should be avoided." Actually, with few tribal exceptions, our race has a triumphant alcoholic history and no definite symptoms of degeneracy can be attributed to it. The theory that alcohol is a poison was too easily and too blindly accepted. So it is to some individuals; sugar is poison to others and meat to others. But to the race in general, alcohol has been an anodyne, a warmer of the soul, a strengthener of muscle and spirit. It has given courage to cowards and has made very ugly people attractive. There is a story told of a Swedish tramp, sitting in a ditch on Midsummer Night. He was ragged and dirty and drunk, and he said to himself softly and in wonder, "I am rich and happy and perhaps a little beautiful."

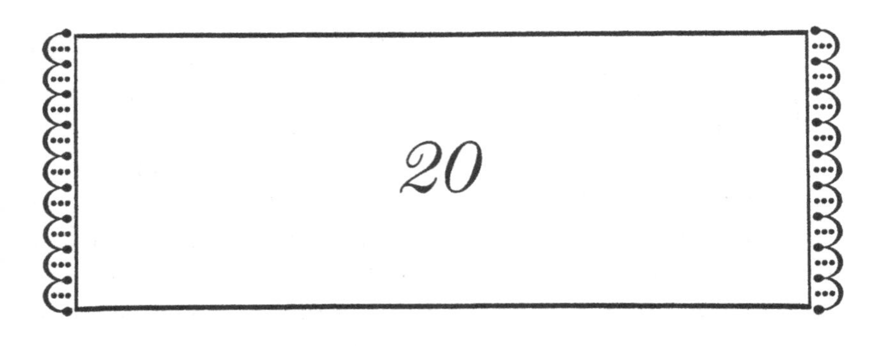

20

March 30

AT EIGHT-THIRTY in the morning the tide was ebbing, uncovering the sand-bar and a great expanse of tidal sand-flat. This flat was made up to a large extent of the broken shells of mollusks. In digging, we found many small clams and a few smooth *Venus*-like clams. We took one very large male fiddler crab. "Sloppy-guts," the *Cerianthus,* was very common here. There were numbers of hermit crabs and many of the swimming crabs with bright-blue claws. These crabs [1] are eaten by Mexicans and are delicious. They swim very rapidly through the water. When we pursued them to the shallows they tried to escape for a time, but soon settled to the bottom and raised their claws to a position not unlike that of a defensive boxer. Their pinch was very painful. When captured and put into a collecting bucket they vented their fury on one another; pinched-off legs and claws littered the buckets on our return. These crabs do not seem to come out of the water as the grapsoids do. Removed from the water, they very soon weaken and lose their fight. Moreover, they do not die as rapidly in fresh water as do most other crabs. Perhaps, living in the lagoons which sometimes must be almost brackish, they have achieved a tolerance for fresh water greater than that of other crabs; greater indeed, although it is not much of a trick, than that of a certain biologist who shall be nameless.

---

[1] *Callinectes bellicosus.*

This varying threshold of tolerance is always an astonishing thing.

Amphioxus ordinarily lives on the seaward side of a sandbar and in sub-tidal water; or, at least, in sand bared by only the lowest of tides. We dug for them here and took only a few weak ones. It was not a very low tide and these were very possibly stragglers. It is probable that an extremely low tide would expose a level in which a great many of them live. The capture of these animals is exciting and requires speed. They are perfectly streamlined and partly transparent. Also, they are extremely nervous. Sometimes if the sand is struck with a shovel they will jump out and then frantically wriggle to get under the sand again—which they readily do. They are able to move through sand and even under it with great rapidity. We turned over the sand and leaped at them before they could escape. There used to be very many of them at Balboa Beach in Southern California, but channel dredging and perhaps the great number of motor boats have made them rare. And they are very interesting animals, being almost the dividing point between vertebrates and invertebrates. Usually one to three inches long and shuttle-shaped, they are perfectly built to slip through the sand without resistance.

The bar was rich with clams, many small *Chione*, and some small razor clams. We extracted the *Cerianthus* from their sloppy casings and found a great many tiny commensal sipunculid worms [2] in the smooth inner linings of the cases. These were able to extend themselves so far that they seemed like hairs, or to retract until they were like tiny peanuts. We did not find commensal pea crabs in the linings, as we had thought we would.

San Lucas Cove is nearly slough-like. The water gets very warm and probably very stale. It is exposed to a deadly sun and is so shallow that the water is soupy. This very quality of proba-

---

[2] *Phascolosoma hesperum.*

ble high salinity and warmth made it very difficult to preserve the *Cerianthus* in an expanded state. The small bunodids are easily anesthetized in Epsom salts, but *Cerianthus,* after six to eight hours of concentrated Epsom-salts solution, and even standing in pans under the hot sun, were able to retract rapidly and violently by expelling water from the aboral pore when the preserving liquid touched them. Sooner or later we will find the perfect method for anesthetizing anemones, but it has not yet been found. There is hope that cold may work as the anesthetic, if we can force absorption of formalin while the animals are relaxed with dry ice. But a great deal of experimenting is necessary, for if too cold they do not receive the formalin, and if too warm they retract on contact with it.

Back on board at about eleven-thirty we sailed for San Carlos Bay. We did not plan to stop at Santa Rosalia. It is a fairly large town which has long been supported by copper mines in the neighborhood, under the control of a French company. A little feeling of hurry was creeping upon us, for by now we had begun to see the magnitude of the job we had undertaken, and to realize that with the limited time and the more than limited equipment and personnel, we could not make much of a job of it. Our time was going fast. Much as Sparky and Tiny wished to continue their research and shore collecting at Santa Rosalia, we sailed on past it. And it looked, from the sea at least, to be less Mexican than other towns. Perhaps that was because we knew it was run by a French company. A Mexican town grows out of the ground. You cannot conceive its never having been there. But Santa Rosalia looked "built." There were industrial works of large size visible, loading trestles, and piles of broken rock. The mountains rose behind the town, burned almost white, and the green about the houses and the red roofs were in startling contrast. Sparky had the wheel as we went by, and his left hand was heavy. It required a definite effort of will for him to keep the course off shore.

At about six P.M. we came to San Carlos Bay, a curious land-locked curve with an inner shallow bay. There is good anchorage for small craft in the outer bay, with five to seven fathoms of water. The inner bay, or lagoon, has a sand beach on all sides. We intended to collect on the heavy boulders on the inner, or eastern, shore. There might be, we thought, a contrasting fauna to that of the tide flats of morning. This beach was piled high with rotting seaweed, left by some fairly recent storm perhaps. Or possibly this beach is at the end of some current-cycle, so that a high tide deposits great amounts of torn weed. There is such a beach at San Antonio del Mar on the western shore of Lower California, about sixty miles south of Ensenada. The debris from ships from hundreds of miles around is piled on this beach—mountains of sea-washed boxes and crates, logs and lumber, great whitened piles of it, mixed in with bottles and cans and pieces of clothing. It is the termination of some great sweeping in the Pacific.

Here at San Carlos there was little human debris; so very few boats pass up the Gulf this far and the people so prize planed wood and cans that such things would be picked up very quickly. In the decaying weed were myriads of flies and beach-hoppers working on this endless food supply. But in spite of their incredible numbers, we were able to catch only a few of the hoppers; they were too fast for us. Again we felt that here in the Gulf a little extra is added to the protection of animals. They are extra-fast, they are extra-armored, they seem to sting and pinch and bite worse than animals in other places. In the sand we found some clams rather like the Pismo clams of California, but shiny brown to black; also some ribbed mussel-like clams.[3] On the rocks we took two species of chitons and some new snails and crabs. There were blue, sharp-spined urchins and a number of flatworms. The flatworms are hard to catch, for they flow over the rocks like quicksilver. Also they are im-

[3] *Carditamera affinis.*

possible to preserve well; many of them simply dissolve in the preservative, while others roll up tightly. *Heliaster,* the sun-star, was here, but he had continued to shrink and was quite small this far up in the Gulf. Under the sand there were a great number of heart-urchins.

That night, using the shaded lamp hung over the side, we had a great run of transparent fish, including a type we had not seen before. We took another squid, a larval mantis-shrimp, and the usual *heteronereis* and crustacea.

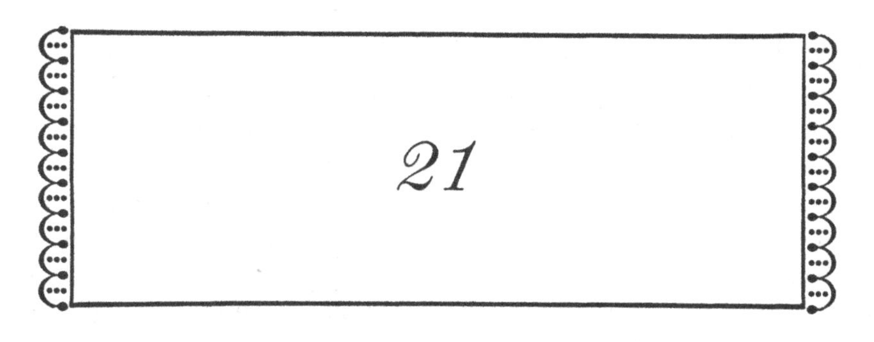

March 31

THE tide was very poor this morning, only two and a half to three feet below the uppermost line of barnacles. We started about ten o'clock and had a little collecting under water, but soon the wind got up and so ruffled the surface that we could not see what we were doing. To a certain extent this was a good thing. Not being able to get into the low littoral, where no doubt the spectacular spiny lobsters would have distracted us, we were able to make a more detailed survey of the upper region. One fact increasingly emerged: the sulphury-green and black cucumber[1] is the most ubiquitous shore animal of the Gulf of California, with *Heliaster,* the sun-star, a close second. These two are found nearly everywhere. In this region at San Carlos, Sally Lightfoot lives highest above the ordinary high tide, together with a few *Ligyda occidentalis,* a cockroach-like crustacean. Attached to the rocks and cliffsides, high up and fully exposed to this deadly sun, were barnacles and limpets, so placed that they must experience only occasional immersion, although they may be often dampened by spray. Under rocks and boulders, in the next association lower down, were the mussel-like ruffled clams and the brown chitons, many cucumbers, a few *Heliasters,* and only two species of brittle-stars—another common species, *Ophiothrix spiculata,* we did not find here although we had seen it everywhere else. In this zone verrucose

[1] *Holothuria lubrica.*

anemones were growing under overhangs on the sides of rocks and in pits in the rocks. There were also a few starfish [2]; garbanzo clams were attached to the rock undersides by the thousands together with club urchins. Farther down in a new zone was a profusion of sponges of a number of species, including a beautiful blue sponge. There were octopi [3] here, and one species of chiton; there were many large purple urchins, although no specimens were taken, and heart-urchins in the sand and between the rocks. There were some sipunculids and a great many tunicates.

We found extremely large sponges, a yellow form (probably *Cliona*) superficially resembling the Monterey *Lissodendoryx noxiosa,* and a white one, *Steletta,* of the wicked spines. There were brilliant-orange nudibranchs, giant terebellid worms, some shell-less air-breathing (pulmonate) snails, a ribbon-worm, and a number of solitary corals. These were the common animals and the ones in which we were most interested, for while we took rarities when we came upon them in normal observation, our interest lay in the large groups and their associations—the word "association" implying a biological assemblage, all the animals in a given habitat.

It would seem that the commensal idea is a very elastic thing and can be extended to include more than host and guest; that certain kinds of animals are often found together for a number of reasons. One, because they do not eat one another; two, because these different species thrive best under identical conditions of wave-shock and bottom; three, because they take the same kinds of food, or different aspects of the same kinds of food; four, because in some cases the armor or weapons of some are protection to the others (for instance, the sharp spines of an urchin may protect a tide-pool johnny from a larger preying fish); five, because some actual commensal partition of activi-

---

[2] *Astrometis sertulifera.*
[3] *Octopus bimaculatus.*

ties may truly occur. Thus the commensal tie may be loose or very tight and some associations may partake of a real thigmotropism.

Indeed, as one watches the little animals, definite words describing them are likely to grow hazy and less definite, and as species merges into species, the whole idea of definite independent species begins to waver, and a scale-like concept of animal variations comes to take its place. The whole taxonomic method in biology is clumsy and unwieldy, shot through with the jokes of naturalists and the egos of men who wished to have animals named after them.

Originally the descriptive method of naming was not so bad, for every observer knew Latin and Greek well and was able to make out the descriptions. Such knowledge is fairly rare now and not even requisite. How much easier if the animals bore numbers to which the names were auxiliary! Then, one knowing that the phylum Arthropoda was represented by the roman figure *VI,* the class Crustacea by a capital *B,* order by arabic figure *13,* and genus and species by a combination of small letters, would with little training be able to place the animals in his mind much more quickly and surely than he can now with the descriptive method tugged bodily from a discarded antiquity.

As we ascended the Gulf it became more sparsely inhabited; there were fewer of the little heat-struck *rancherias,* fewer canoes of fishing Indians. Above Santa Rosalia very few trading boats travel. One would be really cut off up here. And yet here and there on the beaches we found evidences of large parties of fishermen. On one beach there were fifteen or twenty large sea-turtle shells and the charcoal of a bonfire where the meat had been cooked or smoked. In this same place we found also a small iron harpoon which had been lost, probably the most valued possession of the man who had lost it. These Indians do not seem to have firearms; probably the cost of them is beyond even crazy dreaming. We have heard that in some of the houses are the treasured weapons of other times, muskets, flintlocks, old long muzzle-loaders kept from generation to generation. And one man told us of finding a piece of Spanish armor, a breastplate, in an Indian house.

There is little change here in the Gulf. We think it would be very difficult to astonish these people. A tank or a horseman armed cap-a-pie would elicit the same response—a mild and dwindling interest. Food is hard to get, and a man lives inward, closely related to time; a cousin of the sun, at feud with storm and sickness. Our products, the mechanical toys which take up so much of our time, preoccupy and astonish us so, would be considered what they are, rather clever toys but not related to very real things. It would be interesting to try to explain to one of these Indians our tremendous projects, our great drives, the fantastic production of goods that can't be sold, the clutter of possessions which enslave whole populations with debt, the worry and neuroses that go into the rearing and educating of neurotic children who find no place for themselves in this complicated world; the defense of the country against a frantic nation of conquerors, and the necessity for becoming frantic to do it; the spoilage and wastage and death necessary for the retention of the crazy thing; the science which labors to acquire knowledge, and the movement of people and goods contrary to the knowledge obtained. How could one make an Indian understand the medicine which labors to save a syphilitic, and the gas and bomb to kill him when he is well, the armies which build health so that death will be more active and violent. It is quite possible that to an ignorant Indian these might not be evidences of a great civilization, but rather of inconceivable nonsense.

It is not implied that this fishing Indian lives a perfect or even a very good life. A toothache may be to him a terrible thing, and a stomachache may kill him. Often he is hungry, but he does not kill himself over things which do not closely concern him.

A number of times we were asked, Why do you do this thing, this picking up and pickling of little animals? To our own people we could have said any one of a number of meaningless things, which by sanction have been accepted as meaningful. We could have said, "We wish to fill in certain gaps in the knowledge of the Gulf fauna." That would have satisfied our people, for knowledge is a sacred thing, not to be questioned or even inspected. But the Indian might say, "What good is this knowledge? Since you make a duty of it, what is its purpose?" We could have told our people the usual thing about the advancement of science, and again we would not have been questioned further. But the Indian might ask, "Is it advancing, and toward what? Or is it merely becoming complicated? You save the lives of children for a world that does not love them. It is our practice," the Indian might say, "to build a house before we move into it. We would not want a child to escape pneumonia, only to be hurt all its life." The lies we tell about our duty and our purposes, the meaningless words of science and philosophy, are walls that topple before a bewildered little "why." Finally, we learned to know why we did these things. The animals were very beautiful. Here was life from which we borrowed life and excitement. In other words, we did these things because it was pleasant to do them.

We do not wish to intimate in any way that this hypothetical Indian is a noble savage who lives in logic. His magics and his techniques and his teleologies are just as full of nonsense as ours. But when two people, coming from different social, racial, intellectual patterns, meet and wish to communicate, they must do so on a logical basis. Clavigero discusses what seems to our people a filthy practice of some of the Lower California Indians. They were always hungry, always partly starved. When they had meat, which was a rare thing, they tied pieces of string to each mouthful, then ate it, pulled it up and ate it again and again, often passing it from hand to hand. Clavigero found this a disgusting practice. It is rather like the Chinese being ridiculed for eating twenty-year-old eggs who said, "Your cheese is rotten milk. You like rotten milk—we like rotten eggs. We are both silly."

* * *

Costume on the *Western Flyer* had degenerated completely. Shirts were no longer worn, but the big straw hats were necessary. On board we went barefoot, clad only in hats and trunks. It was easy then to jump over the side to freshen up. Our clothes never got dry; the salt deposited in the fibers made them hygroscopic, always drawing the humidity. We washed the dishes in hot salt water, so that little crystals stuck to the plates. It seemed to us that the little salt adhering to the coffee pot made the coffee delicious. We ate fish nearly every day: bonito, dolphin, sierra, red snappers. We made thousands of big fat biscuits, hot and unhealthful. Twice a week Sparky created his magnificent spaghetti. Unbelievable amounts of coffee were consumed. One of our party made some lemon pies, but the quarreling grew bitter over them; the thievery, the suspicion of favoritism, the vulgar traits of selfishness and perfidy those pies brought out saddened all of us. And when one of us who, from being the most learned should have been the most self-controlled, took to hiding pie in his bed and munching it secretly when the lights were out, we decided there must be no more lemon pie. Character was crumbling, and the law of the fang was too close to us.

One thing had impressed us deeply on this little voyage: the great world dropped away very quickly. We lost the fear and fierceness and contagion of war and economic uncertainty. The matters of great importance we had left were not important. There must be an infective quality in these things. We had lost the virus, or it had been eaten by the anti-bodies of quiet. Our pace had slowed greatly; the hundred thousand small reactions of our daily world were reduced to very few. When the boat was moving we sat by the hour watching the pale, burned mountains slip by. A playful swordfish, jumping and spinning, absorbed us completely. There was time to observe the tremendous minutiae of the sea. When a school of fish went by, the gulls followed closely. Then the water was littered with feathers and the scum of oil. These fish were much too large for the gulls to kill and eat, but there is much more to a school of fish than the fish themselves. There is constant vomiting; there are the hurt and weak and old to cut out; the smaller prey on which the school feeds sometimes escape and die; a moving school is like a moving camp, and it leaves a camp-like debris behind it on which the gulls feed. The sloughing skins coat the surface of the water with oil.

At six P.M. we made anchorage at San Francisquito Bay. This cove-like bay is about one mile wide and points to the north. In the southern part of the bay there is a pretty little cove with a narrow entrance between two rocky points. A beach of white sand edges this cove, and on the edge of the beach there was a poor Indian house, and in front of it a blue canoe. No one came out of the house. Perhaps the inhabitants were away or sick or dead. We did not go near; indeed, we had a strong feeling of intruding, a feeling sharp enough even to prevent us from collecting on that little inner bay. The country hereabouts was stony and barren, and even the brush had thinned out. We anchored in four fathoms of water on the westerly side of the bay, then went ashore immediately and set up our tide stake at the water's edge, with a bandanna on it so we could see it from the boat. The wind was blowing and the water was painfully cold. The tide had dropped two feet below the highest line of barnacles. Three types of crabs [4] were common here. There were many barnacles and great limpets and two species of snails, *Tegula* and a small *Purpura*. There were many large smooth brown chitons, and a few bristle-chitons. Farther down under the rocks were great anastomosing masses of a tube-worm with rusty red gills,[5] some tunicates, *Astrometis*, and the usual holothurians.

Tiny found the shell of a fine big lobster,[6] newly cleaned by isopods. The isopods and amphipods in their millions do a beautiful job. It is common to let them clean skeletons designed for study. A dead fish is placed in a jar having a cap pierced with holes just large enough to permit the entrance of the isopods. This is lowered to the bottom of a tide pool, and in a very short time the skeleton is clean of every particle of flesh, and yet is articulated and perfect.

The wind blew so and the water was so cold and ruffled that we did not stay ashore for very long. On board, we put down the baited bottom nets as usual to see what manner of creatures were crawling about there. When we pulled up one of the nets, it seemed to be very heavy. Hanging to the bottom of it on the outside was a large horned shark.[7] He was not caught, but had gripped the bait through the net with a bulldog hold and he would not let go. We lifted him unstruggling out of the water and up onto the deck, and still he would not let go. This was at about eight o'clock in the evening. Wishing to preserve him, we did not kill him, thinking he would die quickly. His eyes were barred, rather like goat's eyes. He did not struggle at all, but lay quietly on the deck, seeming to look at us with a baleful, hating eye. The horn, by the dorsal fin, was clean and

---

[4] *Pachygrapsus crassipes, Geograpsus lividus,* and, under the rocks, *Petrolisthes nigrunguiculatus,* a porcelain crab.

[5] *Salmacina.*

[6] Apparently the northern *Panulirus interruptus.*

[7] *Gyropleurodus* of the Heterodontidae.

white. At long intervals his gill-slits opened and closed but he did not move. He lay there all night, not moving, only opening his gill-slits at great intervals. The next morning he was still alive, but all over his body spots of blood had appeared. By this time Sparky and Tiny were horrified by him. Fish out of water should die, and he didn't die. His eyes were wide and for some reason had not dried out, and he seemed to regard us with hatred. And still at intervals his gill-slits opened and closed. His sluggish tenacity had begun to affect all of us by this time. He was a baleful personality on the boat, a sluggish, gray length of hatred, and the blood spots on him did not make him more pleasant. At noon we put him into the formaldehyde tank, and only then did he struggle for a moment before he died. He had been out of the water for sixteen or seventeen hours, had never fought or flopped a bit. The fast and delicate fishes like the tunas and mackerels waste their lives out in a complete and sudden flurry and die quickly. But about this shark there was a frightful quality of stolid, sluggish endurance. He had come aboard because he had grimly fastened on the bait and would not release it, and he lived because he would not release life. In some earlier time he might have been the basis for one of those horrible myths which abound in the spoken literature of the sea. He had a definite and terrible personality which bothered all of us, and, as with the sea-turtle, Tiny was shocked and sick that he did not die. This fish, and all the family of the Heterodontidae, ordinarily live in shallow, warm lagoons, and, although we do not know it, the thought occurred to us that sometimes, perhaps fairly often, these fish may be left stranded by a receding tide so that they may have developed the ability to live through until the flowing tide comes back. The very sluggishness in that case would be a conservation of vital energy, whereas the beautiful and fragile tuna make one frantic rush to escape, conserving nothing and dying immediately.

Within our own species we have great variation between these two reactions. One man may beat his life away in furious assault on the barrier, where another simply waits for the tide to pick him up. Such variation is also observable among the higher vertebrates, particularly among domestic animals. It would be strange if it were not also true of the lower vertebrates, among the individualistic ones anyway. A fish, like the tuna or the sardine, which lives in a school, would be less likely to vary than this lonely horned shark, for the school would impose a discipline of speed and uniformity, and those individuals which would not or could not meet the school's requirements would be killed or lost or left behind. The overfast would be eliminated by the school as readily as the overslow, until a standard somewhere between the fast and slow had been attained. Not intending a pun, we might note that our schools have to some extent the same tendency. A Harvard man, a Yale man, a Stanford man—that is, the ideal—is as easily recognized as a tuna, and he has, by a process of elimination, survived the tests against idiocy and brilliance. Even in physical matters the standard is maintained until it is impossible, from speech, clothing, haircuts, posture, or state of mind, to tell one of these units of his school from another. In this connection it would be interesting to know whether the general collectivization of human society might not have the same effect. Factory mass production, for example, requires that every man conform to the tempo of the whole. The slow must be speeded up or eliminated, the fast slowed down. In a thoroughly collectivized state, mediocre efficiency might be very great, but only through the complete elimination of the swift, the clever, and the intelligent, as well as the incompetent. Truly collective man might in fact abandon his versatility. Among school animals there is little defense technique except headlong flight. Such species depend for survival chiefly on tremendous reproduction.

The great loss of eggs and young to predators is the safety of the school, for it depends for its existence on the law of probability that out of a great many which start some will finish.

It is interesting and probably not at all important to note that when a human state is attempting collectivization, one of the first steps is a frantic call by the leaders for an increased birth rate—replacement parts in a shoddy and mediocre machine.

Our interest had been from the first in the common animals and their associations, and we had not looked for rarities. But it was becoming apparent that we were taking a number of new and unknown species. Actually, more than fifty species undescribed at the time of capture will have been taken. These will later have been examined, classified, described, and named by specialists. Some of them may not be determined for years, for it is one of the little by-products of the war that scientific men are cut off from one another. A Danish specialist in one field is unable to correspond with his colleague in California. Thus some of these new animals may not be named for a long time. We have listed in the Appendix those already specified and indicated in so far as possible those which have not been worked on by specialists.

Dr. Rolph Bolin, ichthyologist at the Hopkins Marine Station, found in our collection what we thought to be a new species of commensal fish which lives in the anus of a cucumber, flipping in and out, possibly feeding on the feces of the host but more likely merely hiding in the anus from possible enemies. This fish later turned out to be an already named species, but, carrying on the ancient and disreputable tradition of biologists, we had hoped to call it by the euphemistic name *Proctophilus winchellii*.

There are some marine biologists whose chief interest is in the rarity, the seldom seen and unnamed animal. These are often wealthy amateurs, some of whom have been suspected of wishing to tack their names on unsuspecting and unresponsive invertebrates. The passion for immortality at the expense of a little beast must be very great. Such collectors should to a certain extent be regarded as in the same class with those philatelists who achieve a great emotional stimulation from an unusual number of perforations or a misprinted stamp. The rare animal may be of individual interest, but he is unlikely to be of much consequence in any ecological picture. The common, known, multitudinous animals, the red pelagic lobsters which litter the sea, the hermit crabs in their billions, scavengers of the tide pools, would by their removal affect the entire region in widening circles. The disappearance of plankton, although the components are microscopic, would probably in a short time eliminate every living thing in the sea and change the whole of man's life, if it did not through a seismic disturbance of balance eliminate all life on the globe. For these little animals, in their incalculable numbers, are probably the base food supply of the world. But the extinction of one of the rare animals, so avidly sought and caught and named, would probably go unnoticed in the cellular world.

Our own interest lay in relationships of animal to animal. If one observes in this relational sense, it seems apparent that species are only commas in a sentence, that each species is at once the point and the base of a pyramid, that all life is relational to the point where an Einsteinian relativity seems to emerge. And then not only the meaning but the feeling about species grows misty. One merges into another, groups melt into ecological groups until the time when what we know as life meets and enters what we think of as non-life: barnacle and rock, rock and earth, earth and tree, tree and rain and air. And the units nestle into the whole and are inseparable from it. Then one can come back to the microscope and the tide pool and the aquarium. But the little animals are found to be changed, no longer set apart and alone. And it is a strange thing that most of the feeling we call religious, most of the mystical outcrying which is one of the most prized and used and desired reactions of our species, is really the understanding and the attempt to say that man is related to the whole thing, related inextricably to all reality, known and unknowable. This is a simple thing to say, but the profound feeling of it made a Jesus, a St. Augustine, a St. Francis, a Roger Bacon, a Charles Darwin, and an Einstein. Each of them in his own tempo and with his own voice discovered and reaffirmed with astonishment the knowledge that all things are one thing and that one thing is all things—plankton, a shimmering phosphorescence on the sea and the spinning planets and an expanding universe, all bound together by the elastic string of time. It is advisable to look from the tide pool to the stars and then back to the tide pool again.

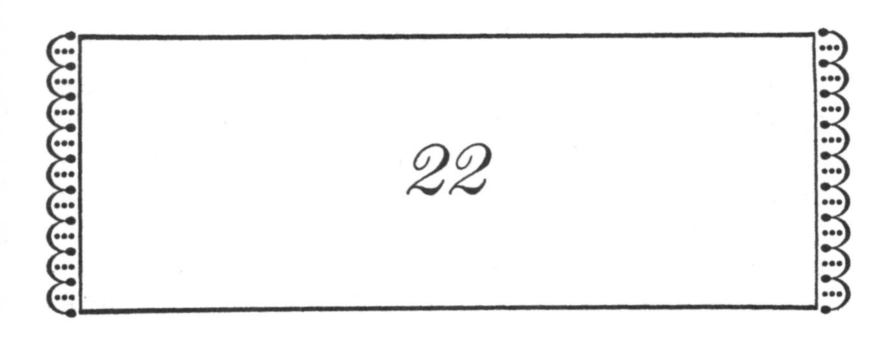

April 1

WITHOUT the log we should have lost track of the days of the week, were it not for the fact that Sparky made spaghetti on Thursdays and Sundays. We think he did this by instinct, that he could come out of a profound amnesia, and if he felt an impulse to make spaghetti, it would be found to be either Thursday or Sunday. On Monday we sailed for Angeles Bay, which was to be our last station on the Peninsula. The tides were becoming tremendous, and while the tidal bore of the Colorado River mouth was still a long way off, Tony was already growing nervous about it. During the trip between San Francisquito and Angeles Bay, we worried again over the fact that we were not taking photographs. As has been said, no one was willing to keep his hands dry long enough to use the cameras. Besides, none of us knew much about cameras. But it was a constant source of bad conscience to us.

On this day it bothered us so much that we got out the big camera and began working out its operation. We figured everything except how to put the shutter curtain back to a larger aperture without making an exposure. Several ways were suggested and, as is often the case when more than one method is possible, an argument broke out which left shutters and cameras behind. This was a good one. Everyone except Sparky and Tiny, who had the wheel, gathered on the hatch around the camera, and the argument was too much for the steersmen. They sent down respectful word that either we should bring the camera up where they could hear the argument, or they would abandon their posts. We suggested that this would be mutiny. Then Sparky explained that on an Italian fishing boat in Monterey mutiny, far from being uncommon, was the predominant state of affairs, and that he and Tiny would rather mutiny than not. We took the camera up on the deckhouse and promptly forgot it in another argument.

Except for a completely worthless lot of 8-mm. movie film, this was the closest we came to taking pictures. But some day we shall succeed.

Angeles Bay is very large—twenty-five square miles, the *Coast Pilot* says. It is land-locked by fifteen islands, between several of which there is entrance depth. This is one of the few harbors in the whole Gulf about which the *Coast Pilot* is willing to go out on a limb. The anchorage in the western part of the bay, it says, is safe from all winds. We entered through a deep channel between Red Point and two small islets, pulled into eight fathoms of water near the shore, and dropped our anchor. The *Coast Pilot* had not mentioned any settlement, but here there were new buildings, screened and modern, and on a tiny airfield a plane sat. It was an odd feeling, for we had been a long time without seeing anything modern. Our feeling was more of resentment than of pleasure. We went ashore about three-thirty in the afternoon, and were immediately surrounded by Mexicans who seemed curious and excited about our being there. They were joined by three Americans who said they had flown in for the fishing, and they too seemed very much interested in what we wanted until they were convinced it was marine animals. Then they and the Mexicans left us severely alone. Perhaps we had been hearing too many rumors: it was said that many guns were being run over the border for the trouble that was generally expected during the election. The fishermen did not look like fishermen, and Mexicans and Americans were *too* interested in us until they discovered what we were doing and too uninterested after they had found out. Perhaps we imagined it, but we had a strong feeling of secrecy about the place. Maybe there really were gold mines there and new buildings for recent development. A road went northward from there to San Felipe Bay, we were told. The country was completely parched and desolate, but half-way up a hill we could see a green spot where a spring emerged from a mountain. It takes no more than this to create a settlement in Lower California.

We went first to collect on a bouldery beach on the western side of the bay, and found it fairly rich in fauna. The highest rocks were peopled by anemones, cucumbers, sea-cockroaches and some small porcellanids. There were no Sally Lightfoots visible, in fact no large crabs at all, and only a very few small members of *Heliaster.* The dominant animal here was a soft marine pulmonate which occurred in millions on and under the rocks. We took several hundred of them. There were some chitons, both the smooth brown *Chiton virgulatus* and the fuzzy *Acanthochitona exquisitus.* We saw fine big clusters of the minute tube-worm, *Salmacina,* and there were a great many flatworms oozing along on the undersides of the rocks like drops of spilled brown sirup. Under the rocks we found two octopi, both *Octopus bimaculatus.* They are very clever and active in escaping, and when finally captured they grip the hand and arm with their little suckers, and, if left for any length of time, will cause small blood blisters or, rather, what in another field are called "monkey bites." Under the water and apparently below the ordinary tidal range were brilliant-yellow *Geodia* and many examples of another sponge of magnificent shape and size and color. This last (erect colonies of the cosmopolitan *Cliona celata,* more familiar as a boring sponge) was a reddish pink and stood high and vase-like, some of them several feet in diameter. Most of them were perfectly regular in shape. We took a number of them, dried some out of formaldehyde, and preserved others. The algal zonation on this slope was sharp and apparent—a *Sargassum* was submerged two or three feet at ebb. The rocks in the intertidal were perfectly smooth and bare but below this *Sargassum johnstonii,* in deeper water, there was a great zone of flat, frond-like alga, *Padina durvillaei.*[1] The wind rippled the surface badly but when an occasional lull came we could look down into this deeper water. It did not seem rich in life except for the algae, but then we were unable to turn over the rocks on the bottom.

While we collected, our fishermen rowed aimlessly about, and in our suspicious state of mind they seemed to be more anxious to appear to be fishing than actually to be fishing. We have little doubt that we were entirely wrong about this, but the place breathed suspicion, and no other place had been like that.

We went back on board and deposited our catch, then took the skiff to the sand flats on the northern side of the bay. It was a hard, compact mud sand with a long shallow beach, and it was heavy and difficult to dig into. We took there a number of *Chione* and *Tivela* clams and one poor half-dead amphioxus. Again the tide was not low enough to reach the real habitat of amphioxus, but if there was one stray in the high area, there must be many to be taken on an extremely low tide. We found a number of long turreted snails carrying commensal anemones on their shells. On this flat there were a number of imbedded small rocks, and these were rich with animals. There were rock-oysters on them and large highly ornamented limpets and many small snails. Tube-worms clustered on these rocks with pea crabs commensal in the tubes. One fair-sized octopus (not *bimaculatus*) had his home under one of these rocks. These small stones must have been havens in the shifting sands for many

---

1 Determinations by Dr. E. Yale Dawson of the Department of Botany, University of California.

animals. The fine mud-like sand would make locomotion difficult except for specially equipped animals, and the others clustered to the rocks where there was footing and security.

The tide began to flow rapidly and the winds came up and we went back to the *Western Flyer*. When we were on board we saw a ship entering the harbor, a big green sailing schooner with her sails furled, coming in under power. She did not approach us, but came to anchorage about as far from us as she could. She was one of those incredible Mexican Gulf craft; it is impossible to say how they float at all and, once floating, how they navigate. The seams are sprung, the paint blistered away, ironwork rusted to lace, decks warped and sagging, and, it is said, so dirty and bebugged that if the cockroaches were not fed, or were in any way frustrated or insulted, they would mutiny and take the ship—and, as one Mexican sailor said, "probably sail her better than the master."

Once the anchor of this schooner was down there was no further sign of life on her and there was no sign of life from the buildings ashore either. The little plane sat in its runway and the houses seemed vacant. We had been asked how long we would remain and had said, until the next morning. Now we felt, curiously enough, that we were interfering with something, that some kind of activity would start only when we left. Again, we were probably all wrong, but it is strange that every one of us caught a sinister feeling from the place. Unless the wind was up, or the anchorage treacherous, we ordinarily kept no anchor watch, but this night the boys got up a number of times and were restless. As with the werewolf, we were probably believing our own imaginations. For a short time in the evening there were lights ashore and then they went out. The schooner did not even put up a riding light, but lay completely dark on the water.

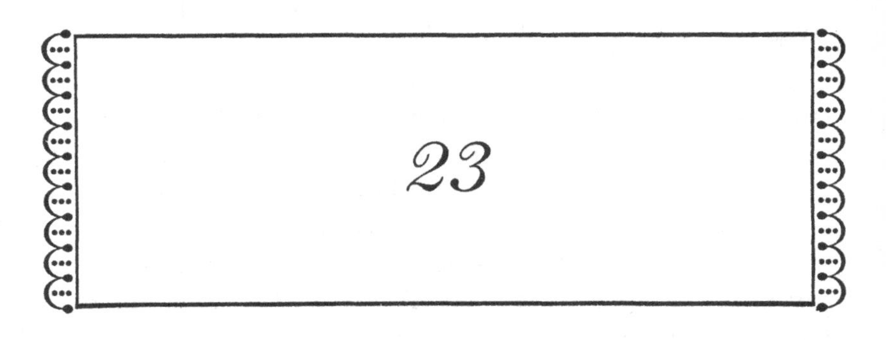

April 2

WE STARTED early and moved out through the channel to the Gulf. It was not long before we could make out Sail Rock far ahead, with Guardian Angel Island to the east of it. Sail Rock looks exactly like a tall Marconi sail in the distance. It is a high, slender pyramid, so whitened with guano that it catches the light and can be seen for a great distance. Because of its extreme visibility it must have been a sailing point for many mariners. It is more than 160 feet high, rises to a sharp point, and there is deep water close in to it. With lots of time, we would have collected at its base, but we were aimed at Puerto Refugio, at the upper end of Guardian Angel Island. We did take some of our usual moving pictures of Sail Rock, and they were even a little worse than usual, for there was laundry drying on a string and the camera was set up behind it. When developed, the film showed only an occasional glimpse of Sail Rock, but a very lively set of scenes of a pair of Tiny's blue and white shorts snapping in the breeze. It is impossible to say how bad our moving pictures were—one film laboratory has been eager to have a copy of the film, for it embodies in a few thousand feet, so they say, every single thing one should not do with a camera. As an object lesson to beginners they think it would be valuable. If we took close-ups of animals, someone was in the light; the aperture was always too wide or too narrow; we made little jerky pan shots back and forth; we have one of the finest sequences of unadorned sky pictures in existence—but when there was something to take about which we didn't care, we got it perfectly. We dare say there is not in the world a more spirited and beautiful picture of a pair of blue and white shorts than that which we took passing Sail Rock.

The long, snake-like coast of Guardian Angel lay to the east of us; a desolate and fascinating coast. It is forty-two miles long, ten miles wide in some places, waterless and uninhabited. It is said to be crawling with rattlesnakes and iguanas, and a persistent rumor of gold comes from it. Few people have explored it or even gone more than a few steps from the shore, but its fine harbor, Puerto Refugio, indicates by its name that many ships have clung to it in storms and have found safety there. Clavigero calls the island both "Angel de la Guardia" and "Angel Custodio," and we like this latter name better.

The difficulties of exploration of the island might be very great, but there is a drawing power about its very forbidding aspect—a Golden Fleece, and the inevitable dragon, in this case rattlesnakes, to guard it. The mountains which are the backbone of the island rise to more than four thousand feet in some places, sullen and desolate at the tops but with heavy brush on the skirts. Approaching the northern tip we encountered a deep swell and a fresh breeze. The tides are very large here, fourteen feet during our stay, and that not an extreme tide at all. It is probable that a seventeen-foot tide would not be unusual here. Puerto Refugio is really two harbors connected by a narrow channel. It is a safe, deep anchorage, the only danger lying in the strength and speed of the tidal current, which puts a strain on the anchor tackle. It was so strong, indeed, that we were not able to get weighted nets to the bottom; they pulled out sideways in the water and sieved the current of weed and small animals, so that catch was fairly worthwhile anyway.

We took our time getting firm anchorage, and at about three-thirty P.M. rowed ashore toward a sand and rubble beach on the southeastern part of the bay. Here the beach was piled with debris: the huge vertebrae of whales scattered about and piles of broken weed and skeletons of fishes and birds. On top of some low bushes which edged the beach there were great nests three to four feet in diameter, pelican nests perhaps, for there were pieces of fish bone in them, but all the nests were deserted —whether they were old or it was out of season we do not know. We are so used to finding on the beaches evidence of man that it is strange and lonely and frightening to find no single thing that man has touched or used. Tiny and Sparky made a small excursion inland, not over several hundred yards from the shore, and they came back subdued and quiet. They had not seen any rattlesnakes, nor did they want to. The beach was alive with hoppers feeding on the refuse, but the coarse sand was not productive of other animal life. The tide was falling, and we walked around a rocky point to the westward and came into a bouldery flat where the collecting was very rich. The receding water had left many small tide pools. The smoothness of the rocks indicated a fairly strong surf; they were dangerously slippery, and Sally Lightfoots and *Pachygrapsus* both scuttled about. As we moved out toward the entrance of the harbor, the boulders became larger and smoother, and then there was a sudden change to unbroken reef, and the smooth rocks gave way to barnacle- and weed-covered stones. The tide was down about ten feet now, exposing the lower tide pools, rich and beautiful with sponges and corals and small pleasant algae. We tried to cover as much territory as possible, but again and again found ourselves fascinated by some small and perfect pool, like a set stage, peopled with broken-back shrimps and small masked crabs.

The point itself was jagged volcanic rock in which there were high mysterious caves. Entering one, we noticed a familiar smell, and a moment later recognized it. For the sound of our voices alarmed myriads of bats, and their millions of squeaks sounded like rushing water. We threw stones in to try to dislodge some, but they would not brave the daylight, and only squeaked more fiercely.

As evening approached, it grew quite cold. Our hands were torn from the long collecting day and we were glad when it was too dark to work any more. We had taken great numbers of animals. There was an echiuroid worm with a spoon-shaped proboscis, found loose under the rocks; many shrimps; an encrusting coral (*Porites* in a new guise); many chitons, some new; and several octopi. The most obvious animals were the same marine pulmonates we had found at Angeles Bay, and these must have been strong and tough, for they were in the high rocks, fairly dry and exposed to the killing sun. The rocky ledge was covered with barnacles. The change in animal sizes on different levels was interesting. In the high-up pools there were small animals, mussels, snails, hermits, limpets, barnacles, sponges; while in the lower pools, the same species were larger. Among the small rocks and coarse gravel we found a great many stinging worms and a type of ophiuran new to us—actually it turned out to be the familiar *Ophionereis* in its juvenile stage. These high tide pools can be regarded as nurseries for more submerged zones. There were urchins, both club- and sharp-spined, and, in the sand, a few heart-urchins. The caverns under the rocks, exposed by the receding tide, were beautiful with many species of sponge, some pure white, some blue, and some purple, encrusting the rock surface. These under-rock caverns were as beautiful as those near Point Lobos in Central California. It was a long job to lay out and list the animals taken; meanwhile the crab-nets meant for the bottom were straining the current. In them we caught a number of very short fat stinging worms (*Chloeia viridis*), a species we had not seen before, probably a deep-water form torn loose by the strength of the current. With a hand-net we took a pelagic nudibranch, *Chioraera leonina,* found also in Puget Sound. The water swirled past the boat at about four miles an hour and we kept the dip-nets out until late at night. This was a strange collecting place. The water was quite cold, and many of the members of both the northern and the southern fauna occurred here. In this harbor there were conditions of stress, current, waves, and cold which seemed to encourage animal life. And it is reasonable that this should be so, for active, churning water means not only a strong oxygen content, but the constant movement of food. And in addition, the very difficulties involved in such a position—necessity for secure footing, crowding, and competition—seem to encourage a ferocity and a tenacity in the animals which go past survival and into successful reproduction. Where there is little danger, there seems to be little stimulation. Perhaps the pattern of struggle is so deeply imprinted in the genes of all life conceived in this benevolently hostile planet that the removal of obstacles automatically atrophies a survival drive. With warm water and abundant food, the animals may retire into a sterile sluggish happiness. This has certainly seemed true in man. Force and cleverness and versatility have surely been the children of obstacles. Tacitus, in the *Histories,* places as one of the tactical methods advanced to be used against the German armies their exposure to a warm climate and a soft rich food supply. These, he said, will ruin troops quicker than anything else. If these things are true in a biologic sense, what is to become of the fed, warm, protected citizenry of the ideal future state?

The classic example of the effect of such protection on troops is that they invariably lost discipline and wasted their energies in weak quarrelsomeness. They were never happy, never contented, but always ready to indulge in bitter and bloody personal quarrels. Perhaps this has no emphasis. So far there has been only one state that we know of which protected its people *without* keeping them constantly alert and organized against a real or imaginary outside enemy. This was the pre-Pizarro Inca state, whose people were so weakened that a little band of fierce, hard-bitten men was able to overcome the whole nation. And of them the converse is also true. When the food supply was wasted and destroyed by the Spaniards, when the fine economy which had distributed clothing and grain was overturned, only then, in their hunger and cold and misery, did the Peruvian people become a dangerous striking force. We have little doubt that a victorious collectivist state would collapse only a little less quickly than a defeated one. In fact, a bitter defeat would probably keep a fierce conquest-ideal alive much longer than a victory, for men can fight an enemy much more successfully than themselves.

Islands have always been fascinating places. The old storytellers, wishing to recount a prodigy, almost invariably fixed the scene on an island—Faëry and Avalon, Atlantis and Cipango, all golden islands just over the horizon where anything at all might happen. And in old days at least it was rather difficult to check up on them. Perhaps this quality of potential prodigy still lives in our attitude toward islands. We want very much to go back to Guardian Angel with time and supplies. We wish to go over the burned hills and snake-ridden valleys, exposed to heat and insects, venom and thirst, and we are willing to believe almost anything we hear about it. We believe that great gold nuggets are found there, that unearthly animals make their homes there, that the mountain sheep, which is said never to drink water, abounds there. And if we were told of a race of troglodytes in possession, we should think twice before disbelieving. It is one of the golden islands which will one day be toppled by a mining company or a prison camp.

Thus far, there had been no illness on board the *Western Flyer*. Tiny drooped a little at Puerto Refugio and confessed that he didn't feel very well, and we held a consultation on him in the galley, explaining to him that consultation was more pleasant to us, as well as to him, than autopsy. After a great many questions, some of which might have been considered personal but which Tiny used as a vehicle for outrageous boasting, we concocted a remedy which might have cured almost anything—which was apparently what he had. Tiny emerged on deck some hours later, shaken but smiling. He said that what he had been considering love had turned out to be simple flatulence. He said he wished all his romantic problems could be solved as easily.

It was now a long time since Sparky and Tiny had been able to carry out the good-will they felt toward Mexico, and they grew a little anxious about getting to Guaymas. There was no actual complaining, but they spoke tenderly of their intentions. Tex was inhibited in his good-will by his engagement to be married, which he wouldn't mention any more for fear we would diet him again. As for Tony, the master, he had no nerves, but the problems of finding new and unknown anchorages seemed to fascinate him. Tony would have made a great exploration captain. There would be few errors in judgment where he was concerned. The others of us were very busy all the time. We mention the health of the crew because we truly believe that the physical condition, and through it the mind, has reins on the actual collecting of animals. A man with a sore finger may not lift the rock under which an animal lives. We are likely to see more through our indigestion than through our eyes, and it seems to us that the ulcer-warped viewpoint is very often evident in animal descriptions. The man best fitted to observe animals, to understand them emotionally as well as intellectually, would be a hungry and libidinous man, for he and the animals would have the same preoccupations. Perhaps we fulfilled these requirements as well as most.

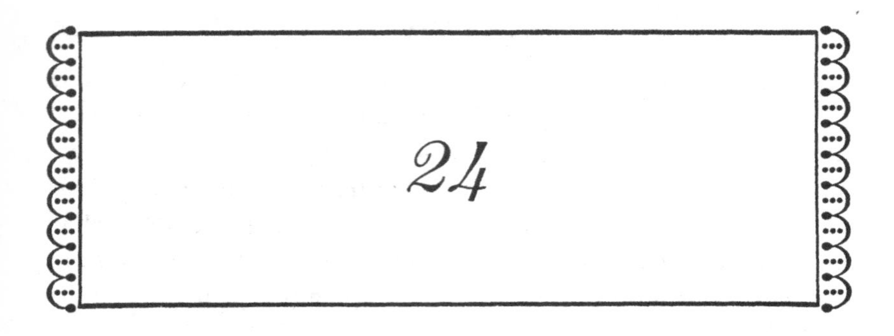

April 3

WE SAILED around the northern tip of Guardian Angel and down its eastern coast. The water was clear and blue, and a large swell flowed past us. About noon we moved through a great group of Zeppelin-shaped jellyfish, ctenophores or possibly siphonophores. They were six to ten inches long, and the sea was littered with them. We slowed down and tried to scoop them up, but the tension of their bodies was not sufficient to hold them together out of water. They broke up and slithered in pieces through the dip-nets. Soon after, a school of whales went by, one of them so close that the spray from his blow-hole came over our deck. There is nothing so evil-smelling as a whale anyway, and a whale's breath is frightfully sickening. It smells of complete decay. Perhaps the droplets were left on the boat, for it seemed to us that we could smell him for a long time after he had gone by. The great schools of tuna, so evident in the Lower Gulf, were not seen here, but a few seals lazed through the water, and on one or two occasions we nearly ran over one asleep on the surface. We felt deeply the loneliness of this sea; no ships, no boats, no canoes, no little ranches on the shore nor villages. Now we would have welcomed a fishing Indian to come aboard and eat canned fruit salad, but this is a deserted sea.

The queer shoulder of Tiburón showed to the southeast of us, and we ran down on it with the wind behind us and probably the tidal current too, for we made great speed. We went down the western coast of Tiburón and watched its high cliffs through the glasses. The cliffs are fairly sheer, and the mountains are higher than those on Guardian Angel Island. This is the island where the Seri Indians come during parts of the year. It is said of them that they are or have been cannibals, a story which has been firmly denied again and again. It is certain that they have killed many strangers, but whether or not they have eaten them does not seem to be documented. Cannibalism is a fascinating subject to most people, and in some way a sin. Possibly the deep feeling is that if people learn to eat one another the food supply would be so generous and so available that no one would be either safe or hungry. It is very curious the amount of hatred and fear that cannibalism inspires. These poor Seri Indians would not be so much feared for their murdering habits, but if in their hunger they should cut a steak from an American citizen a panic arises. Swift's quite reasonable suggestion concerning a possible use for Irish babies aroused a storm of emotionalism out of all proportion to its feasibility. There were not, it is thought, enough well-conditioned babies at that time to have provided anything like an adequate food supply. Swift without a doubt meant it only as an experiment. If it had been successful, there would have been time enough then to think of raising more babies. It has generally been found that starvation is the greatest single cause for cannibalism. In other words, people will not eat each other if they can get anything else. To some extent this reluctance must be caused by an unpleasant taste in human flesh, the result no doubt of our rather filthy eating habits. This need not be a future deterrent, for, if other barriers are removed, such as a natural distaste for eating relatives, or a man's gallant dislike for eating women, who in turn are inhibited by a romantic tendency—if all these difficulties should be solved it would be easy enough to improve the flavor of human flesh by special diets before slaughter and carefully prepared sauces and condiments afterwards. If this should occur, the Seri Indians, if indeed they do be cannibals, far from being loaded with our hatred, must be considered pioneers in a new field and honored as such.

Clavigero, in his *History of* [Lower] *California,*[1] has an account of these Seris.

The vessel, *San Javier* [he says], which had left Loreto in September 1709 with three thousand *scudi* to buy provisions in Yaqui, was carried 180 miles above the port of its destination by a furious storm and grounded on the sand. Some of the people were drowned; the rest saved themselves in the small boat; but after landing they were exposed to another not less serious danger because that coast was inhabited by the Series who were warlike gentiles and implacable enemies of the Spaniards. For this reason they hastened to bury the money and all the possessions which were on the boat; and after embarking again in the small boat they continued with a thousand dangers and hardships to Yaqui, from where they sent the news to Loreto. In a little while the Series came to the place where the Spaniards had buried those possessions, and they dug them up and carried them away. They even removed the rudder from the vessel and they destroyed it in order to get the nails.

As soon as Father Salvatierra learned of that misfortune, he left in the unseaworthy vessel, the *Rosario,* and went to the port of Guaymas. From there he sent this vessel to the place where the *San Javier* was grounded, and he himself went with fourteen Yaqui Indians in that direction over a very bad road which absolutely lacked potable water, and for this reason they suffered great thirst for two days. During the two months which he lived there, exposed to hunger and hardships and to the great danger of all their lives (while the vessel was being repaired), he won the good-will of the Series in such manner that he not only recovered all the cargo of the boat which they had stolen but induced them also to make peace with the Pimas, who were Christian neighbors of theirs and enemies whom they most hated. He baptized many of their children, he catechized the adults and inspired so much affection in them for Christianity that they immediately wanted a missionary to instruct them regularly and to baptize them and govern them in all respects.

So the dominating sweetness of the character of Father Salvatierra, aided by the grace of the Master, triumphed over the ferocity of those barbarians who were so feared, not only by the other Indians,

---

[1] Lake and Gray translation, 1937, pp. 217–18.

but also by the Spaniards. He wept tenderly on seeing their unexpected docility and their good inclinations, thanking God for having had that much good come from the misfortune of the vessel.

The "dominating sweetness" of the character of Father Salvatierra did not, however, change them completely, for they have gone right on killing people until recently. In this account it is also interesting to notice Clavigero's statement that Father Salvatierra took the "unseaworthy" *Rosario.* In the long record of wrecks blowing off course, of marine disaster of every kind, it was wonderful how they were able to judge whether or not a ship was unseaworthy. A little reading of contemporary records of voyaging by these priestly and soldierly navigators indicates that they put more faith in prayer than in the compass. We think the present-day navigators of the Gulf have learned their seamanship in the same school. Some of the ships we saw at Guaymas and La Paz floated in violation of every law of physics. There must be in Heaven a small pilot-house where a worried and distraught St. Christopher spends a good deal of his time looking after the shipping of the Gulf of California with a handful of miracles.

Tiburón looked red to us, and the brush seemed stronger and greener than any we had seen in a long time. In some of the creases between the hills there were growths of small ground-hugging trees like our scrub-oaks. What they were, of course, we do not know. About five-thirty in the afternoon we rounded Red Bluff Point on the southwesterly corner of Tiburón and came to anchor in the lee of the long point, protected from northerly winds. The "corner" of the island is a chosen word, for Tiburón is a rough square lying plumb with the points of the compass. We searched the shore for Seris and saw none. In our usual condition of hunger, it would have been a toss-up whether Seris ate us or we ate Seris. The one who got in the first bite would have had the dinner, but we never did see a Seri.

The coast at this station was interesting; off Red Bluff Point low flat rocks shelved gradually seaward—fine collecting rocks, with many of the pot-holes which make such beautiful natural aquaria at low tide. Southward of this were long reef-like stone fingers extending outward, with shallow sand-bottom baylets between them, almost like boat slips. Next to this was a bouldery beach with stones imbedded in sand; and finally a coarse sand beach. Here again was nearly every kind of environment except mud-flat and lagoon. We began our collecting on the reef, and found the little pot-holes lovely with hydroids and coral and colored sponge and little bright algae. There were many broken-back shrimps in these pools, difficult little fellows to catch, for they are so transparent as to be almost invisible and they move with great speed by flipping their tails like lobsters. Only their stomachs and flickering gills are visible, and one can watch their insides work as though they were little glass models. We caught many of them by working our hands very slowly under them and raising them gradually to the surface.

On the reef, there were the usual *Heliasters,* anemones, and cucumbers, urchins, and a great number of giant snails,[2] of which we collected many hundreds. High up in the intertidal were many *Tegula*-like snails of the kind we found at Cape San Lucas, although here the water was clear and very cold whereas at Cape San Lucas it had been warm. There were very few Sally Lightfoots here; *Pachygrapsus,* the northern crabs, had taken their place. We took abundant solitary corals and laid in a large supply of plumularian hydroids, gathered carefully and preserved so that they might not be crushed or broken. These animals, in appearance at least, are so like plants that they indicate to the imagination a bridge between flora and fauna, just as some plants, like the tropical sensitive plants and the insect-eating plants, indicate by their apparent nervous and muscular versatility an approach from the other side.

---

[2] *Callopoma fluctuosum.*

On the reef, we took a number of barnacles, many *Phataria* and *Linckia,* sponges, and tunicates. Moving from the reef to the stone fingers, we saw and captured a most attenuated spider crab,[3] all legs and little body. On the sand bottom between the fingers were many sting-rays lying quietly, and near the edge of one little harbor there were two in copulation, male (or female) lying on its back with its mate on top of it and the heads together. We wanted these two, and so after a moment in which we toughened the fibers of our romantic feelings, we put a light harpoon through both of them at once and brought them up, angry and disillusioned. We had hoped that they might remain fastened so that they might be preserved in coition, but their softer feelings were offended and they disengaged.

Meanwhile Tiny, moving in the little slip-like bays with the skiff, harpooned several more sting-rays. On the beach we took several sand-living cucumbers, and in the bottom of a mud pool searched long and unsuccessfully for a furry crab which had been seen scuttling into its burrow. This was a rich field for collecting, but the horizon markers were true to their position in the Gulf, and except for the profusion on Red Bluff Point reef, where the footing was excellent, there was nothing novel.

When it grew dark, we turned on the deck lights and saw numbers of a barracuda-like fish coming to eat the small fishes that gathered to the light. We put a fish-line on a small trident spear and began throwing it at them. About every tenth cast we struck one and brought him to the deck. And now a curious thing happened. From the shore came a swarm of very large bats. Their bodies were small but they had a twelve- to fifteen-inch wing-spread. They circled restlessly around the boat, although there were no insects about. Sparky was on the rail, spearing barracuda, and he is very much afraid of bats. Suddenly one swooped near him, and he struck at it with the harpoon. By one of those strange accidents, the barbs went into the bat and captured it,

[3] *Stenorhynchus debilis.*

and now four or five more dived straight at Sparky's head and he dropped the harpoon and ran for the galley. The dead bat fell over the side into the water, where we later picked it up.

Then an even stranger thing happened. As though at a signal, every bat of the hundreds suddenly turned and flew away to shore and not another one was seen. We have not yet a report on the one taken, so we do not know what kind of bats they were. There are reports of fish-eating bats, and these may have been that kind. We warned Sparky seriously to keep very quiet about the incident. "Sparky," we said, "we know that your reputation for truthfulness in Monterey is as good as most. In other words, it is not above reproach. If we were you, when you get back to Monterey, we would never mention to anyone that we had harpooned a bat. We would make up stories and adventures, but there is no reason for straining an already shaky reputation." Sparky promised he would never tell, but back in Monterey he couldn't resist and, just as we supposed, a roar of laughter went up. In Monterey they said, "You know what that Sparky said? He swears he harpooned a bat."

And as punishment to Sparky, when we were questioned we said, "Bat? What bat?" Sparky is a little touchy about the whole subject, and he dislikes bats very intensely now.

Meanwhile we had twelve of the barracuda-like fish. We preserved some of them but did not try to eat any. The sierras and tuna were too delicious to justify making experiments with strong fish.

The mountains of Tiburón were very black against the stars and the sea was calm. On the deck, Tiny made a little noise washing a shirt, for we were not far from Guaymas and Tiny was growing anxious. We discussed bats, and the horror they create in people and the myths about them—in his *Caribbean Treasure,* page 56, Ivan Sanderson makes some very interesting remarks about vampire bats as carriers of rabies, and their whole tie-in with the vampire tradition, so intimately related to werewolfism in the popular mind. A man with rabies, one might infer, could well be the werewolf which occurs all over the world, and vampire and werewolf very often go together. It is a fascinating speculation, and surely the unreasoning and almost instinctive fear of bats might indicate another of those memory-like patterns, some horrible recollection of the evil bats can do.

We find after reading many scientific and semi-scientific accounts of exploration that we have two strong prejudices: the first of these arises where there is a woman aboard—the wife of one of the members of the party. She is never called by her name or referred to as an equal. In the account she emerges as "the shipmate," the "skipper," the "pal." She is nearly always a stringy blonde with leathery skin who is included in all photographs to give them "interest." Our second prejudice concerns a hysteria of love which manifests itself in an outcry against parting and is usually written in Spanish. This outburst comes at the end of the book. It goes, "And so——." Always, "and so," for some reason. "And so we said good-by to Tiburón, vowing to come back again. *Adiós, Tiburón, amigo,* friend." For some reason this stringy shipmate and this rush of emotion are slightly obscene to us. And so we said good-by to Tiburón and trucked on down toward Guaymas.

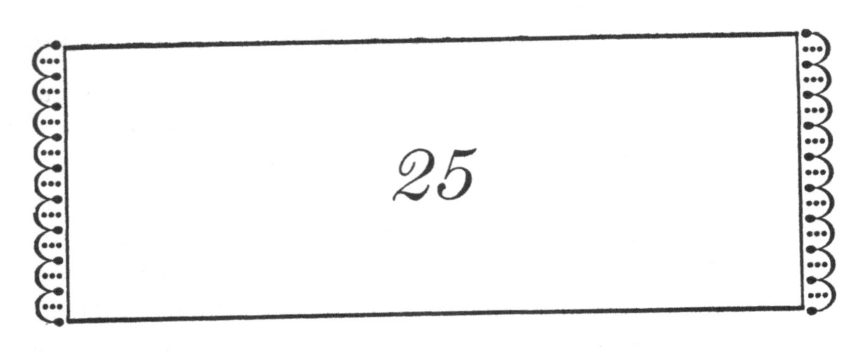

25

April 22

THE trolling jigs picked up two fine sierras on the way. Our squid jigs had gone to pieces from much use, and had to be repaired with white chicken feathers. We were under way all day, and toward evening began to see the sport-fishing boats of Guaymas with their cargoes of sportsmen outfitted with equipment to startle the fish into submission. And the sportsmen were mentally on tiptoe to out-think the fish—which they sometimes do. We thought it might be fun some time to engage in this intellectual approach toward fishing, instead of our barbarous method of throwing a line with a chicken-feather jig overboard. These fishermen in their swivel fishing chairs looked comfortable and clean and pink. We had been washing our clothes in salt water, and we felt sticky and salt-crusted; and, being less comfortable and clean than the sportsmen, we built a whole defense of contempt. With no effort at all on their part we had a good deal of dislike for them. It is probable that Sparky and Tiny had a true contempt, uncolored by envy, for they are descended from many generations of fishermen who went out for fish, not splendor. But even they might have liked sitting in a swivel chair holding a rod in one hand and a frosty glass in the other, blaming a poor day on the Democrats, and offering up prayers for good fishing to Calvin Coolidge.

We could not run for Guaymas that night, for the pilot fees rise after hours and we were getting a little low on money.

Instead, about six P.M. we rounded Punta Doble and put into Puerto San Carlos. This is another of those perfect little harbors with narrow rocky entrances. The entrance is less than eight hundred yards wide, and steep rocks guard it. Once inside, there is anchorage from five to seven fathoms. The head of the bay is bordered by a sand beach, changing to boulders near the entrance. There was still time for collecting.

We went to the bouldery beach and took some snails new to us and two echiuroids. But nothing on or under the rocks was different from the Tiburón animals. The water was warm here and it was soupy with shrimps, of which we took a number in a dip-net. We made a quick survey of the area, for darkness was coming. As soon as it was dark we began to hear strange sounds in the water around the *Western Flyer*—a periodic hissing and many loud splashes. We went to the deckhouse and turned on the searchlight. The bay was swarming with small fish, apparently come to eat the shrimps. Now and then a school of six- to ten-inch fish would drive at the little fish with such speed and in such numbers that they made the sharp hissing we had heard, while farther off some kind of great fish leaped and splashed heavily. Without a word, Sparky and Tiny got out a long net, climbed into the skiff, and tried to draw their net around a school of fish. We shouted at them, asking what they would do with the fish if they caught them, but they were deaf to us. The numbers of fish had set off a passion in them—they were fishermen and the sons of fishermen—let businessmen dispose of the fish; their job was to catch them. They worked frantically, but they could not encircle a school, and soon came back exhausted.

Meanwhile, the water seemed almost solid with tiny fish, one and one-half to two inches long. Sparky went to the galley and put the biggest frying pan on the fire and poured olive oil into it. When the pan was very hot he began catching the tiny fish with the dip-nets, a hundred or so in each net. We passed the nets through the galley window and Sparky dumped them into the frying pan. In a short time these tiny fish were crisp and brown. We drained, salted, and ate them without any cleaning at all and they were delicious. Probably no fresher fish were ever eaten, except perhaps by the Japanese, who are said to eat them alive, and by college boys, who are photographed doing it. Each fish was a curled, brown, crisp little bite, delicate and good. We ate hundreds of them. Afterwards we went back to the usual night practice of netting the pelagic animals which came to the light. We took shrimps and larval shrimps, numbers of small swimming crabs, and more of the transparent fish. All night the hissing rush and splash of hunters and hunted went on. We had never been in water so heavily populated. The light, piercing the surface, showed the water almost solid with fish—swarming, hungry, frantic fish, incredible in their voraciousness. The schools swam, marshaled and patrolled. They turned as a unit and dived as a unit. In their millions they followed a pattern minute as to direction and depth and speed. There must be some fallacy in our thinking of these fish as individuals. Their functions in the school are in some as yet unknown way as controlled as though the school were one unit. We cannot conceive of this intricacy until we are able to think of the school as an animal itself, reacting with all its cells to stimuli which perhaps might not influence one fish at all. And this larger animal, the school, seems to have a nature and drive and ends of its own. It is more than and different from the sum of its units. If we can think in this way, it will not seem so unbelievable that every fish heads in the same direction, that the water interval between fish and fish is identical with all the units, and that it seems to be directed by a school intelligence. If it is a unit animal itself, why should it not so react? Perhaps this is the wildest of speculations, but we suspect that when the school is studied as an animal rather than as a sum of unit fish, it will be found that certain units are assigned special functions to perform; that weaker or slower units may even take their places as placating food for the predators for the sake of the security of the school as an animal. In the little Bay of San Carlos, where there were many schools of a number of species, there was even a feeling (and "feeling" is used advisedly) of a larger unit which was the interrelation of species with their interdependence for food, even though that food be each other. A smoothly working larger animal surviving within itself—larval shrimp to little fish to larger fish to giant fish—one operating mechanism. And perhaps *this* unit of survival may key into the larger animal which is the life of all the sea, and this into the larger of the world. There would seem to be only one commandment for living things: Survive! And the forms and species and units and groups are armed for survival, fanged for survival, timid for it, fierce for it, clever for it, poisonous for it, intelligent for it. This commandment decrees the death and destruction of myriads of individuals for the survival of the whole. Life has one final end, to be alive; and all the tricks and mechanisms, all the successes and all the failures, are aimed at that end.

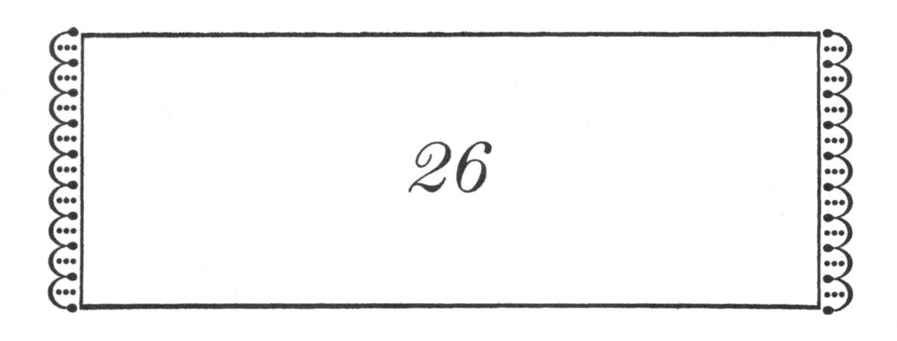

26

April 5

WE SAILED in the morning on the short trip to Guaymas. It was the first stop in a town that had anything like communication since we had left San Diego. The world and the war had become remote to us; all the immediacies of our usual lives had slowed up. Far from welcoming a return, we rather resented going back to newspapers and telegrams and business. We had been drifting in some kind of dual world—a parallel realistic world; and the preoccupations of the world we came from, which are considered realistic, were to us filled with mental mirage. Modern economies; war drives; party affiliations and lines; hatreds, political, and social and racial, cannot survive in dignity the perspective of distance. We could understand, because we could feel, how the Indians of the Gulf, hearing about the great ant-doings of the north, might shake their heads sadly and say, "But it is crazy. It would be nice to have new Ford cars and running water, but not at the cost of insanity." And in us the factor of time had changed: the low tides were our clock and the throbbing engine our second hand.

Now, approaching Guaymas, we were approaching an end. We planned only two or three collecting stations beyond, and then the time of charter-end would be crowding us, and we would have to run for it to be back when the paper said we would. The charter at least fixed our place in time. And already our crew was trying to think of ways to come back to the Gulf.

This trip had been like a dreaming sleep, a rest from immediacies. And in our contacts with Mexican people we had been faced with a change in expediencies. Perhaps—even surely—these people are expedient, but on some other plane than our ordinary one. What they did for us was without hope or plan of profit. We suppose there must have been some kind of profit involved, but not the kind we are used to, not of material things changing hands. And yet some trade took place at every contact—something was exchanged, some unnamable of great value. Perhaps these people are expedient in the unnamables. Maybe they bargain in feelings, in pleasures, even in simple contacts. When the Indians came to the *Western Flyer* and sat timelessly on the rail, perhaps they were taking something. We gave them presents, but it is sure they had not come for presents. When they helped us, it was with no idea of material payment. There were material prices for material things, but one couldn't buy kindness with money, as one can in our country. It was so in every contact, and they were so used to the spiritual transaction that they had difficulty translating material things into money. If we wanted to buy a harpoon, there was difficulty immediately. What was the price? An Indian had paid three pesos for the harpoon several years ago. Obviously, since that had been paid, that was the price. But he had not yet learned to give time a money value. If he had to go three days in a canoe to get another harpoon, he could not add his time to the price, because he had never thought of time as a medium of exchange. At first we tried to explain the feeling we all had that time is a salable article, but we had to give it up. Time, these Indians said, went on. If one could stop time, or take it away, or hoard it, then one might sell it. One might as well sell air or heat or cold or health or beauty. And we thought of the great businesses in our country—the sale of clean air, of heat and cold, the scrabbling bargains in health offered over the radio, the boxed and bottled beauty, all for a price. This was not bad or good, it was only different. Time and beauty, they thought, could not be captured and sold, and we knew they not only *could* be, but that time could be warped and beauty made ugly. And again it was not good or bad. Our people would pay more for pills in a yellow box than in a white box—even the refraction of light had its price. They would buy books because they should rather than because they wanted to. They bought immunity from fear in salves to go under their arms. They bought romantic adventure in bars of tomato-colored soaps. They bought education by the foot and hefted the volumes to see that they were not short-weighted. They purchased pain, and then analgesics to put down the pain. They bought courage and rest and had neither. And they are vastly amused at the Indian who, with his silver, bought Heaven and ransomed his father from Hell. These Indians were far too ignorant to understand the absurdities merchandising can really achieve when it has an enlightened people to work on.

One can go from race to race. It is coming back that has its violation. As we feel greatness, we feel that these people are very great. It seems to us that the repose of an Indian woman sitting in the gutter is beyond our achievement. But even these people wish for our involvement in temporal and material things. Once we thought that the bridge between cultures might be through education, public health, good housing, and through political vehicles—democracy, Nazism, communism—but now it seems much simpler than that. The invasion comes with good roads and high-tension wires. Where those two go, the change takes place very quickly. Any of the political forms can come in once the radio is hooked up, once the concrete highway irons out the mountains and destroys the "localness" of a community. Once the Gulf people are available to contact, they too will come to consider clean feet more important than clean minds. These are the factors of civilization and their paths, good roads, high-voltage wires, and possibly canned foods.

A local 110-volt power unit and a winding dirt road may leave a people for a long time untouched, but high-voltage operating day and night, the network of wires, will draw the people into the civilizing web, whether it be in Asiatic Russia, in rural England, or in Mexico. That *Zeitgeist* operates everywhere, and there is no escape from it.

Again, this is not to be considered good or bad. To us, a little weary of the complication and senselessness of a familiar picture, the Indian seems a rested, simple man. If we should permit ourselves to remain in ignorance of his complications, then we might long for his condition, thinking it superior to ours. The Indian on the other hand, subject to constant hunger and cold, mourning a grandfather and set of uncles in Purgatory, pained by the aching teeth and sore eyes of malnutrition, may well envy us our luxury. It is easy to remember how, when we were in the terrible complication of childhood, we longed for easy and uncomplicated adulthood. Then we would have only to reach into our pockets for money, then all problems would be ironed out. The ranch-owner had said, "There is no poverty in your country and no misery. Everyone has a Ford."

* * *

We arrived early at Guaymas, passed the usual tests of customs, got our mail at the consulate, and then did the various things of the port. Some of those things are amusing, but they are out of drawing for this account. Guaymas was already in the pathway of the good highway; it was no longer "local." At La Paz and Loreto the Gulf and the town were one, inextricably bound together, but here at Guaymas the railroad and the hotel had broken open that relationship. There were gimcracks for tourists everywhere. This is no criticism of the change, but Guaymas seems to us to be outside the boundaries of the Gulf. We had good treatment there, met charming people, did good and bad things, and left with reluctance.

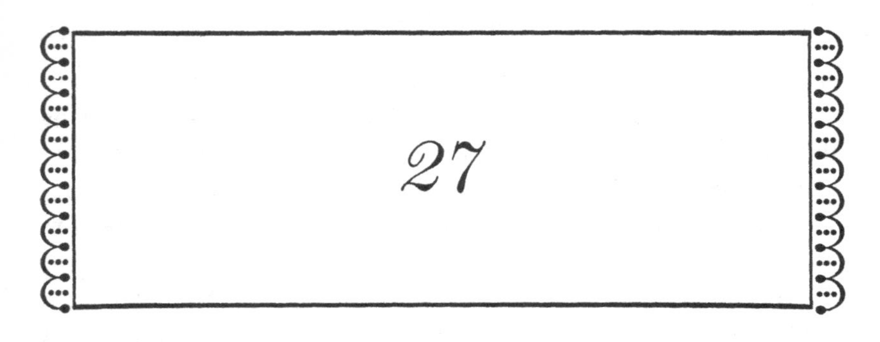

27

April 8

WE SAILED out on Monday, a little tattered and a little tired. Captain Corona, pilot and shrimp-boat owner, who had been kind and hospitable to us, piloted us out and stopped one of his incoming boats for us to inspect. It was a poor small boat, and had not much of a catch of shrimps. Everyone in this neighborhood had complained of the Japanese shrimpers who were destroying the shrimp fisheries. We determined to pay them a visit on the next day. The moment we dropped the pilot, just outside of Guaymas, the Gulf was local again and part of the design it had put in our heads. The mirage was over the land and the sea was very blue. We sailed only a short distance and dropped our anchor in a little cove opposite the Pajaro Island light. That night we caught a number of fish that looked and felt like catfish. Tex skinned them and prepared them, and we did not eat them. A little gloom hung over all of us; Sparky and Tiny had fallen in love with Guaymas and planned to go back there and live forever. But Tex and Tony were gloomy and a little homesick.

We were awakened well before daylight by the voices of men paddling out for the day's fishing, and it was with some relief that we pulled up our anchor and started out to continue the work we had come to do. The day was thick with haze, the sun came through it hot and unpleasant, and the water was oily and at the same time choppy. The sticky humidity was on us.

In about an hour we came to the Japanese fishing fleet. There were six ships doing the actual dredging while a large mother ship of at least 10,000 tons stood farther offshore at anchor. The dredge boats themselves were large, 150 to 175 feet, probably about 600 tons. There were twelve boats in the combined fleet including the mother ship, and they were doing a very systematic job, not only of taking every shrimp from the bottom, but every other living thing as well. They cruised slowly along in echelon with overlapping dredges, literally scraping the bottom clean. Any animal which escaped must have been very fast indeed, for not even the sharks got away. Why the Mexican government should have permitted the complete destruction of a valuable food supply is one of those mysteries which have their ramifications possibly back in pockets it is not well to look into.

We wished to go aboard one of the dredge boats. Tony put the *Western Flyer* ahead of one of them, and we dropped the skiff over the side and got into it. It was not a friendly crew that looked at us over the side of the iron dredge boat. We clung to the side, almost swamping the skiff, and passed our letter from the Ministry of Marine aboard. Then we hung on and waited. We could see the Mexican official on the bridge reading our letter. And then suddenly the atmosphere changed to one of extreme friendliness. We were helped aboard and our skiff was tied alongside.

The cutting deck was forward, and the great dredge loads were dumped on this deck. Along one side there was a long cutting table where the shrimps were beheaded and dropped into a chute, whether to be immediately iced or canned, we do not know. But probably they were canned on the mother ship. The dredge was out when we came aboard, but soon the cable drums began to turn, bringing in the heavy purse-dredge. The big scraper closed like a sack as it came up, and finally it deposited many tons of animals on the deck—tons of shrimps, but also tons of fish of many varieties: sierras; pompano of several species; of the sharks, smooth-hounds and hammer-heads; eagle rays and butterfly rays; small tuna; catfish; *puerco*—tons of them. And there were bottom-samples with anemones and grass-like gorgonians. The sea bottom must have been scraped completely clean. The moment the net dropped open and spilled this mass of living things on the deck, the crew of Japanese went to work. Fish were thrown overboard immediately, and only the shrimps kept. The sea was littered with dead fish, and the gulls swarmed about eating them. Nearly all the fish were in a dying condition, and only a few recovered. The waste of this good food supply was appalling, and it was strange that the Japanese, who are usually so saving, should have done it. The shrimps were shoveled into baskets and delivered to the cutting table. Meanwhile the dredge had gone back to work.

With the captain's permission, we picked out several representatives of every fish and animal we saw. A stay of several days on the boat would have been the basis of a great and complete collection of every animal living at this depth. Even going over two dredgeloads gave us many species. The crew, part Mexican and part Japanese, felt so much better about us by now that they brought out their treasures and gave them to us: bright-red sea-horses and brilliant sea-fans and giant shrimps. They presented them to us, the rarities, the curios which had caught their attention.

At intervals a high, chanting cry arose from the side of the ship and was taken up and chanted back from the bridge. From the upper deck a slung cat-walk extended, and on it the leadsman stood, swinging his leadline, bringing it up and swinging it out again. And every time he read the markers he chanted the depth in Japanese in a high falsetto, and his cry was repeated by the helmsman.

We went up on the bridge, and as we passed this leadsman he said, "Hello." We stopped and talked to him a few moments before we realized that that was the only English word he knew. The Japanese captain was formal, but very courteous. He spoke neither Spanish nor English; his business must all have been done through an interpreter. The Mexican fish and game official stationed aboard was a pleasant man, but he said that he had no great information about the animals he was overseeing. The large shrimps were *Penaeus stylirostris,* and one small specimen was *P. californiensis.*

The shrimps inspected all had the ovaries distended and apparently, as with the Canadian *Pandalus,* this shrimp had the male-female succession. That is, all the animals are born male, but all become females on passing a certain age. The fish and game man seemed very eager to know more about his field, and we promised to send him Schmitt's fine volume on *Marine Decapod Crustacea of California* and whatever other publications on shrimps we could find.

We liked the people on this boat very much. They were good men, but they were caught in a large destructive machine, good men doing a bad thing. With their many and large boats, with their industry and efficiency, but most of all with their intense energy, these Japanese will obviously soon clean out the shrimps of the region. And it is not true that a species thus attacked comes back. The disturbed balance often gives a new species ascendancy and destroys forever the old relationship.

In addition to the shrimps, these boats kill and waste many hundred of tons of fish every day, a great deal of which is sorely needed for food. Perhaps the Ministry of Marine had not realized at that time that one of the good and strong food resources of Mexico was being depleted. If it has not already been done, catch limits should be imposed, and it should not be permitted that the region be so intensely combed. Among other things, the careful study of this area should be undertaken so that its potential could be understood and the catch maintained in balance with the supply. Then there might be shrimps available indefinitely. If this is not done, a very short time will see the end of the shrimp industry in Mexico.

We in the United States have done so much to destroy our own resources, our timber, our land, our fishes, that we should be taken as a horrible example and our methods avoided by any government and people enlightened enough to envision a continuing economy. With our own resources we have been prodigal, and our country will not soon lose the scars of our grasping stupidity. But here, with the shrimp industry, we see a conflict of nations, of ideologies, and of organisms. The units of the organisms are good people. Perhaps we might find a parallel in a moving-picture company such as Metro-Goldwyn-Mayer. The units are superb—great craftsmen, fine directors, the best actors in the profession—and yet due to some overlying expediency, some impure or decaying quality, the product of these good units is sometimes vicious, sometimes stupid, sometimes inept, and never as good as the men who make it. The Mexican official and the Japanese captain were both good men, but by their association in a project directed honestly or dishonestly by forces behind and above them, they were committing a true crime against nature and against the immediate welfare of Mexico and the eventual welfare of the whole human species.

The crew helped us back into our skiff, handed our buckets of specimens down to us, and cast us off. And Tony, who had been cruising slowly about, picked us up in the *Western Flyer*. We had taken perhaps a dozen pompano as specimens when hundreds were available. Sparky was speechless with rage that we had brought none back to eat, but we had forgotten that. We set our course southward toward Estero de la Luna—a great inland sea, the borders of which were dotted lines on our maps. Here we expected to find a rich estuary fauna. In the scoop between Cape Arco and Point Lobos there is a fairly shallow sea which makes a deep ground-swell. It was Tiny up forward who noticed the great numbers of manta rays and suggested that we hunt them. They were monsters, sometimes twelve feet between the "wing" tips. We had no proper equipment, but finally we rigged one of the arrow-tipped harpoons on a light line. This harpoon was a five-inch bronze arrow slotted on an iron shaft. After the stroke, the arrow turns sideways in the flesh and the shaft comes out and floats on its wooden handle. The line is fastened to the arrow itself.

The huge rays cruised slowly about, the upturned tips of the wings out of water. Sparky went to the crow's-nest, where he could look down into the water and direct the steersman. A hundred feet from one of the great fish we cut the engine and coasted down on it. It lay still on the water. Tiny poised prettily on the bow. When we were right on it, he drove the harpoon into it. The monster did not flurry, it simply faded for the bottom. The line whistled out to its limit, twanged like a violin string, and parted. A curious excitement ran through the boat. Tex came down and brought out a one-and-one-half-inch hemp line. He ringed this into a new harpoon-head, and again Tiny took up his position, so excited that he had his foot in a bight of the line. Luckily, we noticed this and warned him. We coasted up to another ray—Tiny missed the stroke. Another, and he missed again. The third time, his arrow drove home, the line sang out again, two hundred feet of it. Then it came to the end where it was looped over a bitt, vibrated for a moment, and parted. The breaking strain of this rope was enormous. But we were doing it all wrong and we knew it. A ton and a half of speeding fish is not to be brought up short. We should have thrown a keg overboard with the line and let the fish fight the keg's buoyancy until exhausted. But we were not equipped. Tiny was almost hysterical by this time. Tex brought a three-inch line with an extremely high breaking strain. We had no more arrow harpoons. Tex made his hawser fast to a huge trident spear. When he finished, the assembly was so heavy that one man could hardly lift it. This time, Tex took the harpoon. He did not waste his time with careless strokes; he waited until the bow was right over one of the largest rays, then drove his spear down with all his strength. The heavy hawser ran almost smoking over the rail. Then it came to the bitt and struck with a kind of groaning cry, quivered, and went limp. When we pulled the big harpoon aboard, there was a chunk of flesh on it. Tiny was heart-broken. The wind came up now and so ruffled the water that we could not see the coasting monsters any more. We tried to soothe Tiny.

"What could we do with one if we caught it?" we asked.

Tiny said, "I'd like to pull it up with the boom and hang it right over the hatch."

"But what could you do with it?"

"I'd hang it there," he said, "and I'd have my picture taken with it. They won't believe in Monterey I speared one unless I can prove it."

He mourned for a long time our lack of foresight in failing to bring manta ray equipment. Late into the night Tex worked, making with file and emery stone a new arrow harpoon, but one of great size. This he planned to use with his three-inch hawser. He said the rope would hold fifteen to twenty tons, and this arrow would not pull out. But he never had a chance to prove it; we did not see the rays in numbers any more.

We came to anchorage that night south of Lobos light and about five miles from the entrance to Estero de la Luna. In this shallow water Tony did not like to go closer for fear of stranding. It was a strange and frightening night, and no one knew why. The water was glassy again and the deck soaked with humidity. We had a curious feeling that a stranger was aboard, some presence not seen but felt, a dark-cloaked person who was with us. We were all nervous and irritable and frightened, but we could not find what frightened us. Tex worked on the Sea-Cow and got it to running perfectly, for we wanted to use it on the long run ashore in the morning. We had checked the tides in Guaymas; it was necessary to leave before daylight to get into the estuary for the low tide. In the night, one of us had a nightmare and shouted for help and the rest of us were sleepless. In the darkness of the early morning, only two of us got up. We dressed quietly and got our breakfast. The light on Lobos Point was flashing to the north of us. The decks were soaked with dew. Climbing down to the skiff, one of us fell and wrenched his leg. True to form, the Sea-Cow would not start. We set off rowing toward the barely visible shore, fixing our course by Lobos Light. A little feathery white shape drifted over the water and it was joined by another and another, and very soon a dense white fog covered us. The *Western Flyer* was lost and the shore blotted out. With the last flashing of the Lobos light we tried to judge the direction of the swell to steer by, and then the light was gone and we were cut off in this ominous glassy water. The air turned steel-gray with the dawn, and the fog was so thick that we could not see fifteen feet from the boat. We rowed on, remembering to quarter on the direction of the swells. And then we heard a little vicious hissing as of millions of snakes, and we both said, "It's the *cordonazo*." This is a quick fierce storm which has destroyed more ships than any other. The wind blows so that it clips the water. We were afraid for a moment, very much afraid, for in the fog, the *cordonazo* would drive our little skiff out into the Gulf and swamp it. We could see nothing and the hissing grew louder and had almost reached us.

It seems to be this way in a time of danger. A little chill of terror runs up the spine and a kind of nausea comes into the throat. And then that disappears into a kind of dull "what the hell" feeling. Perhaps this is the working of some mind-to-gland-to-body process. Perhaps some shock therapy takes control. But our fear was past now, and we braced ourselves to steady the boat against the impact of the expected wind. And at that moment the bow of the skiff grounded gently, for it was not wind at all that hissed, but little waves washing strongly over an exposed sand-bar. We climbed out, hauled the boat up, and sat for a moment on the beach. We had been badly frightened, there is no doubt of it. Even the sleepy dullness which follows the adrenal drunkenness was there. And while we sat there, the fog lifted, and in the morning light we could see the *Western Flyer* at anchor offshore, and we had landed only about a quarter of a mile from where we had intended. The sun broke clear now, and true to form when there was neither danger nor much work the Sea-Cow started easily and we rounded the sand-hill entrance of the big estuary. Now that the sun was up, we could see why there were dotted lines on the maps to indicate the borders of the *estero*. It was endless—there were no borders. The mirage shook the horizon and draped it with haze, distorted shapes, twisted mountains, and made even the bushes seem to hang in the air. Until every foot of such a shore is covered and measured, the shape and extent of these estuaries will not be known.

Inside the entrance of the estuary a big canoe was anchored and four Indians were coming ashore from the night's fishing. They were sullen and unsmiling, and they grunted when we spoke to them. In their boat they had great thick mullet-like fish, so large that it took two men to carry each fish. They must have weighed sixty to one hundred pounds each. These Indians carried the fish through an opening in the brush to a camp of which we could see only the smoke rising, and they were definitely unfriendly. It was the first experience of this kind we had had in the Gulf. It wasn't that they didn't like *us*—they didn't seem to like each other.

The tide was going down in the estuary, making a boiling current in the entrance. Biologically, the area seemed fairly sterile. There were numbers of small animals, several species of large snails and a number of small ones. There were burrowing anemones[1] with transparent, almost colorless tentacles spread out on the sand bottom. And there were the flower-like *Cerianthus* in sand-tubes everywhere. On the bottom were millions of minute sand dollars of a new type, brilliant light green and having holes and fairly elongate spines. Farther inside the estuary we took a number of small heart-urchins and a very few larger ones. On the sand bottoms there were large burrows, but dig as we would, we could never find the owners. They were either very quick or very deep, but even under water their burrows were always open and piles of debris lay about the entrances. Some large crustacean, we thought, possibly of the fiddler crab clan.

The commonest animal of all was the enteropneust, an "acorn-tongued" worm presumably about three feet long that we had found at San Lucas Cove and at Angeles Bay. There were hundreds of their sand-castings lying about. We were not convinced that with all our digging we had got the whole animal even once (and the specialist subsequently confirmed our opinion with regret!).

Deep in the estuary we took several large beautifully striped *Tivela*-like clams and a great number of flat pearly clams. There were hundreds of large hermit crabs in various large gastropod shells. We found a single long-armed sand-burrowing brittle-star which turned out to be *Ophiophragmus marginatus,* and our listing of it is the only report on record since Lütken, in Denmark, erected the species from Nicaraguan material nearly a hundred years ago. In the uneasy footing of the sand, every stick and large shell and rock was encrusted with barnacles; even one giant swimming crab carried a load of barnacles on his back.

The wind had been rising a little as we collected, rippling the water. We cruised about in the shallows trying to see the

---

1 *Harenactis.*

bottom. There were great numbers of sand sharks darting about, but the bottom was clean and sterile and not at all as well populated as we would have supposed. The mirage grew more and more crazy. Perhaps these sullen Indians were bewildered in such an uncertain world where nothing half a mile away could maintain its shape or size, where the world floated and trembled and flowed in dream forms. And perhaps the reverse is true. Maybe these Indians dream of a hard sharp dependable world as an opposite of their daily vision.

We had not taken riches in this place. When the tide turned we started back for the *Western Flyer.* Perhaps the Sea-Cow too had been frightened that morning, for it ran steadily. But the tide was so strong that we had to help it with the oars or it would not have been able to hold its own against the current. It took two hours of oars and motor to get back to the *Western Flyer.*

We felt that this had not been a good nor a friendly place. Some quality of evil hung over it and infected us. We were not at all sorry to leave it. Everyone on board was quiet and uneasy until we pulled up the anchor and started south for Agiabampo, which was, we thought, to be our last collecting station.

It was curious about this Estero de la Luna. It had been a bad place—bad feelings, bad dreams, and little accidents. The look and feel of it were bad. It would be interesting to know whether others have found it so. We have thought how places are able to evoke moods, how color and line in a picture may capture and warp us to a pattern the painter intended. If to color and line in accidental juxtaposition there should be added odor and temperature and all these in some jangling relationship, then we might catch from this accident the unease we felt in the *estero.* There is a stretch of coast country below Monterey which affects all sensitive people profoundly, and if they try to describe their feeling they almost invariably do so in musical terms, in the language of symphonic music. And perhaps here the mind and the nerves are true indices of the reality neither segregated nor understood on an intellectual level. Boodin remarks the essential nobility of philosophy and how it has fallen into disrepute. "Somehow," he says, "the laws of thought must be the laws of things if we are going to attempt a science of reality. Thought and things are part of one evolving matrix, and cannot ultimately conflict." [2]

And in a unified-field hypothesis, or in life, which is a unified field of reality, everything is an index of everything else. And the truth of mind and the way mind is must be an index of things, the way things are, however much one may stand against the other as an index of the second or irregular order, rather than as a harmonic or first-order index. These two types of indices may be compared to the two types of waves, for indices are symbols as primitive as waves. The first wave-type is the regular or cosine wave, such as tide or undulations of light or sound or other energy, especially where the output is steady and unmixed. These waves may be progressive—increasing or diminishing—or they can seem to be stationary, although deeply some change or progression may be found in all oscillation. All terms of a series must be influenced by the torsion of the first term and by the torsion of the end, or change, or stoppage of the series. Such waves as these may be predictable as the tide is. The second type, the irregular for the while, such as graphs of rainfall in a given region, falls into means which are the functions of the length of time during which observations have been made. These are unpredictable individually; that is, one cannot say that it will rain or not rain tomorrow, but in ten years one can predict a certain amount of rainfall and the season of it. And to this secondary type mind might be close by hinge and "key-in" indices.

We had had many discussions at the galley table and there

---

[2] *A Realistic Universe*, p. xviii. 1931. Macmillan, New York.

had been many honest attempts to understand each other's thinking. There are several kinds of reception possible. There is the mind which lies in wait with traps for flaws, so set that it may miss, through not grasping it, a soundness. There is a second which is not reception at all, but blind flight because of laziness, or because some pattern is disturbed by the processes of the discussion. The best reception of all is that which is easy and relaxed, which says in effect, "Let me absorb this thing. Let me try to understand it without private barriers. When I have understood what you are saying, only then will I subject it to my own scrutiny and my own criticism." This is the finest of all critical approaches and the rarest.

The smallest and meanest of all is that which, being frightened or outraged by thinking outside or beyond its pattern, revenges itself senselessly; leaps on a misspelled word or a mispronunciation, drags tricky definition in by the scruff of the neck, and, ranging like a small unpleasant dog, rags and tears the structure to shreds. We have known a critic to base a vicious criticism on a misplaced letter in a word, when actually he was venting rage on an idea he hated. These are the suspicious ones, the self-protective ones, living lives of difficult defense, insuring themselves against folly with folly—stubbornly self-protective at too high a cost.

Ideas are not dangerous unless they find seeding place in some earth more profound than the mind. Leaders and would-be leaders are so afraid that the *idea* "communism" or the *idea* "Fascism" may lead to revolt, when actually they are ineffective without the black earth of discontent to grow in. The strike-raddled businessman may lean toward strikeless Fascism, forgetting that it also eliminates him. The rebel may yearn violently for the freedom from capitalist domination expected in a workers' state, and ignore the fact that such a state is free from rebels. In each case the idea is dangerous only when planted in unease and disquietude. But being so planted, growing in such earth, it ceases to be idea and becomes emotion and then religion. Then, as in most things teleologically approached, the wrong end of the animal is attacked. Lucretius, striking at the teleology of his time, was not so far from us. "I shall untangle by what power the steersman nature guides the sun's courses, and the meanderings of the moon, lest we, percase, should fancy that of own free will they circle their perennial courses round, timing their motions for increase of crops and living creatures, or lest we should think they roll along by any plan of gods. For even *those* men who have learned full well that godheads lead a long life free of care, if yet meanwhile they wonder by what plans things can go on (and chiefly yon high things observed o'erhead on the ethereal coasts), again are hurried back unto the fears of old religion and adopt again harsh masters, deemed almighty,—wretched men, unwitting what can be and what cannot, and by what law to each its scope prescribed, its boundary stone that clings so deep in Time." [3]

* * *

In the afternoon we sailed down the coast carefully, for the sand-bars were many and some of them uncharted. It was a shallow sea again, and the blueness of deep water had changed to the gray-green of sand and shallows. Again we saw manta rays, but not on the surface this day, and the hunt had gone out of us. Tex did not even get out his new harpoon. Perhaps the crew were homesick now. They had seen Guaymas, they were bloated with stories, and they wanted to get back to Monterey to tell them. We would stop at no more towns, see no more people. The inland water of Agiabampo was our last stop, and then quickly home. The shore was low and hot and humid, covered with brush and mangroves. The sea was sterile, or populated with sharks and rays. No algae adhered to the sand bot-

---

[3] Lucretius, *On the Nature of Things,* W. E. Leonard translation, Everyman's Library, 1921, p. 190.

tom, and we were sad in this place after the booming life of the other side. We sailed all afternoon and it was evening when we came to anchorage five miles offshore in the safety of deeper water. We would edge in with the leadline in the morning.

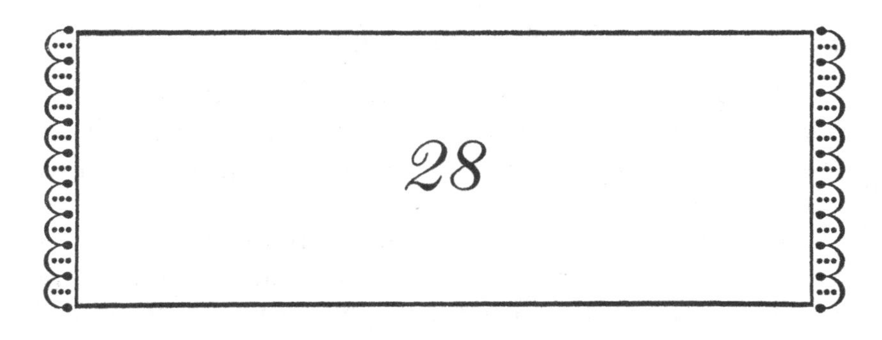

28

April 11

AT TEN O'CLOCK we moved toward the northern side of the entrance of Agiabampo estuary. The sand-bars were already beginning to show with the lowering tide. Tiny used the leadline on the bow while Sparky was again on the crow's-nest where he could watch for the shallow water. Tony would not approach closer than a mile from the entrance, leaving as always a margin of safety.

When we anchored, five of us got into the little skiff, filling it completely. Any rough water would have swamped us. Sparky and Tiny rowed us in, competing violently with each other, which gave a curious twisting course to the boat.

Agiabampo is a great lagoon with a narrow seaward entrance. There is a little town ten miles in on the northern shore which we did not even try to reach. The entrance is intricate and obstructed with many shoals and sand-bars. It would be difficult without local knowledge to bring in a boat of any draft. We moved in around the northern shore; there were dense thickets of mangrove with little river-like entrances winding away into them. We saw great expanses of sand flat and the first extensive growth of eel-grass we had found.[1] But the eel-grass, which ordinarily shelters a great variety of animal life, was here not very rich at all. We saw the depressions where

[1] The true *Zostera marina* according to Dr. Dawson, botanist at the University of California, who remarks that it had not been reported previously so far south.

botete, the poison fish, lay. And there were great numbers of sting-rays, which made us walk very carefully, even in rubber boots, for a slash with the tail-thorn of a sting-ray can easily pierce a boot.

The sand banks near the entrance were deeply cut by currents. High in the intertidal many grapsoid crabs [2] lived in slanting burrows about eighteen inches deep. There were a great many of the huge stalk-eyed conchs and the inevitable big hermit crabs living in the cast-off conch shells. Farther in, there were numbers of *Chione* and the blue-clawed swimming crabs. They seemed even cleverer and fiercer here than at other places. Some of the eel-grass was sexually mature, and we took it for identification. On this grass there were clusters of snail eggs, but we saw none of the snails that had laid them. We found one scale-worm,[3] a magnificent specimen in a *Cerianthus*-like tube. There were great numbers of tube-worms in the sand. The wind was light or absent while we collected, and we could see the bottom everywhere. On the exposed sand-bars birds were feeding in multitudes, possibly on the tube-worms. Along the shore, oyster-catchers hunted the burrowing crabs, diving at them as they sat at the entrances of their houses. It was not a difficult collecting station; the pattern, except for the eel-grass, was by now familiar to us although undoubtedly there were many things we did not see. Perhaps our eyes were tired with too much looking.

As soon as the tide began its strong ebb we got into the skiff and started back to the *Western Flyer.* Collecting in narrow-mouthed estuaries, we are always wrong with the currents, for we come in against an ebbing tide and we go out against the flow. It was heavy work to defeat this current. The Sea-Cow gave us a hand and we rowed strenuously to get outside.

That night we intended to run across the Gulf and start for

---

[2] *Ocypode occidentalis.*

[3] *Polyodontes oculea.*

home. It was good to be running at night again, easier to sleep with the engine beating. Tiny at the wheel inveighed against the waste of fish by the Japanese. To him it was a waste complete, a loss of something. We discussed the widening and narrowing picture. To Tiny the fisherman, having as his function not only the catching of fish but the presumption that they would be eaten by humans, the Japanese were wasteful. And in that picture he was very correct. But all the fish actually were eaten; if any small parts were missed by the birds they were taken by the detritus-eaters, the worms and cucumbers. And what they missed was reduced by the bacteria. What was the fisherman's loss was a gain to another group. We tried to say that in the macrocosm nothing is wasted, the equation always balances. The elements which the fish elaborated into an individuated physical organism, a microcosm, go back again into the undifferentiated macrocosm which is the great reservoir. There is not, nor can there be, any actual waste, but simply varying forms of energy. To each group, of course, there must be waste—the dead fish to man, the broken pieces to gulls, the bones to some and the scales to others—but to the whole, there is no waste. The great organism, Life, takes it all and uses it all. The large picture is always clear and the smaller can be clear—the picture of eater and eaten. And the large equilibrium of the life of a given animal is postulated on the presence of abundant larvae of just such forms as itself for food. Nothing is wasted; "no star is lost."

And in a sense there is no over-production, since every living thing has its niche, *a posteriori,* and God, in a real, non-mystical sense, sees every sparrow fall and every cell utilized. What is called "over-production" even among us in our manufacture of articles is only over-production in terms of a status quo, but in the history of the organism, it may well be a factor or a function in some great pattern of change or repetition. Perhaps some cells, even intellectual ones, must be sickened before others can be well. And perhaps with us these production climaxes are the therapeutic fevers which cause a rush of curative blood to the sickened part. Our history is as much a product of torsion and stress as it is of unilinear drive. It is amusing that at any given point of time we haven't the slightest idea of what is happening to us. The present wars and ideological changes of nervousness and fighting seem to have direction, but in a hundred years it is more than possible it will be seen that the direction was quite different from the one we supposed. The limitation of the seeing point in time, as well as in space, is a warping lens.

Among men, it seems, historically at any rate, that processes of co-ordination and disintegration follow each other with great regularity, and the index of the co-ordination is the measure of the disintegration which follows. There is no mob like a group of well-drilled soldiers when they have thrown off their discipline. And there is no lostness like that which comes to a man when a perfect and certain pattern has dissolved about him. There is no hater like one who has greatly loved.

We think these historical waves may be plotted and the harmonic curves of human group conduct observed. Perhaps out of such observation a knowledge of the function of war and destruction might emerge. Little enough is known about the function of individual pain and suffering, although from its profound organization it is suspected of being necessary as a survival mechanism. And nothing whatever is known of the group pains of the species, although it is not unreasonable to suppose that they too are somehow functions of the surviving species. It is too bad that against even such investigation we build up a hysterical and sentimental barrier. Why do we so dread to think of our species as a species? Can it be that we are afraid of what we may find? That human self-love would suffer too much and that the image of God might prove to be a mask? This could be only partly true, for if we could cease to wear the image of a kindly, bearded, interstellar dictator, we might find ourselves true images of his kingdom, our eyes the nebulae, and universes in our cells.

The safety-valve of all speculation is: *It might be so.* And as long as that *might* remains, a variable deeply understood, then speculation does not easily become dogma, but remains the fluid creative thing it might be. Thus, a valid painter, letting color and line, observed, sift into his eyes, up the nerve trunks, and mix well with his experience before it flows down his hand to the canvas, has made his painting say, "It might be so." Perhaps his critic, being not so honest and not so wise, will say, "It is not so. The picture is damned." If this critic could say, "It is not so with me, but that might be because my mind and experience are not identical with those of the painter," that critic would be the better critic for it, just as that painter is a better painter for knowing he himself is in the pigment.

We tried always to understand that the reality we observed was partly us; the speculation, our product. And yet if somehow, "The laws of thought must be the laws of things," one can find an index of reality even in insanity.

* * *

We sailed a compass course in the night and before daylight a deep fog settled on us. Tony stopped the engine and let us drift, and the dawn came with the thick fog still about us. Tiny and Sparky had the watch, and as the dawn broke, they heard surf and reported it. We came out of our bunks and went up on the deckhouse just as the fog lifted. There was an island half a mile away. Then Tony said, "Did you keep the course I gave you?" Tiny insisted that they had, and Tony said, "If that is so, you have discovered an island, and a big one, because the chart shows no island here." He went on delicately, "I want to congratulate you. We'll call it 'Colletto and Enea Island.' " Tony continued silkily, "But you know Goddamn well you didn't keep the course. You know you forgot, and are a good many miles off course." Sparky and Tiny did not argue. They never claimed the island, nor mentioned it again. It developed that it was Espíritu Santo Island, and would have been a prize if they had discovered it, but some Spaniards had done that several hundred years before.

San Gabriel Bay was near us, its coral sand dazzlingly white, and a good reef projecting and a mangrove swamp along part of the coast. We went ashore for this last collecting station. The sand was so white and the water so clear that we took off our clothes and plunged about. The animals here had been affected by the white sand. The crabs were pale and nearly white, and all the animals, even the starfish, were strangely colored. There were stretches of this blinding sand alternating with bouldery reef and mangrove. In the center of the little bay, a fine big patch of green coral almost emerged from the water. It was green and brown coral in great heads, and there were *Phataria* and many club-spined urchins on the heads. There were multitudes of the clam *Chione* just under the surface of the sand, very hard to find until we discovered that every clam had a tiny veil of pale-green algae growing on the front of each valve and sticking up above the sand. Then we took a great number of them.

Near the beds of clams lived heart-urchins with vicious spines.[4] These too were buried in the sand, and to dig for the clams was to be stabbed by the heart urchins, and to be stung badly. There were many hachas here with their clustered colonies of associated fauna. We found solitary and clustered zoanthidean anemones, possibly the same we had been seeing in many variations. We found light-colored *Callinectes* crabs and one of the long snake-like sea-cucumbers [5] such as the ones we

---

[4] *Lovenia cordiformis.*
[5] *Euapta godeffroyi.*

had taken at Puerto Escondido. On the rocky reef there were anemones, limpets, and many barnacles. The most common animal on the reef was a membranous tube-worm [6] with tentacles like a serpulid's. These tentacles were purple and brown, but when approached they were withdrawn and the animal became sand-colored. The mangrove region here was rich. The roots of the trees, impacted with rocks, maintained a fine group of crabs and cucumbers. Two large, hairy grapsoid crabs [7] lived highest in the littoral. They were very fast and active and difficult to catch, and when caught, battled fiercely and ended up by autotomizing.

There was also a *Panopeus*-like crab, *Xanthodius hebes,* but dopey and slow. We found great numbers of porcelain crabs and snapping shrimps. There were barnacles on the reef and on the roots of the mangroves; two new ophiurans and a large sea-hare, besides a miscellany of snails and clams. It was a rich haul, this last day. The sun was hot and the sand pleasant and we were comfortable except for mosquito bites. We played in the water a long time when we were tired of collecting.

When once the engine started now, it would not stop until we reached San Diego. We were reluctant to go back. This balance in time is one of the very few occasions when we have the right of "yes" and "no," and even now the cards were stacked against "yes."

At last we picked up the collecting buckets and the little crowbars and all the tubes, and we rowed slowly back to the *Western Flyer*. Even then, we had difficulty in starting. Someone was overboard swimming in the beautiful water all the time. Tony and Tex, who had been eager to get home, were reluctant now that it was upon them. We had all felt the pattern of the Gulf, and we and the Gulf had established another pat-

---

[6] *Megalomma mushaensis.*
[7] *Geograpsus* and *Goniopsis.*

tern which was a new thing composed of it and us. At last, and with sorrow, Tex started the engine and the anchor came up for the last time.

All afternoon we stowed and lashed equipment, set the corks in hundreds of glass tubes and wrapped them in paper toweling, screwed tight the caps of jars, tied down the skiffs, and finally dropped the hatch cover in place. We covered the bookcase with triple tarpaulin, and one last time overcame the impulse to throw the Sea-Cow overboard. Then we were under way, sailing southward toward the Cape. The swordfish jumped in the afternoon light, flashing like heliographs in the distance. We took back our old watches that night, and the engine drummed happily and drove us through a calm sea. In the morning the tip of the Peninsula was on our right. Behind us the Gulf was sunny and calm, but out in the Pacific a heavy threatening line of clouds hung.

Then a crazy literary thing happened. As we came opposite the Point there was one great clap of thunder, and immediately we hit the great swells of the Pacific and the wind freshened against us. The water took on a gray tone.

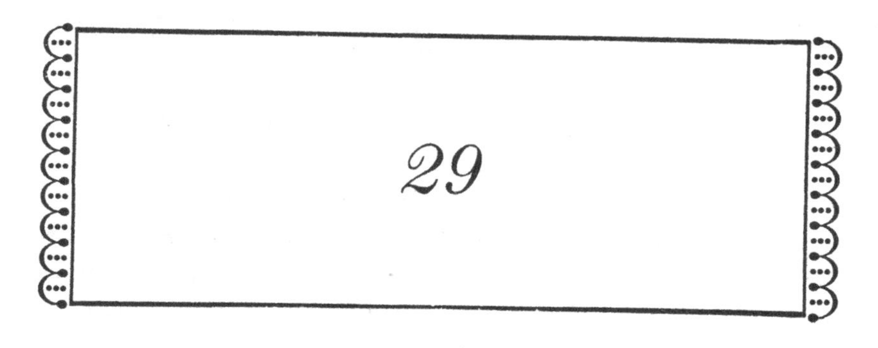

29

April 13

AT THREE A.M. Pacific time we passed the light on the false cape and made our new course northward, and the sky was gray and threatening and the wind increased. The Gulf was blotted out for us—the Gulf that was thought and work and sunshine and play. This new world of the Pacific took hold of us and we thought again of an unseen person on the deck-house, some kind of symbol person—to a sailor, a ghost, a pre-monition, a feeling in human form.

We could not yet relate the microcosm of the Gulf with the macrocosm of the sea. As we went northward the gray waves rolled up and the *Western Flyer* stubbed her nose into them and the white spray flew over us. The day passed and a new night came and the sea grew more stern. Now we plunged like a nervous horse, and no step could be taken without a steadying hand. The galley was in confusion, for a can of olive oil had leaped from its stand and flooded the floor. On the stove, the coffee pot slipped back and forth between its bars.

Over the surface of the heaving sea the birds flew landward, zigzagging to cover themselves in the wave troughs from the wind. The man at the wheel was the lucky one, for he had a grip against the pitching. He was closest to the boat and to the rising storm. He was the receiver, but also he was the giver and his hand was on the course.

What was the shape and size and color and tone of this little expedition? We slipped into a new frame and grew to be a part of it, related in some subtle way to the reefs and beaches, related to the little animals, to the stirring waters and the warm brackish lagoons. This trip had dimension and tone. It was a thing whose boundaries seeped through itself and beyond into some time and space that was more than all the Gulf and more than all our lives. Our fingers turned over the stones and we saw life that was like our life.

On the deckhouse we held the rails for support, and the blunt nose of the boat fought into the waves and the gray-green water struck us in the face. Some creative thing had happened, a real tempest in our small teapot minds. But boiling water still produces steam, whether in a watch-glass or in a turbine. It is the same stuff—weak and dissipating or explosive, depending on its use. The shape of the trip was an integrated nucleus from which weak strings of thought stretched into every reachable reality, and a reality which reached into us through our perceptive nerve trunks. The laws of thought seemed really one with the laws of things. There was some quality of music here, perhaps not to be communicated, but sounding clear and huge in our minds. The boat plunged and shook herself, and rivers of swirling water ran down the scuppers. Below in the hold, packed in jars, were thousands of little dead animals, but we did not think of them as trophies, as things cut off from the tide pools of the Gulf, but rather as drawings, incomplete and imperfect, of how it had been there. The real picture of how it had been there and how we had been there was in our minds, bright with sun and wet with sea water and blue or burned, and the whole crusted over with exploring thought. Here was no service to science, no naming of unknown animals, but rather —we simply liked it. We liked it very much. The brown Indians and the gardens of the sea, and the beer and the work, they were all one thing and we were that one thing too.

The *Western Flyer* hunched into the great waves toward

Cedros Island, the wind blew off the tops of the whitecaps, and the big guy wire, from bow to mast, took up its vibration like the low pipe on a tremendous organ. It sang its deep note into the wind.

A NOTE ON PREPARING SPECIMENS

Following are the methods of preparing, anesthetizing, and preserving specimens which we have followed. As previously remarked, some are by no means entirely satisfactory, but they are the best we know.

The chitons, which otherwise curl up tightly during preservation, can be removed directly from rock or collecting bucket and tied firmly to glass plates with string or strips of cloth. When they grip the plate, they are dropped into a container of ten-percent formalin; whereupon they are preserved relaxed.

Sipunculids (or peanut worms), to be preserved with the introvert extended, must first be anesthetized by sprinkling powdered menthol crystals on the sea water in a scrupulously clean glass dish. These worms are so delicate that metal or metal salts will cause them to draw in their heads immediately. The more refractory forms, after a few hours' treatment with menthol, are killed by letting fresh water gradually replace the sea water. This must be done over a period of several hours. They are then preserved in formalin solution.

Alcyonaria, such as *Renilla* (the sea-pansy), *Stylatula* (the sea-pen), etc., are treated with Epsom salts for several hours until the expanded polyps are completely inert, then preserved in formalin if they are intended for display where color is important (although even thus the color is not successfully retained) or in alcohol if for study or identification.

Hydroids are relaxed with menthol for the small forms, with Epsom salts for the large gymnoblasts.

Anemones are very difficult to handle, and a perfect method for expanding them has not been developed. They are sensitive both chemically and physically. A specimen expanded and relaxed with Epsom salts will still draw in its tentacles if touched. Some results have been achieved by introducing pure oxygen directly into the stomach with a tiny hollow glass needle—a process which seems to intoxicate them. But it is at best a delicate and over-arduous operation. A fairly good method is gradually to introduce a saturated solution of Epsom salts into the pan where the animals are expanded, using a drip-string. This is followed later with novocain, or best of all, but usually unavailable, with cocaine, which is put into the water directly over the animal. Finally, formalin is introduced with a drip-string. Any shock, either chemical or physical, will cause immediate retraction of the tentacles. For permanent use a running seawater system which supplies O_2 to the circulation of the animals being narcotized has proven most effective of all.

Crabs and shrimps of most types, and brittle-stars, are best killed by dropping them into fresh water as soon after they are captured as possible. After a brief struggle the crabs die without casting their legs, and the brittle-stars (with the exception of *Ophioderma teres*) will not curl up if this method is used. All these should subsequently be preserved in alcohol, since formalin will disintegrate the calcareous portions of the brittlestars and soften the chitin of the crabs. For color notes only, specimens may be preserved in formalin.

Holothurians (or sea-cucumbers) are very delicate and have a tendency to eviscerate if kept in stale or over-warm water. They should be got as soon as possible into trays of cool, clean sea water, allowed to expand, then relaxed with Epsom salts used in considerable quantity. They should be preserved in alcohol. Most holothurians deteriorate badly in a few months if formalin is used on them, owing to the dissolution of the calcareous plates by which specialists determine them. Their neurotic tendency of deliberately casting out their viscera when they are sick or shocked or unhappy makes them difficult to handle. The cucumber itself, if times get better and a pleasant environment is restored, is able to grow a new set.

Starfish relax and puff out in a life-like manner when placed for several hours in fresh water. They can ordinarily be preserved in formalin, but museum specimens for determination should be placed in alcohol for a time and dried in the air. This method also applies to sea-urchins.

Sponges must not be placed in formalin even for a few moments. Entire colonies can be dried after a preliminary immersion in alcohol; small portions of the colonies so treated ought, however, to be preserved permanently in a vial of alcohol.

Flatworms such as the turbellarians are hard to collect, hard to handle, and hard to preserve. They are so delicate that the bodies are easily injured in picking them up. When they are crawling on a rock it is satisfactory to place a thin-bladed knife in their path; when the flatworm oozes onto the knife-blade he can be lifted into a container. Very small specimens may be lifted from the rock with a camel's-hair brush and transported to a glass plate. This is flooded with hot Bouin's solution and immediately covered with another glass plate. If one could devise smooth plates permeable by Bouin's solution it would be even better.

Several methods are applicable to the preservation of such pelagic invertebrates as jellyfish. Menthol crystals are a successful anesthetic for the contractile forms. However, the difficulties are likely to relate to the extreme softness of the more delicate forms. Hardening solutions of chromic acid and formalin, or osmic acid (which, however, is very expensive—five to six dollars a gram) in formalin, are most effective. Often there are mechanical difficulties involved in attempts to retain the original shapes of animals. The jellyfish *Beroë* tends to buckle up. We have had some success by inserting the closed end of a test-tube in the body cavity during the hardening process. It is really impossible to preserve jellyfish aboard a boat that is rolling even slightly. Formalin is usually a satisfactory preservative, and is often indicated exclusively.

It seems from these notes that all animals are difficult to preserve, and it is true that all of them require care which is not often enough given. The almost universal cry of specialists engaged in species determination is that specimens arrive in such bad condition that their work is made doubly difficult. Only extreme care can rectify this.

Some of the annelid worms are extremely difficult to handle. Anesthetizing methods are useful, but the delicacy of the animals and their constricting traits make special procedure necessary. Very long specimens may be wrapped around glass rods or test-tubes and suspended in formalin or alcohol, depending on the species. The chief difficulty with worms is to get the entire animal in the first place, especially the extreme anterior and posterior segments without which identification is difficult if not impossible. Some of the worms sting very badly and should not be handled at all.

For the preservation of fishes, formalin in general gives better results than alcohol. Small specimens are hardened by putting them directly into trays of formalin solution; while larger fishes must have the body cavities injected with a twenty- to twenty-five-percent solution of formalin to which a little glycerin has been added.

Formalin should be used as follows: U.S.P. formaldehyde solution of thirty-eight to forty percent is a gas dissolved in water, and this percentage represents saturation. One part of this to sixteen parts of water for small specimens is successful when the amount of the solution is many times the bulk of the animal. Large specimens are hardened in trays of ten- to fifteen-percent solution. One must check by experiment on the particular animal involved, feeling it after a few hours to note the consistency of the tissue. Formaldehyde is very irritating to nose, lungs, and skin. Rubber gloves should always be used and the solution worked with in large rooms or in the open air. The tolerance of a person working with formaldehyde sometimes decreases with time, so that he is sickened at the odor or even breaks out with allergy eruptions on contact or association.

With alcohol the ultimate preservation is usually in seventy-percent strength. For crabs and so forth in the tropics or in very hot weather, glycerin should be added. This keeps the animals flexible and less brittle—less likely to break up—and also prevents poor preservation due to bubbles forming in the solution. For ideal preservation the specimen should be brought immediately after killing into a twenty-five-percent solution, then to fifty percent and finally to seventy percent. Large specimens, where the amount of liquid is small in proportion to the amount of tissue involved, may require ninety-percent solution for a few days of hardening before they are placed in a new seventy-percent solution.

Labeling is easy, simple, and necessary. Yet the failure to label clearly and immediately has led to many ridiculous situations. One expedition, which need not be named, labeled Atlantic animals as coming from the Panamic regions. And another completely lost track of its collection, to the disgust of the specialists who later tried to determine the species. Labels are best made on slips of good drawing paper and printed with a drawing pen in India ink. Each label should include the date, the exact place, the depth, and a number added which will agree with the number in the collecting notes. In the collecting notes, under this number, should occur any remarks covering ecological factors or observed action of the living animal which would be impossible to put on the label. The label should be placed *inside* the jar with the animal, and it should be done immediately, before a new lot of specimens comes in. There has not, to our knowledge, been any single expedition or extended trip which failed to turn out some unlabeled, or mislabeled material, so that the records are full of obviously incorrect reports. Some Panamic animals have been reported from Puget Sound, and our common California shore crab, *Pachygrapsus,* was originally described as from the Sandwich Islands. Immediate labeling, on the same day as the collection, is the only way to reduce these errors to a minimum. This cannot be over-emphasized.

Fig. 1. Phyllonotus bicolor (Valenciennes) 1852 § S-358
Pink Murex

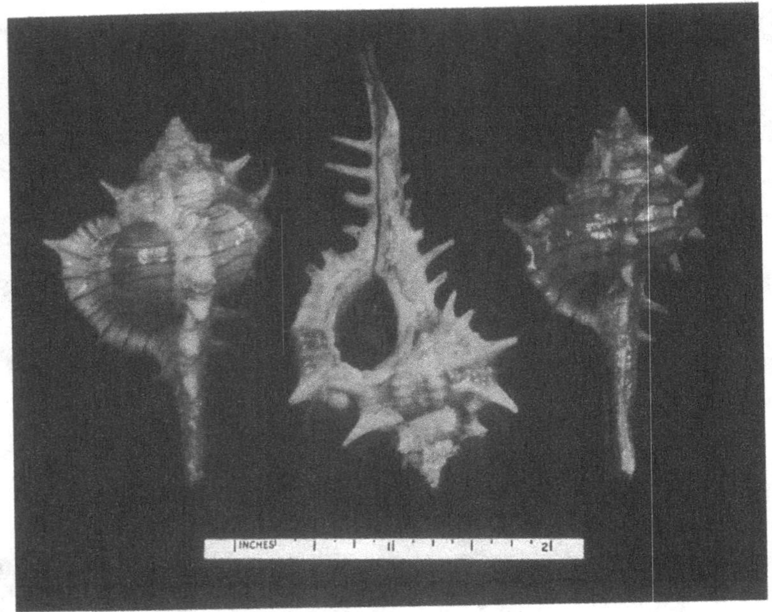

Fig. 1. Murex rectirostris Sowerby. Gulf spiny Murex § S-355

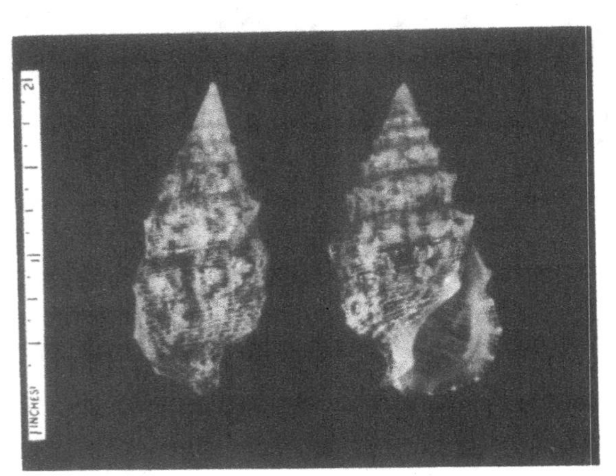

Fig. 2. Cerithium maculosum Kierner § S-374
The Spotted Cerithium

PLATE 2

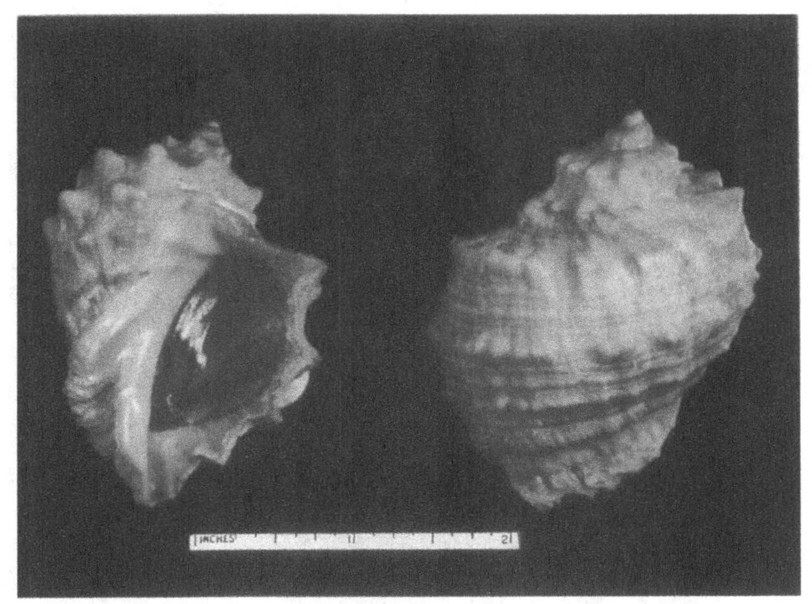

Fig. 1. Thais tuberculata Gray. Knobby Thais § S-367

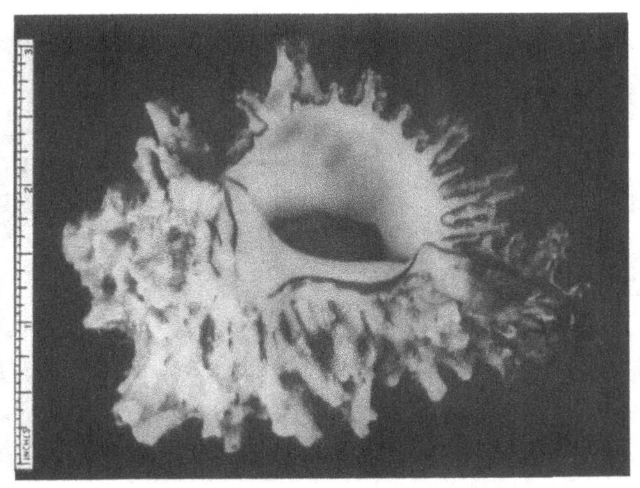

Fig. 2. Phyllonotus princeps (Broderip) § S-361
Regal Murex

Fig. 1. Protothaca grata (Sowerby) § S-252
Varicolored Edible Clam

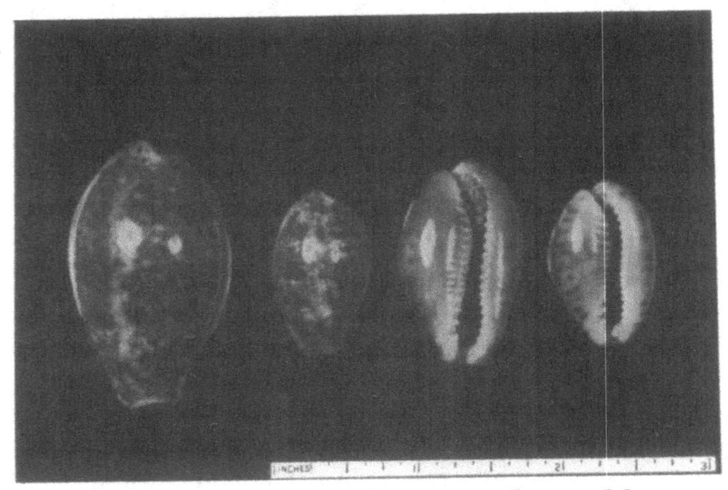

Fig. 2. Cypraea annettae Dall 1909. Brown Cowry § S-370

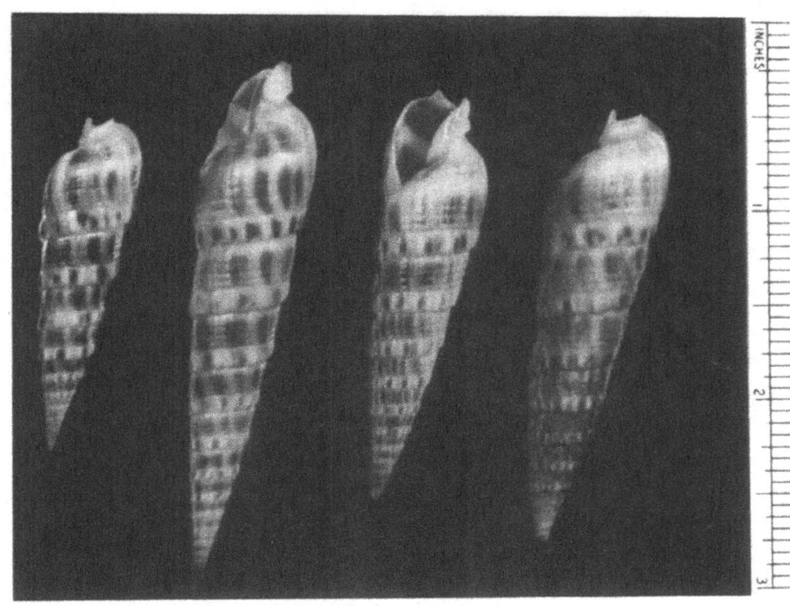

Fig. 1. Terebra variegata Gray 1834 § S-326
Variegated Augur-shell

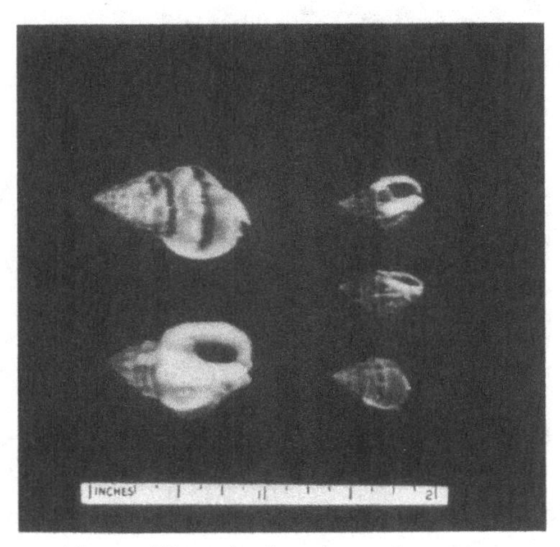

Fig. 2. Nassarius ioaedes and
luteostoma § S-340 and § S-341

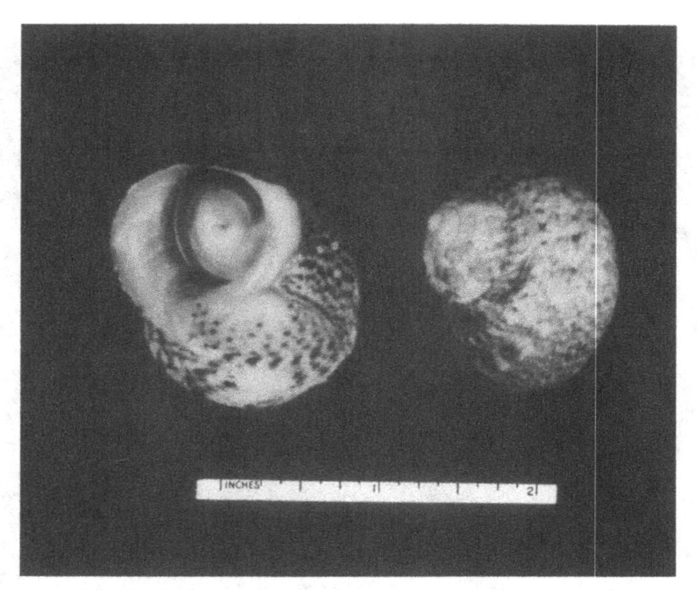

Fig. 1. Callopoma fluctuosum (Wood) 1828 § S-404
Common Gulf Turbine-shell

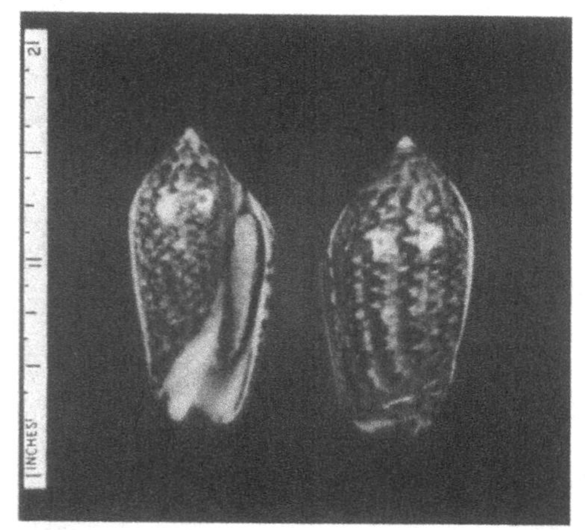

Fig. 2. Oliva venulata Lamarck § S-331
Gulf Olive-shell

Fig. 1. Gulf Olive-shell. Albino color variety § S-331

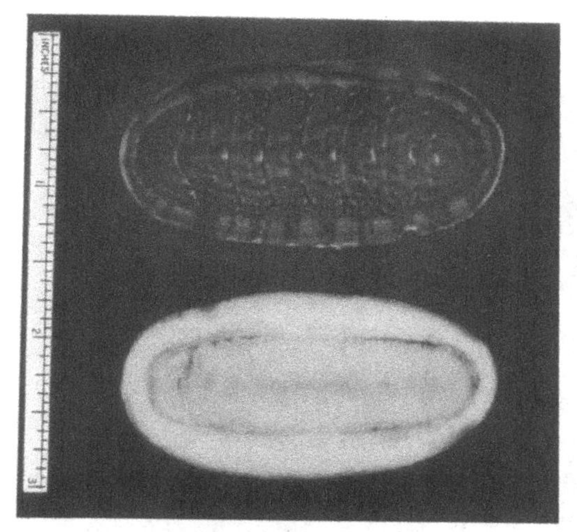

Fig. 2. Chiton virgulatus Sowerby 1840 § U-17
Common Gulf Chiton

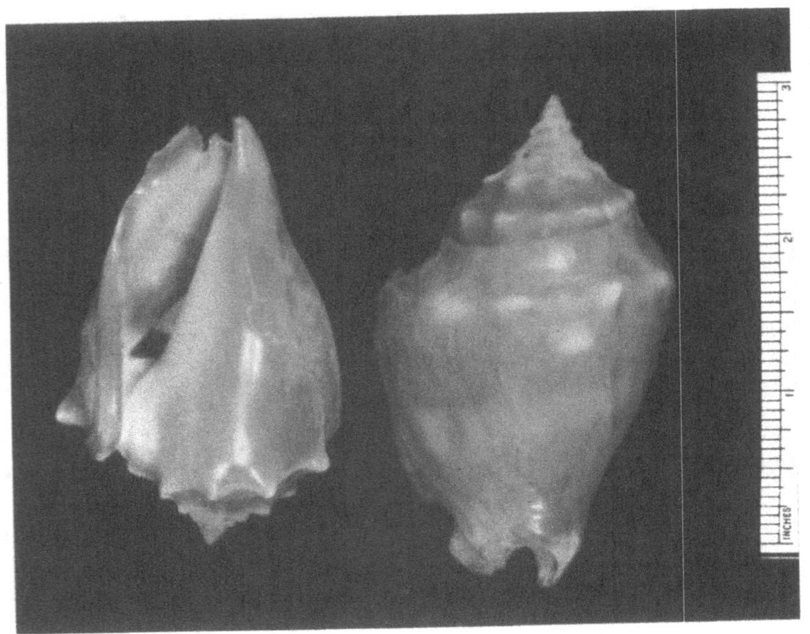

Fig. 1. Strombus gracilior Sowerby 1825 § S-372
Smooth Conch

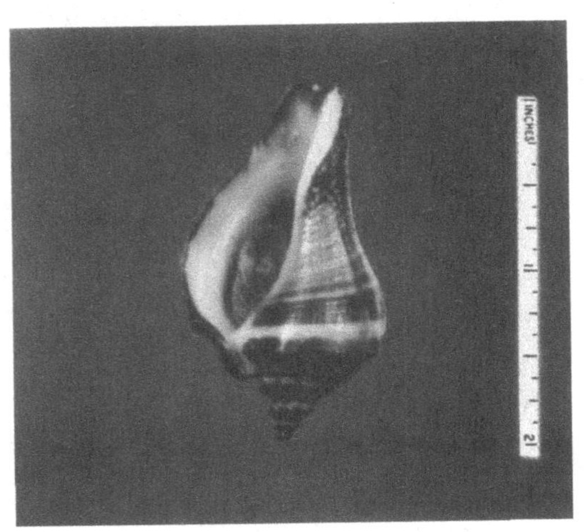

Fig. 2. Melongena patula Broderip and Sowerby § S-338

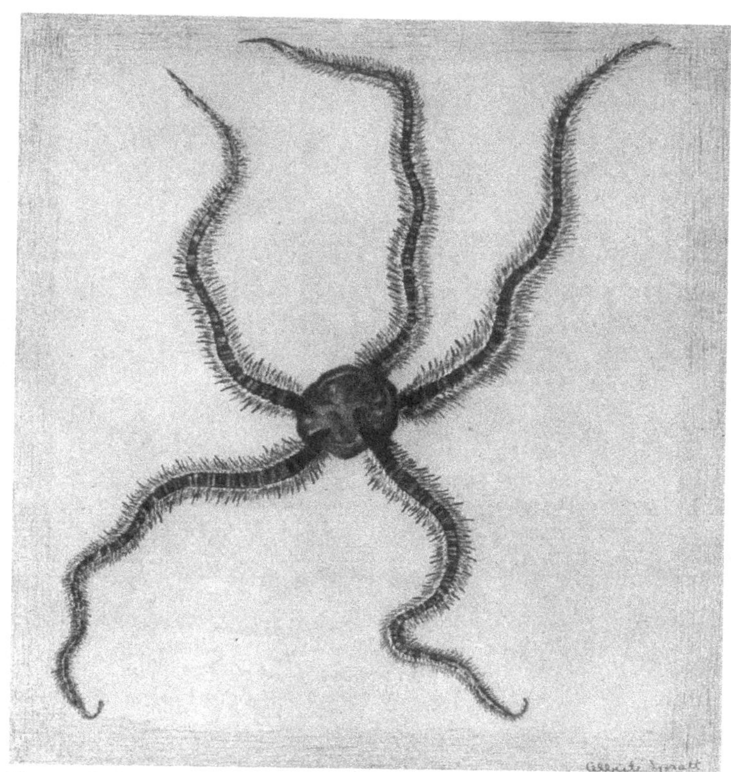

Fig. 1. Ophiocoma alexandri Lyman 1860 **§ K-215**
½ natural size

The drawings on plates 9-24 inclusive are by Alberté Spratt

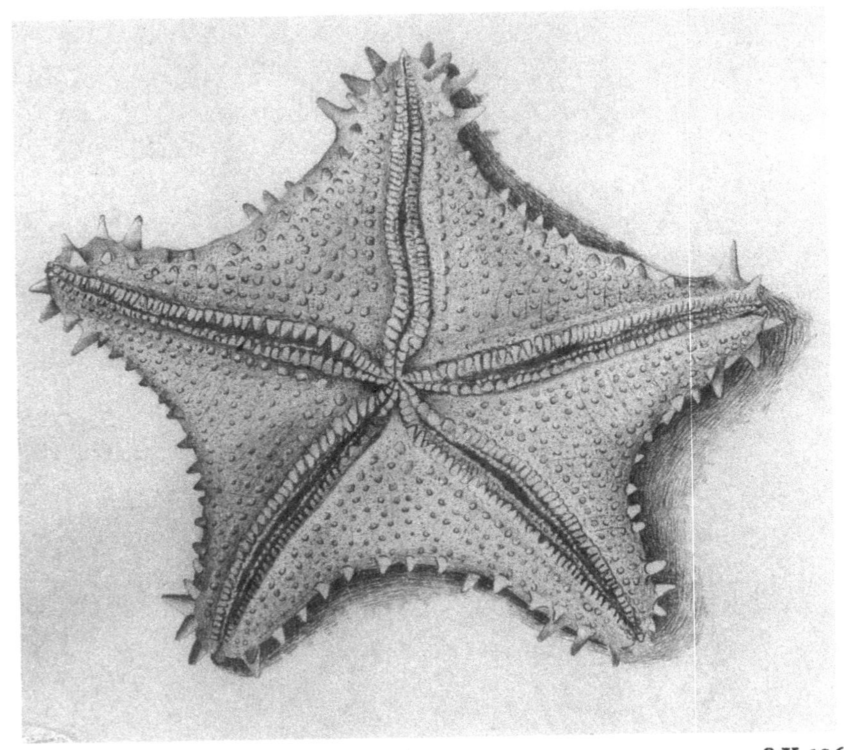

Fig. 1. Nidorellia armata (Gray) 1840 — § K-126
natural size, 5½"-6" — oral surface

Fig. 2. Holothuria impatiens (Forskål) 1775 — § L-21
½ natural size

Fig. 1. Oreaster occidentalis Verrill 1867 **§ K-125**
natural size, 6″

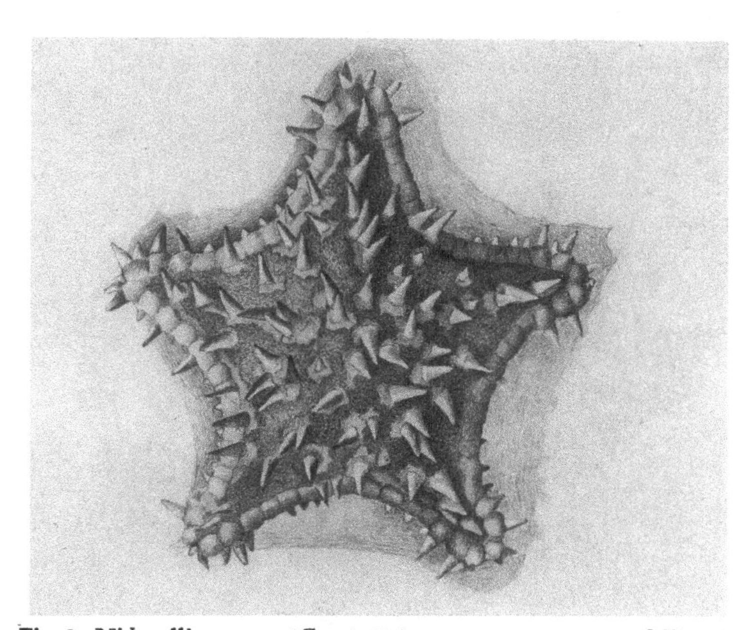

Fig. 2. Nidorellia armata (Gray) 1840 **§ K-126**
natural size, 5½″-6″ **aboral surface**

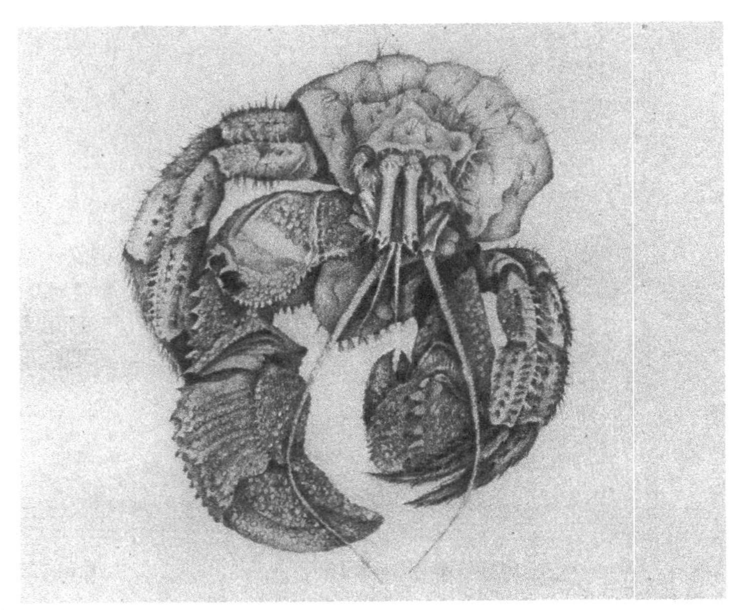

Fig. 1. Petrochirus californiensis Bouvier 1895 **§ Q-16**
¾ natural size

Fig. 2. Tedania ignis (Duchassaing and Michelotti) 1864 **§ A-23**
½ natural size

PLATE 12

Fig. 1. Ophiocoma aethiops Lütken 1859 § K-214
aboral surface. ½ natural size

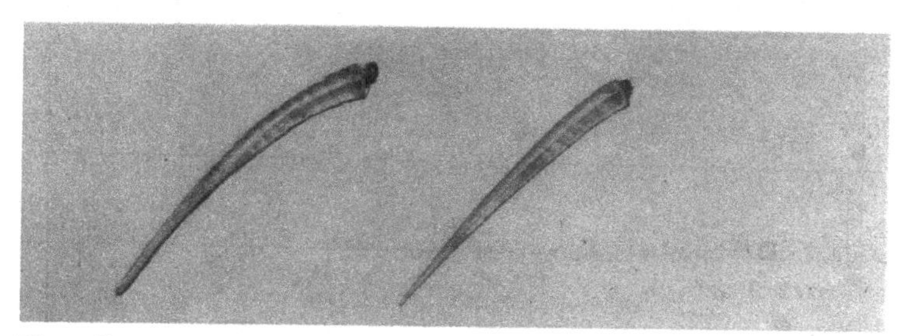

Fig. 2. Dentalium semipolitum Broderip and Sowerby § S-103 and 104
and D. splendidum Sowerby 1832 1¼ natural size

Fig. 1. Ocypode occidentalis Stimpson 1860 § R-75
½ natural size

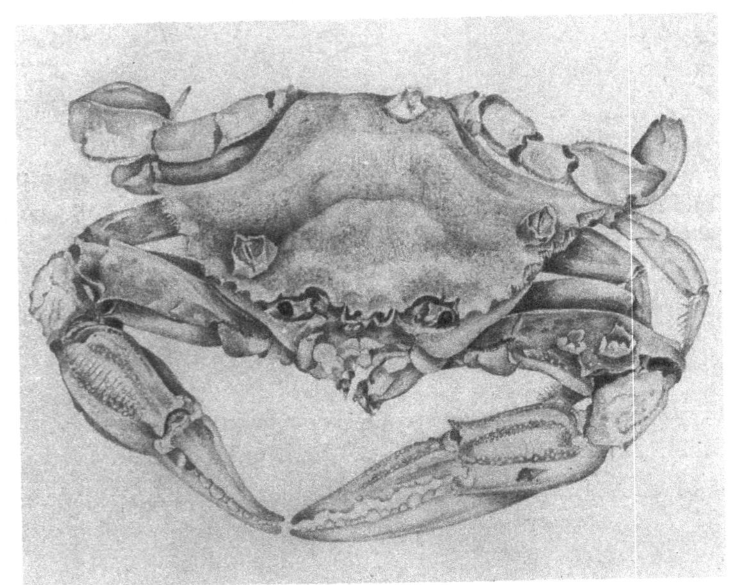

Fig. 2. Callinectes bellicosus Stimpson 1859 § R-42
½ natural size

Fig. 1. Phataria unifascialis (Gray) 1840 § K-120
½ natural size

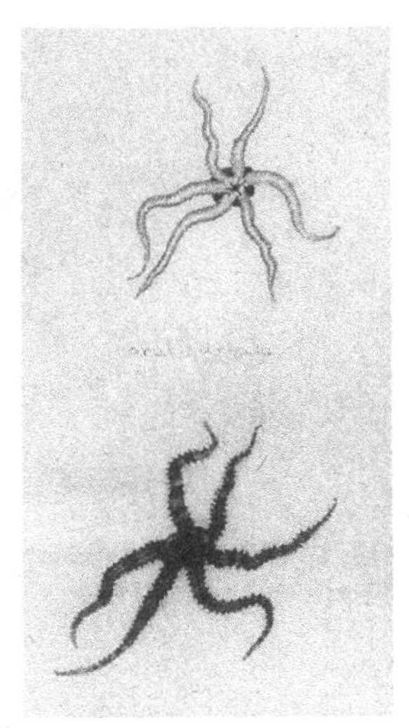

Fig. 2. Ophiactis sp.
1½ natural size
§ K-218 (219?)

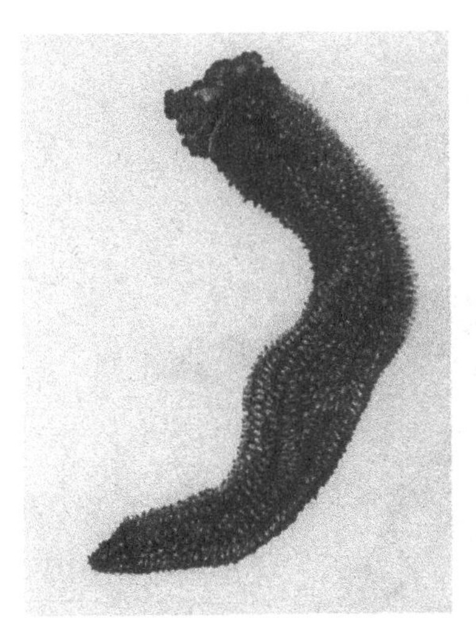

Fig. 3. H. lubrica Selenka 1867
Sulphur cucumber § L-25
½ natural size

Fig. 4. Luidia phragma Clark 1910 § K-128
natural size, 4½″-5″

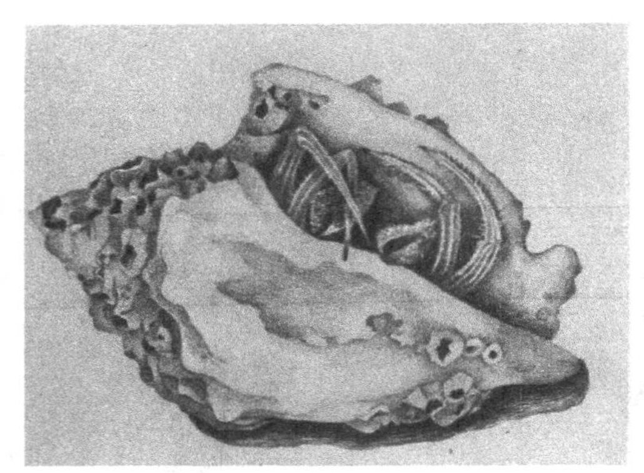

Fig. 1. Clibanarius panamensis Stimpson (1859) 1862 § Q-14 natural size

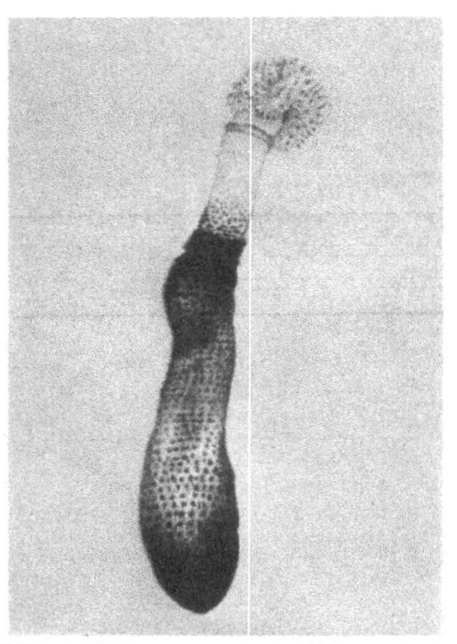

Fig. 2. Phycosoma antillarum (Grube and Örsted) 1859 § I-10 natural size

Fig. 3. Othilia tenuispina (Verrill) 1871 as Echinaster § K-124

PLATE 16

Fig. 1. Clypeaster rotundus (A. Agassiz) 1863 § K-322
½ natural size

Fig. 2. Echinometra vanbrunti A. Agassiz 1863 § K-314
½ natural size

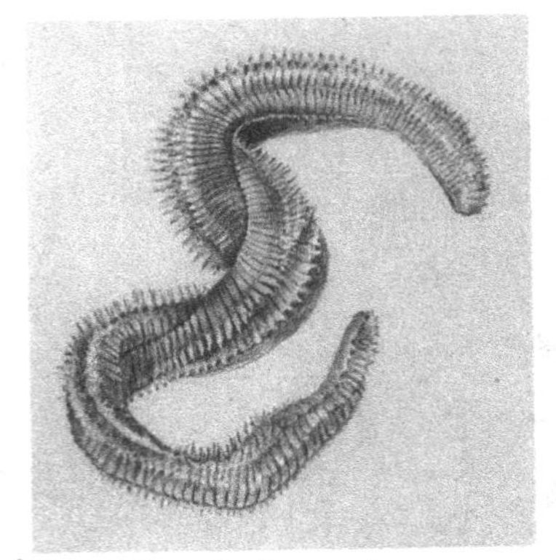

Fig. 1. Eurythoë complanata
(Pallas) 1766 § J-35
⅔ natural size

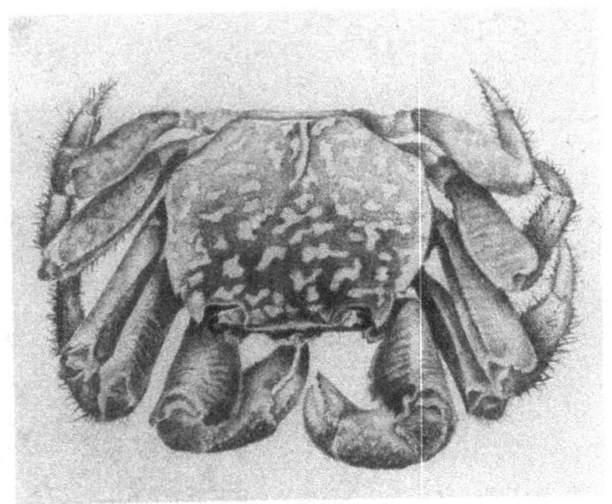

Fig. 2. Geograpsus lividus § R-68
(Milne-Edwards) 1837 ½ natural size

Fig. 3. Portunus minimus § R-40
Rathbun 1898 natural size

Fig. 4. Eriphia squamata § R-58
Stimpson 1859 1½ natural size

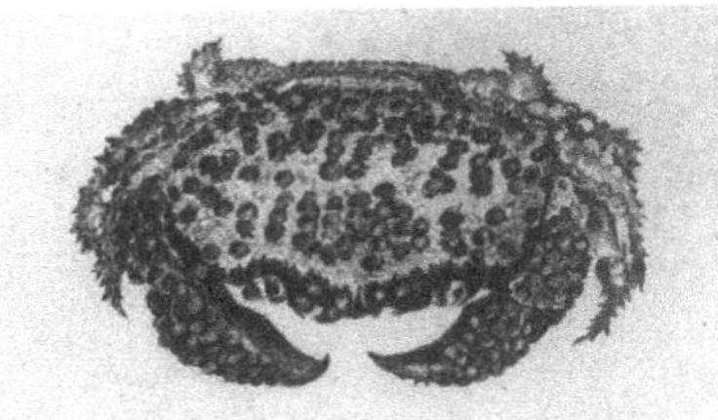

Fig. 5. Daira americana § R-45
Stimpson 1860 natural size

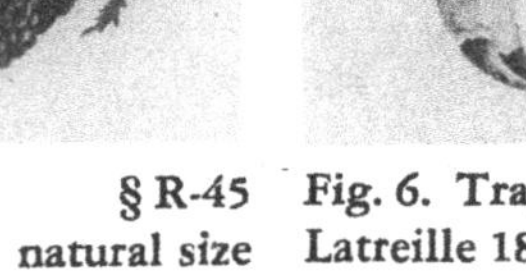

Fig. 6. Trapezia cymodoce ferruginea
Latreille 1825 1½ natural size § R-60

PLATE 18

Fig. 1. H. paraprinceps § L-26
Deichmann 1937 ½ natural size

Fig. 2. Meoma grandis Gray 1852 § K-324
½ natural size

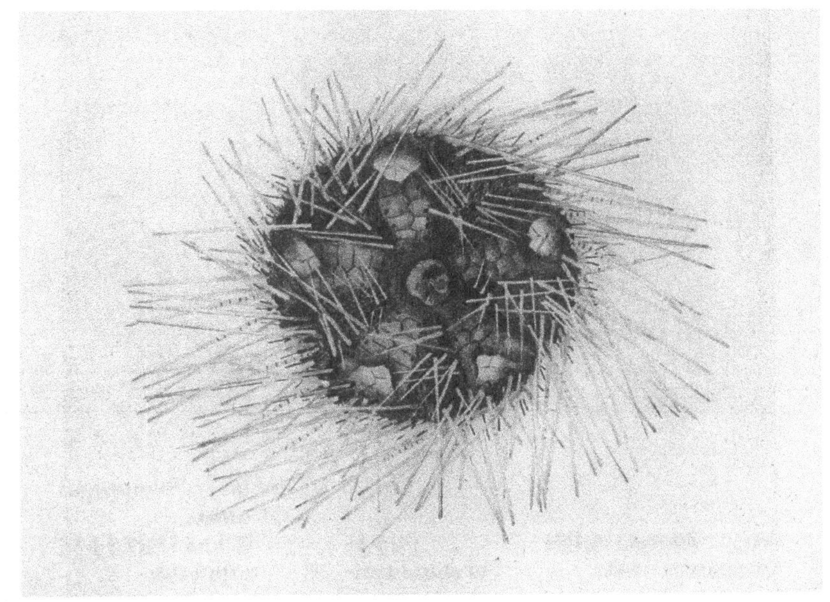

Fig. 3. Astropyga pulvinata (Lamarck) 1816 § K-318

Fig. 1. Pharia pyramidata (Gray) 1840 § K-119

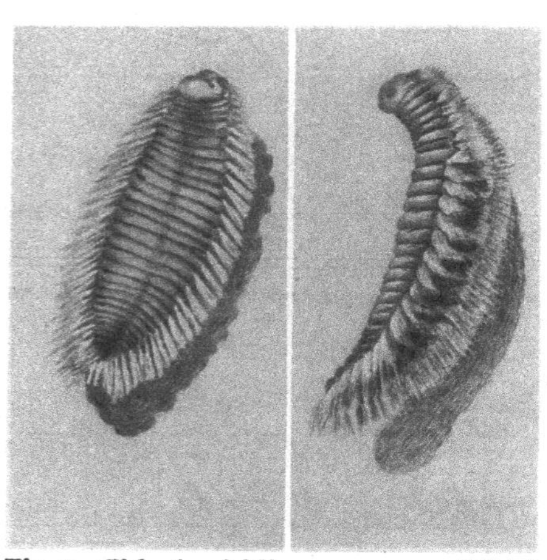

Fig. 2. Chloeia viridis (Schmarda) 1861 § J-34
⅔ natural size

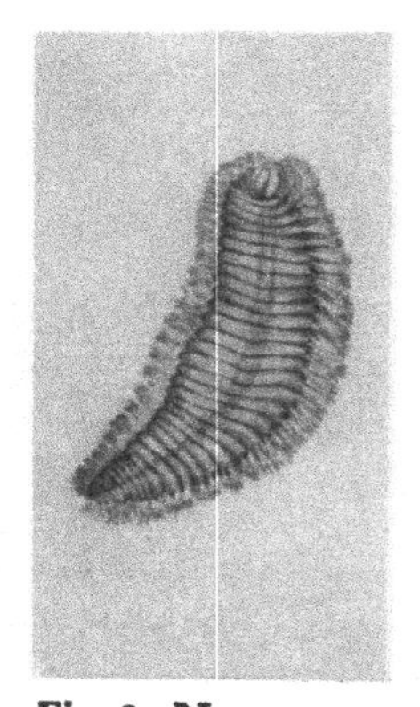

Fig. 3. Notopygos ornata Grube 1856 § J-36
natural size

PLATE 20

Fig. 1. Evibacus princeps Smith 1869 § P-105 ¾ natural size

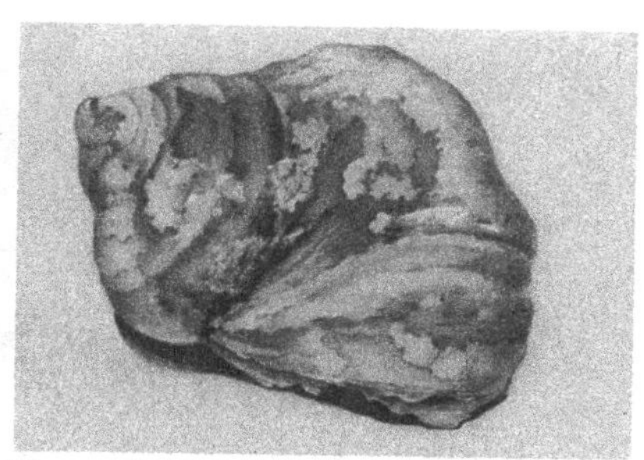

Fig. 2. Callopoma fluctuosum (Wood) 1828 § S-404 natural size

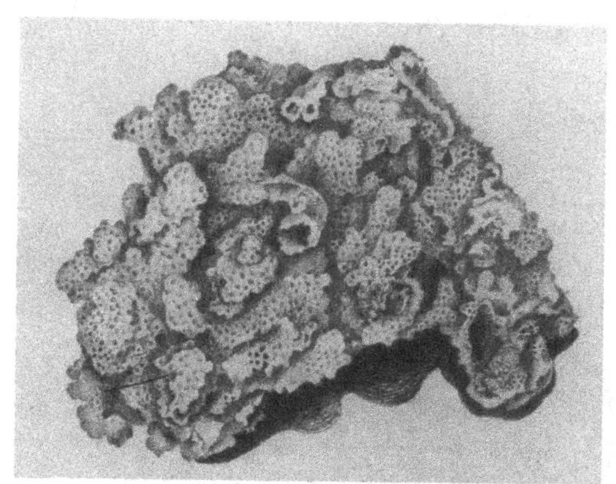

Fig. 3. Pocillopora capitata Verrill 1864 § E-7 ¾ natural size

Fig. 4. Hypoconcha panamensis Smith in Verrill 1869 § R-25 ¾ natural size

Fig. 5. Pontonia pinnae Lockington (1878) 1879 natural size § P-25

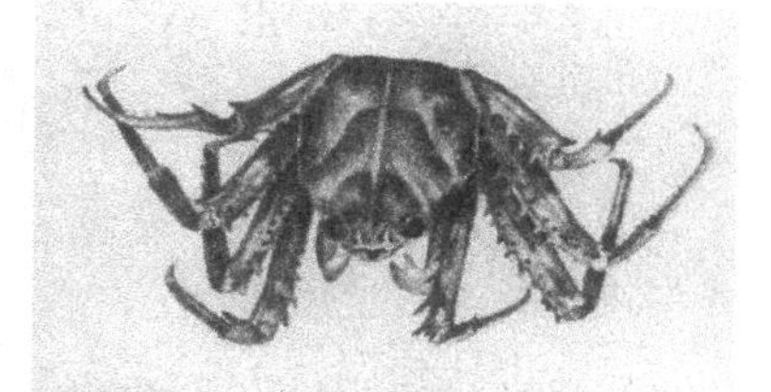

Fig. 6. Percnon gibbesi (Milne-Edwards) 1853 § R-74 natural size

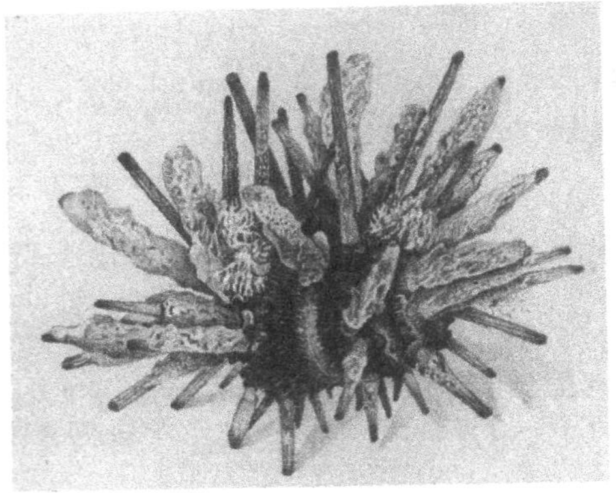

Fig. 1. Eucidaris thouarsii § K-313
(Valenciennes in L. Agassiz and Desor) 1846
½ natural size

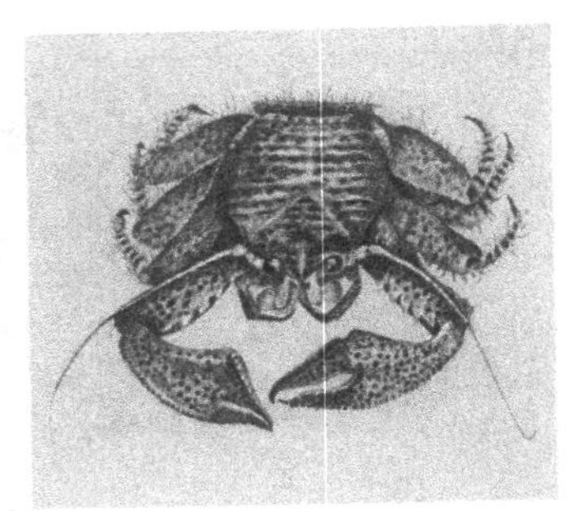

Fig. 2. Petrolisthes edwardsii
(Saussure) 1853 § Q-27
½ natural size

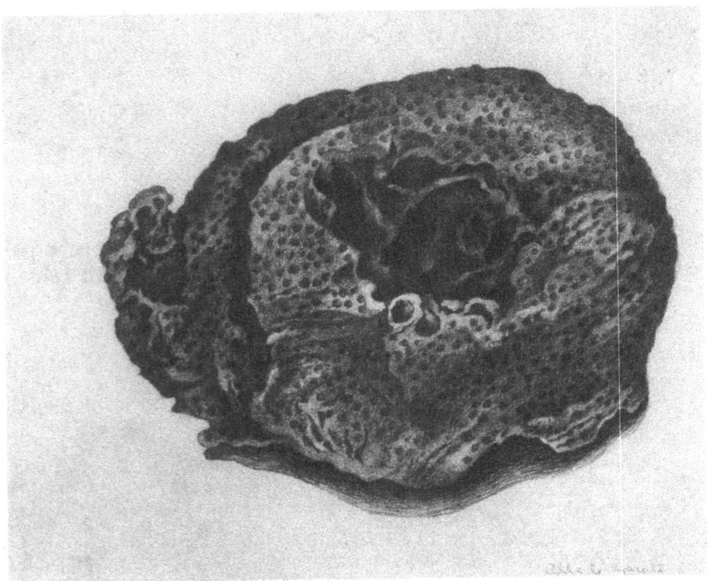

Fig. 3. Cliona celata Grant 1826 § A-13
⅓ natural size

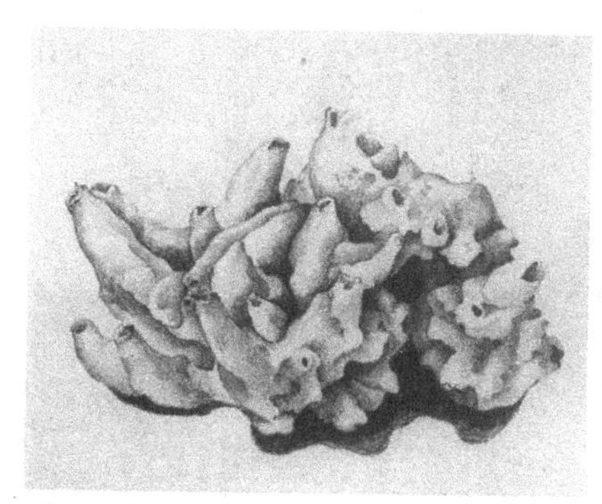

Fig. 1. Leucetta losangelensis § A-18
(de Laubenfels) 1930 ½ natural size

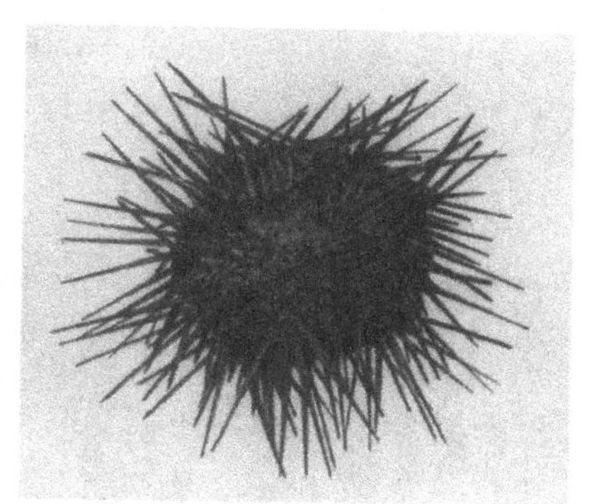

Fig. 2. Arbacia incisa § K-315
A. Agassiz 1863 ½ natural size

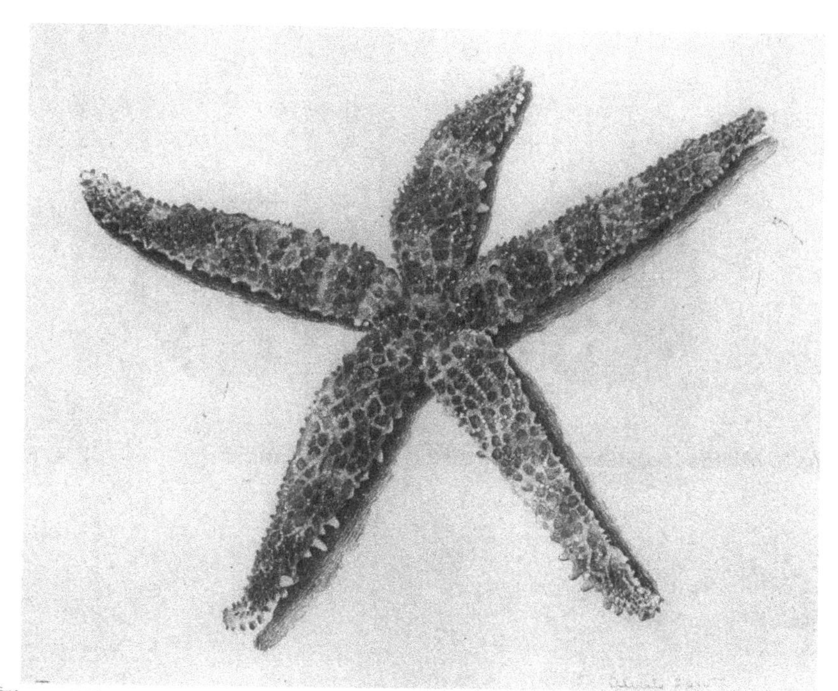

Fig. 3. Mithrodia bradleyi Verrill 1867 ½ natural size § K-122

Fig. 1. Heliaster kubiniji Xantus 1860 § K-116

Fig. 2. Thoë sulcata § R-32
Stimpson 1860
PHOTO BY HARLAND L. SWIFT

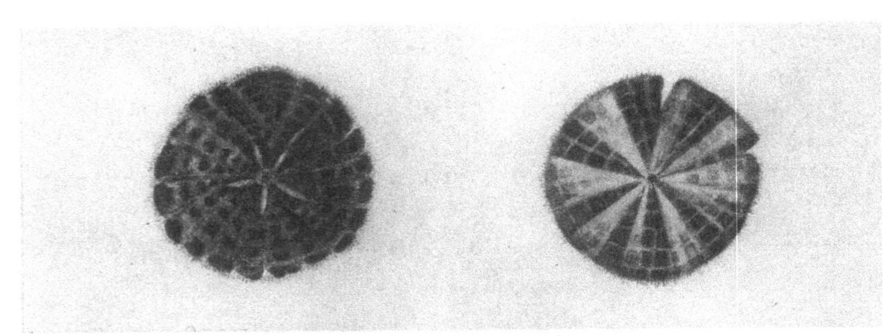

Fig. 3. Mellita longifissa Michelin 1858 1½ natural size § K-323

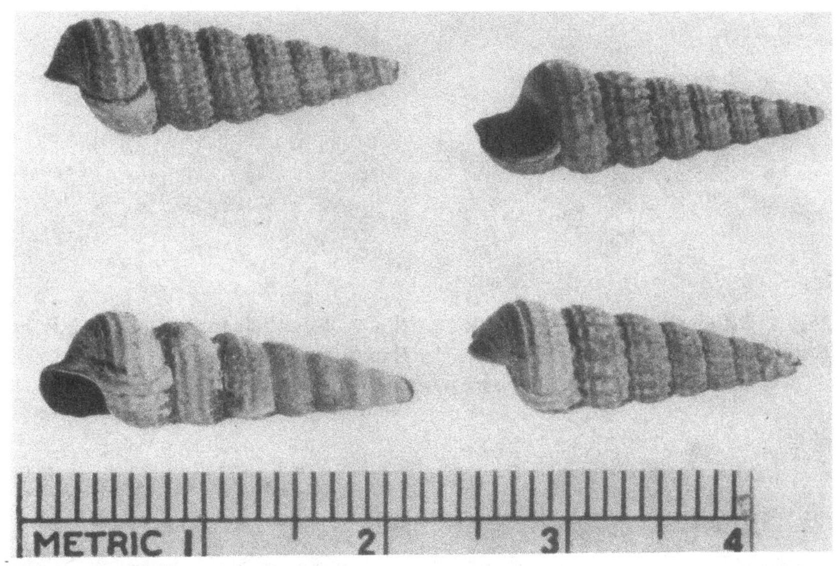

Fig. 1. Cerithidea mazatlanica Carpenter 1857 **§ S-373**
Mazatlan Horn-shell

PHOTO BY HARLAND L. SWIFT

Fig. 2. Octopus sp. upper surface
Diameter of body 49.1 mm.

Fig. 3. Octopus sp. under surface
§ U-116

PHOTOS BY WILLIAM G. VESTAL

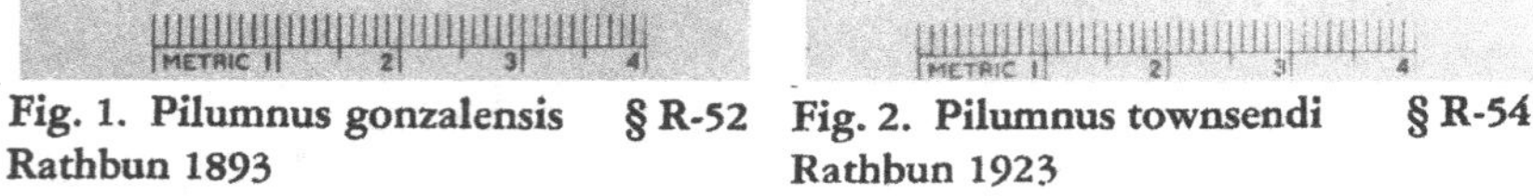

Fig. 1. Pilumnus gonzalensis § R-52
Rathbun 1893

Fig. 2. Pilumnus townsendi § R-54
Rathbun 1923

PHOTO BY HARLAND L. SWIFT

Fig. 3. Siphonaria pica Sowerby. Starry Pulmonate Limpet, under surface § S-325

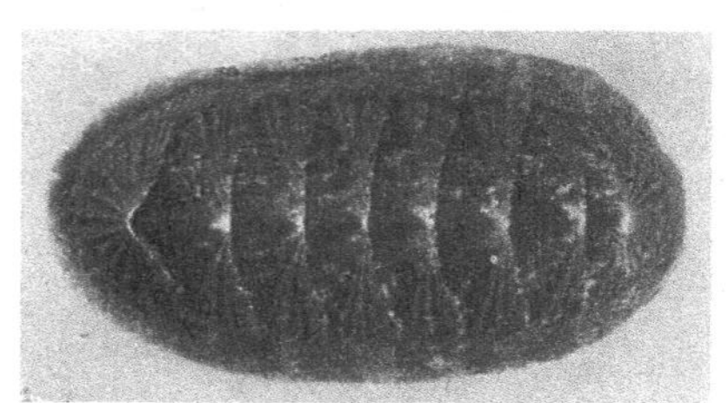

Fig. 5. Ischnochiton (Lepidozona) clathratus (Reeve) 1847 § U-18
length 45.8 mm.

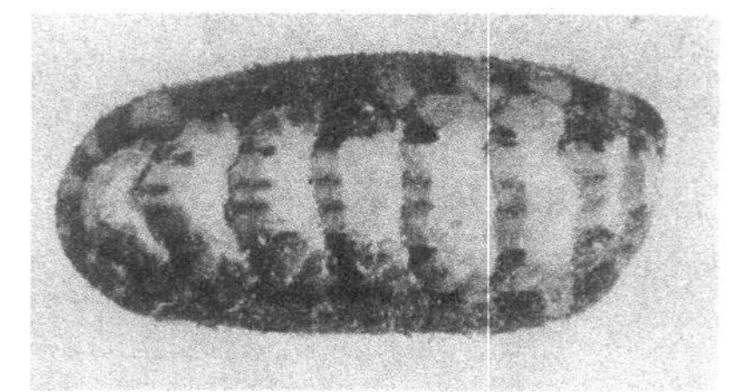

Fig. 6. Nuttalina sp. cf. allantophora Dall 1919 § U-22
length, curled, 17.3 mm.

PHOTOS BY WILLIAM G. VESTAL

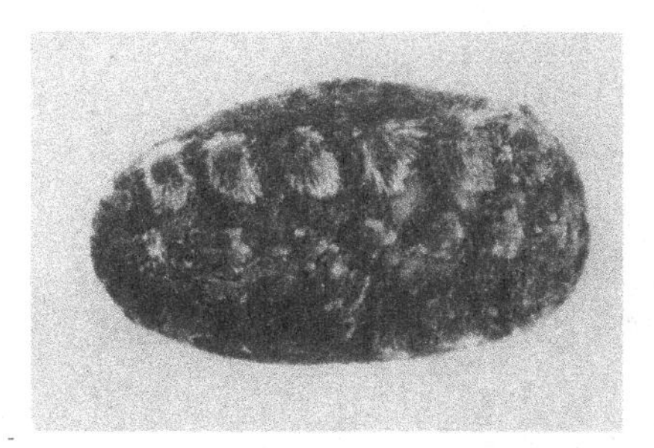

Fig. 1. Acanthochitona exquisita (Pilsbry) 1893 § U-14

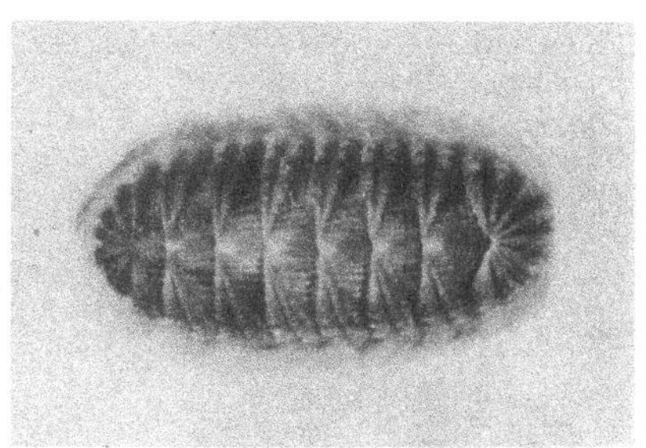

Fig. 2. Callistochiton infortunatus Pilsbry 1893 § U-15

Fig. 3. Chaetopleura aff. lurida (Sowerby) 1832 § U-16

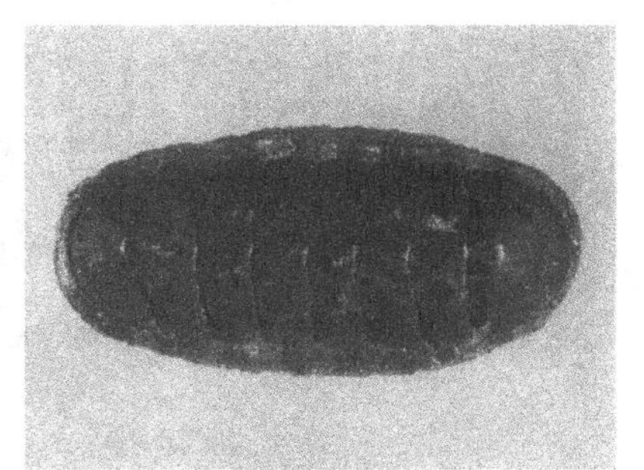

Fig. 4. Chiton virgulatus Sowerby 1840 § U-17

Fig. 5. Ischnochiton (Radsiella) tridentatus Pilsbry 1892 § U-19

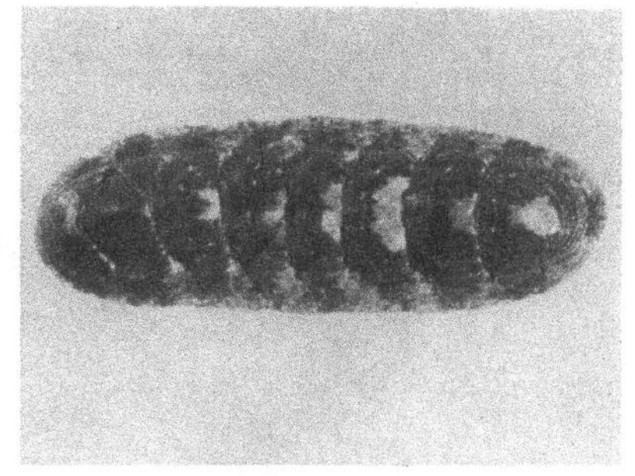

Fig. 6. Ischnochiton (Stenoplax) limaciformis (Sowerby) 1832 § U-20

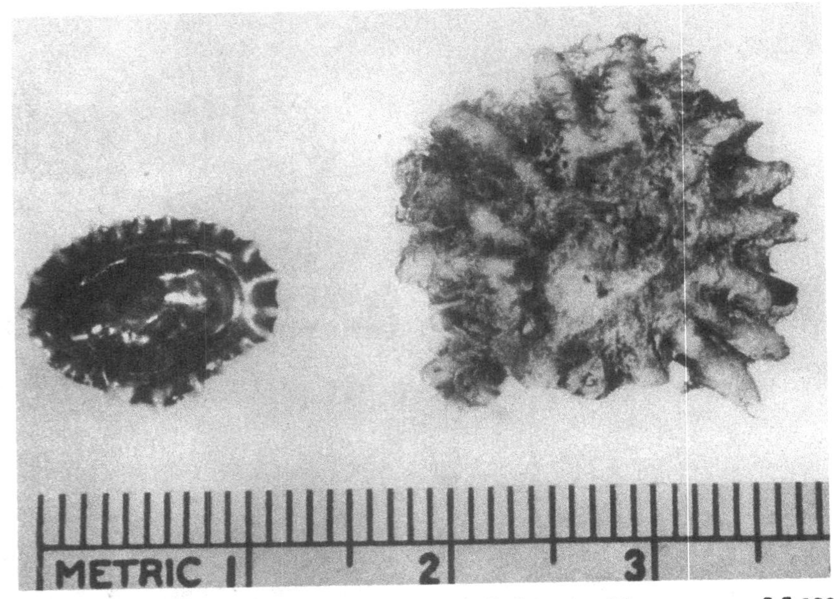

Fig. 1. Siphonaria aequilirata Reeve. Digitate Pulmonate Limpet **§ S-323**

PHOTO BY HARLAND L. SWIFT

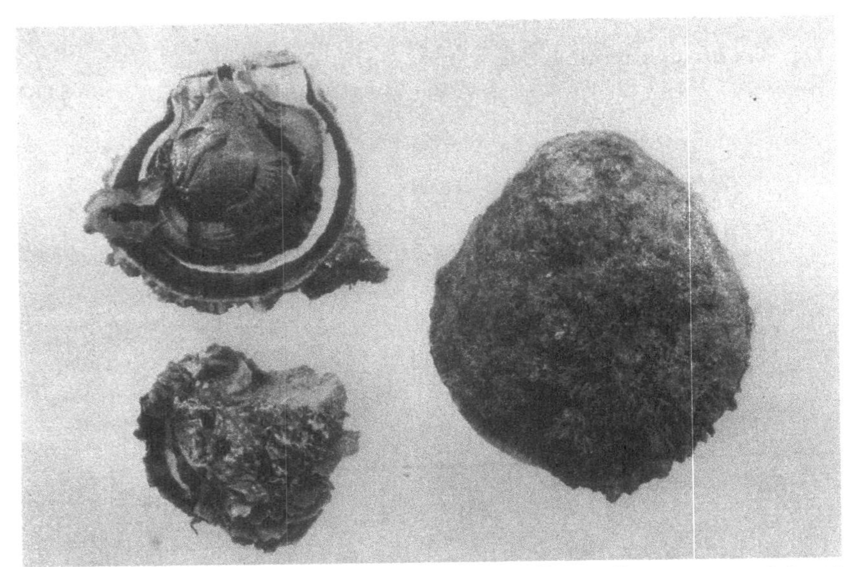

Fig. 2. Spondylus sp. probably *limbatus* Sowerby. Thorny Oyster **§ S-234**

Fig. 1. Mithrax areolatus (Lockington) 1876 (1877) § R-36

Fig. 2. Pachycheles panamensis Faxon 1895 § Q-23

Fig. 3. Petrolisthes hirtipes Lockington 1878 § Q-29 dorsal view

Fig. 4. Petrolisthes gracilis Stimpson 1859 (1862) § Q-28 dorsal view

Fig. 5. Petrolisthes nigrunguiculatus Glassell 1936 dorsal view § Q-31

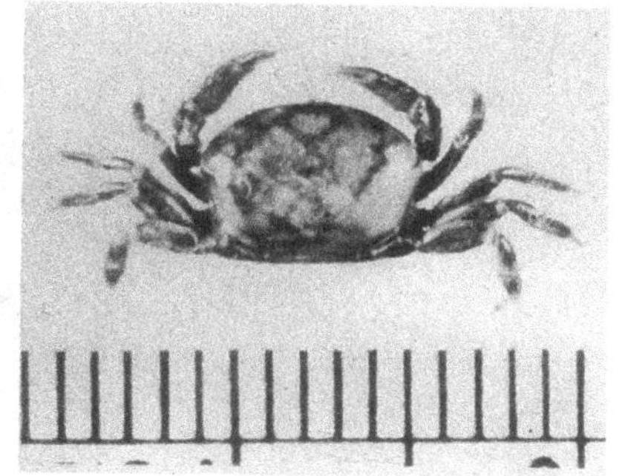

Fig. 6. Dissodactylus xantusi Glassell 1936 § R-63

PHOTOS BY HARLAND L. SWIFT

Fig. 1. Diodon hystix or holocanthus. Pufferfish. Length about 6″. § W-20

Fig. 2. Fodiator acutus (Cuvier and Valenciennes) § W-22

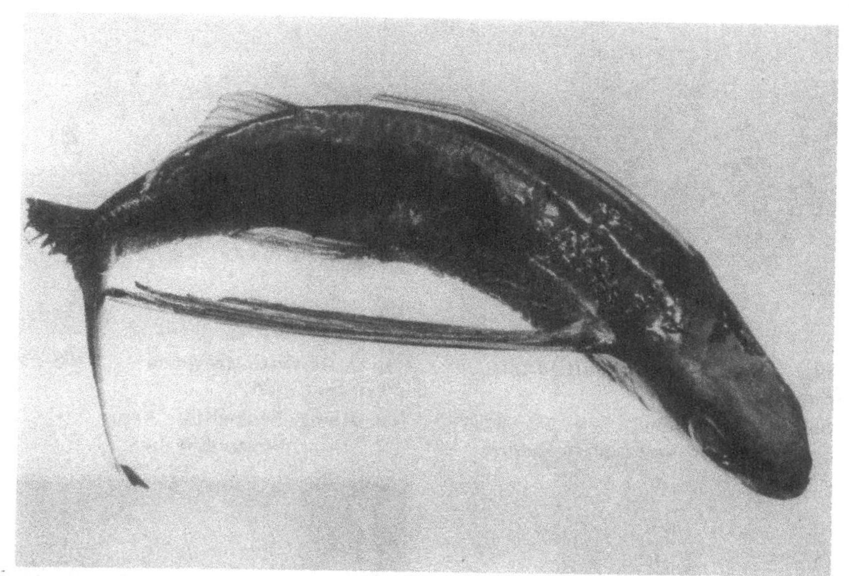

Fig. 1. Cypselurus californicus (Cooper) **§ W-19**
California Flying Fish

Fig. 2. Balistes polylepis Steindachner 1876 **§ W-18**
Trigger Fish or Puerco

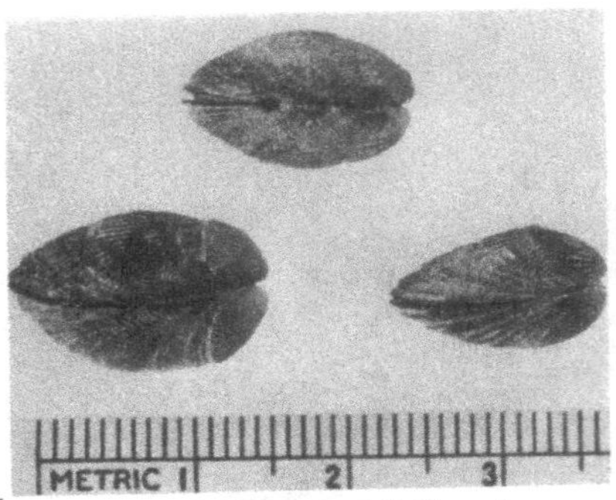

Fig. 1. Brachidontes multiformis Carpenter
Small Shore Mussel § S-237
PHOTO BY HARLAND L. SWIFT

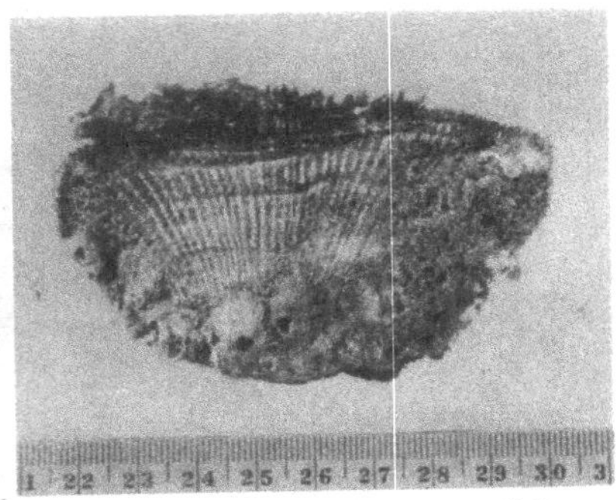

Fig. 2. Barbatia reeveana d'Orbigny 1846 § S-219
The Bristly Mussel-like Arca
PHOTO BY FISHER

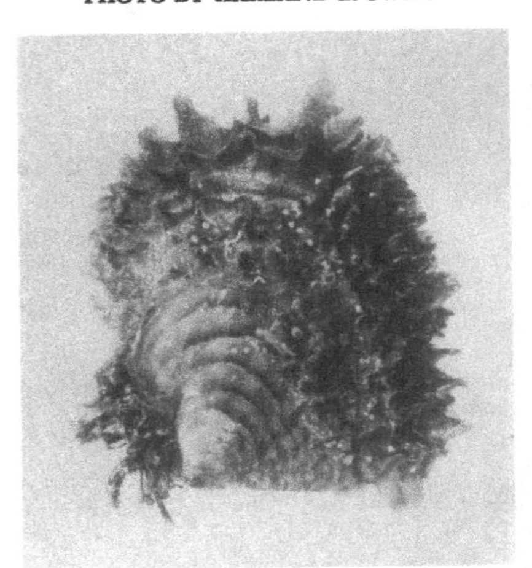

Fig. 3. Pinctada mazatlanica (Hanley) 1855. The Gulf Pearl Oyster § S-230
Length about 3½″
PHOTO BY FISHER

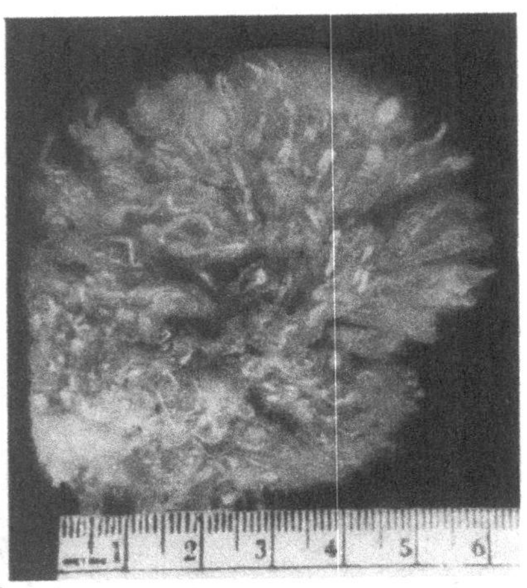

Fig. 4. Clavelina sp. § V-115
Semi-compound Tunicate
PHOTO BY FISHER

Fig. 5. Emerita rathbunae Schmitt 1935 § Q-43

Fig. 6. Microphrys platysoma (Stimpson) 1860 § R-39

PHOTOS BY HARLAND L. SWIFT

PLATE 32

Fig. 1. Chione succincta § S-248
(Valenciennes) 1821. Hard-shell Cockle

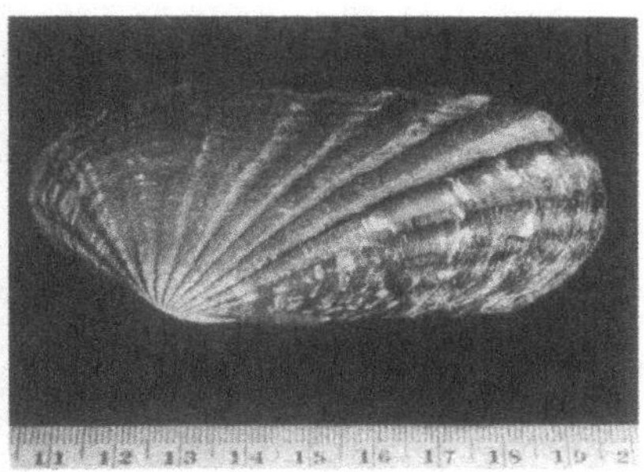

Fig. 2. Carditamera affinis californica
Deshayes 1852. Ruffled Clam § S-240

Fig. 3. Arca multicostata Sowerby 1833
The Cockle-like Arca § S-217
Diagonal length 3¼″

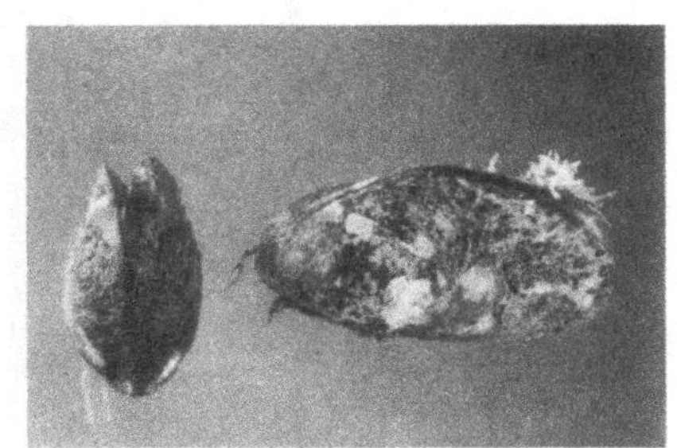

Fig. 4. Volsella capax § S-239
(Conrad) 1837 The Horse Mussel
Length of larger specimen 3½″

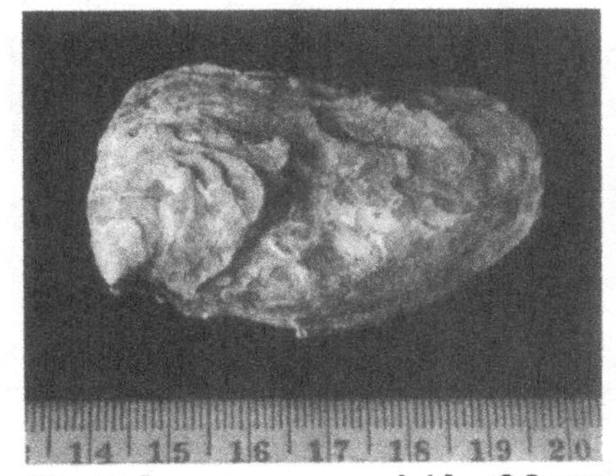

Fig. 5. Isognomon anomioides § S-227
Reeve Paper-shell Clam

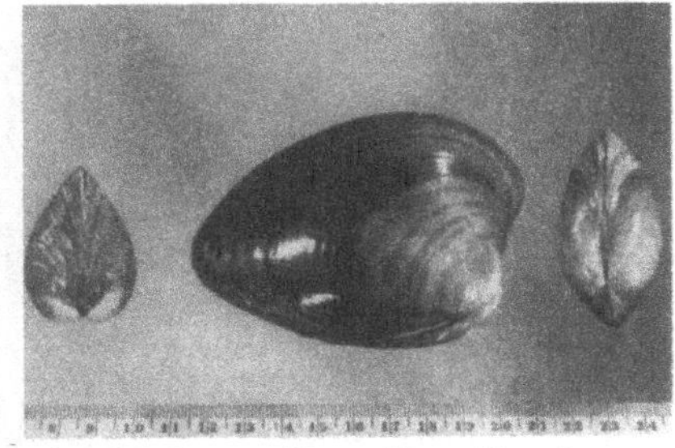

Fig. 6. Macrocallista (Paradione)
squalida (Sowerby) 1835 § S-251
The Gulf Pismo Clam

PHOTOS BY FISHER

PLATE 33

Fig. 1. Conus princeps Linné 1758 § S-329 Royal Cone

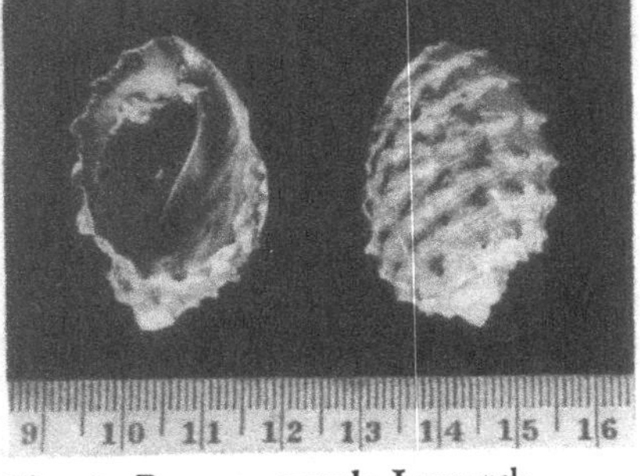

Fig. 2. Purpura patula Lamarck § S-357

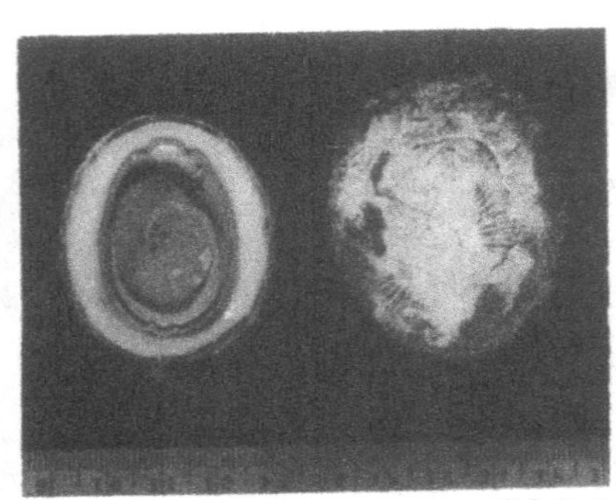

Fig. 3. Acmaea discors Philippi § S-398 Eroded Limpet

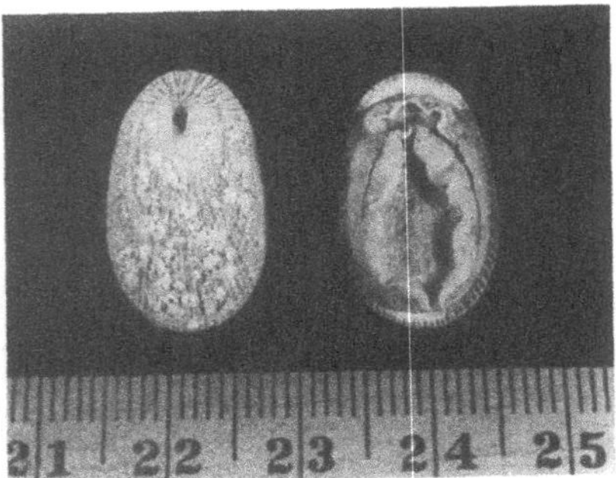

Fig. 4. Diadora inequalis (Sowerby) 1835 § S-410
Asymmetrical Keyhole Limpet with tubes of serpulid worms § J-72 or 76

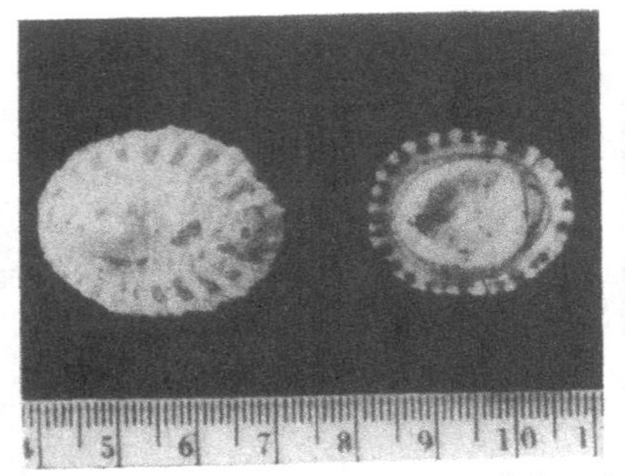

Fig. 5. Acmaea atrata Carpenter § S-396 Coolie-hat Limpet

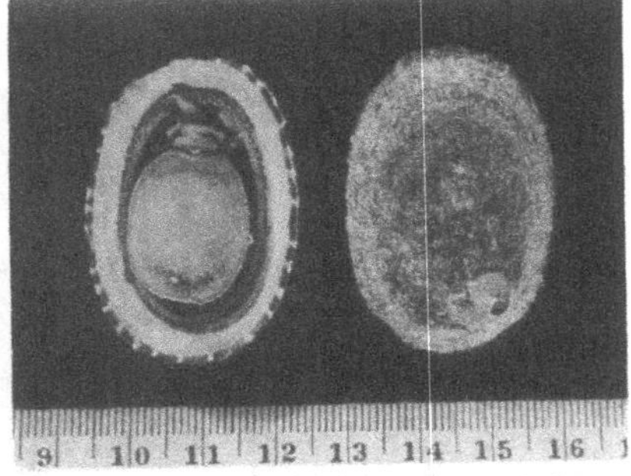

Fig. 6. Acmaea dalliana Pilsbry § S-397 Dall's Limpet

PHOTOS BY BEAUFORT FISHER

PLATE 34

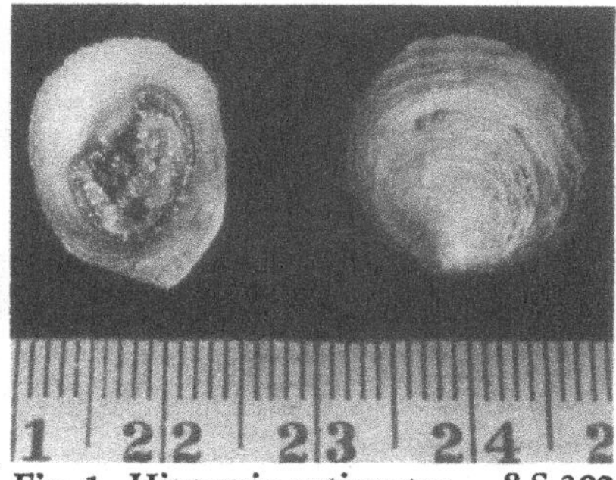

Fig. 1. Hipponix antiquatus § S-382
(Linné) Ancient Hoof-shell

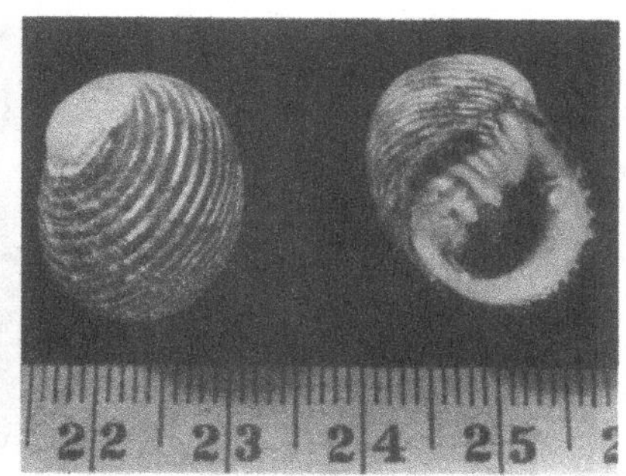

Fig. 2. Nerita scabricostata § S-394
Lamarck Black and white Whorl-shell

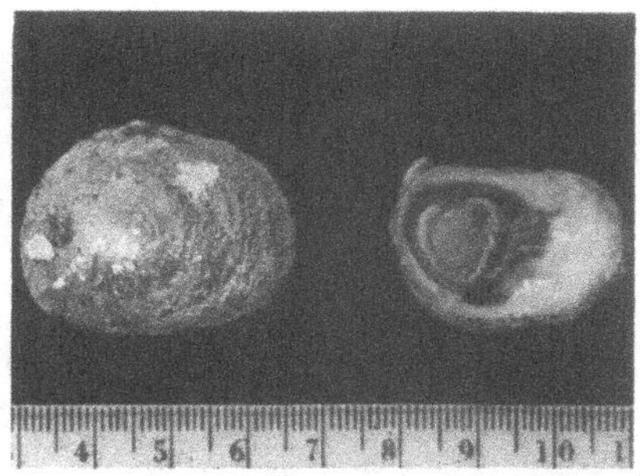

Fig. 3. Crepidula onyx § S-384
Sowerby 1824 Onyx Slipper-shell

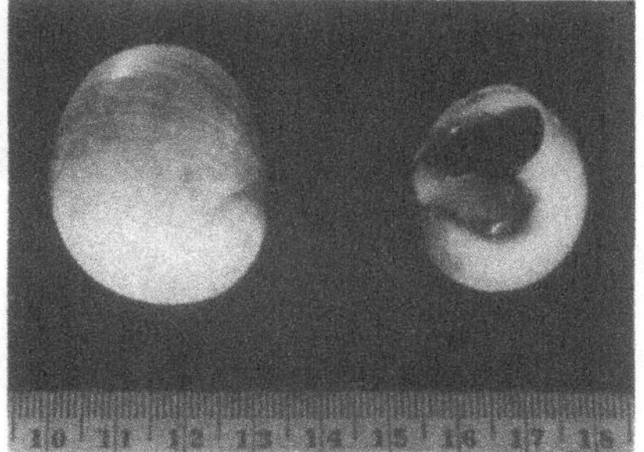

Fig. 4. Polinices reclusianus § S-390
(Deshayes) 1839
Recluz's Moon-shell or Bull's-eye

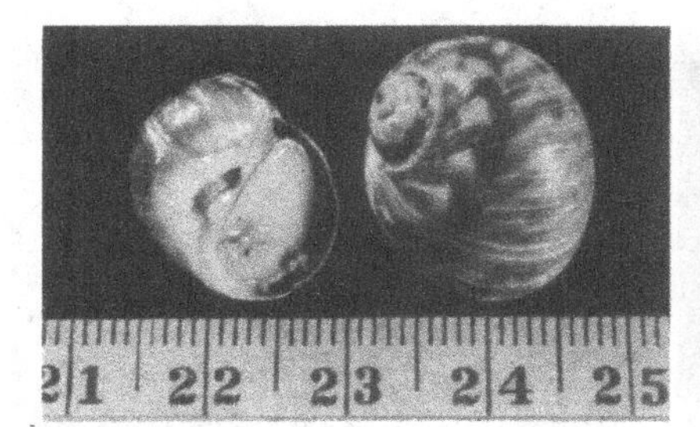

Fig. 5. Natica chemnitzii § S-388
Pfeiffer Variegated Moon-shell,
or Chemnitz's Bull's-eye

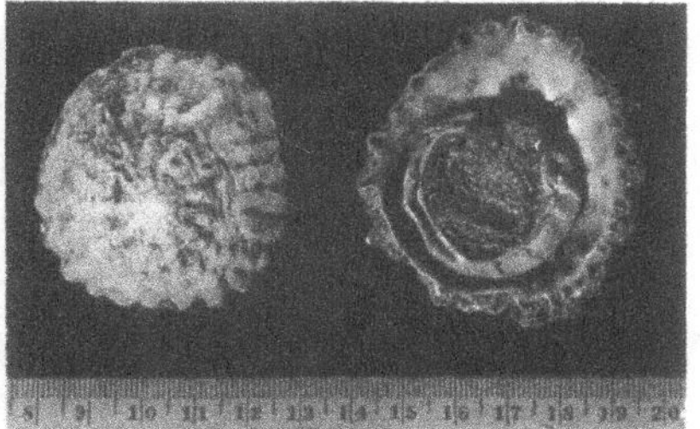

Fig. 6. Crucibulum imbricatum § S-386
(Sowerby) 1824
Imbricated Cup and Saucer Limpet

PHOTOS BY BEAUFORT FISHER

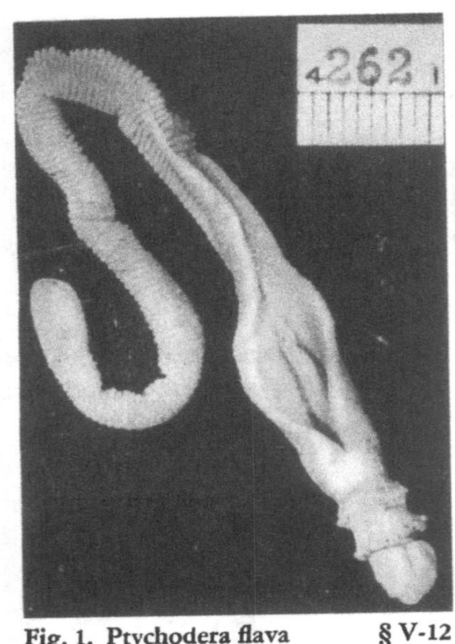

Fig. 1. Ptychodera flava § V-12
Eschscholtz, cf.

PHOTO BY T. H. BULLOCK OF A COMPLETELY EXTENDED SPECIMEN

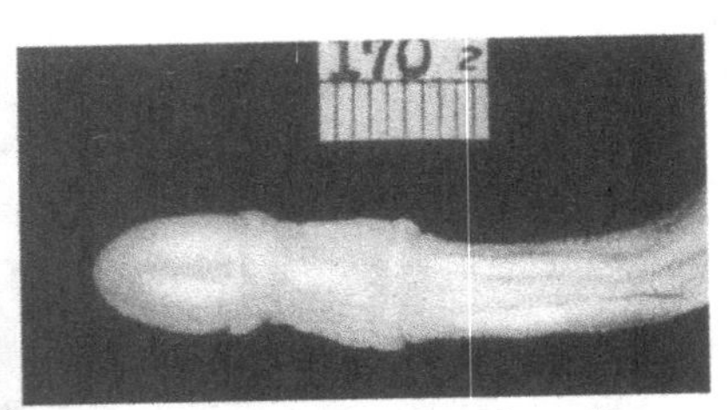

Fig. 2. Balanoglossus occidentalis
Ritter (MS) § V-11
(Portion of anterior end)

PHOTO BY T. H. BULLOCK FROM MISSION BAY (SAN DIEGO) SPECIMEN

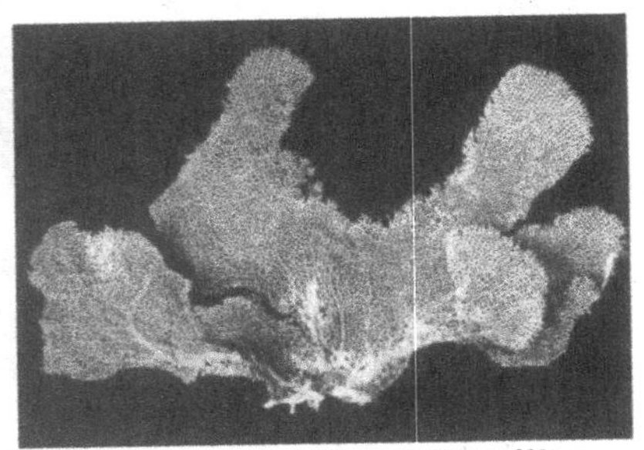

Fig. 3. Gorgonia adamsi (Verrill)
Sea-Fan (undetermined) *not treated*

PHOTO BY BEAUFORT FISHER

Fig. 4. Axius (neaxius) vivesi § P-201
(Bouvier) 1895. La Paz Ghost Shrimp

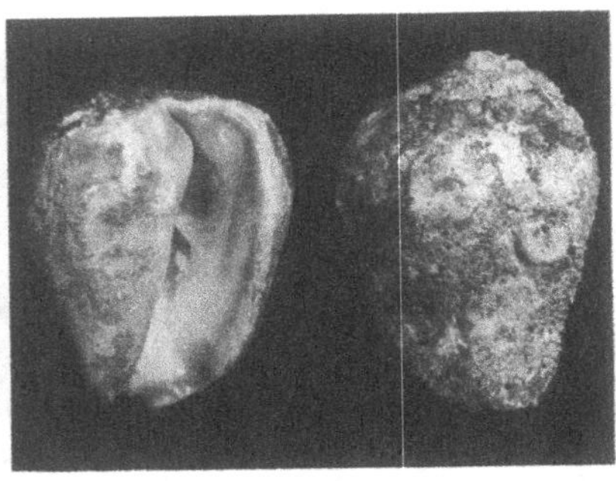

Fig. 5. Strombus galeatus § S-371
Swainson 1823 Giant Conch

PHOTOS BY FISHER

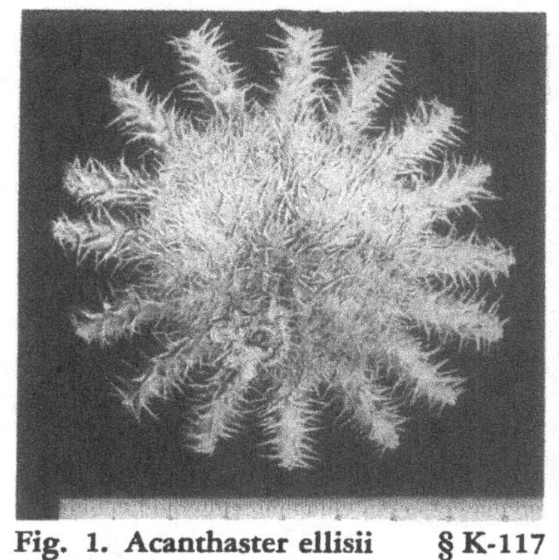

Fig. 1. Acanthaster ellisii (Gray) 1840 § K-117

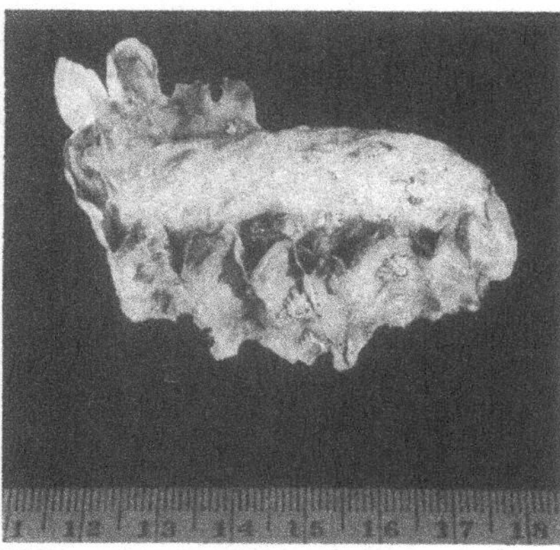

Fig. 2. Ostrea mexicana Sowerby 1871 The Mangrove Oyster § S-232

Fig. 3. Tegula rugosa A. Adams Variegated Turban § S-407

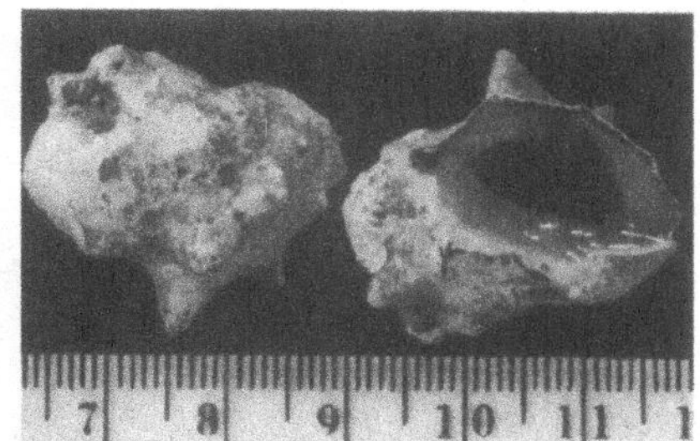

Fig. 4. Thais centriquadrata Duclos Four-pronged Rock-shell § S-363

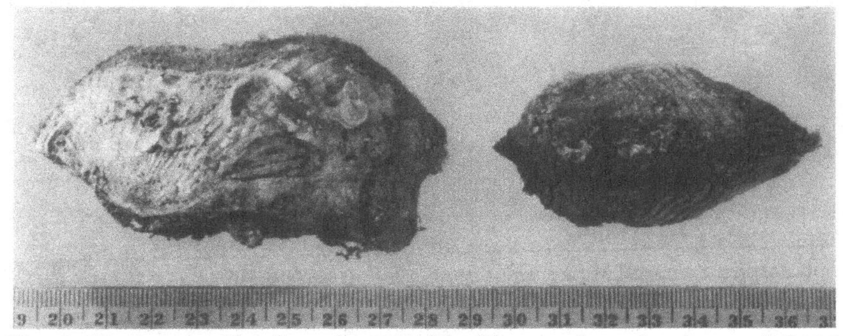

Fig. 5. Navicula pacifica Sowerby 1833. The Elongate Irregular Arca § S-224

PHOTO BY FISHER

Fig. 1. Acanthina lugubris § S-351
(Sowerby) 1821 Gulf Unicorn-shell

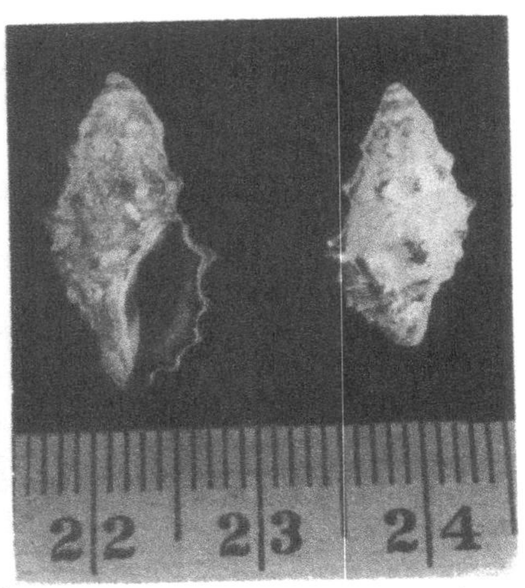

Fig. 2. Engina ferruginosa § S-339
Reeve Rusty Engina

Figs. 3 and 4. Thais (Tribulus) § S-364
planospira Lamarck Masked Flat Snail

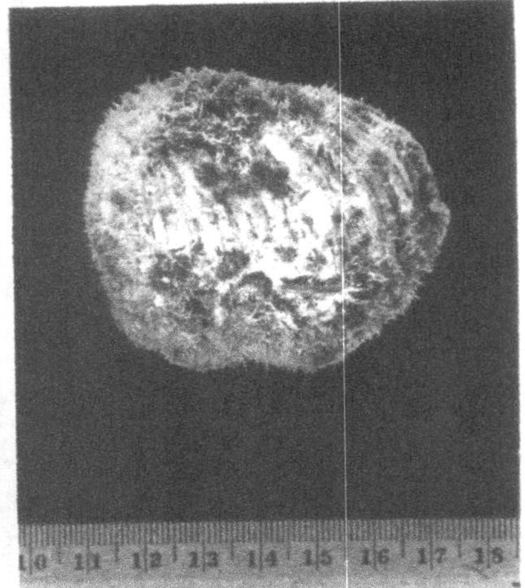

The annelid worm is:
Eunice antennata § J-50

PHOTOS BY BEAUFORT FISHER

Fig. 1. Penaeus stylirostris § P-22
Stimpson 1871
Guaymas Edible Shrimp

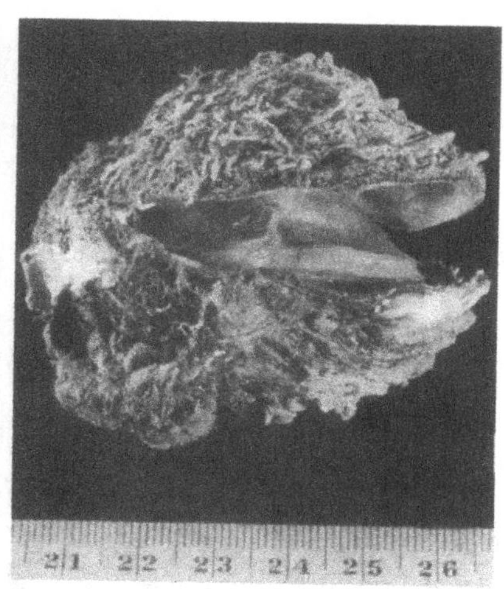

Fig. 2. Chama mexicana § S-242
Carpenter The Mexican Rock Oyster

Fig. 3. Felaniella sericata § S-245
Reeve 1850 The Satin Diplodon

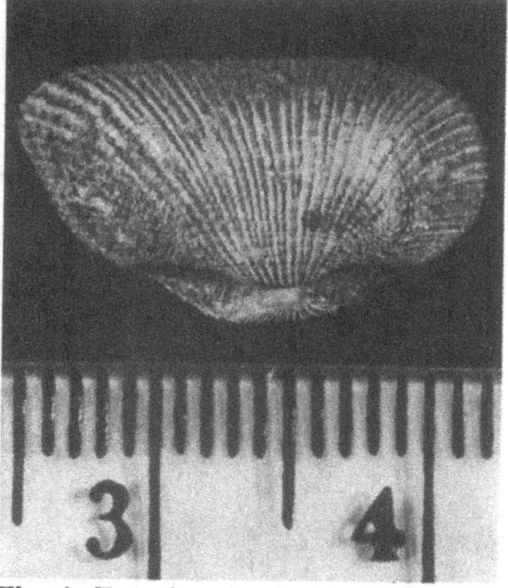

Fig. 4. Fossularca solida § S-220
Sowerby 1833 Garbanzo Clam
In typical specimens, the ribs have been worn smooth

PHOTOS BY BEAUFORT FISHER

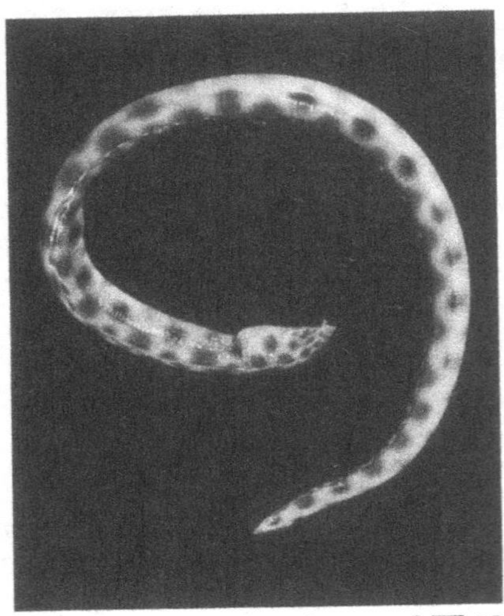

Fig. 1. Myrichthys tigrinus Girard 1859 (spread about 4½") § W-25

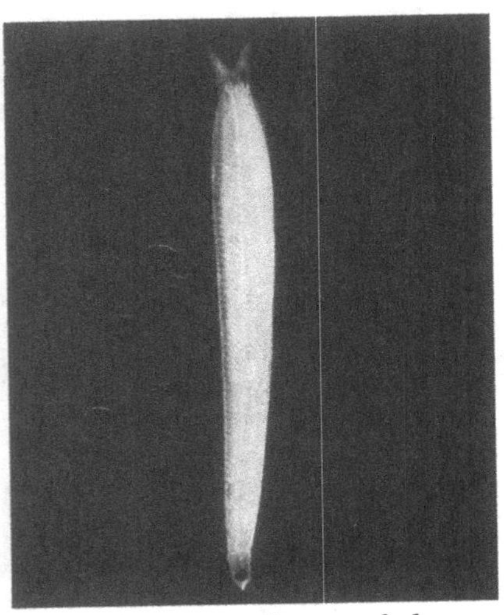

Fig. 2. Transparent Leptocephalus larva of Albula vulpes (Linn.) Ladyfish. (length about 2½") § W-17

Fig. 3. Fasciolaria princeps Sowerby (about 11½" long in life) § S-336

Fig. 4. Cerithium sculptum (Sowerby) 1855 Sculptured Cerithium § S-375

PHOTOS BY BEAUFORT FISHER

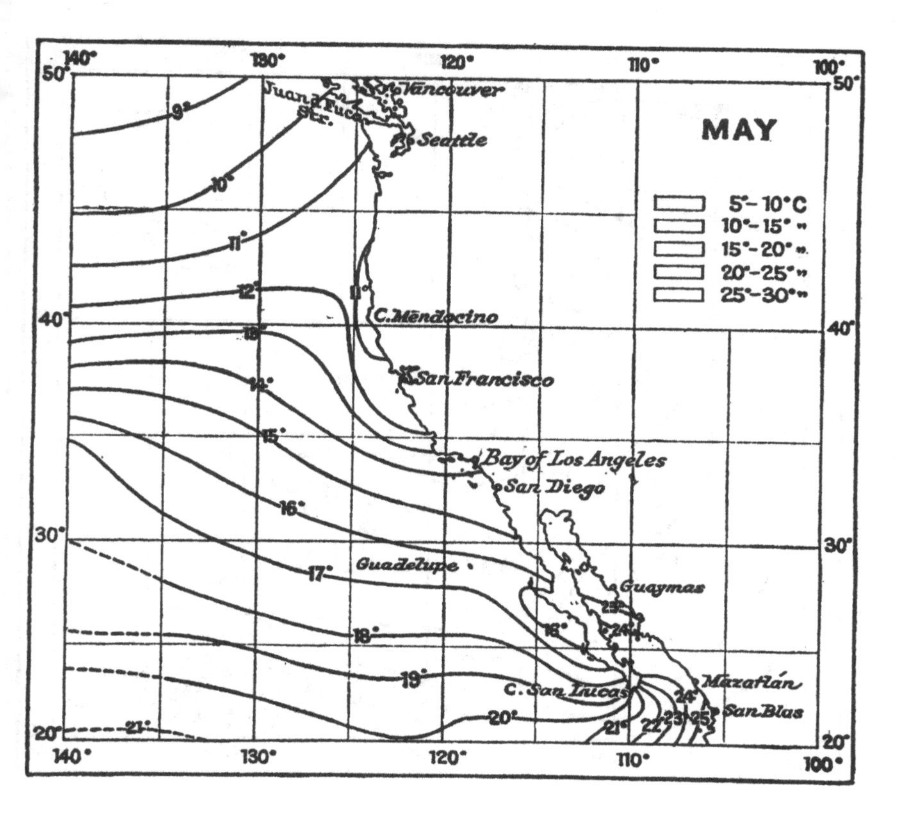

ISOTHERMS FOR MAY

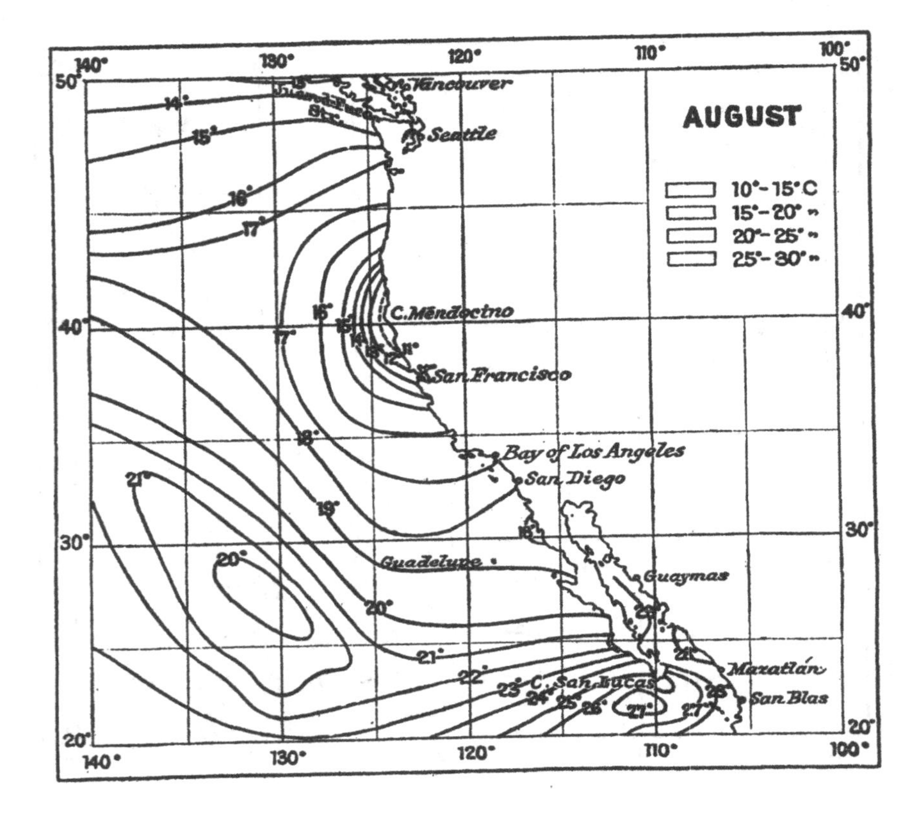

ISOTHERMS FOR AUGUST

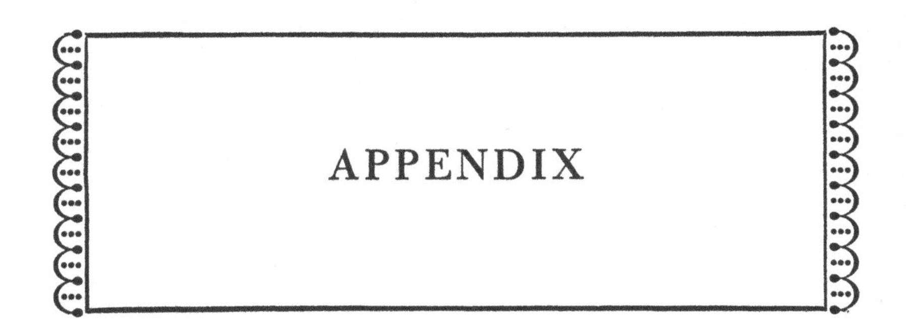

APPENDIX

Consisting of an annotated catalogue of the species encountered, a bibliography and résumé of the literature, and a summary of the present state of our knowledge with regard to the littoral natural history of the Gulf of California, together comprising materials for a source book on the marine invertebrates of the Panamic Faunal Province.

Annotated Phyletic Catalogue and Bibliography

INTRODUCTION AND SUMMARY

I:A

THE following acknowledgments, being first in importance, are given very gladly. We are indebted particularly to two of the institutions affiliated with the Universidad Nacional Autónoma de México, for the free use of facilities extended to one of us during the summer of 1940, and for the warm and friendly personal co-operation which would seem to be a national characteristic of Mexico: First, the Institute of Biology in Chapultepec Park and its director, Dr. I. Ochoterena, its professor of invertebrate zoology, Dr. Enrique Rioja, and its excellent library headed by Señor Crisóforo Vega. For its service to Mexico, and, more widely, for co-ordinating biological literature, both North and South American, in the Spanish language, this Institute and library merit the heartiest support of all United States scientific bodies. Second, to the more general scientific library of the Academia Nacional de Ciencias Antonio Alzate, which extended, equally freely, its not inconsiderable facilities.

We are grateful also to the Hopkins Marine Station at Pacific Grove, California, and especially to its director, Professor W. K. Fisher, for assistances too numerous to specify, and for access to many rare items in his personal library which must have remained unavailable to us otherwise. Acknowledgments are due also to the following: to the U. S. National Museum, particularly to Dr. Waldo L. Schmitt; to the libraries of the University of California, and of Stanford University, especially to Librarian M. J. Abbott in Life Sciences at Berkeley for library privileges granted; to Virginia Scardigli, who was an efficient associate and a pleasant companion during some of the bibliographic research; but most of all to Toni Jackson who, in addition to manuscript revising and glossary constructing, transcribed the bulk of the difficult text, both journal and appendix, the latter phase involving that most vexing of all secretarial tasks—the editing of scientific copy interlarded with handwriting of a doubtful legibility.

The services of specialists in the various fields, which, as in a previous survey (Ricketts and Calvin 1939, § Y–3), have been willingly bestowed, are acknowledged in appropriate sections of the ensuing catalogue.

The drawings—and pictorially they are among the finest of this sort that we have ever seen—are, almost without exception, the work of Alberté Spratt of Carmel, California, who, with no specialized zoological knowledge, has nevertheless constructed painstakingly accurate illustrations, and who, with little spare time, has given abundantly of that little. For a group of color photographs, equally painstakingly done in perhaps a still more specialized field, we are indebted to Mr. Russell Cummings of Pacific Grove. The very welcome contribution of a color drawing of one of the holothurians by Dr. Fisher is acknowledged with thanks. Photographs have been contributed by Dr. Bullock of Yale University, and by Dr. MacFarland of the California Academy of Sciences. Other photographs are by specially trained professionals such as Mr. Beaufort Fisher of Pacific Grove, Mr. Harland L. Swift of West Los Angeles, and by Mr. William G. Vestal of Redlands, to all of whom much credit is due.

I:B

THIS is an attempt to bring order to a subject previously unordered, and to shed light onto a field that has been dark.

Before the publication of Johnson and Snook and of Ricketts and Calvin, it was difficult for students or travelers or even for scientists themselves to obtain any ready information on the animals of the Pacific shores of temperate America. A considerable amount of work had been done, but there were many gaps, and the literature was sparse, scattered, and unco-ordinated. This has proved to be the case even more seriously with reference to the western shores of tropical America.

In the citations, therefore, an attempt is being made to compile a fairly comprehensive bibliography of the marine invertebrates of the Panamic Faunal Province, particularly with reference to the Gulf of California. Occasional mentions of gulf invertebrates in large general taxonomic accounts, or in papers on the fauna of other regions, were not specifically investigated; there was merely a cursory examination of the literature prior to 1860 (when the standards were in any case less finely drawn than those at present; and in conchology the listing is confessedly fragmentary. But otherwise a serious effort has been made to search out, examine, and list, if apropos, all the large and important papers concerned with Panamic invertebrates.

We are heartily in accord with the p. 5 remarks of Keen 1937 (§ S–19): "Because a bibliography is the foundation upon which adequate organization of information rests, no small part of this paper is devoted to the listing of titles." Too often, careful and original effort needed elsewhere has been expended on material worked over equally carefully years ago in another country and recorded in a publication not thought worthy of examination until pointed out.

Among contributions toward the toto-understanding of a given fauna, collation of the literature is at least as important as the field work—which should include the general collation of all groups of animals (1) in the field, (2) in the reporting of field notes, and (3) in the construction of the final report.

Interestingly, the most useful references have proved to be the very oldest and the very newest. Seventy to a hundred years ago the invertebrates of the Panamic Province were known as well as, or possibly better than they are today, and by a larger proportion of zoologists. And the examination and delineation of that fauna attracted a far greater quota of the total biological energy available then, than it attracts even now. So in constructing this report we have used frequently the 1867-71 coelenterate and echinoderm papers of Verrill, and those of Stimpson, Lockington, Streets, and others, on crustacea; the monographs of Lyman and Agassiz, the shell catalogues of Carpenter and Adams, and the reports of the *Vettor Pisani* and of other expeditions. From then until recently, except for work on the crabs and shells dredged by the *Albatross,* comparatively little has been added to our knowledge of the marine animals of this area. But since 1915, several modern investigations, notably those of Dr. Mortensen; of the steam yacht *St. George;* of the New York Zoological Society; and of the Hancock Pacific Expeditions, have again penetrated this long neglected field.

Aside from papers of the Smithsonian Institution and of the U. S. National Museum, the most frequently consulted periodical turned out to be no local nor even western hemisphere publication, strangely enough, but the *Videnskabelige*

Meddelelser fra Dansk naturhistorisk Forening, published in Denmark. This contains, in addition to some of the descriptions of echinoderms, etc., collected almost a hundred years ago on the west coast of Central America, more than sixty papers issued thus far as a result of *Dr. Th. Mortensen's 1914-16 Pacific Expedition.*

The publications of the *Hancock Pacific Expeditions,* now in progress, would seem to be the most important of all, despite their considerable cost, were it not for their unavailability in even the largest libraries. Some were on hand at the Institute of Biology in Mexico City, but the Hopkins Marine Station of Stanford University has no file of this Pacific coast series published only a few hundred miles away. Even at the libraries of the University of California, only Volume Two was available in the spring of 1941, and we were forced to purchase some of the most needed numbers and to consult others as separates wherever they might be seen, usually in the libraries of specialists. Publications of this sort, furthermore, are too expensive for the private worker to purchase—the separates comprising Volume Two, alone, come to $17.50. Though the publications are as costly as those of the Museum of Comparative Zoology at Harvard, they lack the large-size page, the splendid lithographic illustrations, and, in some cases, the recognized scientific authority characteristic especially of the *Memoir* volumes of that series. Considering all this, the biologist who has to finance his own library may justly regret the advent of the rich man in science.

A really formidable list of references accrued during the construction of this bibliography. In selecting items to be included, we were guided by certain considerations which made the work more difficult than if we had listed indiscriminately all apropos citations, but our care has reduced the bulk to more usable proportions. Preference has been given to papers with comprehensive bibliographies of their own. By indicating in the annotation that a given citation is listed in the bibliography of an account which we cite, we have been able to avoid duplication and to conserve space, while at the same time mentioning all the important references. The original description of a species has not been cited *necessarily*, this ordinarily having been covered in the more monographic papers cited for the group. It has been considered more important to list references which have full descriptions and synonymies, and, most of all, which carry adequate photographs or drawings of the *entire* animal. The method of identification by pictures may be both superficial and primitive, but it still remains not only the most popular, but by far the fastest method for the layman, or even for the specialist out of his own field. For most minds, a lifetime is too short for studying more than a few of the many groups which the collector or ecologist will surely encounter even in the most delimited area, and the non-specialist who has recourse to adequately illustrated literature for the identification of his catch saves time and energy for other considerations.

An incidental dissertation on the art and science of scientific writing would seem to be indicated at this point—a dissertation arising from the unusual circumstance of a non-specialist having had occasion to review rather carefully a large section of specialized literature.

The gulf between the general zoologist, the ecologist, or the layman and the specialist is very great and is becoming greater. The situation among polyclad-worm specialists is illustrative. They *may* be the opposite of the field naturalists. They needn't be, and the only ones with whom we are acquainted definitely are not, but there is nothing in the nature of the work as such to prevent them from remaining far removed from the living material. Identifications of polyclad-worms are made from serial sections of preserved material—minute slices of tissue, so thin as to show clearly the cellular structure. In the case of polychaet worms, identifications are made chiefly from the detached bristles. And as a rule, the interest of the specialist remains in identification and nomenclature alone. Therefore, the illustrations in the specialized literature are usually drawings of the diagnostic anatomical details, often microscopic or at least highly magnified, and separated from the animal itself. Thus the average zoologist or the layman, consulting the special literature in order to identify a worm, finds the description and illustration so wrapped in technicalities and so unrecognizable that he is likely to become discouraged by its remoteness from the animal itself.

And this situation is true also of the work on sponges, holothurians, alcyonaria, and to a surprising extent of the work on conchology. It might be well at this point for zoological workers to recognize that specialization, although desirable and necessary, has, too, its hazards and that liaison writers are much needed here. Co-ordination of this type would not only benefit zoology, but would render its findings more intelligible to workers in other fields.

Setting aside any evaluations as to the thoroughness, integrity, or competence brought to a given task, there are several kinds of papers. In working up a collection of animals from a given region, the specialist has three choices. He may do the job in hand most simply by listing the species taken, with citations in the synonymy only, by suitable descriptions and by diagnostic illustrations especially of such species as appear to be new to science, and by repeating any collecting or natural history observations which may have been noted on the label by the collector. Or, he may find that with a little extra exertion he can also review all the published work on that particular group in that area, evaluating and collating the literature to date, and citing in a bibliography all the apropos items with short abstracts. By so doing, he will have made it unnecessary for future workers to dig back into scattered and often inaccessible literature; or, if they must do this, at least he will have simplified the task by having provided a foundation. The Nielsen 1932 ophiuran paper based on Dr. Mortensen's expedition is a good example. As a third possibility, our hypothetical specialist may discover that with still more effort, he may be able to have suitable toto-illustrations prepared (as contrasted to figures of diagnostic details) of all the commoner forms, so that laymen using the paper may quickly familiarize themselves with that particular fauna without having to make a special study of the highly specialized subject matter. Miss Rathbun's "Stalk-Eyed Crustacea of Peru," but especially her four volume monograph on the Brachyura of North America, are cases in point.

But the "field eye" is as different from the "laboratory eye" and the "library eye" as these are from each other, and they all contribute to the whole and to each other. So the ideal situation would be for the specialist to have been a member of the expedition, or for him otherwise to have familiarized himself largely and at first hand with the living animals in the field. Opportunities of this sort must be rare, but they provide rich rewards in bringing together, for evaluation by a single mind, all the known logical factors in the specialist's field. The subsequent report will be an expression of this co-ordination, and should prove an increasing joy to work with in the field, laboratory, or library. Darwin's account of the barnacles is a classic example; in recent years Fisher's monograph on the asteroids and Ashworth's Arenicola paper are good illustrations, along with some of the Glassell decapod papers, among the shorter accounts. Even by the informed general public, expressions of this type are coming to be recognized as monuments to real achievement, to be contrasted with the annoyance and inconvenience of less thorough works as time shows up their inherent oversights and shortcomings. The British conchologist Philip P. Carpenter summarized this situation well for his day, by saying (p. vi, § S–8): "My principal object in the preparation of these works has been to make out and compare the writing of previous naturalists, so that it might be possible for succeeding students to begin where I left off, without being obliged to waste so large an amount of time as I have been compelled to do in analyzing the (often inaccurate) work of their predecessors."

In these remarks—the complaint of one who unwittingly has been led through some pretty bad morasses of indifferent work and of poorer reporting of it—there may be sensed a frank propaganda for research and write-ups of the most thorough sort. Work of this type would seem to be, in its own right, a satisfying task and an honorable avocation not unconnected with what Keats would have termed pure "joy," and, from the viewpoint of functionalism, a hearty and lasting contribution to the "increase and diffusion of knowledge" which is one of the fundamental aims of science.

I:C

In connection with the phyletic catalogue, it should be emphasized that, although we seem to have collected a good representation of the *littoral* species, no general survey of the entire fauna has been attempted. Dredged animals will be lacking from our list unless it happens that they occur also along shore. Therefore any conclusions arrived at through the analyses and syntheses of these data must be regarded as valid for shore conditions only.

Experience elsewhere has indicated that depth is an important, although by no means the only, or even the most important, factor of distribution. A typical beach fauna, for instance, will consist in part only of true shore forms. The majority will derive obviously from the continuously submerged zones, as more or less successful shore colonizers, with migrations still going on, and with possibly a few motile stragglers from land—the whole welded into a fairly coherent, mutually interrelated society in a more or less stable state of equilibrium. Although this particular zone, centering around the line of lowest tides, is thought to be the richest area in the entire ocean, both for species and for individuals, there will be no attempt to suggest a delineation of the whole region from a study of the one part.

II:A

THERE has been mention of a Panamic Fauna. It may be well to characterize it at this point more clearly. A faunal province for mollusks has been defined by Schenck and Keen, (1936, p. 923) as a region "populated by a distinctive assemblage of species. Distinctiveness is not simply a matter of ecologic situation but of spatial extent or range. . . ." Just how distinctive the assemblage must be would seem to vary with the province. What is regarded as the Panamic Fauna seems to be pretty definite beside that which has been regarded traditionally as the Californian, the Oregonian, the Aleutian, the Sitkan, etc.

A more generalized definition would substitute the word "forms" for "species" so as to read: "populated by a distinctive assemblage of forms." There are indications that the word "species" may have different connotations in conchology, in carcinology, and in echinodermology. Quite possibly the amount of differential which to Fisher indicates a subspecies or no more than a forma among the starfish would, on the basis of shell alone, constitute a full species or even a subgenus for the conchologist. The standards, the mean, may be different in different groups. The further advantage of using a more generalized term in attempting to evaluate provinces is illustrated by the following situation: In comparing lists of distinctive and abundant animals of the southern California area with those of similar exposure, depth, and type of bottom in the Monterey Bay area, it turns out that a sea pansy, *Renilla,* characterizes tidal sand flats of a certain sort in the south, and there is nothing comparable in the northern lists. Sea pansies which the veriest amateur would recognize as such, and which would be differentiated only by specialists, occur at Newport Bay in southern California, at Estero de Punta Banda in northern Lower California, at Cape San Lucas, at Panama, in northern Peru, and probably elsewhere between and beyond. The fact that the *Renilla* of southern California and northern Lower California is identified by specialists as *R. köllikeri,* and that of Panama Bay as *R. amethystina,* is secondary in a zoogeographical sense to the fact that there are abundant sea pansies—a readily recognizable "form" of animal—south of Pt. Conception, and none whatsoever to the north.

Transcending species differentiations, there is a deeply real distinctiveness which is valid both scientifically and in factual common sense, although difficult to state. There is a profound gulf between an assemblage which comprises, say, the sea pansies (which are obviously separated from their nearest relatives), the sea hare *Navanax inermis,* the tubed anemone *Cerianthus*—a distinctive form—the fiddler crab *Uca* (as distinguished from the related grapsoids), the swimming crabs *Portunus* or *Callinectes* (as distinguished from the Cancridae), the clams *Chione* spp. and *Dolichoglossus,* and an ecologically similar assemblage only a few hundred miles removed spatially, which comprises the more northern *Hemigrapsus oregonensis, Schizothaerus, Cardium corbis, Zirfaea pilsbryi, Saxidomus* spp. and *Macoma* spp., *Callianassa, Upogebia,* the large northern *Tethys* (p. 737, MacGinitie 1935, § Y–26), and the dog whelk, *Nassarius fossatus.* And that gulf persists even though it may be shown that *Haminoea, Urechis,* etc., and dozens of less abundant or less dominant species are common to both assemblages.

One's personal criterion of a zoogeographical province is

difficult to put into words. If a field zoologist fairly well acquainted with the communities of shore invertebrates in a given region, X, should find familiar animals in comparable communities and in about the same proportions (depth, type of bottom, and conditions of exposure to wave shock being equal), at the widely separated points X-3 and X-1, and if, by traveling only a little farther in a given direction, he should find at the point X-2, markedly different animals proportioned differently or in different communities (the above mentioned factors being equal), he will likely assume that in the first case he was working entirely in one province, and that in the second he has passed the bounds of that province into another.

So the Monterey Bay worker must include in one zoogeographical province all the *exposed rocky shore* fauna between Sitka, Alaska, and Boca de la Playa (some 40 miles south of Ensenada), at least. In this 2600 mile stretch, many animals will have disappeared gradually (except at Pt. Conception where the changes are fairly sudden) to be replaced by comparable forms often scarcely distinguishable as different species. The large picture is found to be the same, the chief constituents are the same, even their proportions are similar. *Balanus glandula, Littorina scutulata, Acmaea digitalis, Mitella, Mytilus, Pisaster, Thais emarginata,* the lined *Lepidochiton,* the large solitary green *Cribrina, Katherina,* the large and abundant forms of *Nereis, Tegula funebrale, Pagurus samuelis* in *Tegula* shells, *Mopalia, Strongylocentrotus franciscanus* and *purpuratus, Pycnopodia, Patiria, Cryptochiton* (north of Pt. Conception especially), and *Parastichopus californicus* (along shore only in *quiet* waters north of Pt. Conception), all occur unchanged from Alaska to northern Mexico, although most of them become notably scarcer south of Pt. Conception. If we add to this list *Littorina planaxis* (recorded range intermediate), and several shore crabs, sponges, and compound tunicates, we shall have considered most of the common shore forms which make up 95% of the total production on exposed rocky shores. Some are commoner to the north, or to the south, or to the center of this great traverse, but on the whole our collector would find himself equally at home in either of the extremes or anywhere between, ecological conditions being the same. He would have to be experienced and discriminating indeed to remark any difference whatsoever down at least to Pt. Conception, in landscape, weather, summer water temperature, rocky shore topography, or in the animals themselves. Such an assemblage of distinctive forms would seem to constitute a zoogeographical province, or at least some sort of subdivision whatever the terminology, and regardless of whatever else might be indicated by the *total* records of *all* the species involved.

These *total* records, on which zoogeographical provinces have been based, include both rare and common forms, from shore to several hundred fathoms. There are reasons for believing that *total* records of *all* the species provide a distorted picture. Distribution records of this sort give equal weight to such dominant forms as *Mytilus californianus* and *Pisaster ochraceus* (which for every linear mile of rocky coastline in the temperate zone will produce hundreds of pounds of animal tissue and shell per year), and to such rarities as certain minute snails or worms which have never been seen by any save a few specialists and museum attendants. Records of deep water animals are included also, although obviously the zoogeographical provinces of the ocean floor, if any exist, must differ from those along shore. The deeper an animal lives, the more it is subjected to oceanic conditions which are more uniform than anything else on earth, so that abyssal animals are cosmopolitan. In deep water there is little differential geographic distribution, the only measurable differential is bathymetric. In the Monterey Bay area at least, and probably throughout the Pacific coastline with its great depths close offshore, identical and similar animals will be found indiscriminately at depths varying from 20 to several hundred fathoms, so that any attempt to limit the shore fauna by imposing the 100 fathom limit, generally presumed to mark the continental shelf, is seen to be highly artificial. Monterey divers insist there is a sharp threshold at 12 fathoms—72 feet—and it may be that somewhere around this level the limit of true littoral animals will be found, but documenting such a statement would require much work. However there can be no doubt that the only animals not subjected to uniform oceanic conditions are those restricted to the shore. And the only way to be certain of that, since here again no definite line of demarcation can be set at present, is to take animals from between the tides and from the few feet just below the line of low water. These surely can be considered as subject to shore conditions and not to the uniform conditions of ocean bottoms. Littoral zoogeographical provinces can be based only on such conditions (which are largely a function of ocean currents in controlling air temperatures and humidities), and on the ocean currents themselves which have not yet been adequately plotted. And, just as surely, the animals important in the marine sociology of an area are not museum rarities, but the common and obvious forms which make up the bulk of the population.

II:B

TRAVELING only a few hundred miles farther south, our Monterey Bay observer, at Cape San Lucas, for instance (where the rocks are similar in type and exposure to those at Boca de la Playa below Ensenada, at Pt. Lobos south of Monterey, or at the Sitka outer islands), would find himself in a territory wholly unfamiliar zoologically. He would be unlikely to see at first glance even one single familiar form. Unfamiliar animals would be found inhabiting familiar ecological niches. *Pisaster* would be replaced by the many rayed *Heliaster* which clings equally tightly; *Strongylocentrotus* spp. by *Echinometra van brunti*. There would be littorines, and *Thais*-like forms, all adapted to withstand heavy surf as are those in the north, but of different species, usually of entirely different groups. The limpets, the barnacles, the few chitons would be obviously different. Pendent gorgonians and (in more quiet waters) stony corals would occupy the overhangs and ledges inhabitated in the north by a complex of varied sponges, anemones, hydroids, and encrusting Bryozoa and tunicates. Unfamiliar crabs would scuttle about in the uppermost zone, just as hard to catch—harder in fact—than those in the north, but utterly different. All this would be evidence, even to the untrained observer, of an entirely new fauna.

The fact of the matter is that these different animals of the exposed rocky shore, and *Renilla* and others of the estuarine soft bottoms, characterize what has been called the Panamic Fauna, which in distinctiveness would seem to fulfill even the strictest definition of a zoogeographical province. Even its boundaries may be drawn with some degree of sharpness. The southern limits seem to be pretty well marked at latitude 4° 30′ S., bisecting the bulge in northern Peru near Paita, north of Sechura Bay. A short overlap area thereabout is formed by the variation in the limits of the cold Humboldt current.

The northern limit has been set generally at Cape San Lucas, latitude 23° N. However, there can be no doubt that the Panamic Province in its entirety extends at least to Magdalena Bay, and quite probably to the Pt. San Eugenio complex inshore from Cedros Island, latitude 28° N. In reporting on hydroid distribution of the 1934 Hancock Expedition, Fraser, 1938 (§ B–2) p. 5, states: "In passing northward along the west coast of Lower California, the most decided break in continuity of distribution in both fauna and flora appears to be in the vicinity of Thurloe Point. There the large kelps so characteristic of the coasts of United States and Canada make their first striking appearance, and other species appear, coincidentally with these." Thurloe Head is about 25 miles south of Pt. San Eugenio. Also the northward limit of such important horizon-marking tropical shore forms as the Sally Lightfoot crab is known to be at Cedros Island, and many of the northern animals make their last stand there. The whole coastline between San Diego and Magdalena Bay has been very little worked over, and it can be expected confidently that future collecting will establish many northern limits of Panamic animals in that area.

Actually, a considerable number of Panamic invertebrates, especially the quiet water forms, reach their extreme frontier in southern California. A few of these have already become extinct there within our own experience, coincident with changes due to commerce, dredging, and the construction of breakwaters. The following come to mind as both common and representative of present or immediately past conditions: the pendent gorgonians *Muricea* spp., *Renilla,* the burrowing anemones *Cerianthus* and *Harenactis;* the starfish, *Astropecten armatus, Linckia,* and *Astrometis;* two of the three commonest shore ophiurans south of Pt. Conception, *Ophioderma* and *Ophionereis* (*Ophiothrix* occurs occasionally at Pacific Grove), the urchins *Lovenia* and *Lytechinus; Sipunculus nudus,* the spiny lobster *Panulirus, Synalpheus* spp., *Callianassa affinis, Penaeus brevirostris* (slightly sublittoral; the record for San Francisco must be the result either of downright error, or of the finding of strays), *Pilumnus, Paraxanthias taylori, Uca crenulata,* Amphioxus, the cockles *Chione* spp., most of the cowries and the murices and the cones, etc. None of these extend importantly north of Pt. Conception, which seems to be a barrier more effective for southern than for northern forms, although many of the rocky shore inhabitants are fairly common at Santa Barbara which is just a few miles to the south.

The fact is worth passing mention that competent collecting has turned out some of these southern animals as strays in the north. With increased collecting intensity, more and more ranges will be extended in both directions, so that the *qualitative* zoogeographical records of the future will read, almost literally: "everything—everywhere." But the great numbers of given animals will still remain in their optimum areas, with differential distribution observable *quantitatively* only—which is in any case the way the competent field zoologist will see it. In the large picture, the rarities will be regarded as rarities only, and weight will be given chiefly to the common forms which comprise 99% of the scene.

II:C

WITH reference to faunal barriers in general, but particularly to Pt. Conception as a northern limit for Panamic forms, it will be argued that many species are common to areas both north and south of this alleged line of demarcation. But it can be shown equally that a greater number of commoner and more important *shore* animals are limited by this complex than by any similar stretch of coastline many times its length. Actually there is more divergence between the common shore animals fifty or a hundred miles north and south of this point than there is within the whole area between St. Sal (just to the north) and Cape Flattery (in the latitude of Victoria)—or, for that matter, clear up to Sitka, ecologic conditions being equal.

As attenuated extensions of the Panamic fauna can be traced all the way to southern California where Pt. Conception is thought to function as a last barrier to their farther northward migrations, so the more northern fauna fades out gradually and terminates finally somewhere in the almost unexplored stretch of coastline in northern or middle Lower California. (An unexplored area, incidentally, can be a very effective barrier in zoogeography!) This might aptly be termed the North Temperate Fauna, extending as it does almost throughout the entire temperate zone as far as to Sitka without any sudden breaks, as can be seen by any shore observer who is more concerned with common animals than with rarities (which most experienced collectors definitely are *not*). There is a fairly obvious break at Cape Flattery, but this seems to be ecological rather than geographic, since most of the littoral distribution records immediately north of there are based on quiet water collecting in the Straits of Juan de Fuca, in the Straits of Georgia, and in Puget Sound, as opposed to open coast conditions to the south. So marked is this uniformity in the distribution records, on the whole, that Fraser, noting not even a Pt. Conception break in his *total* records, writes (1938, "Hydroid Distribution in the North Eastern Pacific," Trans. Roy. Soc. Canada, Sec. V: p. 42) : "There is nothing to indicate that there is any especially effective barrier to hydroid distribution at any point between Bering Sea and the boundary between United States and Mexico."

But for these northern animals also, Pt. Conception is something of a real barrier, because, as has been noted above, such of those as have dispersed past this boundary to the south will be found there sparingly or stuntedly, as compared to their lush and healthy abundance in Monterey Bay and to the north. But Cedros Island, opposite the mainland Pt. San Eugenio complex, marks the extreme southern limit of such few as have come thus far all the way from Alaska: *Strongylocentrotus franciscanus, S. purpuratus* (Clark 1913, § K–5, p. 222) , and *Parastichopus californicus* (in 35–45 fathoms, Deichmann 1937, § L–5, p. 163, 1938, § L–6, p. 362) , to name a few. And *Pisaster, Mitella, My-*

tilus, and other northern horizon markers will have disappeared somewhere in the poorly-known stretch between the Ensenada area, where they all occur sparingly, and Cedros Island which has finally barred their farther southward dispersings.

II:D

If then the region between Pt. San Eugenio and Pt. Conception can be looked upon correctly as an overlap area, the Panamic Province can be defined as extending from about latitude 4° 30′ S., just below Ecuador, to about 28° N., opposite Cedros Island in Lower California, with an attenuating extension to the north, limited finally by Pt. Conception, 34° 30′ N. A north temperate fauna can be conceived as extending from an unknown point in the north (possibly around 57° at Icy Strait north of Sitka, but possibly not short of the passes into the Bering Sea between the Aleutian Islands) down to Pt. Conception, 34° 30′ N., with an overlap extension to Cedros Island at about 28° N.

Hence for purposes of zoogeographical evaluation and classification, we have adopted the following possibly arbitrary rules: *north temperate* forms are those which, having a southern limit at or north of Cedros Island, range materially north of Pt. Conception. Species restricted to the Cedros Island—Pt. Conception area are probably *overlap* forms, although they may be North Temperate or Panamic animals not yet discovered in their northern or southern extensions. *Panamic* animals have their northern limits at or south of Pt. Conception, but range materially south of Cedros Island. Animals known to occur in several provinces are rated as *cosmopolitan, tropicopolitan,* etc.

The Panamic area so defined is then the subject of this bibliography, and, as illustrated by the Gulf of California as a typical and fairly virginal collecting ground representing its north extension, the subject also of the phyletic catalogue and of the book proper. A previous account, Ricketts and Calvin 1939, has been concerned with the common littoral animals of the temperate province, there considered to extend from Sitka to San Quentin Bay. The overlap animals are treated bibliographically in both books. There are references herein to apropos parts of *Between Pacific Tides,* and cross references are considered desirable there, and are projected for possible future editions.

III

ALTHOUGH our purpose was, primarily, to get an understanding of the region as a whole, and to achieve a toto-picture of the animals in relation to it and to each other, rather than to amass a great collection of specimens, we nevertheless procured more than 550 different species in the pursuit of this objective, and almost 10% of these will prove to have been undescribed at the time of capture. And we merely scratched the surface.

Even this comprises a considerable amount of material. It would have been best—since a more or less intuitive feeling of familiarity with the bulk of the animals in their surroundings was a desideratum—if we could have kept track of every one of these. But the whole elaborate process detailed here, of which collecting and final write-up are merely first and last steps, comprises a labyrinth in which even the finest memory may falter. Hence, right at the start we deliberately restricted our attention to the obviously common forms, while still in the field. In this way we shall probably have missed most of the obscure, minute, or subtly differentiated species, even where common. And conversely we shall have remarked any spectacular animals we may have found, however rare.

Any random group of collected materials will consist 95 to 99% of common items, with a sprinkling of rarities, and there is no way of determining at the time of collection whether a given form is rare or common. There it is. It may never have been seen before. It may never be found again. But there, at that time, it looks ordinary enough. This difficulty can never be surmounted in a short survey. All one can do is to operate on the assumption that the bulk of the total number seen will be prevalent and characteristic forms. Human nature being what it is, the big plum-blue starfish at Puerto Escondido is far more likely to stand out in one's mind, even though it may turn out to be one of the three known examples of its type on record, than the nondescript sponges which cover the rocks at every suitable location.

The plan of procedure followed these lines: After the preliminary field discrimination, we tried to differentiate our species, loosely at least, until a decently correct name could be found by which to handle them, to list them in the collecting report for the day, noting their position and abundance, and to follow this information through the specialist's determination. Then, having a definitive name, to search out the original description of the animal, if important and accessible, and to trace its history through the literature so as to determine where it may have been found, by whom, and under what circumstances. In fact, generally to become familiar with the separate species, even in the field with its multitude of diverse individuals, without at the same time losing sight of the whole picture.

The achievement of this objective was fortunately not impossible, thanks to a unique and very favorable set-up. Most expeditions have been so large that there could be no summing up for many years, if at all. Or, at best, the report was merely a rehash of field notes, with determinations which were almost wholly tentative, or written by experts who had had only secondary contact with the actual field work. In our case, everything from the original planning to the final write-up was closely co-ordinated. One or the other of us has been able to keep a finger on the pulse of the whole situation, having had a hand in the direction of the trip; in the collecting, anesthetizing, and preserving of the specimens; in the recording of field notes; in the sorting, labeling, and packaging of the material, with the later examining, departmentalizing, and re-labeling attendant upon our retaining duplicates and records of examples dispatched to specialists; through the subsequent allocation of final names to the withheld specimens, through the bibliographic research, and finally in erecting an accurate conceptual and verbal structure of the published report. And all this, through intense concentration and directionalism, has fortunately been accomplished in less than two years, before the original freshness shall have dimmed.

Although more than 500 species have been determined, the work is still far from complete. Scarcely any information is available on the anemones and little more on the Alcyonaria. These two important groups have been omitted in the phyletic catalogue and bibliography which follow. Only a fraction of the fishes have been determined, and the information on flatworms and naked opisthobranchs is incomplete.

But despite incompleteness, some generalizations emerge. Now it can be affirmed positively that, except for a few scattered exceptions, such as sponges and tunicates, the color of the Gulf littoral as a whole is distinctly tropical. With these few exceptions, the entire fauna represents an extreme northerly extension of the Panamic Province. Also, though this may be untrue of certain separate groups, there is little distinctiveness to the fauna of the Gulf. For the Panamic Faunal Province, however, as exemplified by its northerly extension in the Gulf, considerable distinctiveness can be claimed. There are but slight affiliations with animals of the north temperate zones, again excepting the sponges and tunicates, and comparatively few species are identical with those inhabiting the West Indian or Indo-Pacific areas. A good many, however, are tropicopolitan or cosmopolitan. The faunal differences between the northern and southern portions of the Gulf are well marked in most groups. For analytical purposes, we divide the region into two halves by means of a line drawn from Guaymas to Santa Rosalia. Actually, there is evidence that the line should be drawn farther south, and that a quantitative investigation might disclose that this southern fauna ranges south from La Paz and Espiritu Santo Island to Panama and northern Peru. There are even some species which occur only from Cape San Lucas south.

Of the 507 to 520 plus species treated herein, geographic ranges have been compiled for 415, not always completely. 72 seem to be limited to the Gulf. This division ordinarily includes the area contained within a line drawn from Mazatlan to Cape San Lucas but, in some of the group totals, Magdalena Bay has been included. 160 species not reported from north of the Gulf (or in some cases, Magdalena Bay) range only to the south. 42 range northwardly, 68 *both* north and south. The range to the northward is usually limited by Magdalena Bay or Cedros Island, and rarely extends beyond southern California, although some species may occur at Monterey, and a few extend even to Puget Sound or Alaska. 39 are tropicopolitan or cosmopolitan. 20 are West Indian. Only 14 are Indo-Pacific.

These figures are thought to be fairly significant. They probably reflect a representative cross section of the common littoral species, whereas studies of a single group in this region ordinarily include all species recorded from the area—rare and common, littoral and dredged. They point to a high degree of distinctiveness, not of the Gulf fauna, but of the Panamic fauna as represented by its most northerly extension.

Although only 18 of these common species occur identically in the West Indies and 14 in the Indo-Pacific, this must not be construed as reflecting slight affiliations with these districts. The generic relationships may be very close, many genera being common to both, sometimes to all three areas. And in many cases analogous species are related so closely that only an expert can determine whether a given animal belongs to the West Indian or to the Panamic species. In this connection, the § S–413 remarks are applicable. Conchologists have, in the past, arranged lists to show up to 50 species of mollusks alone, common to the Panamic and West Indies regions, and half that many to the Panamic and Indo-Pacific, but modern differentials have become so delicate that the species in question are now generally considered distinct.

Zoogeographically significant information would accrue from the compiling of percentages to show how large a portion of the Gulf fauna is limited in northward extension by Magdalena Bay, how much by the Pt. San Eugenio complex, how much by Cape San Quentin, and, finally, how much by Pt. Conception. Percentages should be adduced to indicate what proportion of the common littoral species are restricted to the northern, and what to the southern half of the Gulf. We should determine how many of the extra-limital and tropicopolitan species have been able to migrate beyond the zoogeographical barriers which restrict Panamic forms, so as to populate North Temperate or Peruvian waters. All this must await a more favorable opportunity.

A few of the more general observations might be summarized here. Both the sponge affiliations and the tunicate affiliations, curiously, are with more northern (Californian) waters, although each has species in common with the West Indies, and among the sponges, the most spectacular, large, and common forms are West Indian; smaller species are cosmopolitan or southern Californian. The Panamic affiliations of both groups are entirely nil so far as known. There are few shore hydroids; their scarcity, beside the lushness reported from the Galápagos and easily observable in southern California, is all the more remarkable. These few have mostly northern Pacific affiliations, with faint Caribbean and cosmopolitan tinges. So far as can be seen from cursory observation, except for a single *Stylatula,* there are no Alcyonarian relationships with California or even with southern California. No *Renilla* are in the collection. There is one fleshy sea-pen, a *Ptilosarcus* or *Leioptilus,* but surely not *quadrangularis* of the north Pacific. A purple pendent gorgonian, a brown fleshy gorgonian, and a sea-fan are all common. Some of the commonest anemones are 2 species of *Epizoanthus* superficially indistinguishable from West Indian forms. Bunodid anemones similar to the aggregated north Pacific *Cribrina* are very common, and there are *Cerianthus* and *Harenactis* not separable from southern California species in appearance. Corals are very abundant, and it would seem that at Pulmo we have one of the few true reefs reported from the American Pacific. The polyclad fauna is abundant and varied, there are highly colored Cotyleans in the south, and mostly nondescript Leptoplanids in the north. One Nemertean, the Panamic *Basiodiscus mexicanus,* was widely obtained. Both sipunculids and echiurids are common—commoner on the whole than they are on the California coast, and one species, *Physcosoma agassizii,* occurs in both areas. Annelids are abundant, varied, and obvious, the tropicopolitan stinging worm *Eurythoë* being one of the commonest invertebrates of the region. Encrusting Bryozoa are abundant, especially in the north, and the purple *Bugula neretina* is widespread. A many-rayed starfish, *Heliaster,* is, with the club urchin *Eucidaris,* and the cucumber *Holothuria lubrica,* and several species of ophiurans including the giant *Ophiocoma,* among the commonest of shore animals. Among the 60 species of Echinoderms, holothurians are most abundant, including some giant creeping forms and a great worm-like synaptid, with 15 species of Echinoids next in importance. Sand flats are paved with huge cake urchins and keyhole urchins. The poisonous *Centrechinus mexicanus,* fortunately, occurs sparingly. Starfish are ubiquitous and abundant. The acorn barnacle fauna is varied and abundant, mostly with cosmopolitan, northern, and unique species.

107 species of decapods were taken. Sally Lightfoot is everywhere on high rocks. Sand bugs, Xanthids, Majids, hermits, the large swimming crab *Callinectes bellicosus,* and many small decapods are abundant, although in its crab fauna the Gulf fulfills the proverbial tropical formula: many species, few individuals per species. Sand flats have hordes of fiddlers and some *Ocypode.* Spiny lobsters are occasional, the Panamic *P. inflatus* in the south, the southern Californian *P. interruptus* in the north. An important shrimp industry involving *Penaeus stylirostris* centers at Guaymas. Snapping shrimps are very abundant. The sand flats near La Paz constitute one of the few known regions where *Dentalium* can be taken intertidally. The Pelecypod fauna, with its pearl oysters and edible oysters, its huge *Pinna,* with its commensal shrimp, *Pontonia,* its great pearly and ornamented *Spondylus* and its edible clams, *Chione* and *Macrocallista,* is world famous. Among its 40 common species, the garbanzo clam *Fossularca,* the paper-shell clam, *Isognomon anomonioides,* the heavy ribbed *Carditamera affinis,* and the small mussel, *Brachidontes multiformis,* the two last especially in the north, are very common. Among the naked Gastropods, a marine pulmonate, *Onchidium,* is characteristic of the northern part, occurring in overlapping hordes high on the rocky shore. Pleurobranchids and sea-hares are spotty; a large *Dolabella* may be found. Among other nudibranchs, a highly colored swimming species, with its undulating margins resembling a Cotylean flatworm, occurs commonly in the south. More than 90 species of shelled Gastropods are common enough to have been taken along shore in our short survey. It is to this group, especially in the south, that the Gulf's more spectacular inhabitants belong. Some of the conches and murices are very large and highly colored. Olives, cones, cowries, limpets, and keyhole limpets are abundant. Pulmonate limpets were taken at several stations. A rather handsome hairy chiton, and the large *Chiton virgulatus* represent the Amphineura, along with 8 or 10

less common species. Octopi, said to be abundant, we found only occasionally. There is an important Pro-chordate fauna, all three groups being well represented. A huge *Balanoglossus* covers the sand flats with its castings. Although we collected only a few, Amphioxus confidently can be expected in the sub-littoral sand bars. Tunicates are common, mostly compound. Fishes are very noticeable. Several hundred species are known to inhabit these waters, from the swordfish, tuna, and huge manta rays, down to the sardines and to minute *Fierasfer* living in the cloaca of holothurians. We took more than 40 species without being interested in the group, and while attempting to avoid the temptation of collecting them. *Botete,* the poison-liver fish, and the puffer *Diodon,* are common especially in the south. There are important associations of pelagic animals, attracted to the boat especially at night if a light is hung over side. Under these circumstances, assemblages of worms, minute crustaceans, ribbon fish, and snake eels are likely to collect, with occasional squid, *Squilla* larvae, etc. Hordes of *Pleuroncodes* may be swept into the Gulf by currents from the open Pacific. Whales, porpoises, sharks, and great sea turtles are likely to be encountered. On one occasion, although this is outside our scope, we were visited by a great swarm of bats, a quarter of a mile or more offshore.

But beyond these summaries, the chief aim in constructing this account has been to provide a source book for material on the Panamic invertebrates. Herein also will be found a statement of the common shore animals of the Gulf, of their range —whence zoogeographers may compile their own statistics and draw their own conclusions as to how the animals arrived and from where—and a list of references from which added information can be obtained by lay reader or biologist.

Finally, this résumé of an effort to cast light onto a scene small and specific in the breadth of the world, but large in our minds, and equally large, no doubt, in the eyes of travelers and scientists concerned with it, can be closed with the remarks of the conchologist Carpenter, who touched universals in his study of a special group (1857, § S–6, p. 92): "The object of this Report has been so to condense and arrange the existing materials that those who consult it may know what has been done, and may have the means of deciding on the value to be attached to the different sources of information. Thus they may be enabled to begin where the writer leaves off, and not spend precious time in working out afresh what has already been ascertained." So now (p. 71), "we proceed . . . to the details . . . merely premising that the student must bear in mind the very unsatisfactory nature of most of our materials, and must therefore receive what follows simply as the approximation partially attainable in the present state of the science and not as absolute truth."

TABLE OF REFERENCES

PHYLUM PORIFERA

A-1 to A-26—Sponges.

PHYLUM COELENTERATA

B-1 to B-13—Hydroids.

C—Alcyonaria (not yet reported upon), Gorgonians (sea-fans, etc.), sea-pens, etc.

D—Anthozoa other than corals (not yet reported upon). Anemones.

E-1 to E-9—Corals.

PHYLUM PLATYHELMINTHES

F-1 to F-16—Polyclads. Marine flat-worms.

G-1 to G-3—Nemerteans. Ribbon worms.

PHYLUM POLYZOA OR BRYOZOA

H-1 to H-21—Bryozoa. Moss animals and certain of the corallines.

PHYLA SIPUNCULOIDEA AND ECHIUROIDEA

I-1 to I-15—Sipunculids and Echiuroids. Peanut worms, etc.

PHYLUM ANNELIDA

J-1 to J-77—Polychaets. Worms.

PHYLUM ECHINODERMATA

K-1 to K-16—References to the group as a whole.

K-101 to K-130—Asteroidea. Starfish.

K-201 to K-226—Ophiuroidea. Brittle stars.

K-301 to K-328—Echinoidea. Sea urchins, etc.

L-1 to L-42—Holothurioidea. Sea cucumbers.

PHYLUM ARTHROPODA, MOSTLY CLASS CRUSTACEA

M-1 to M-17—References to the group as a whole.

M-101 to M-200—Copepods (not yet reported upon; we got only one or two).

M-201 to M-222—Amphipods.

M-301 to M-320—Isopods.

M-401 to M-411—Schizopods and Cumacea (we got none of the latter).

M-501 to M-600—Pantopods (none taken).

M-601 to M-609—Stomatopods.

N-1 to N-19—Cirripedia, the barnacles.

O-1 to O-22—Decapods. References to the group as a whole.

P-1 to P-36—Macrurous decapods. All the Natantia.

P-101 to P-105—All the macrurous Reptantia except the Family Galatheidae. Shrimps, etc.

P-201 to P-205—Superfamily Thalassinidae of the Tribe Anomura.

Q-1 to Q-44—Anomurous decapods. All the Anomura except the subfamily Thallasinidea. Hermit crabs, etc.

R-1 to R-78—Brachyura. True Crabs.

PHYLUM MOLLUSCA

S-1 to S-47—References to the group as a whole.

S-101 to S-104—Scaphopods. Tooth shells.

S-201 to S-256—Pelecypods. Bivalves, clams, mussels, oysters, etc.

S-301 to S-414—Shelled gastropods (not including shelled Tectibranchs). Snails, etc.

T-1 to T-24—Naked Gastropods. Sea-hares, nudibranchs, and Onchiids.

T-101 to T-106—Shelled Tectibranchs. The bubble-shells.

U-1 to U-27—Amphineura. Chitons.

U-101 to U-118—Cephalopods. Squid, Octopi, etc.

PHYLUM CHORDATA

V-1 to V-13—Enteropneusts. Acorn-tongue worms.

V-101 to V-119—Tunicates. Sea squirts, etc.

V-201 to V-203—Cephalochordates. Lancelets (Amphioxus, etc.).

W-1 to W-27—Fishes.

MISCELLANEOUS

X—Groups not considered in detail. (Not yet reported upon. We got only a few. One medusa jellyfish, a bat, a gordian worm considered in the text only, etc.)

Y—Non-specific books and accounts.

Z—No references.

LIST OF ABBREVIATIONS

The uniform method of bibliographic citation employed herein involves abbreviations according to the following list, and volume, page, etc., numbers according to a definite schedule. If there is a series number which should be shown, this is indicated in parentheses before the volume number. If there is a Part, or Section, or Fascicule, or Number, which should be shown as an important direction, this is indicated in parentheses after the volume number. This symbol is followed by a colon which precedes the number of pages, plates, and figures, thus: (5) Vol. 1 (2): 212-216, Pl. 1; which indicates: Series 5, Volume 1, Part (or Number) 2, pages 212-216, Plate 1.

Following are the standardized abbreviations used to indicate periodicals. The letters shown in roman face type are those which are used as abbreviations.

Albatross abbreviations, see end of this section.

A*llan* Hancock Paci*fic* Exp*editions,* Allan Hancock Foundation, University of Southern California, Los Angeles, California

Amer*ican* Journ*al of* Sci*ence* (American Journal of Science and Art, 1820-79), New Haven, Connecticut

Amer*ican* Mid*land* Nat*uralist,* University of Notre Dame, Notre Dame, Indiana

Amer*ican* Mus*eum* Journ*al,* American Museum of Natural History, New York, N. Y.

Amer*ican* Mus*eum* Nov*itates,* American Museum of Natural History, New York, N. Y.

Amer*ican* Nat*uralist,* Boston (published at various times elsewhere: Salem, Mass., New York, N. Y., etc.)

An*ales del* Inst*ituto de* Biolog*ía de la Universidad de México*

Ann*ales de l'*Inst*itut* Océano*graphique,* Monaco (Paris)

Ann*als and* Mag*azine of* Nat*ural* Hist*ory,* London

Ann*als of the* Lyc*eum of* Nat*ural* Hist*ory,* New York, N. Y. (now New York Academy of Sciences)

Ann*ual* Rep*ort,* Smith*sonian* Inst*itution,* Washington, D. C.

Ann*ual* Rep*ort,* Tortugas Lab*oratory,* Carn*egie* Inst*itution of Washington,* Washington, D. C.

Ark*iv* för Zool*ogi utgivet av* K. Svenska Vetens*kapsakademien,* Stockholm

Bos*ton* Journ*al of* Nat*ural* Hist*ory*

Boll*ettino dei* Mus*ei di* Zool*ogia ed* Anat*omia* comp*arata della R.* Univ*ersità di* Torino, Italy

Boll*ettino della* Soc*ietà di* Nat*uralista in* Napoli, Italy

Bronn's Klass*en* un*d* Ord*nungen* de*s* Tier-Reichs, Leipzig

Bull*etin du* Mus*éum d'*Hist*oire* Nat*urelle,* Paris

Bull*etin de la* Soc*iété* Phil*omathique de Paris,* Paris

Bull*etin of the* Amer*ican* Mus*eum of* Nat*ural* Hist*ory,* New York, N. Y.

Bull*etin of the* Bingham Oceanog*raphic* Coll*ection,* Peabody Museum of Natural History, Yale University, New Haven, Connecticut

Bull*etin of the* Bur*eau of* Fish*eries,* Washington, D. C. *See also,* Bull. U. S. Fish Comm.

Bull*etin of the* Mus*eum of* Comp*arative* Zoöl*ogy at Harvard College,* Cambridge, Massachusetts. *See also, Albatross* abbreviations at end of section.

Bull*etin of the* Scripps Inst*itution of* Ocean*ography of the University of California,* La Jolla, California

Bulletin of the United States Fish Commission, Washington. See also, U. S. Fish Comm. Bull.

Bulletin of the United States National Museum, Washington, D. C.

Bulletin of the Vanderbilt Marine Museum, Huntington, Long Island, New York

Bulletin of the New York Zoological Society. See also, Zool. Soc. Bull.

Bulletins of the Laboratories of Natural History, State University of Iowa, Iowa City

Ciencia, Mexico City, Mexico

Comptes Rendus du Académie Scientifique, Paris

Field Museum of Natural History, Zoological Series, Chicago, Illinois

First (1st) Annual Report, Laguna Marine Laboratory, Pomona College, Claremont, California

Jenaische Zeitschrift für Naturwissenschaft, Jena, Germany

John Murray Expedition, *1933-34*, Scientific Reports, The British Museum (N. H.), London

Journal of Entomology and Zoology, Pomona College, Claremont, California. Also, Pomona College Journal of Entomology and Zoology

Journal of Malacology, London

Journal of the Washington Academy of Sciences, Washington, D. C.

Kongliga Svenska Vetenskaps-Akademiens Handlingar, Stockholm, Sweden

Meddelelser fra det Zoologiske Museum, Oslo, Norway

Memoirs of the National Academy of Sciences, Washington, D. C.

Mémoires de l'Académie Impériale des Sciences de St. Pétersbourg, now Akademïia nauk, Leningrad

Mémoires de la Société de Physique et d'Histoire Naturelle de Genève, Switzerland

Memoirs of the Boston Society of Natural History

Memoirs of the Museum of Comparative Zoölogy at Harvard College, Cambridge, Massachusetts. *See also, Albatross* abbreviations at end.

Mission Scientifique au Mexique et dans l'Amérique Centrale, Paris

Mitteilungen des Naturhistorisches Museum, Hamburg

Nautilus, *Philadelphia.* Now being published through the Department of Zoology at the University of Pennsylvania

New York Academy of Sciences, Scientific Survey of Porto Rico and the Virgin Islands

Philippine Journal of Science, Manila, P. I.

Pomona College Journal of Entomology and Zoology. See Journal

Proceedings of the Academy of Natural Sciences, Philadelphia, Pennsylvania

Proceedings of the American Philosophical Society, Philadelphia, Pennsylvania

Proceedings of the Biological Society of Washington

Proceedings of the Boston Society of Natural History

Proceedings of the California Academy of Sciences, San Francisco, California

Proceedings of the Cambridge Philosophical Society, Cambridge, England

Proceedings of the Essex Institute, Salem, Massachusetts

Proceedings of the Malacological Society, London

Proceedings of the New England Zoological Club, Cambridge, Massachusetts

Proceedings of the United States National Museum, Washington, D. C.

Proceedings of the Washington Academy of Sciences, Washington, D. C.

Proceedings of the Zoological Society of London

Publications of the University of California at Los Angeles in Mathematical and Physical Sciences

Report of the . . . Meeting of the British Association for the Advancement of Science

Revista de la Sociedad Mexicana de Historia Natural, Museo de la Fauna y Flora, Mexico

Reisen im Archipel der Philippinen, von C. Semper, Jena. *Also as:* Semper Reisen, etc.

Revue et Magasin de Zoologie pure et appliquée, Paris

Records of the Indian Museum, Calcutta

Smithsonian Miscellaneous Collections, Washington. *Also* Smithsonian Edition

Transactions of the Connecticut Academy of Arts and Sciences, New Haven

Transactions of the Linnaean Society, London

Transactions of the San Diego Society of Natural History, San Diego, California

Transactions of the Royal Canadian Institute, Toronto

Transactions of the Royal Society of Edinburgh

United States Fish Commission Bulletin. *See* Bulletin

University of California Publications in Botany, Berkeley, California

University of California Publications in Zoology, Berkeley, California

Videnskabelige Meddelelser fra Dansk naturhistorisk Forening i Kobenhavn, Denmark

Zeitschrift für wissenschaftliche Zoologie, Leipzig

Zoologica, Scientific Contributions of the New York Zoological Society

Zoologische Jahrbücher, Abtheilung (now Abteilung) für Systematik, ökologie und geographie der Tiere, Jena

Zoologischer Anzeiger, Leipzig

"*Albatross* 1891" will be understood as an abbreviation for: "Reports on an exploration off the west coasts of Mexico, Central, and South America, and off the Galápagos Islands, in charge of Alexander Agassiz, by the U. S. Fish Commission Steamer *Albatross*, during 1891. . . ."

"*Albatross* 1899-1900" will be understood as an abbreviation for: "Reports on the scientific results of the expedition to the tropical Pacific, in charge of Alexander Agassiz, by the U. S. Fish Commission Steamer *Albatross*, from August 1899, to March 1900. . . ."

"*Albatross* 1904-1905" will be understood as an abbreviation for: "Reports on the scientific results of the expedition to the eastern tropical Pacific, in charge of Alexander Agassiz, by the U. S. Fish Commission Steamer *Albatross*, from October 1904, to March 1905. . . ."

(Glossary will be found on page 587.)

PHYLETIC CATALOGUE

Phylum Porifera (Sponges)

REFERENCES

§ A–1 Carter, H. J. 1882. "Some sponges from the West Indies and Acapulco in the Liverpool Free Museum described, with general and introductory remarks." Ann. Mag. Nat. Hist. (5) Vol. 9: 266-301, Pls. 11 and 12, and 346-368.

An interesting account embodying a pleasant and dignified critique of scientific writing. Classification obsolete. Bibliographic citations include Duchassaing et Michelotti 1864 which describes sponges from the Caribbean now known to occur also in the Pacific. The following Panamic species are treated from Acapulco: *Tuba acapulcaensis* n. sp., p. 279; *Reneira fibulata* Sdt., p. 284; *Halichondria isodictyalis* n. sp., p. 285; *Donatia multifida* n. sp., p. 358 with natural size toto-illustration as fig. 22, Pl. 12; all dredged in 4–9 fathoms.

§ A–2 Hyatt, A. 1877. "Monograph of North American Poriferae, Part II," Mem. Boston Soc. Nat. Hist., Vol. 2 (18): 481-554, Pls. 15-17.

The only Panamic sponge is from Zorritos, northern Peru: *Spongelia ligneana* Hyatt, p. 539, with no data on spiculation, and no illustration, although there is a fair toto-illustration of *dubia* (from Bahamas) and of *dubia* var. *excavata,* which it much resembles, from Cape Florida (p. 534 and 535) as figs. 9-11, Pl. 17.

§ A–3 Laubenfels, M. W. de 1932. "The marine and fresh-water sponges of California." Proc. U.S.N.M., Vol. 81 (4): 1-140, 79 text figs.

Monographic. 5 of the sponges we found in the Gulf are treated as occurring here also. Extensive bibliography cites: Schulze 1899 which deals with Hexactinellids of the Galápagos to Alaska area, and which is beautifully illustrated, but chiefly concerned with species from deep and from very deep water; Lendenfeld 1915 which, again, is concerned only with very deep to abyssal Hexactinellids, and other apropos papers.

§ A–4 .... 1935. "Some sponges of Lower California (Mexico)." Am. Mus. Novit. No. 779, 14 pp., 9 text figs.

REFERENCES

Material collected on the *Albatross* 1911 cruise, presumably in the Gulf, since most of the shallow-water invertebrate collections were taken there. 22 species and varieties, of which 9 are new. No data as to station or depth was available to the author. Some of the material is obviously intertidal. Bibliography with de Laubenfels 1930, etc. The faunal affiliations are found to be with California and with the Indo-Pacific.

§ A–5 .... 1936. "A discussion of the Sponge Fauna of the Dry Tortugas in particular and the West Indies in general, with material for a revision of the families and orders of the Porifera." Papers from the Tortugas Laboratory, Carnegie Institution, Vol. 30: 225 pp., 22 pls.

Key paper for the sponges. Diagnoses of all families and orders. Mention of all known genera allocated into a new classification. Treatment of all genera and species known from the West Indies, and hence including several forms we found in the Gulf, often with excellent toto-illustrations, sometimes of the living animals. Comprehensive bibliography.

§ A–6 .... 1936a. "A comparison of the shallow-water sponges near the Pacific end of the Panama Canal with those at the Caribbean end." Proc. U.S.N.M., Vol. 83: 441-466, text figs. 40-45.

Ecological and geographical considerations on p. 441-42. Reports great differences between the sponges of California and Lower California, to neither of which the Panama-Pacific species are related. 15 Pacific species of which 5 are new. Collecting at Panama City "yielded nine species and proved astonishingly similar to collecting near Plymouth, England. Four out of the nine . . . are forms common to both localities." Three other species occurring near by are Arctic and Antarctic! Six species are common to the Atlantic and Pacific entrances, of which only two are cosmopolitan. Bibliography.

§ A–7 .... 1939. "Sponges collected on the Presidential Cruise of 1938." Smith. Misc. Coll., Vol. 98 (15): 7 pp., 1 fig.

9 species taken at Galápagos, only one of which occurred at Magdalena Bay—the only sponge collected there. One of the Galápagos sponges represented a new species, *Merriamium roosevelti*, p. 1. Bibliography.

§ A–8 Lendenfeld, Robert von 1910. "The Sponges I. The Geodidae" and "The Sponges II. The Erylidae." *Albatross* 1904-05, and *Albatross* 1888-1904. Mem. Mus. Comp. Zool., Vol. 41 (1): 260 pp., 48 pls., and (2): 265-323, 8 pls.

Many of the dredging stations at which these Tetraxonian sponges were collected were in shallow water, 50 to 100 meters; the extremes

REFERENCES

were 18 to 205 fathoms. *Geodia ataxastra* sp. nov., and G. *media* Bowerbank, p. 194, were taken on shore at Panama, the latter being reminiscent of some of the specimens we saw in the Gulf. Altogether, some 13 new species are described and two more considered within our region, but de Laubenfels regards most of them as belonging to a single species, and warns us against placing too much dependency on any of von Lendenfeld's work. Excellent lithographic illustrations, including photographs of the entire sponges.

§ A–9 Wilson, H. V. 1904. "The Sponges." *Albatross* 1891. Mem. Mus. Comp. Zool., Vol. 30 (1): 164 pp., 26 pls.

47 species and subspecies, of which 33 are new. There are 26 Hexactinellida, 7 Tetractinellida, and 14 Monaxonida. No calcareous or horny sponges were taken, mostly deep water material being involved, although the station at which the greatest number of species was taken, 3405, was also the shallowest, 53 fathoms, and one other was scarcely deeper. Sponges from these two fairly shallow stations are described, with excellent toto-photographs, on pp. 77, 105, 115, 118, 121, 136, 139, 146, 149, 151, 155, and 158. Bibliography.

§ A–10 ACKNOWLEDGMENT: For determining the following species, for suggestions and help with the literature, and for checking over this portion of the phyletic catalogue, we are grateful to Dr. M. W. de Laubenfels of Pasadena, California.

LIST OF SPECIES TAKEN:

§ A–11 *Aaptos van namei* de Laubenfels 1935
de Laubenfels 1935 (§ A–4), p. 8, no station data, presumably from the Gulf.

FAMILY SUBERITIDAE of the ORDER HADROMERINA. As "a very dark sponge that often assumes a domed shape" at Pt. Lobos on Espiritu Santo Island. A smooth black encrusting sponge at Coronado Island. A domed or flat top very dark form, yellow inside, at Puerto Escondido. Taken also at Puerto Refugio. Related to the gigantic *Spheciospongia* (p. 48, de Laubenfels 1932, § A–3) which is treated in Ricketts and Calvin 1939 (§ Y–3).

§ A–12 *Chondrosia reniformis* Schulze 1877
Schulze, F. E. 1877. "Untersuchingen über den Bau und die Entwicklung der Spongien, III. Die Familie der Chondrosidae." Zeit. f. wiss. Zool., 29: 87-122, Pls. 8 and 9. Page 97, with toto-figures on Pl. 8. Reported to be cosmopolitan by de Laubenfels in a personal communication.

FAMILY CHONDROSIIDAE of the ORDER CARNOSA. These are the very smooth, slate-colored encrustations which are almost indistinguishable from the compound tunicate *Cystodytes* (§ V–109) except on

REFERENCES

careful study. Pulmo Reef. A similar form, superficially almost identical with the tunicate *Phallusia nigra* in the Tortugas, is treated by de Laubenfels 1936 (§ A–5) on p. 183.

§ A–13 *Cliona celata* Grant 1826 PL. 22 FIG. 3

de Laubenfels 1932 (§ A–3) treats the subspecies *californiana* on p. 47.

FAMILY CLIONIDAE of the ORDER HADROMERINA. This is the very obvious, massive vase-like sponge found in the lowest intertidal as an important feature, at Angeles Bay, on gravel flats with boulders and *Sargassum*. Noted also at Port San Carlos, Sonora. The yellow *Cliona*-like sponge which honeycombs the rocks east of La Paz and at Puerto Escondido may belong here.

§ A–14 *Geodia mesotriaena* von Lendenfeld 1910

von Lendenfeld 1910 (§ A–8), p. 96 with toto-figs. 1 and 2, Pl. 21 (equals also his *agassizii, mesotrianella, breviana,* and *ovis,* all 1910, and mostly with good toto-illustrations), southern California, 20 to 205 fathoms.

de Laubenfels 1932 (§A–3), p. 25, Alaska to southern California, all dredged.

FAMILY GEODIIDAE of the ORDER EPIPOLASIDA. The commonest sponge encountered on shore in the Gulf. Curious that a California dredged animal should be found intertidally here. De Laubenfels notes, however, that our specimens may turn out to be the cosmopolitan *Geodia gibberosa* Lamarck, reported from both Atlantic and Pacific sides of Panama (de Laubenfels 1936a, § A–6, p. 454), and treated in greater detail on p. 172 of de Laubenfels 1936 (§ A–5), from the West Indies. Taken at Pt. Lobos on Espiritu Santo Island, east of La Paz on the drowned coral flats, on Pt. Marcial Reef, at Coronado Island, at Concepcion Bay, at Puerto Refugio, and at San Gabriel Bay on Espiritu Santo Island. Typically as a white encrusting amorphous to massive form, a similar species, or possibly a variety, has black, dome-like excrescences, slightly concave on the top.

§ A–15 *Haliclona ecbasis* de Laubenfels 1930

de Laubenfels 1932 (§ A–3), p. 117, from southern California. Reported also by de Laubenfels 1935 (§ A–4), p. 2, from Lower California, probably from the Gulf.

FAMILY HALICLONIDAE of the ORDER HAPLOSCLERINA. Noted in the collecting reports as "small cratered sponge, looks like our *Haliclona permollis*" (at Pt. Lobos on Espiritu Santo Island), and as a "lavender erect sponge on hacha shells, like the Monterey *Haliclona*" (Concepcion Bay). Another example of anomalous distribution. The cosmopolitan *H. permollis* occurs at Monterey (*cinerea,* p. 120, § A–3 becomes *permollis,* p. 444, § A–6), at Panama (p. 444, § A–4, very

REFERENCES

common on the rocky tidal flats), and in the Galápagos (p. 1, § A–7), but specimens taken in-between, where ordinarily a tropical fauna can be expected, comprise a more northern species.

§ A–16 *Hircinia variabilis* (Schmidt) Schulze 1862
de Laubenfels 1936 (§ A–5), p. 19, excellent toto-photograph of dried specimen as fig. 1, Pl. 4. "One of the most abundant species in the vicinity of Tortugas." West Indies, Isthmus of Panama, possibly Mediterranean.

FAMILY SPONGIDAE of the ORDER KERATOSA. Taken only once, on the drowned coral flats east of La Paz, where it occurred very commonly in arborescent clusters in the coral heads. Closely related to the commercial sponges. Brick red base and large blackish oscula.

§ A–17 *Hymeniacidon* spp.
Several species are treated by de Laubenfels 1936 (§ A–5), p. 137, and 1932 (§ A–3), p. 57, especially *H. sinapium* from southern California.

FAMILY HYMENIACIDONIDAE of the ORDER HALICHONDRINA. A thin, encrusting, often highly colored sponge with minute volcano-like oscules. One of a group of genera formerly lumped under the collective and better known name of *Axinella*. Our only representative was taken as a high tide pool sponge at Puerto Refugio.

§ A–18 *Leucetta losangelensis* (de Laubenfels) 1930 PL. 23 FIG. 1
de Laubenfels 1932 (§ A–3), p. 13, southern California, low intertidal, especially where there is strong wave action. de Laubenfels 1935 (§ A–4), p. 2, on oyster shells, presumably from the Gulf.

FAMILY LEUCETTIDAE of the ORDER ASCONOSA. A white calcareous sponge, growing in great mildly encrusting to flat-massive clusters which are very much a feature of the intertidal landscape under rocks and in overhangs at Cape San Lucas, at Pt. Lobos on Espiritu Santo Island, at Amortajada Bay on San José Island, at Coronado Island, at San Francisquito Bay, at Puerto Refugio, and at San Gabriel Bay on Espiritu Santo Island. Very abundant and widespread, perhaps outranking *Geodia* as the commonest sponge encountered.

§ A–19 *Leuconia heathi* (Urban) 1905
de Laubenfels 1932 (§ A–3), p. 12, Monterey Bay and southern California, intertidal and dredged to 78 meters in the south, intertidal only at Monterey.

FAMILY GRANTIIDAE of the ORDER ASCONOSA. As a Grantia-like sponge in the tide pools at Puerto Refugio.

REFERENCES

§ A–20 *Leucosolenia coriacea* Montagu
Topsent 1936, p. 2. Bull. Inst. Ocean. Monaco, No. 711. Reported to be cosmopolitan by de Laubenfels in a personal communication.

FAMILY LEUCOSOLENIIDAE of the ORDER ASCONOSA. Recorded only once, from Amortajada Bay on San José Island.

§ A–21 *Spirastrella* sp.
The genus is discussed on p. 142 of de Laubenfels 1936 (§ A–5).

FAMILY CHOANITIDAE of the ORDER HADROMERINA. A white encrusting sponge from the south end of Tiburon Island.

§ A–22 *Steletta estrella* de Laubenfels 1932
de Laubenfels 1932 (§ A–3), p. 31, southern California, intertidal and dredged to 41 meters.

FAMILY ANCORINIDAE of the ORDER CHORISTIDA. One of the stinging sponges, the spicules of which are very sharp and bristly, enter the skin very readily, and can be extracted only with difficulty. Taken only once, on the rocky shore at Angeles Bay.

§ A–23 *Tedania ignis* (Duchassaing and Michelotti) 1864
PL. 12 FIG. 2
de Laubenfels 1936 (§ A–5), p. 89, "exceedingly abundant in shallow water in the vicinity of the Dry Tortugas—it is quite possible that literally bushels of this sponge could be accumulated by a few hours' collecting . . . The color in life is a brilliant vermilion." A red, erect, and large sponge which at Concepcion Bay was very much a feature of the slightly sublittoral gravel flats, on rocks, on shells of *Pinna,* etc. The tissue is very spongy, the flamboyant color fades rapidly to a coppery green. De Laubenfels remarks, in a personal communication, that this may turn out to be the cosmopolitan *T. anhelans.* FAMILY TEDANIDAE of the ORDER POECILOSCLERINA.

§ A–24 *Tethya aurantia* (Pallas) 1766
de Laubenfels 1932 (§ A–3), p. 44, in the var. californiana, from the Monterey Bay area intertidal, and from southern California intertidal and dredged.
de Laubenfels 1935 (§ A–4), p. 11, from Lower California, presumably from the Gulf. Stated to be cosmopolitan. The Panama specimens, however, are *T. diploderma* (p. 451, de Laubenfels 1936a, § A–6), noted also as cosmopolitan.

REFERENCES

FAMILY TETHYIDAE of the ORDER EPIPOLASIDA. Usually hemispherical, yellow, woody, with surface "superficially warty with mushroom-shaped elevations about 2 mm. high, crowded over the upper portion." (p. 44, § A-3) Very common at Puerto Escondido, at Angeles Bay on the rocks, and at the south end of Tiburon Island.

§ A-25 Partly because of improperly preserved material, resulting from inadequate laboratory facilities and insufficient stocks of alcohol, some of our sponges were undetermined, and are probably indeterminable. A cake-like specimen brought up by the shrimp trawler south of Guaymas may be a species of *Cliona*. So also may be massive chunks found at Puerto Escondido, in appearance very distinct from the *C. celata* at Angeles Bay. A distinctive *Geodia* with slightly concave flat black top, very common at Puerto Escondido, may be other than *G. mesotriaena*. Collecting directions, calculated to insure identifiable material, have been supplied by de Laubenfels. Mere drying, without any attempt at preservation whatsoever, results in material superior to that preserved in formalin. Alcohol only should be used. The approved method is to excise a small portion of the colony, while still in the field, to preserve it in alcohol then and there, and to store it with the remainder of the colony which can be dried.

SUMMARY

§ A-26 A total of 14 species, all intertidal or very slightly subtidal, has been identified from our sponge collections. Other material which may have comprised additional species could not be determined. Two of the large species, the cosmopolitan *Cliona celata,* and the common West Indies *Tedania ignis,* were among the most abundant and spectacular of the shore invertebrates at two only stations. A cosmopolitan *Tethya,* and two southern California forms (the former known only from dredgings)—*Geodia mesotriaena* and *Leucetta losangelensis*—were among the more ubiquitously abundant species. Geographical data is lacking for *Hymeniacidon* sp. and *Spirastrella* sp. Of the remainder, five are known from southern California (one only extending so far north as Monterey), two are West Indian, four are cosmopolitan. Only one seems to be restricted to the Gulf. The affiliations with southern California therefore are very close, with the West Indies less so, with the North Pacific or with Panama itself practically zero. Only one species, *Leuconia heathi,* aside from the cosmopolitan *Tethya aurantia* is common both to Monterey and the Gulf. But between the 16 species reported from along shore at Panama by de Laubenfels, and the 14 we took in the Gulf, there is not one single species in common, although among the Echinoderms and worms the conformity may run up to 50%. Both in their northern affinities and in their lack of uniqueness, the sponges differ from all other groups so far examined in the area. Just why this fauna should refute the generalizations applicable to other Gulf animals remains something of an enigma. Interesting is the fact that between certain sponges and tunicates, or between certain tunicates and sponges, there is resemblance so striking as to suggest mimicry. *Didemnum* (§ V–111) may be mistaken for a white sponge. *Cystodytes* (§ V–109) specifically resembles *Chondrosia.*

Phylum Coelenterata

CLASS HYDROZOA—HYDROIDS ONLY

REFERENCES

§ B–1 Fraser, C. McLean 1937. Hydroids of the Pacific Coast of Canada and the United States. pp. 207, pls. 44, Univ. Toronto Press.

Key paper for Pacific hydroids. Some 236 species are briefly characterized and illustrated, each with a natural-size toto-drawing, and with one or more detail drawings x20 to show taxonomic features. Section on geographic and bathymetric distribution. Complete bibliography.

§ B–2 .... 1938. "Hydroids of the 1934 Allan Hancock Pacific Expedition." A. Hancock Pac. Exp., Vol. 4 (1): 1-104, 15 pls.

Results mostly of fairly shallow dredging, usually under 100 fathoms, but of the 173 species taken in the area from middle Lower California to Ecuador, including the oceanic islands, 60 were found also on shore, chiefly in the south. No collections were made in the Gulf. Dr. Fraser was a member of this expedition, and personally attended to most of the hydroid collecting, so that the report emerges from that rare and most perfect combination: a specialist who is to write the account doing the actual work in the field. 73 species are described as new. Of the remaining one hundred, 77 have been reported also from the north Atlantic, 52 also from the northeastern Pacific, and 13 from the southeastern Pacific including the waters south of the tip of South America; 42 of these having been reported from two or more of the areas, Toto- and detail drawings of all new species. Good bibliography including: S. F. Clarke 1894 and 1907; Agassiz *Albatross* reports (total of 20 species entirely from deep water); the Nutting 1900 monograph (which mentions *Aglaophenia octocarpa* and *Lytocarpus philippinus* since taken by the Hancock Expeditions from the Panamic area): and the A. B. Hastings 1930 report of the Galápagos gymnoblast *Zanclea protecta,* not since taken.

The following is from p. 5. "The distribution of species in the whole area, with the possible exception of the northern portion of the coast of Lower California, indicates strong affinities between the hydroid fauna here and that in the North Atlantic, more particularly in the West Indian area. There is evidence of some continuity of distribution in the California-Lower California area but practically no indication of such continuity at the southern extremity."

§ B–3 .... 1938a. "Hydroids of the 1936 and 1937 Allan Hancock Pacific Expeditions." A. Hancock Pac. Exp., Vol. 4 (2): 107-126, 3 pls.

47 species collected in the Gulf of which 45 were dredged only. 10 species on the Pacific coast of northern Lower California, all dredged.

REFERENCES

Only one of these was common to both regions. Of the 56 species considered, 10 are new, with x20 and x40 detail drawings.

§ B-4 .... 1938b. "Hydroids of the 1932, 1933, 1935, and 1938 Allan Hancock Pacific Expeditions." A. Hancock Pac. Exp., Vol. 4 (3): 129-152, 3 pls.

Mostly dredged forms, from Peru to Lower California (no collecting in the Gulf), but 11 species were taken along shore, 2 in the Peruvian, and 9 in the Panamic area. However, of the total of 99 species, many were found in water under 20 fathoms, and some in 5 fathoms or less. 12 new species are described with detailed figures x20 and x40.

§ B-5 .... 1939. "Distribution of the hydroids in the collections of the Allan Hancock Expeditions." A Hancock Pac. Exp., Vol. 4 (4): 155-178.

Brings the situation to date in the area involved, from the southern boundary of the United States through Peru. Analytical tables of distribution to show occurrence within the area, and to show the distribution elsewhere of such species as occur also outside the area involved. Index to Vol. 4.

§ B-6 ACKNOWLEDGMENT: For determination of the hydroids, for the gift of literature, and for many other assistances at various times, we have to thank Dr. C. McLean Fraser, Professor Emeritus of Zoology at the University of British Columbia, Vancouver, B. C.

LIST OF SPECIES TAKEN:

§ B-7 *Aglaophenia longicarpa* Fraser 1938a

Detail figures 2a etc., Pl. 16, 24 fathoms, Espiritu Santo Island, p. 112, Fraser 1938a (§ B-3). Recorded otherwise only from 10-15 fathoms, Isabel Island, Fraser 1938b, p. 135 (§ B-4).

A coarse, large, dark-colored, colonial ostrich-plume hydroid, apparently the gulf analogue of the north temperate *A. struthionides.* Taken only once, at Puerto Escondido, where there were pure culture stands many square feet in extent, in the channel and just within the enclosure, immediately below low tide and exposed possibly for a few minutes on the lowest tides.

§ B-8 *Aglaophenia diegensis* Torry

Toto-drawing as figs. 212a on Pl. 40, p. 175, Fraser 1937 (§ B-1). Fraser 1938, p. 56 (§ B-2). Fraser 1938a, p. 111 (§ B-3). Total range: San Francisco to Galápagos, shore to 30 fathoms.

A comparatively delicate, elongate ostrich-plume, about midway between the above *longicarpa* and *Plumularia setacea* in this respect. Solitary or in clumps of a few stems. The commonest littoral hydroid

REFERENCES

in the Gulf, especially in the north. Taken at Marcial Point, Puerto Refugio, south end of Tiburon Island, and at Port San Carlos in Sonora.

§ B–9 *Plumularia setacea* (Ellis)

Toto-drawings as figs. 213a on Pl. 44; Sitka to San Diego, low tide to 90 fathoms, p. 191, Fraser 1937 (§ B–1). Fraser 1938, p. 66 (§ B–2); Clarion Island and Galápagos, low tide to 68 fathoms. Fraser 1938a, p. 111 (§ B–3), Espiritu Santo Island, 24 fathoms.

A delicate, almost transparent form, common also along the California coast (p. 65, fig. 21, Ricketts and Calvin 1939 [§ Y–3]). We collected this only once in the Gulf, on the flats east of La Paz where it is fairly common in grottoes and overhangs of rocks and on the coral *Porites porossa.*

§ B–10 *Obelia dichotoma* (Linn.)

Toto-drawing as figs. 86a on Pl. 17, p. 85, Fraser 1937 (§ B–1). Fraser 1938, p. 36 (§ B–2). Fraser 1938b, p. 133 (§ B–4). Alaska to Ecuador, low tide to 50 fathoms.

A delicate, small, single-stemmed form (not profusely branching). On carapace of tortoise-shell turtle pelagic southwest of Pt. Abrojos. With *O. plicata,* attached to gorgonian skeleton embedded in sandy mud of Estero de la Luna, Sonora. A personal communication from Dr. Fraser notes that this species had not been reported previously from the Lower California region.

§ B–11 *Obelia plicata* Hincks

Toto-drawings as figs. 94a on Pl. 18, p. 90, Fraser 1937 (§ B–1). Fraser 1938, p. 38 (§ B–2). Fraser 1938a, p. 109 (§ B–3). Alaska to San Francisco Bay; thence Santa Maria Bay to Ecuador, Galápagos, etc.

More profusely branching and bushy than the above, with tufts sometimes several inches long. Taken only once, attached with *O. dichotomae* to gorgonian skeleton embedded in Estero de la Luna, Sonora.

§ B–12 *Sertularia versluysi* Nutting

We have been unable to find any toto-drawing. fig. 4, Pl. 1, is fairly diagnostic, referring to the description on p. 63 in Nutting 1904, "American Hydroids. Part II. The Sertularidae," U.S. Nat. Mus. Spec. Bull., 325 pp., 41 pls. Fraser 1938, p. 55 (§ B–2), Galápagos, low tide.

A minute (½") buff-colored form taken under rocks and overhangs at the 0.0′ level and below at Cape San Lucas. A West Indian species. In a personal communication, Dr. Fraser remarks that it has not been reported heretofore from the Lower California region.

SUMMARY

REFERENCES

§ B–13 During our spring 1940 trip, shore hydroids were the rarest animals in the Gulf, especially by contrast to the north Pacific region with which we were more familiar. This situation is borne out by the Hancock reports which list only two littoral Gulf hydroids, *Aglaophenia diegensis* (which we found to be fairly widespread and common), and *A. pinguis* which we failed to find (p. 111, [§ B–3] Fraser 1938a). The six species found there reverse, in their affinities, the conclusions reached through the study of several other groups. There are no Indo-Pacific affiliations whatsoever. The commonest form (*A. diegensis*) can be considered Panamic since it has not been reported outside the California-Galápagos area, and it seems to be far more common in the Gulf than in California, certainly than it is at Monterey Bay where we have never knowingly taken it. *A. longicarpa* would seem to be a local form, limited to the Gulf and to the area just to the south. *Plumularia setacea* is a Pacific north temperate species; *Sertularia versluysi* is West Indian; and the two species of *Obelia* are more or less cosmopolitan. Fraser, through his studies of the shore and dredged hydroids from Ecuador to Cedros Island, concludes that their affiliations are with the North Atlantic, especially with the West Indian region, except for those of the northern part of Lower California, where the affiliations are with the north temperate Pacific coast. It seems obvious, from our findings, (only 6 littoral species) and from an examination of the literature (wherein only 2 littoral species are reported) that the hydroid fauna of the Gulf is largely sublittoral—in the sense of subtidal.

SUB-ORDER MADREPORARIA, Stony Corals

§ E–1 Joubin, M. L. 1912. "Carte des bancs et récifs de Coraux (Madrépores)." Ann. Inst. Ocean., Vol. 4 (2): 7 pp., 5 maps.

On Map number 5, which shows the area on the Pacific coast from Mazatlan south, the exact location of all Panamic coral banks known at that time is indicated. Submarine banks, indicated by red cross hatching, are shown at Mazatlan, Isabel Island, Cape Corrientes, Maldonato, and to the south. Exposed banks, indicated by solid red, are shown at Tehuantepec, Acajutla, and to the south.

REFERENCES

§ E–2 Palmer, R. H. 1928. "Fossil and recent corals and coral reefs of Western Mexico." Proc. Amer. Phil. Soc., Vol. 67 (1): 21-31, 3 pls.

Mention of the deficiency of American coral reefs in the Pacific as compared with Atlantic shores. Ecology and temperature limits of rock-living shore forms. Occurrence of Miocene or Pliocene reef at the head of the Gulf of California, composed of genera now living in the Caribbean. Marked differences between recent coral fauna of the Atlantic and Pacific. Small true reef of *Pocillopora capitata* var. *robusta* Verrill at Puerto Angelito, Oaxaca, "closely related to Samoan and also Hawaiian species. So close is this relationship that the species are separated on very small and unimportant details." (p. 25) Description of species includes *Porites panamensis* Verrill 1866, the Pleistocene *Oculina* sp., *Astrangia browni* n. sp., *A. oaxacensis* n. sp. (which Wells considers synonymous with *A. pedersenii* Verrill 1864), and *Pocillopora palmata* n. sp., all with good toto-illustrations.

§ E–2a Verrill, A. E. 1866. "On the corals and polyps of Panama, with descriptions of new species." Proc. Bost. Soc. Nat. Hist., Vol. 10: 323-333, no figs.

Report on collection sent by Bradley, with statement of the Anthozoa situation to date at Panama. 25 species are described or cited in the genera *Renilla, Gorgonia, Muricea, Echinogorgia, Sympodium, Zoanthus* (with reference to a LeConte 1851 Proc. Phil. Acad. Nat. Sci. article): *Porites, Stephanocora* (new), *Astrangia, Phyllangia,* and *Ulangia.*

§ E–2b .... 1869. "Review of the corals and polyps of the west coast of America." Trans. Conn. Acad., Vol. 1 (6): 377-558, 6 pls.

With the following, comprises a résumé of all the Anthozoa then known from the whole Pacific coast, pennatulids, gorgonians, alcyonarians, actinians, zoanthideans, antipatharians, and madreporarians.

§ E–2c .... 1869a. "On the geographic distribution of the polyps of the west coast of America." Ibid. (7): 558-570.

Lists of species from provinces which he characterizes as: Arctic, Sitkan, Oregonian, Californian, and Panamanian. The last was at that time the best known, with 104 species recorded. These lists (except for the last) have little modern value, due to synonymy and to added information (especially in the form of collecting reports from what were then unexplored areas), and the deductions are equally invalid. Detailed lists of species collected at La Paz, at Guaymas, at Nicaragua, and in the three South American provinces considered.

§ E–3 Verrill, A. E. 1869b. "On some new and imperfectly known echinoderms and corals." Proc. Bos. Soc. Nat. Hist., Vol. 12: 381-396.

REFERENCES

Four new species of madrepores are described: *Astropsammia pedersenii* from La Paz, a *Dendrophyllia* and two *Pavonia* from Panama, low water and 6–8 fathoms. No illustrations; no natural history data.

§ E–4 See also Madrepore section of Verrill papers cited as our (§ K–12 and K–13).

§ E–5 ACKNOWLEDGMENT: For determination of the stony corals, for literature references, and for data on geographic distribution, we have to thank Dr. John W. Wells, Dept. of Geology, Ohio State University, Columbus, Ohio.

LIST OF SPECIES TAKEN:

§ E–6 *Astrangia pedersenii* Verrill 1864
Verrill, 1869, § E–2b, p. 529. From La Paz and Guaymas. Never previously illustrated. Dr. Wells regards as conspecific: *A. dentata* Verrill 1866, § E–2a, p. 332, described from the Bay of Panama, also at Acajutla (San Salvador), Acapulco, and La Paz; and *A. oaxacensis* Palmer 1928, § E–2, p. 29 from Puerto Angelito. FAMILY ASTRANGIDAE VERRILL (ASTRAEIDAE) of the APOROSA.

The zooids are minute, in cups up to 3 or 4 mm. across at the base, a few mm. high, and a few mm. across at the top. Bases often confluent, forming a loose encrusting colony. Fawn brown to dark gray. About 36 septa.

The only solitary coral we found in the gulf. Intertidal and subtidal on sea fans, on undersides and sides of rocks: on sea fan from Pulmo Reef or Pt. Marcial Reef; Coronado Island, Puerto Escondido, on shell of living Pelecypod at Concepcion Bay; San Carlos Bay, and south end of Tiburon Island. Suitable large rocks, as at Puerto Escondido, may be covered with dozens of specimens.

§ E–7 *Pocillopora capitata* Verrill 1864 PL. 21 FIG. 3
Verrill 1869, p. 520 (§ E–2b). Never previously figured. Originally described from La Paz, but presumably merely because Capt. Pedersen made shipments from that point. Also reported by Verrill from Panama, Acapulco, and Socorro Island. FAMILY POCILLOPORIDAE VERRILL of the APOROSA.

Although forming great massive colonies, the outlines are delicate and sharp, with zooids more widely spaced than in *Porites*. Colonies may be foliosely encrusting, but are often erect and palmately branching. 12 septa.

A true reef building coral, and the form involved in Pulmo Reef,

REFERENCES

which seems to be a true fringing reef. We found this form nowhere else, but at Pulmo there is a linear mile or more of it, 10 to 30 feet wide, and protruding well above the low tide mark.

§ E–8 *Porites cf. nodulosa* Verrill 1869

Verrill 1869, p. 505 (§ E–2b). Never illustrated; La Paz. Known also from Carmen Island in the Gulf (Wells, personal communication). FAMILY PORITIDAE of the perforate corals.

The colonies that we took are definitely but coarsely branching. Dr. Wells writes: "may represent a new species, the branches are smaller and more discrete than Verrill's described type." Septa usually 12.

We took this form only once, in a crab net cast overboard at the ship's anchorage in 4–5 fathoms, south of Coronado Island.

§ E–9 *Porites porosa* Verrill 1869

Verrill 1869, p. 504, (§ E–2b). Never illustrated. Described from near La Paz. Wells in a personal communication considers possibly conspecific *P. californica* Verrill 1869, p. 504 (§ E–2b). FAMILY PORTIDA of the perforate corals.

Coarse, with rounded outlines; hemispherical, nodular, or encrusting, but rarely if ever branching. Zooids closely adjacent.

We found this to be the commonest coral of Gulf shores. In the south and on the Lower California shores of the Gulf, it occurred in dome-shaped heads up to 8″ in diameter, or in variously shaped masses and clusters. But at Puerto Refugio and on the southern end of Tiburon Island, the thin encrustation on tide pool rocks was scarcely recognizable as the same species; we took it for a hydro-coral. The massive form was very much a feature of the tidal landscape east of La Paz, at Pt. Lobos (Espiritu Santo Island), and on the rocky reef at Pt. Marcial. It occurred thus also at Concepcion Bay.

SUMMARY

§ E–10 The relationships would seem to be Indo-Pacific (see Palmer's remarks cited in our § E–2), or West Indian. However, several species of *Astrangia,* a large, almost world wide genus, occur elsewhere in the Panamic area. The solitary coral *A. insignifica* Nomland is reported to be common along shore in southern California (Johnson and Snook 1927, § Y–2, p. 108), but we have never knowingly found this, which should not be mistaken for the common tide pool coral of more northern waters, *Balanophyllia elegans* Verrill of the FAMILY EUPSAMMIDAE.

Phylum Platyhelminthes

TURBELLARIA (FLATWORMS)

REFERENCES

§ F–1 Bock, Sixten 1913. "Studien über Polycladen." Zoologiska Bidrag från Uppsala Vol. 2: 31-344, Pls. 4-10, 66 text figs.

A revision of the group complete and monographic to its date, cited partly for its orientation value. All the orders, families, and genera known to that time are diagnosed, and all known species are cited, with their geographic distribution. Methods are stated, there is information on terminology and classification, and a comprehensive bibliography. *Latocestus viridis* n. sp. (p. 63) from Panama and *Emprosthopharynx opisthoporus* n. sp. (p. 161) from Galápagos are described for the first time, although the specimens were taken in 1852 on the *Eugénie* Expedition. Three other Panamic species are treated or mentioned in the genus *Stylochoplana* (*panamensis*, p. 179, *plehni* and *californica*, p. 180).

§ F–2 .... 1925. "Planarians, Pts. I to III" and "Pt. IV." (Nos. 25 and 27, Papers from Dr. Th. Mortensen's Pac. Exped. 1914-16). Vidensk. Medd. fra Dansk naturh. Foren., Vol. 79: 1-84, 3 pls. and 22 text figs., and Vol. 79: 97-184, 2 pls., 31 text figs.

Material in the other parts is extra-limital, but No. III considers two polyclads living commensally with the hermit crab *Petrochirus californiensis* Boucrez (misprint for Bouvier) at Panama, 9 M. *Euprosthiostomum adherens*, n. sp. (p. 49, toto-fig. 1 and 2, Pl. 2a). *Emprosthopharynx opisthoporus* Bock 1913, on p. 61 with toto-fig. 7 on Pl. 2a. The relationship between the polyclads and their host is considered. Bibliography.

§ F–3 Hyman, Libbie H. 1939. "Polyclad worms collected on the Presidential Cruise of 1938." Smith. Misc. Coll., Vol. 98 (17): 13 pp., 15 text figs.

Two new species from the Pacific, *Euplana clippertoni*, p. 4, figs. 9-12, and *Prosthiostomum parvicelis*, p. 6, figs. 13-15, plus some Caribbean material.

§ F–4 .... 1940. "The polyclad flatworms of the Atlantic coast of the United States and Canada." Proc. U.S.N.M., Vol. 89: 444-495, text figs. 24-31.

Cited here because it seems to comprise the only easily available and modern conspectus of this group, with definitions of sections, families, and genera. These diagnoses are applicable to other than Atlantic polyclads. Bibliography of 38 titles includes Hyman 1939a (Sargassum) and Lang 1884 (Naples) which describe specimens related to certain of our finds.

REFERENCES

§ F-5 Plehn, Marianne 1896. "Neue Polycladen gesammelt von Herrn Kapitän Chierchia bei der Erdumschiffung der Korvette *Vettor Pisani*. . . ." Jena Zeit. f. Nat. Vol. 30: 137-176, Pls. 8-13.

Allioplana delicata n. g., n. sp. p. 142 with toto-illustration as fig. 1, Pl. 13, under stones, Payta, Peru. *Leptoplana panamensis* n. sp., p. 151, toto-illustration of cleaned, mounted specimen on Pl. 10; under stones in sand, two localities in the Gulf of Panama; a *Leptoplana* from Peru, and two pelagic species. Bibliography.

§ F-6 Woodworth, W. McM. 1894. "Report on the Turbellaria." *Albatross* 1891. Bull. Mus. Comp. Zool., Vol. 25 (4): 49-52, 1 pl.

Prostheceraeus panamensis n. sp. From the reef at Panama, and two pelagic species, one *Stylochoplana californica* n. sp., with excellent toto-illustration, pelagic in the Gulf at the surface.

§ F-7 ACKNOWLEDGMENT: For provisional determinations of the following material, for literature, and for assistance and information, we have to thank Dr. L. H. Hyman of the American Museum of Natural History in New York City.

LIST OF SPECIES TAKEN:

FAMILY LATOCESTIDAE (p. 62 Bock 1913 [§ F-1] and Hyman 1940 [§ F-4] p. 457.)

§ F-8 *Latocestus* sp.

2 smooth slate-colored specimens taken at Puerto Escondido; very abundant dark, smooth flatworms from the rocks at Angeles Bay, and at San Carlos Bay, Sonora.

FAMILY STYLOCHIDAE (Bock [§ F-1] p. 108 and Hyman [§ F-4] p. 459.)

§ F-9 *Stylochus* sp. or spp.

As 2 large, brown, dark-speckled forms at Pulmo Reef. As a large, buff-colored "bumpy" specimen at Pt. Lobos on Espiritu Santo Island.

§ F-10 *Stylochus* (?) sp.

Representatives taken on the reef at Pt. Marcial, as large thin flatworms with nuchal tentacles.

FAMILY LEPTOPLANIDAE (Bock [§ F–1] p. 167 and Hyman [§ F–4] p. 467.)

REFERENCES

§ F–11 *Stylochoplana plehni* Bock 1913 cf.
Equals *Leptoplana californica* Plehn 1897, per p. 179-180 Bock 1913 (§ F–1), from Monterey Bay and Galápagos Island.

Taken with shelled Gastropods, rocky littoral at Tiburon Island.

§ F–12 One or more additional species, reminiscent of the superficially nondescript genus *Leptoplana,* were taken at El Mogote, on the reef at Pt. Marcial, and at Port San Carlos, Sonora.

FAMILY PLANOCERIDAE (Bock [§ F–1] p. 228 and Hyman [§ F–4] p. 477.)

§ F–13 Specimens labeled "*Planocera*-like forms" were taken at Cape San Lucas (under rock below 0.0′ tide level).

FAMILY PSEUDOCERIDAE (Bock [§ F–1] p. 252 and Hyman [§ F–4] p. 484.)

§ F–14 *Pseudoceros* sp.

Large fimbriated, dark-purple specimen with white digestive tract showing through on the underside. Reminded one of a black *Gyrocotyle.*

§ F–15 Entirely undetermined material possibly comprising one or more additional species was taken at San Carlos Bay in Lower California, at Puerto Refugio, and possibly at other stations with some of the flatworms mentioned above.

SUMMARY

§ F–16 Probably no less than 7 species and possibly up to 10 or 12 (only one of which has been identified with previously known material) resulted from our short survey of the common and representative shore fauna in this group. Since a large proportion of these will prove to be new species, little can be indicated in the way of biological or zoogeographical summarizing.

Species of *Latocestus* and of one or more Leptoplanids were among the most common and characteristic under

REFERENCES

rock inhabitants, the former in the south, the latter especially in the northern part of the Gulf and high up in the intertidal. In the south also, the "bumpy" *Stylochus* and a striking *Pseudoceros* were large and noticeable.

The difficulties of preparing materials of this sort in the field, or with the inadequate laboratory facilities incident to travel in a small boat, are very considerable. On the whole, we are conscious of having handled this phase of our survey poorly—a situation all the more to be regretted since so little is known of this group, although abundant and varied in the Panamic area. Few collectors, furthermore, will be likely to take the trouble of capturing and preparing these usually rather fragile and often nondescript animals. Work of this sort needs the full-time attention of a specialist who actually goes out into the field, observes the living animals in their natural ecology, and correlates this information with the finally identified material which by this time will have become remote in the form of serial-section microscopic slides. It is in groups such as this, where identifications are made microscopically from material altered radically from the living animal, that liaison work is most needed.

CLASS NEMERTEA

§ G–1 Coe, W. R. 1940. "Revision of the Nemertean Fauna of the Pacific coasts of North, Central, and northern South America." A. Hancock Pac. Exp., Vol. 2 (13): 247-323, Pls. 24-31, 2 of which are colored.

Including a report on the 13 species collected by the Hancock expeditions through 1939. Four of these are from Peru, the remainder constituting Panamic forms. Seven of these were taken in the Gulf. There is in many cases no data as to depth, but both dredged and littoral forms presumably are included. In all, 16 species seem to have been reported to date from the Panamic area. 10 occur elsewhere also, 5 to the north, 1 in Peru. 2 are tropicopolitan, 2 are cosmopolitan. Only 6 of the total 16 are restricted to the Panamic as here considered. With its listing of 98 species, with synonymy, descriptions, data on habitat and distribution, this comprises a key paper and summary of the Pacific coast nemerteans to date. The bibliography lists Joubin 1905 (report of M. Diguet's Lower Californian collections) and the Coe 1926 *Albatross* report, the only other items in which nemerteans of the west American tropics are mentioned, in addition to the Coe 1904 and 1905 etc. considerations of Pacific American forms.

LIST OF SPECIES TAKEN:

REFERENCES

§ G–2 ACKNOWLEDGMENT: For identifying the specimens mentioned below, for the gift of literature, and for much previous co-operation, we have to thank Dr. W. R. Coe of Yale University and of Scripps Institution.

§ G–3 We took only a single species in the Gulf:

Baseodiscus mexicanus (Bürger) 1895.

p. 261, Coe 1940 (§ G–1), good toto-photograph as fig. 24, Pl. 26; Gulf of California to Colombia and Galápagos, shore to 100 M. or more.

Representatives of this large, spectacular, and common form were collected at Pt. Lobos on Espiritu Santo Island under and among the boulders, on the sand beach at El Mogote, and on the rocky shore at Angeles Bay, and were seen elsewhere at several points. From our experience and from the reports of Joubin and Coe, it would seem that this is the most abundant Panamic nemertean, indeed the only one that can be counted upon as occurring ubiquitously, and that it is equally common in the northern and southern portions of the Gulf. Coe records specimens up to 4 M. in length. Most of ours were under 12". The color is variable, usually vivid, with numerous narrow white rings encircling the body at irregular or regular intervals.

Phylum Bryozoa

REFERENCES

§ H-1 Bassler, R. S. 1922. "The Bryozoa, or Moss Animals." Ann. Rep. Smith. Inst., 1920: 339-380, 4 pls., text figs.

A non-technical account highly valuable for orientation. The emphasis is, however, on fossil material. General characters, classification, methods, a systematic section with consideration of some of the typical species, anatomy. The illustrations are largely paleontological.

§ H-2 Canu, Ferdinand, and Ray S. Bassler 1923. "North American later Tertiary and Quaternary Bryozoa." Bull. U.S.N.M., 125: 301 pp., 47 pls.

Treating only such forms as are known fossily—which includes a good many living species in the genera *Membranipora, Stylopoma,* etc. Fine bibliography including Levinsen 1909.

§ H-3 .... 1930. "The Bryozoan Fauna of the Galápagos Islands." Proc. U.S.N.M., Vol. 76 (13): pp. 78, pls. 14.

53 species, of which 29 are new, all resulting from 3 *Albatross* dredge hauls in 33½, 40, and 684 fathoms, 51 of them recorded from the two shallow stations. 10 species occur also in the Gulf of Mexico. "Another remarkable phenomenon is the persistence in this region of archaic forms known hitherto only as fossils and in which naturally the anatomic structure was unknown." (p. 1)

§ H-4 Hastings, Anna B. 1930. "Cheilostomatous Polyzoa from the vicinity of the Panama Canal collected by Dr. C. Crossland in the cruise of the S. Y. *St. George*. Proc. Zool. Soc. London, 1929: 697-740, 17 pls.

A most important paper. Reviews what little had been done previously in the region, considers geographic distribution, treats over 60 species and varieties, several of them new. Fine bibliography including Folke Borg 1926, Levinsen 1909, etc. Of the 59 species and 2 varieties found at the Pacific end, 42 are tropicopolitan, etc., 12 are west American, and 5 are anomalous.

§ H-5 A group of papers by Alice Robertson as follows:

1900. "Studies in Pacific Coast Entoprocta." Proc. Calif. Acad. Sci., (3) Vol. 2 (4): 323-348, 1 pl., including p. 345 mention of one southern Californian species.

1902. "Some observations on *Ascorhiza occidentalis* Fewkes, and related Alcyonidia." Proc. Calif. Acad. Sci., (3) Vol. 3 (3): 99-108, Pl. 14. 45 fathoms, southern California.

1905. "Non-encrusting Chilostomatous Bryozoa of the West Coast of North America." Univ. Calif. Publ. Zool., Vol. 2 (5): 235-322, Pls. 4-16. 12 southern

REFERENCES

California species are treated, several of which were new, 10 from shore, 2 from 125 fathoms.

1908. "The incrusting Chilostomatous Bryozoa of the West Coast of North America." Univ. Calif. Publ. Zool., Vol. 4 (5): 253-344, Pls. 14–24. 8 southern California shore species, and 8 from dredging to 99½ fathoms. The p. 265 *Membranipora tehuelcha* later became *M. tuberculata* which we found abundantly in the Gulf.

1910. "The Cyclostomatous Bryozoa of the west coast of North America." Univ. Calif. Publ. Zool., Vol. 6 (120): 225-284, Pls. 18–25. 11 southern California species are treated, 5 of them new, from shore and from shallow dredgings.

§ H–6 There are other important references, a modern revision by Folke Borg 1933, in Bd. 14, Zoologiska Bidrag från Uppsala, etc. The exigencies of time and circumstance, and human inadequacies of energy have precluded a more thorough review of the literature in this group.

§ H–7 ACKNOWLEDGMENT: For determinations of the following species, and for checking over this portion of the phyletic catalogue, we are grateful to Dr. R. S. Bassler of the United States National Museum at Washington, D. C.

LIST OF SPECIES TAKEN:

§ H–8 *Bugula neretina* (Linn.) 1758

p. 266, Robertson 1905, with good toto-fig. 97 on Pl. 16. Southern California, from Monterey where it is scarce, to San Diego, etc., where it is extremely abundant.

p. 704, Hastings 1930 (§ H–4), Panama. Detailed geographic range which indicates its cosmopolitan distribution.

Brown or reddish brown, occurring in great lush growths. Taken at Pt. Marcial Reef; abundantly on the rocks at Angeles Bay; but most commonly on dead shells in the channel entrance at Agiabampo Estuary, Sonora, April 12, 1940.

§ H–9 *Cellepora* sp.

Taken once only, at Coronado Island, on rocks in dazzling white coral sand.

REFERENCES

§ H–10 *Crisia* sp.

Common on and under rocks at and below the 0.0′ tide level at Cape San Lucas.

§ H–10a *Flustra* (?) sp.

This also was taken only once, on low tide rocks at Puerto Refugio on Angel de la Guardia Island.

§ H–11 *Lagenipora erecta* O'Donoghue 1923

p. 33, O'Donoghue, C. H. and Elsie, 1923. "A preliminary list of Polyzoa (Bryozoa) from the Vancouver Island region." Contributions to Canadian Biology (N.S.), Vol. 1 (10): 143-201, 4 pls. From Georgia Strait, B. C., 10-20 fathoms, and from the Queen Charlotte Islands.

Taken with *Bugula neretina* on Pt. Marcial Reef, March 24, 1940.

§ H–12 *Lichenspora* sp.

Marcial Point, half tide rocks, March 24, 1940.

§ H–13 *Membranipora tuberculata* Bosc 1802

p. 22, Canu and Bassler 1923 (§ H–2), with good x20 illustrations of both living and fossil forms as figs. 3–5, Pl. 33. Distribution: cosmopolitan. Includes *M. tehuelca,* p. 265 Robertson 1908 (§ H–5): ". . . obtained from San Francisco southward. One of the most abundant species on the rock weed at La Jolla, San Diego, and San Pedro."

On the upper valve of a sessile Pelecypod at Pulmo Reef. Encrusting on an *Ulva*-like alga growing on a rock in the estuary at San Lucas Cove, Lower California.

§ H–14 *Membranipora* sp.

Puerto Escondido, encrusting on rock.

§ H–15 *Porella* sp.

On the rocky beach at the south end of Tiburon Island.

§ H–16 *Scrupocellaria diegensis* Robertson 1905

p. 261, Robertson 1905 (§ H–5), with good toto-fig. 96, Pl. 16. The most abundant Bryozoan in the southern California region. Less abundant to the northward; occurs sparingly at San Francisco Bay which marks its northern limit.

San Francisquito Bay, on rocks; abundant.

§ H–17 *Scrupocellaria scruposa* (Linn.) 1758

p. 703, Hastings 1930 (§ H–4), Galápagos, 5 fathoms. Distribution: practically cosmopolitan.

On rocks at Marcial Point, half tide, March 24, 1940.

REFERENCES

§ H–18 *Scrupocellaria* sp.

Rocky shores at Angeles Bay.

§ H–19 *Stylopoma spongites* (Pallas) 1766

p. 102, Canu and Bassler 1923 (§ H–2), with illustrations of fossil specimens, mostly x20, figs. 1–12, Pl. 12. Habitat: Gulf of Mexico, Bermuda, etc.

Encrusting on old coral clusters on the sand flats east of La Paz, very abundant.

§ H–20 *Thalamoporella californica* Levinsen 1909

p. 716, Hastings 1930 (§ H–4), dredged and along shore at the Galápagos. Previous distribution: California.

Gravelly beach at Angeles Bay; on green algae.

SUMMARY

§ H–21 Although common and widely distributed within the Gulf, none of the Bryozoans were highly noticeable excepting *Bugula neretina,* great colonies of which were seen at several stations. 14 species were taken. Of the 7 for which we have adequate data, 3 are cosmopolitan, 1 is Caribbean, 2 are Californian, and 1 British Columbian.

Phyla Sipunculoidea and Echiuroidea

REFERENCES

§ I–1 Chamberlin, Ralph V. 1919 (1920). "Notes on the Sipunculida of Laguna Beach." Pomona Coll. Journ. Ent. and Zool., Vol. 12: 30-31.

Notes on *Sipunculus nudus, Physcosoma agassizii,* and terse descriptions of three new species of *Dendrostoma,* and of *Phascolosoma hesperum,* as *hespera*), sp. nov. Scant ecological information, no illustrations, no bibliography.

§ I–2 Gerould, J. H. 1913. "The sipunculids of the eastern coast of North America." Proc. U.S. Nat. Mus., Vol. 44: 373-437, Pls. 58–62.

Although concerned with Atlantic forms, there is a listing on p. 374 of the species reported as occurring on both sides of the Isthmus (*Physcosoma pectinatum, P. antillarum, Sipunculus nudus* and *S. titubans*) and those known only from the Pacific side (*Physcosoma agassizii, Aspidosiphon truncatus,* and *Dendrostoma peruvianum*). Comprehensive accounts of *Phascolosoma gouldii* (Pourtalès) 1851, p. 380; and of *Physcosoma antillarum,* p. 420; mention of *Sipunculus nudus* Linn., and of *S. titubans* described originally by Selenka (§ I–4) from Puntarenas (Punta Arenas). Bibliography.

§ I–3 Keferstein, W. 1867. "Untersuchungen über einige amerikanische Sipunculiden." Zeitschr. f. Wiss. Zool., Vol. 17: 44-55, Pl. 6.

Based on a collection received from A. Agassiz at Harvard, mostly described briefly during the previous year. p. 45, *Sipunculus nudus* from Panama, considered to be identical with that noted in Grube 1859, as *S. phalloides* Pallas from Puntarenas. *Phascosolosoma* (now *Physcosoma) agassizii,* p. 46, etc., Panama, etc. *P. pectinatum,* p. 47 etc., Panama. *P. puntarenae* is synonymized into *P. varians* of the West Indies, p. 48. P. (*Aspidosiphon*) *truncatum,* p. 50, Panama. *P. antillarum* G. and O., Panama. No bibliography.

§ I–4 Selenka, E., J. G. de Man, and C. Bülow 1883, etc. "Die Sipunculiden." Reis. Arch. Phil. Semper., Teil II, Bd. IV, Abt. 1: XXXII, 131 pp., 14 plates, mostly colored.

A monographic account, not only of the Philippine sipunculids brought back by Semper, but of everything known of the group to that date. The table of geographic distribution on pp. XXIV ff. shows the following Panamic species: *pectinatum, agassizii, antillarum, truncatus, varians, titubans,* all of which are described with some detail-illustrations in the body of the work. *Dendrostoma blandum* sp. nov. is described on p. 85, with fine but small colored lithograph toto-figure, from Japan only, [subsequently mentioned 1939 by Satô, as Californian) and *S. phalloides* Pallas is mentioned on p. 99 as having been found in the Grube collection labeled from Punta

REFERENCES

Arenas (Puntarenas). Good bibliography, including Baird 1868, etc., which records material from the Pacific American coast.

§ I–5 Satô, Hayao 1939. "Studies on Echiuroidea, Sipunculoidea, and Priapuloidea of Japan." Sci. Rep. Tôhoku Imp. Univ., (4) Biol. Vol. 14 (4): 339-460, Pls. 19–23, 60 text figs. and 16 tables.

Because Japanese shores are rich in sipunculids and echiuroids, with 69 and 19 species respectively, it is to Japanese literature we must refer for comprehensive modern treatment of these groups. The genus characterizations include all genera represented in the Panamic littoral in our Gulf findings and in the literature to date. *Physcosoma nigrescens* is listed from Costa Rica, and *Dendrostoma blandum* from California, p. 387 and 411. *Aspidosiphon truncatum* is listed from Panama. There is information on complete geographic distribution for all species found in Japan, analytical keys, and a comprehensive bibliography, including the above-mentioned Baird 1868.

§ I–6 ACKNOWLEDGMENT: For determinations of the sipunculids and echiuroids, for help with and access to the literature, and for reading critically this portion of the phyletic index, we are indebted to Prof. W. K. Fisher, Director, Hopkins Marine Station of Stanford University, Pacific Grove, California.

LIST OF SPECIES TAKEN: (Sipunculids)

§ I–7 *Phascolosoma hesperum* Chamberlin 1919.
p. 31, sand and eel grass, Balboa, California, Chamberlin 1919 (§ I–1). Also mentioned, p. 212 Ricketts and Calvin 1939 (§ Y–3), southern California.

Exceedingly abundant as commensals in the tubes of the sand flat anemone *Cerianthus* at San Lucas Cove, March 29. A minute, thread-like worm with introvert wrinkled, elongate, and much thinner than the balance of the body from which it is noticeably set off.

§ I–8 *Phascolosoma* sp., similar to *gouldii*.

Probably a new species. Under boulders, with *Physcosoma agassizii*, and *P. antillarum*, at Pt. Lobos on Espiritu Santo Island; several specimens. The Atlantic *P. gouldii* is considered on p. 380, Gerould 1913 (§ I–2). The genus is characterized on p. 401 by Satô 1939 (§ I–5).

§ I–9 *Physcosoma agassizii* (Keferstein) 1866
Keferstein 1867 p. 46 (§ I–3). Ricketts and Calvin 1939 p. 47 (§ Y–3), with good toto-photograph on Pl. VIII.

REFERENCES

Ranges from northern California (at least, the more northern records in the past seem to have been confused with *P. japonicum*) to Panama and possibly to South America.

One specimen only of this exceedingly common and characteristic Monterey Bay under rock intertidal form was taken under similar conditions at Pt. Lobos, Espiritu Santo Island, with *Physcosoma antillarum* and *Phascosoloma* sp.

§ I–10 *Physcosoma antillarum* (Grube and Örsted) 1859.
PL. 16 FIG. 2

Described originally from Punarenas (Punta Arenas), west coast of Costa Rica in the "Annulata Örstediana" published in the Vidensk. etc. There is a description on p. 420 and a couple of fair toto-photographs as figs. 19 and 20 on Pl. 62, Gerould 1913 (§ I–2) who synonymizes with it, Baird's 1868 *nigriceps* from Chile. Reported by Satô 1939 (§ I–5), p. 394, from Japan, who reports the range also as from West Indies and Brazil and from the west coast of Central America.

A large specimen which was preserved well expanded, was taken with *Phascolosoma* sp. and *Physcosoma agassizii* at Pt. Lobos, Espiritu Santo Island, under boulders. An additional specimen was taken in the interstices of dead coral east of La Paz.

§ I–11 *Physcosoma* sp., similar to *agassizii*

Would seem to represent the commonest sipunculid in the Gulf. A small edition of *agassizii,* taken at Pulmo Reef (in interstices of the living coral *Pocillopora*), under boulders at Puerto Escondido, among rocks at San Carlos Bay March 30-31, and under rocks at Puerto Refugio. May turn out to be the true *agassizii.* The genus is characterized on p. 381 by Satô 1939 (§ I–5).

§ I–12 *Sipunculus nudus* Linnaeus

There is a short description, not illustrated, in Gerould 1913, p. 428, (§ I–2), and a comprehensive account and description with toto-photographs (as figs. 3 and 31) in Satô 1939 (§ I–5). p. 365 of the latter gives the range as almost cosmopolitan in temperate and tropical seas. Ricketts and Calvin 1939 (§ Y–3) p. 223 report material from southern California and northern Lower California, with a good toto-photograph on p. 43.

We took this twice in the Gulf, on the muddy sand flats at El Mogote, and again on sand flats at Angeles Bay.

LIST OF SPECIES TAKEN: (Echiuroids)

REFERENCES

§ I–13 A new species of *Thalassema,* to be named *T. steinbecki* by Dr. Fisher, was taken in a single example at El Mogote, opposite La Paz, in the sandy mud, apparently burrowing, and in association with *Dentalium, Sipunculus nudus,* and close to a colony of *Holothuria paraprinceps.* Appears to be related to the species to be named *T. pelodes* by Dr. Fisher, found under similar conditions at Newport Bay in southern California. The genus is characterized on p. 352 of Satô 1939 (§ I–5).

§ I–14 A new species of *Ochetostoma*

(one of the genera characterized on p. 357 of Satô 1939 (§ I–5) resulting from the splitting of *Thalassema*) to be named *O. edax* by Dr. Fisher. Was found abundantly at the following points, as a common and characteristic tide pool inhabitant: Pt. Lobos, Espiritu Santo Island, under tide flat boulders, with an Enteropneust; Coronado Island, under and among slightly subtidal rocks on white sand; Puerto Refugio, under rocks. Specimens were elongate to grape-shaped, smooth and thin-skinned, greenish, with obvious and comparatively large spoon-shaped proboscis.

§ I–15 Six sipunculids and two echiuroids were taken, of which 2 of the former and both of the latter seem to represent new species. With the limited information available, no generalizations can be drawn for these groups. Representatives were fairly common, of *Physcosoma* sp. and of the *Ochetostoma* especially, and both of these must be rated as important in the economy of the rocky shores visited.

Phylum Annelida

CLASS POLYCHAETA

REFERENCES

§ J-1 ACKNOWLEDGMENT: To Dr. Olga Hartman of the Hancock Foundation, University of Southern California, Los Angeles, we are under obligation for the following identifications, for help with the literature, for suggestions and assistance, and most of all for checking this portion of the phyletic catalogue.

When the Hartman Hancock monographs have been completed, the Panamic Polychaet bibliography will be simple, but for the present, many citations have to be shown. The literature, as usual, is scattered, but fortunately fairly voluminous. The Fauvel papers provide good general orientation material, although concerned with European species. The Grube 1877 and Chamberlin 1919 monographs comprise handbooks for tropical forms. Beyond these, the four Monro papers, the Berkeley 1939, and of course the Hartman 1939a and 1940 accounts are essential.

§ J-2 Augener, Hermann 1933. "Polychaeten von den Galápagos-Inseln." No. VI in: "The Norwegian Expedition to the Galápagos Islands 1925 . . ." Medd. Zool. Mus. Oslo, No. 32: 55-66, 1 text fig.

9 species only were taken. *Lepidasthenia minikoënsis* Potts 1909-10 is considered in some detail. The ubiquitous *Erythoë complanata* was taken in 18 examples, with species of *Eulalia, Lumbrinereis, Cirratulus,* etc.

§ J-3 Berkeley, E. and C. 1939. "On a collection of Polychaeta, chiefly from the West Coast of Mexico." Ann. Mag. Nat. Hist., (11) Vol. 3 (15): 321-346, 12 text figs.

Collected by Capt. Lewis of the yacht *Stranger.* 34 species (one of which was taken only in the Caribbean) of which 5 are new, and 7 appear to constitute new Pacific coast records. Bibliography.

§ J-4 Chamberlin, R. V. 1919. "Pacific coast Polychaeta collected by Alexander Agassiz." Bull. Mus. Comp. Zool., Vol. 63 (6): 251-270, 3 pls.

Among other (mostly Californian) species, the following are treated after more than 50 years, having been taken by Agassiz in 1859-60 at Panama: *Notopygos maculatus* (equals *ornata), Eurythoë complanata, Cirratulus exuberans* sp. nov., *Eupolymnia regnans* Chamberlin, and *Pseudopotamilla panamanica* sp. nov.

REFERENCES

§ J–5 .... 1919a. "The Annelida Polychaeta." *Albatross* 1891, 1899-1900, 1904-05. Mem. Mus. Comp. Zool., Vol. 48: 514 pp., 80 pls.

Except for the fine Monro papers and for Hartman's monograph (Hancock, in progress) this is the most important reference for Panamic annelids, and it comprises, furthermore, a handbook of annelid knowledge complete to its publication date, with a résumé of all the families and genera known at that time. However, Hartman warns us against placing too much confidence in its trustworthiness. There are lists to show bathymetric distribution, and a key to the 49 families then recognized. In the comprehensive family characterizations, there are natural history notes, keys to all the genera then recognized, and data on genera synonymy. Résumé of the literature, and bibliography only are lacking.

175 species are treated, 118 of them new, from the Panamic area and from Polynesia, with a few from the Atlantic, and 2 from southwestern Alaska. Although concerned primarily with dredged forms, many littoral species are considered, including 25 from shore collecting: 19 at Panama, 1 from the west coast of Lower California (outside Magdalena Bay), and 4 from the Galápagos. Within the Panamic Province as here defined, some 8 species occur pelagically, and 47 in deep water (8 from the Gulf of California), but all except one are deeper than 100 fathoms, many are below 500 fathoms, and some are abyssal. The lithographic illustrations show details only of Panamic forms, or at most toto-drawings of the anterior segments.

§ J–6 Fauvel, Pierre 1923. "Polychètes errantes." Faune de France, Vol. 5: 488 pp., 181 text figs.

Covering 15 families, APHRODITIDAE to EUNICIDAE, with their sub-families, plus two small aberrant families. See below.

§ J–7 .... 1927. "Polychètes sédentaires." Faune de France, Vol. 16: 494 pp., 152 text figs.

Covering 21 families, STERNASPIDIDAE to SERPULIDAE, with the archiannelids and myzostomes.

These two monographic volumes together comprise an excellent source book, apparently both reliable and complete, transcending its original purpose. Beyond serving as a résumé of French polychaets, it provides a Polychaet handbook, as illustrated by the fauna of France. There are analytical keys to the families, and within each family to the genera, and then to the species. The species descriptions are fairly complete, with dimensions, synonymy, geographic distribution both locally and elsewhere, and natural history notes. Every species is illustrated diagnostically, and two toto-drawings at

REFERENCES

least of the anterior and often of the entire animal. The introduction includes information on external morphology, anatomy, reproduction, ecology, and natural history, methods of preservation, etc. For each volume there is a comprehensive bibliography and systematic index, both alphabetical, and a very usable table of contents. The information is so complete that, for the species listed, one has rarely to seek earlier data. The majority of the families likely to be encountered in the Panamic area are diagnosed and considered here, many of the genera, and not a few of the species. Apparently the very finest type of source book, its completeness and accuracy are astonishing. One wonders how a single worker could find time in a life-span to digest, co-ordinate, and present all this information.

§ J–8 Gravier, Ch. 1905. "Sur un Polynoidien (*Lepidasthenia digueti*, nov. sp.) commensal d'un Balanoglosse de Basse-Californie." Bull. Mus. Hist. Nat. Paris, Vol. 11: 177-181.

The same article, but enlarged, with several figures including a toto-drawing (x⅓) of the enteropneust host with annelid and stomatopod commensals, appears in: Bull. Soc. Phil. Paris (9) Vol. 9: 160-173, 9 text figs. Much natural history information on these very large polynoids and on several other animals found commensally in the same association. "One might remark that the dorsal tube laterally limited by the genital wings of Balanoglossus constitutes a site of great advantage for polynoids. . . ." Startling resemblance (in connection with which the term "mimicry" comes to mind) between the ornamentation of the stomatopod and the scales of the polynoid, with no apparent "purpose" in a survival sense, since the assemblage lives in the total darkness of a burrow.

§ J–9 .... 1907 and 1908. "Sur les Annélides polychètes rapportés par M. le Dr. Rivet, de Payta (Pérou)." Bull. Mus. Hist. Nat. Paris, Vol. 13: 525-530, and Vol. 14, 40-44.

Low tide. Of 19 species taken, 8 were new: *Phyllodoce parvula, Eulalia personata, Marphysa schmardai, Chrysopetalum riveti, Dodecaceria opulens, Scoloplos grubei, Sabellaria fauveli* and *Branchiomma roulei*. Many of these have since been revised or synonymized. Diagnoses apparently of the same species from the same collection were published also elsewhere, in Arch. zool. exp. gén., Paris (4) Vol. 10: 617-659, 3 pls., and 638-641 etc., in 1909; and in 1910, Mission de service géographique de l'armée pour la mesure d'un arc méridien équatorial en Amérique du Sud. Zoologie, Vol. 9 (3): 93-126, 6 pls., not consulted.

§ J–10 Grube, E. 1877. "Annulata Semperiana. Beiträge zur Kenntniss der Anneliden Fauna der Philippinen." Mem. Acad. Imp. Sci. St. Pétersbourg (7) Vol. 25 (8): 299 pp., 15 pls.

A great collection of polychaets brought back by Semper from a sojourn of several years in the Philippines is made herein the basis for a monograph of what was known to date of that group particu-

REFERENCES

larly in the tropics. A number of Panamic species are described and beautifully illustrated; many genera characteristic of the Panamic area are treated. The lithographs rarely show less than a complete anterior end, and they sometimes figure the entire worm.

§ J-11 Hartman, Olga 1939. "The polychaetous annelids collected on the Presidential Cruise of 1938." Smith. Misc. Coll., Vol. 98 (13): 22 pp., 3 text figs.

31 species, 24 of which occurred only in the Pacific collections, 6 only in the Atlantic, and one, *Eurythoë complanata,* in both. *Neanthes roosevelti* and *Polydora tricuspa* are new species, *Cirratulus niger* is newly named, and two indeterminate forms of *Scalisetosus* and *Armandia* may represent new species, all Pacific. There is a list of stations with annelids collected at each, an annotated catalogue with geographic range of all species taken, and a bibliography of 39 titles. The illustrations figure anatomical details, usually magnified, and toto-drawings of the anterior segments and of one larva.

§ J-12 .... 1939a. "Polychaetous annelids. Pt. I. Aphroditidae to Pisionidae." A. Hancock Pac. Exp., Vol. 7 (1): 156 pp., incl. 28 pls.

Of the 59 mostly Panamic species treated, 26 are shore inhabitants of which 2 may be overlap forms, 1 occurs also in the North Pacific, 1 also in the Atlantic, and 2 also in the Indo-Pacific. There are notes on synonymy, data on geographic range and distribution and illustrative material of usually magnified structural diagnostic detail. Good bibliography which seems to represent painstakingly careful work of the most discriminating sort.

§ J-13 .... 1940. "Polychaetous annelids. Part II. Chrysopetalidae to Goniadidae." A Hancock Pac. Exp., Vol. 7 (3): 173-287, incl. 14 pls.

When this series is completed, summing up what is known to date of the Panamic annelids, no other bibliographic citations will need to be shown in this section. 61 chiefly Panamic species are treated here. 23 are shore forms of which 1 is probably overlap, 1 occurs also in the North Pacific, 1 in the Mediterranean, 3 in the Indo-Pacific, and 4 are cosmopolitan. Bibliography.

§ J-14 Monro, C. C. A. 1928. "Polychaeta of the families Polynoidae and Acoëtidae from the vicinity of the Panama Canal, collected by Dr. C. Crossland and Dr. Th. Mortensen." Journ. Linn. Soc. London, Zool., Vol. 36: 553-576, 30 text figs.

16 Pacific species, with dimensions, and often with natural history and color notes recorded from the living material. The collecting was mostly along shore, and largely in the Pacific. Short bibliography.

§ J-15 .... 1928a. "On the Polychaeta collected by Dr. Th. Mortensen off the coast of Panama." No. 45 in

REFERENCES

"Papers from Dr. Th. Mortensen's Pacific Expedition." Vidensk. Medd. fra Dansk. Naturh. Foren., Vol. 85: 75-103, 19 text figs.

40 species and varieties, of which 6 are new, mostly from along shore. Some natural history notes. Good bibliography. "The fauna of the Panama coasts is tropical, and the forms here studied confirm Prof. P. Fauvel's contention that among the Polychaeta the tropical fauna includes many species common to the Atlantic, the Pacific, and the Indian Oceans."

§ J-16 .... 1933. "The Polychaeta Errantia collected by Dr. Crossland at Colón, in the Panama Region, and the Galápagos Islands during the Expedition of the S. Y. *St. George*." Proc. Zool. Soc. London, 1933, Vol. (1): 1-96, 36 text figs.

Collecting was mostly in the Pacific, and at depths less than 30 fathoms, usually along shore. "Crossland made elaborate field notes on the collections and these I have given almost in full. Our knowledge of the appearance in life and habits of the tropical forms amounts almost to nothing, and I therefore regard the field observations by Crossland, who, both as a collector and systematist, has had much experience of the Polychaeta, as of great value." 99 species and varieties were taken, of which 11 were new; 85 of these were Panamic. Good bibliography.

§ J-17 .... 1933a. "The Polychaeta Sedentaria collected by Dr. C. Crossland at Colón, in the Panama region, and the Galápagos Islands during the expedition of the S. Y. *St. George*." Proc. Zool. Soc. London, 1933: 1039-1092, 31 text figs.

68 species and varieties of which 11, including one new genus, are new; 59 were taken in the Pacific. There is a consideration of distribution especially by reference to water temperatures. 37 species are represented on both sides of the Isthmus, but, "The Pacific species in Crossland's collection show considerable affinities with the Indo-Pacific fauna, and I find that 45% of them are represented in the Indo-Pacific area, as against only 22% common to the West Indies and the Panama Region with the Galápagos . . . moreover, the following species are European, [list of 27 from the total collection]. . . . from this it would seem that the drift of species has been in the direction west to east, rather than east to west." Many collecting and color notes, as before. Bibliography. *Stylarioides papillata,* which we have found along shore both in Puget Sound and in the Gulf, is reported from low tide here in coral; along the California coast it occurs only in deep water.

§ J-18 Rioja, E. 1939. "Estudios anelidológicos. 1. Observaciones acerca de varias formas larvarias y post-larvarias pelágicas de Spionidae, procedentes de Acapulco, con descripción

REFERENCES

de una especie nueva del género *Polydora*." An. Inst. Biol., Vol. 10: 297-311, 31 figs.

One of the larvae is mentioned as occurring very abundantly during midsummer as a pelagic form at Acapulco.

§ J–18a Rioja, E. 1941. "Estudios Anelidológicos. II. Especies del género *Hydroides* de las costas mexicanas del Pácifico." An. Inst. Biol., Vol. 12 (1): 161-175, 48 text figs.

4 new species and *H. californicus* Treadwell 1929, from Acapulco and Mazatlan. Fine figures of all species described, with good toto-drawings of *H. ochotereana* and *H. brachyacantha*, spp. nov.

§ J–19 Treadwell, A. L. 1917. "A new species of polychaetous annelid from Panama with notes on an Hawaiian form." Proc. U.S.N.M., Vol. 52: 427-430, 5 text figs.

Phyllodoce panamensis sp. nov. from Chame Point, Panama.

§ J–20 .... 1923. "Polychaetous annelids from Lower California with descriptions of new species." Amer. Mus. Novit., Vol. 74, 11 pp., 8 text figs.

26 species, of which 2 were new, the annelid collection of the *Albatross* 1911 cruise to the Gulf. Although concerned primarily with dredged material, often from great depths, the expedition procured 5 shore species, and *Halosydna brevisetosa* from 13½ fathoms. *Chloeia flava* (Pallas), *Polynoë lordi* are stated as occurring on the beach at Francisquito Bay (probably San Francisquito) and at Pichilinque Bay; *Nereis mediator, Platynereis integer* and *Nereis kobiensis* reported as "from Lower California" are presumably littoral. Bibliography of 16 titles.

§ J–21 .... 1928. "A new polychaetous annelid of the genus *Phyllodoce* from the west coast of Costa Rica." Proc. U.S.N.M., Vol. 74: 1-3, 3 text figs.

Phyllodoce nicoyensis, sp. nov., 3 specimens, one 90 mm. long, from the Gulf of Nicoya.

§ J–22 .... 1928. "Polychaetous annelids, from the *Arcturus* Oceanographic Expedition." Zoologica, Vol. 8 (8): 449-485, 3 figs.

A total of 65 species was collected, Atlantic and Pacific, of which 16 and 1 genus were new. Many were from the Galápagos. Good bibliography.

§ J–23 .... 1929. "New species of polychaetous annelids in the collections of the American Museum of Natural History, from Porto Rico, Florida, Lower California, and British Somaliland." Amer. Mus. Novit., 392, 13 pp., 34 text figs.

REFERENCES

Ceratonereis singularis sp. nov., 15 mm. average, from 200 specimens taken pelagically at night, Carmen and San José Islands, in the Gulf. *Eupomatus similis* sp. nov., and *Hydroides californicus* sp. nov., both collected by C. H. Townsend in 1911, "Lower California," probably in the Gulf.

§ J-24 .... 1931. "New species of polychaetous annelids from California, Mexico, Porto Rico, and Jamaica." Amer. Mus. Novit., 482, 7 pp., 21 text figs.

Oenone brevimaxillata, sp. nov., over 400 mm. long is recorded merely from "Mexico." Since most Mexican marine collecting has been done on the west coast, this will likely prove to be a Panamic species.

§ J-25 .... 1937. "Polychaetous annelids from the west coast of Lower California, the Gulf of California, and Clarion Island." No. 9 in: The Templeton Crocker Expedition. Zoologica, Vol. 22 (pt. 2, No. 9): 139-160, 2 pls.

34 species were taken. *Chloeia entypa* Chamberlin (which Chamberlin and Harman both place in the FAMILY AMPHINOMIDAE, whereas Treadwell treats it as an Aphroditid) was found on shore in the Gulf, 2 species of *Leodice* (undeterminable) are from water shallower than 9 fathoms, and *Cirratulus exuberans* was taken at Station 159 in tide pool. All the remainder were dredged in depths of from 20 to 100 fathoms; 16 of the 30 having been collected in the Gulf. Bibliography.

§ J-26 Among a few others not cited here (family revisions, lists of types in the M.C.Z., U.S.N.M., etc.), the following treat southern California shore and dredged species, some of which are overlap or which may prove to be Panamic species:

Chamberlin 1919b, Journ. Entom. Zool., Vol. 11: 1-23. Laguna Beach polychaets.

Essenberg 1917 U.C. Publ. Zool., Vol. 16 (22): 401-430. California Aphroditids.

.... 1917a U.C. Publ. Zool., Vol. 18 (3): 45-60. California Polynoids.

.... 1917b U.C. Publ. Zool. (4): 61-74. California Amphinomids.

.... 1922 U.C. Publ. Zool., Vol. 22 (6): 379-381. San Diego *Stylarioides*. Under sea-weed roots.

Hamilton 1915. Journ. Entom. Zool., Vol. 7: 234-240. Laguna beach Polynoids only.

REFERENCES

Hartman 1936. U.C. Publ. Zool., Vol. 41: 117-132. California Phyllodocids.

.... 1938. U.C. Publ. Zool., Vol. 43: 93-112. California GLYCERIDAE, EUNICIDAE, STAURONERIDAE and OPHELIDAE.

.... 1939a. A. Hancock Pac. Exp., Vol. 7 (2): 159-170. Southern California hesionid, lumbrinereid, and disomid.

.... 1941. A. Hancock Pac. Exp., Vol. 7 (4): 289-322. California spionids, one of which (p. 308), *Polyodora commensalis,* was taken at Mazatlan.

Moore, J. P., a series of papers on California polychaets in the Proc. Acad. Nat. Sci. Phil for 1904, 1909, 1909a, 1910, 1911, and 1923. The last four are devoted to material dredged by the *Albatross,* largely off the coast of southern California, in very shallow to moderately deep water. The 1909 paper treats 64 intertidal species from San Diego (and Monterey—15 being common to both regions).

Treadwell, A. L. 1914. U.C. Publ. Zool., Vol. 13 (9): 175-234. Pacific coast polychaets.

LIST OF SPECIES TAKEN:

FAMILY POLYNOIDAE: Scale worms (Chamberlin 1919a [§ J–5] p. 35, and Fauvel 1923 [§ J–6] p. 39.)

§ J–27 *Halosydna glabra* Hartman 1939
p. 5, Hartman 1939 (§ J–12), Gulf and Panama, shore to 12 fathoms.

Taken once free-living in the coral at Pulmo Reef; again at Concepcion Bay in tray with hermit crabs, presumably living in umbilicus of shell which harbored hermit, taken at night via crab net at anchorage.

§ J–28 *Iphione ovata* Kinberg 1855
Hartman 1939 (§ J–12) p. 27 reports as abundant along shore down to 15 fathoms in the Panamic area, ranging from the Gulf to Ecuador and Galápagos, and in Hawaii. Monro 1928 (§ J–14) p. 557 repeats Crossland's notes on the living specimens: "Coiba. Short and stiff and elytra are uniformly coloured. They completely cover the animal, and the feet are invisible but for dense tufts of fine

REFERENCES

notopodial chaetae which form a continuous fringe just outside the elytra. The elytra are reticulated honey-comb fashion, but colour is quite uniform. . . . in coral, Gorgona Island. Red-brown colour due to body being filled with scarlet eggs, which colour shows through the brown elytra to some extent. Underside of body rose-red. Neuropodial chaetae stiff and dark coloured; notopodial fine, white, and directed both dorsally and ventrally. Elytra shaped to overlap tile-fashion over the back."

An elongate yellow-brown scale worm taken at Pt. Lobos on Espiritu Santo; at Pt. Marcial Reef; and at Coronado Island, all in the southern half of the Gulf.

§ J–29 *Lepidonotus hupferi* Augener 1918
Hartman 1939 (§ J–12) p. 43, Gulf to Ecuador, shore to 12 fathoms, also west coast of central Africa.

Taken only once, again from the umbilicus of a large snail shell which harbored a hermit crab, at Agiabampo Estuary, Sonora, April 11, 1940.

§ J–30 *Thormora johnstoni* (Kinberg) 1855
Hartman 1939 (§ J–12) p. 50. Southern California to Panama via the Gulf. Galápagos, possibly Colombia, Hawaii. Shore to 20 fathoms. Monro 1928 (§ J–14), p. 556, reporting on 60 Panama specimens, records color notes made by Crossland from the living animals, averaging 15 x 16 mm. and finds it "recognizable by the difference in colour of the first two pairs of elytra from the rest. . . . Body uniformly chestnut coloured, being covered with elytra of that colour, but feet are white. Dorsal cirri are white with a black band below the tip. The anterior end, as far as the first two pairs of feet, whitish with black specks, divided clearly by a black line from the brown part. . . . All chaetae stout and golden or even brown in colour. . . . Later got several specimens from washings. The peculiar coloration of the head end is constant, but some are otherwise coloured bright orange and have white spots at the attachment of the elytra."

Taken only once, as a typical small polynoid worm, at Pt. Marcial Reef.

FAMILY SIGALIONIDAE (Chamberlin 1919a [§ J–5], p. 89, and Fauvel 1923 [§ J–6], p. 101.)

REFERENCES

§ J–31 *Eusigalion lewisii* (Berkeley) 1939
Hartman 1939 (§ J–12) p. 59, as *E. hancocki* sp. nov. Range: Gulf to Ecuador, sublittoral to 32 fathoms.

Taken from the sand flats at San Lucas Cove, March 29, 1940.

FAMILY POLYODONTIDAE (p. 85 as [partly] ACOETIDAE, Chamberlin 1919a [§ J–5] and Fauvel 1923 [§ J–6] p. 95.)

§ J–32 *Polyodontes oculea* (Treadwell) 1902
Hartman 1939 (§ J–12) p. 83, Lower California to Panama, subintertidal to 20 fathoms. West Indies. Monro 1928 (§ J–14) p. 575 reports the following collector's notes by Dr. Crossland: "Port of Spain Roadstead, Trinidad. The species is unmistakably recognized by the brown-black borders to the middle and posterior edges of the elytra. Head prominent, 2 long palps with white dots and a pair of stalked eyes which in life could be protruded from under the first elytra almost like those of *Strombus* from its shell . . . The tubes are irregular masses of slimy mud and mucus rather pointed at each end. Also common in Limon Bay, at the Atlantic end of the Panama Canal. Taboga Island, Panama Bay. Anterior elytral borders dark brown rather than black, posterior elytra with black borders, but unmistakably the same species as those obtained months ago on the other side of the Isthmus of Panama."

A fine example of this spectacular and highly colored tube-dwelling scale worm was dug out of a mid-tidal mud-flat covered with eel grass, in the inner bay at Agiabampo Estuary, Sonora. Additional tubes, presumably of this species, covered a slightly elevated area of several hundred square yards in the center of this bay.

FAMILY CHRYSOPETALIDAE (Chamberlin 1919a [§ J–5] p. 92 and Fauvel 1923 [§ J–6] p. 122.)

§ J–33 *Bhawania riveti* (Gravier) 1908
Monro 1933 (§ J–16) p. 17; Taboga (Panama). The size

REFERENCES

as given by Monro is 47 x 2 mm. the color "dull purple relieved by brilliant sulphur-coloured dorsal cirri and a pair of transverse marks of the same." Range: northern Peru (Gravier), Panama, and Gulf (herewith).

A *Phyllodoce*-like form taken only once, in dead coral clusters east of La Paz.

FAMILY AMPHINOMIDAE (Chamberlin 1919a [§ J–5] p. 23 and Fauvel 1923 [§ J–6] p. 125.)

§ J–34 *Chloeia viridis* (Schmarda) 1861 PL. 20 FIG. 2
Hartman 1940 (§ J–13) p. 205; range: Gulf of California to Panama, Galápagos, and Cocos; West Indies. Monro 1933 (§ J–16) p. 9: "body pinkish-white with distinct black and opaque white and yellow markings."

A giant stinging worm, illustrated herewith, taken at night at Puerto Refugio anchorage in crab nets put over the side in 7 fathoms' depth.

§ J–35 *Eurythoë complanata* (Pallas) 1766 PL. 18 FIG. 1
There is a long account starting on p. 28 of Chamberlin 1919 (§ J–5), based on Galápagos and Polynesian specimens (*complanata* was not then thought to occur on continental American shores). Monro 1933 (§ J–16) p. 4: "salmon-pink all over, this colour being in contrast with the snow-white chaetae. The gills are bright orange. . . ." Hartman 1940 (§ J–13) p. 202; range: tropicopolitan, intertidal to 10 fathoms; occurs on the Pacific North American Coast from southern California (dredged), through the Gulf (shore to 60 fathoms) to the Galápagos, etc.

A stinging worm. The commonest annelid encountered in the Gulf, found at every suitable locality examined (rocky shores, coral, etc.) from Pulmo Reef to Angeles Bay and on the islands.

§ J–36 *Notopygos ornata* Grube 1856 PL. 20 FIG. 3
Hartman 1940 (§ J–13) p. 207: Gulf of California to Colombia and Galápagos, shore to 70 fathoms. Monro 1933 (§ J–16) p. 10: "brownish, with, even for an Amphinomid, conspicuous long bunches of white chaetae. . . ."

A smaller, stouter, stinging worm, in its chunkiness resembling the sea-mouse (*Aphrodita*) dredged in California waters. Taken in coral interstices east of La Paz, and under rocks at Puerto Refugio.

FAMILY PHYLLODOCIDAE: The Paddle worms. (Chamberlin 1919a [§ J–5] p. 97 and Fauvel 1923 [§ J–6] p. 141.)

REFERENCES

§ J–37 *Anaitides madeirensis* Langerhans 1880
Chamberlin 1919a (§ J–5) p. 104; as a synonym of *A. patagonica* (Kinberg) 1865 (from the southwestern Atlantic). Range reported there as Peru, 536 fathoms, and as from Hawaii (Treadwell 1906 etc. as *Phyllodoce sancti-vicentis*). If this range, discontinuous and curious for a shore animal, should be substantiated, it would constitute another zoological anomaly. Dr. Hartman, however, regards this species as more or less cosmopolitan. Monro 1933 (§ J–13) p. 21; Panama and Gorgona Island, Colombia.

An elongate paddle worm taken at Pulmo Reef and at Coronado Island. Similar forms (but different species) are mentioned on p. 110 of Ricketts and Calvin 1939 (§ Y–3), and on p. 314, Vol. 2 (1910) of the *Cambridge Natural History*.

§ J–38 *Eulalia myriacyclum* (Schmarda) 1861
Reported in a personal communication from Dr. Hartman as occurring in Jamaica (p. 24-25, Augener 1925, Publ. Univ. Zool. Mus. Kobenhavn, No. 39. We were unable to consult this reference).

A spectacularly colored paddle worm taken under rocks at Puerto Refugio. Long, flat specimens of this sort, up to several inches in length, will be recognized by their rich dark-green background color, against which the two light longitudinal stripes show up strikingly.

FAMILY HESIONIDAE (Chamberlin 1919 [§ J–5] p. 183 and Fauvel 1923 [§ J–6] p. 231.)

§ J–39 *Podarke pugettensis* Johnson 1901
Hartman 1940 (§ J–13) p. 211; North Pacific (California and Puget Sound) to Peru. Japan. Littoral.

At Pulmo Reef, a single specimen was found in the groove on the underside of one of the rays of the starfish *Oreaster occidentalis* (§ K–125), resembling a commensal scale worm but with longer bristles and lacking elytra, and occurring exactly as the polynoid *Arctonoë fragilis* occurs in *Evasterias* at Puget Sound (Ricketts and Calvin 1939, § Y–3, p. 157). *Podarke pugettensis* has been reported as a commensal of *Luidia,* and of "Asterias" (*Pisaster* or *Evas-*

REFERENCES

terias at Puget Sound, and as a parasite of starfish (type not stated) at San Diego (Moore, 1909, § J–26, p. 243).

FAMILY DORVILLEIDAE (Chamberlin 1919a [§ J–5] p. 328.)

§ J–40 *Stauronereis cerasina* (Ehlers) 1901
Hartman 1940 (§ J–13) p. 214, from 4 stations in the Gulf, shore to 15 fathoms. Range: Gulf of California, and Juan Fernandez off southwestern South America; sublittoral to 15 fathoms.

Small nondescript worms removed from mass of sponge or compound tunicates at Coronado Island.

FAMILY SYLLIDAE (Chamberlin [§ J–5] p. 163 and Fauvel [§ J–6] p. 352.)

§ J–41 *Amblyosyllis* sp.
(genus equals *Pterosyllis,* Fauvel 1923 [§ J–6] p. 279.)

Small white elongate worms taken pelagically at night by light hung overboard at San Lucas Cove, March 29, with *Platynereis agassizi* (§ J–46).

§ J–42 *Odontosyllis* sp.
(genus on p. 180 Chamberlin [§ J–5] and p. 274 Fauvel [§ J–6].)

Atokous form taken as minute elongate specimens in coral interstices at Pulmo Reef, and on the reef at Pt. Marcial. The Epitokous or pelagic stage was taken at night by light hung overboard at Puerto Escondido anchorage.

FAMILY NEREIDAE (Chamberlin [§ J–5] p. 191 and Fauvel [§ J–6] p. 328.)

§ J–43 *Ceratonereis tentaculata* Kinberg 1866
Hartman 1940 (§ J–13) p. 218; range reported as almost tropicopolitan; eastern and western Pacific, Australia, West Indies; shore to 40 fathoms.

The atokous phase was taken as an elongate worm resembling *Lumbrinereis* under rocks at Puerto Escondido, with *Eurythoë* and *Dasybranchus.* The epitokous or heteronereis stage was netted pelagically at night by light hung over side at Concepcion Bay anchorage, associated with the *Platynereis polyscalma* below.

REFERENCES

§ J–44 *Neanthes* sp.
Genus treated as a subgenus under Nereis in Fauvel (§ J–6) p. 346; as a genus in Hartman 1938, Univ. Calif. Publ. Zool., Vol. 43 (4): 79-82.

Two individuals taken pelagically at San Lucas Cove with the *Perinereis* below.

§ J–45 *Perinereis* sp.
(Genus treated on p. 352, Fauvel 1923)

The epitokous stage was taken abundantly with the above *Neanthes* at San Lucas Cove, March 28 or 29, 1940; by light hung over side during the night at anchorage.

§ J–46 *Platynereis agassizi* (Ehlers) 1868
Hartman 1940 (§ J–13) p. 231; numerous specimens as *Uncinereis agassizi*. Range reported as from British Columbia to Peru; shore to 900 fathoms.

Numerous epitokous (heteronereis) forms were taken pelagically at night by light hung over side at the San Lucas Cove anchorage (March 28 or 29, 1940) associated with a small species of *Amblyosyllis*.

§ J–47 *Platynereis polyscalma* Chamberlin 1919
Hartman 1940 (§ J–13) p. 229; numerous specimens. Range reported as tropical Pacific, Indo-Pacific; pelagic to 15 fathoms.

Numerous epitokous or heteronereis forms were taken pelagically at night, by light hung overside at the Concepcion Bay anchorage, associated with the *Ceratonereis* (§ J–43) above.

FAMILY GLYCERIDAE: The Proboscis worms.
(Chamberlin [§ J–5] p. 343 and Fauvel [§ J–6] p. 381.)

§ J–48 *Glycera dibranchiata* Ehlers 1868
Hartman 1940 (§ J–13) p. 246; one specimen from Mission Bay, San Diego. Range: New England coast south to Cape Hatteras; Southern California; shore to 633 fathoms.

Burrowing in the sandy mud-flats of El Mogote, and again in the sand of San Lucas Cove (March 29, 1940). These 2" to 3" proboscis worms appear to be miniature replicas of the very noticeable reddish *G. rugosa* (Ricketts and Calvin [§ Y–3] p. 198) often turned out of the sandy mud in California estuaries, where they roll a wicked-looking introvert in and out, which changes the shape of the head from pointed to club shape.

FAMILY EUNICIDAE: (Chamberlin [§ J–5] p. 229 as LEODICIDAE. Fauvel [§ J–6] p. 488.)

REFERENCES

The Palolo worms, usually tube-building forms, very elongate, with the anterior cirri modified to form feathery gills.

§ J–49 Eunice afra (Peters)
Monro 1933 (§ J–16), p. 66. Taboga Island.

Pt. Lobos, Espiritu Santo Island; one specimen only, with other worms. Reported by Dr. Hartman as an addendum.

§ J–49a E. aphroditois (Pallas)

Common in and around coral clusters at Pulmo Reef, with two other species of *Eunice*. Also under rocks at Pt. Lobos, Espiritu Santo Island. Reported by Dr. Hartman while this section was in press.

§ J–50 *E. antennata* Savigny PL. 38 FIG. 4
Berkeley 1939 (§ J–3) p. 334; from Haiti and from Tiburon Island. Monro 1933 (§ J–16) p. 59; from Colón (Atlantic) Panama, Gorgona Island, Galápagos.

With the above at Pulmo Reef. Also at Coronado Island; reported in our field notes as a *Lumbrinereis*-like form, elongate, and with round cross-section of constant diameter (not tapering).

§ J–51 *E. filamentosa* Grube
Monroe 1933 (§ J–16) p. 65; Galápagos and Gorgona.

Taken only once, under tide pool rocks at Puerto Refugio, Angel de la Guardia Island, with the following:

§ J–52 *E.* (*Palola*) *siciliensis* Grube
Fauvel 1923 (§ J–6) p. 405, circumtropical.

Taken with the above at Puerto Refugio, and among coral heads at Pulmo Reef with two other *Eunice* spp.

§ J–53 *E. schemacephala* Schmarda
Hartman 1939 (§ J–11) p. 15; from Old Providence Island in the Caribbean. Range: West Indies, Florida, Caribbean.

Taken with *E. afra* under rocks on the beach at Pt. Lobos, Espiritu Santo Island.

§ J–54 *Eunice* sp.

Removed from mass of sponge or tunicate on shell of *Pinna* from slightly below low tide level, Concepcion Bay.

§ J–55 *Marphysa aenea* (Blanchard)
Monro 1933 (§ J–16) p. 67; Galápagos.

A worm found commonly at the base of mangroves at El Mogote; recorded at the time of capture as reminding us somewhat of a proboscis worm.

FAMILY LYSARETIDAE (included as a subfamily of LUMBRINEREIDAE in Chamberlin [§ J–5] p. 325 and Fauvel [§ J–6] p. 426.)

REFERENCES

§ J–56 *Aglaurides fulgida* (Savigny)
Monro (§ J–16) p. 91; 5–10 fathoms. Coiba. As *Oenone fulgida.*

One of the "miscellaneous worms" taken under rocks on the beach at Pt. Lobos on Espiritu Santo Island, and again on a rock at Puerto Escondido.

FAMILY LUMBRINERIDAE (Chamberlin [§ J–5] p. 325 and Fauvel [§ J–6] p. 429.)

§ J–57 Fragments of *Lumbrinerids,* lacking those portions which carry identification characters, were taken in the sand at San Lucas Cove, March 29, 1940, and (probably in a different species) under tide pool rocks at Puerto Refugio.

FAMILY SPIONIDAE (Chamberlin [§ J–5] p. 367 and Fauvel [§ J–7] p. 26.)

§ J–58 *Polyodora* sp.
(Genus treated on p. 305 Hartman 1941 [§ J–28].)

(To be described by Dr. Hartman only when additional material comes to hand.) Attached to a tube of *Spirobranchus* from Cape San Lucas.

§ J–59 *Scolelepis* sp.

(To be described only when additional material comes to hand.) A minute sand-living form (these often build membranous tubes in sand) taken at San Lucas Cove, March 29, 1940.

FAMILY CIRRATULIDAE (Chamberlin [§ J–5] p. 370 and Fauvel [§ J–7] p. 88.)

§ J–60 *Cirriformia spirabranchus* (Moore) 1904
Moore 1904 (§ J–26) p. 492; San Diego, presumably on shore.

Recorded in our notes as a tubed cirratulid living in the dazzling white coral sand about rocks at Coronado Island. Dr. Hartman notes it as occurring normally in a burrow. Our notes may be incorrect.

FAMILY FLABELLIGERIDAE (Chamberlin [§ J–5] p. 394 Fauvel [§ J–7] p. 112 equals *Chlorhaemidae.*)

REFERENCES

§ J–61 *Stylarioides capulata* (Moore) 1909
Moore 1909 (§ J–26) p. 284; San Diego Bay, between tides.

Under rocks at Puerto Refugio.

§ J–62 *Stylarioides papillata* (Johnson) 1901
Johnson 1901, Proc. Bost. Soc. Nat. Hist., Vol. 29, p. 416; from Port Orchard, Washington. Monro [§ J–17] p. 1058; "Taboga Island, from dead coral just below tide level (3). . . . Crossland writes of these specimens: 'Whether these live in crannies or actually bore is doubtful. Body slender as in preserved specimens, and a dull green all over. No gelatinous or sandy covering, bodies firm and rather stiff.' "

Taken in dead heads of coral east of La Paz. Remarked in the field notes as "resembling *Stylarioides.*" We were surprised to find that this material belonged to this species which occurs in deep water in Monterey Bay and along shore in the Straits of Juan de Fuca in Puget Sound.

FAMILY OPHELIIDAE (p. 382 Chamberlin [§ J–5] and Fauvel [§ J–7] p. 128.)

§ J–63 *Armandia* sp.
(Genus treated on p. 105 Hartman 1938 [§ J–26] and on p. 135 Fauvel [§ J–7].)

Another sand-living species, taken at San Lucas Cove, March 29, 1940.

§ J–64 *Travasia gigas* Hartman 1938
Juvenile, determination not certain. Hartman, p. 103, 1938 (§ J–26) from San Francisco Bay, Tomales Bay, and San Diego Bay.

Taken with the above at San Lucas Cove.

FAMILY CAPITELLIDAE (Chamberlin [§ J–5] p. 463 and Fauvel [§ J–7] p. 139.)

§ J–65 *Dasybranchus caducus* Grube
Fauvel (§ J–7) p. 148, fig. 52. Atlantic and Mediterranean coasts of Europe; Red Sea; China Sea; Indian Ocean; Pacific Ocean.

REFERENCES

Monro 1928a (§ J–15) p. 97; Panama, sandy shore "about a dozen fragments of this common, almost cosmopolitan species."

Monro 1933a (§ J–17) p. 1059; Trinidad and as *D. c. var. lumbricoides* Grube, from Galápagos shore pools (3 specimens); and Gorgona, low water.

From muddy sand flats at El Mogote.

FAMILY MALDANIDAE (Chamberlin [§ J–5] p. 407 and Fauvel [§ J–7] p. 167.)

§ J–66 The anterior fragments of some maldanid worm, not determinable without additional material, were taken with the above in the sandy mud flats of El Mogote, while we were digging out larger forms. This seemed to be fairly common.

FAMILY OWENIIDAE (Chamberlin 1919 [§ J–5] p. 418 as AMMOCHARIDAE. Fauvel [§ J–7] p. 202.)

§ J–67 *Owenia fusiformis* della Chiaje

Fauvel [§ J–7] p. 203; North Sea; English Channel; Atlantic; Mediterranean; South Atlantic; Indian Ocean; Pacific.

Taken at El Mogote, but whether under occasional rocks (as with the *Eurythoë* that occurred at this point) or dug out of the sand (probably the latter) our notes fail to mention.

FAMILY SABELLARIIDAE (Chamberlin [§ J–5] p. 483 and Fauvel [§ J–7] p. 205.)

§ J–68 *Idanthyrsus pennatus* (Peters)

Monro 1933a (§ J–17) p. 1065; Taboga and Gorgona Island; numerous at low tide and in coral heads brought up by diving. Crossland writes, "Large, not forming colonies. In single tubes or several together under one stone, but hardly more colonial than a Terebellid. Tubes attached to the stones throughout their length and resemble those of a large Terebellid until broken, when they are found to be much more solid. . . . Tentacles in life when not extended 10 mm. long. Fresh water or alcohol brings out a rich green solution immediately." Hartman 1939 (§ J–11), p. 19; from Clipperton Island. Range: cosmopolitan, in the littoral.

REFERENCES

Taken on the reef or on coral clusters at Pulmo Reef. Probably one of the sandy-membranous tube builders similar to the *Sabellaria* which builds great honeycomb colonies, per photo taken at Santa Cruz, California (Pl. XXVI of Ricketts and Calvin 1939 [§ Y-3]).

FAMILY TEREBELLIDAE (Chamberlin [§ J-5] p. 418 and Fauvel [§ J-7] p. 240.)

§ J-69 *Pista elongata* Moore 1909
Moore 1909 (§ J-26) p. 271; between tides, San Diego.
Monro 1933a (§ J-17) p. 1068 "Balboa, rocks and rock pools, low tide in Panama (6)."

A characteristic terebellid taken, with its shelly-membranous tube, at El Mogote.

§ J-70 *Thelepus setosus* (Quatrefages) 1865
Fauvel 1927 [§ J-7] p. 273 with synonymy. Atlantic coast of Europe; Red Sea; Indian Ocean; Pacific (Chile, Australia); Falkland Islands; Antarctic.
Berkeley, 1939 (§ J-3) p. 342, from San Felipe Bay, Sonora.

A tubed worm, readily recognizable in the field as a terebellid, from the Reef at Marcial Point. Taken again (probably on rocks) at Angeles Bay, where it was reported in the field notes as "a giant *Thelepus*-like form"—obviously, since it is the same species (= Johnson's 1901 *Thelepus crispus*) with which we were familiar up north.

FAMILY SABELLIDAE (Chamberlin [§ J-5] p. 466 and Fauvel [§ J-7] p. 293.)

§ J-71 *Megalomma mushaensis* (Gravier) 1908
Monro 1933a (§ J-17) p. 1078; from fine gravel in shore pools, and from 12 fathoms, abundant. "Crossland shows that the colour in this species is very variable, 'body greenish white, further whitened by numerous opaque white blotches. Gills above base, chocolate brown with two deep chocolate lines, distal half of gills whitish.' Elsewhere he says that the gills may be dark greenish brown and white . . . almost black with no white . . . or deep crimson. . . . The tubes are heavily encrusted with coarse shelly sand."

In the white coral sand at the base of rocks, Coronado Island. Also, very noticeably, at Gabriel Bay on Espiritu Santo Island, where,

REFERENCES

again in white coral sand about the bases of rocks, specimens of this tubed worm with highly colored tentacles were so numerous as to cause the sand to change color when the animals were disturbed and snapped back into their white sandy and shell-covered tubes. The majority of the specimens were small, ¼"–½", and could be collected readily by uprooting the longer tubes from the sand. Large specimens however, up to 2"–3" in length, were so intermingled with interstices of rock that they could be got only with great difficulty.

FAMILY SERPULIDAE (Chamberlin [§ J–5] p. 472 and Fauvel [§ J–7] p. 346.)

§ J–72 *Eupomatus* sp. PL. 34 FIG. 4

Minute serpulids, attached to the calcareous tubes of the larger serpulid *Spirobranchus* (below) at Cape San Lucas.

§ J–73 *Protula tubularia* (Montagu) 1803

Fauvel (§ J–7) p. 382; Atlantic and Mediterranean coast of Europe; Persian Gulf; Indian Ocean.

Monro (§ J–17) p. 1088, in var. *balboensis* nov. Balboa; Taboga Island; Gorgona Island; numerous, shore.

Taken at Puerto Refugio, but lacking any field notes to which it may be referred.

§ J–74 *Salmacina dysteri* (Huxley) 1855

Fauvel 1927 (§ J–7) p. 377, North Sea; Atlantic and Mediterranean coast of Europe; Red Sea; Australia.

Monro (§ J–17) p. 1090, in var. *tribranchiata* (Moore) 1923; Galápagos shore, numerous . . . "a large mass of tubes . . . Crossland describes these specimens as being 'conspicuous by their brilliant vermilion red, which colour disappears on the colony being touched.' The pigmentation of the gills is a further difference between this variety and the stem form."

Moore 1923 (§ J–26) p. 250; 38–45 fathoms, Southern California, as *Filograna tribranchiata* n. sp.

Minute grayish-white calcareous tubed forms with rusty reddish gills, which, like *Filograna* sp. along the California coast, change the color of the rocks depending on whether their gills are retracted or extended. Literally encrusting the rocks at San Francisquito and Angeles Bays, and (no doubt) elsewhere in the northern part of the Gulf, high in the intertidal.

§ J–75 *Spirobranchus incrassatus* Mörch 1863

("Revisio critica Serpulidarum" Naturh. Tidsskrift, Copenhagen, Vol. 3 (1): 347-470, Pl. X. Not accessible to us.)

REFERENCES

An important feature of the low intertidal landscape at Cape San Lucas, where the anastomosing calcareous tubes of this large and spectacular worm encrust the rocks. The gills are brilliantly colored in a banded pattern of red, white, and black.

§ J–76 Spirorbis tubes, Agiabampo Bay (on eel grass) and attached to vermetid mollusk tubes at Amortajada Bay.

(See Pl. 34 Fig. 4)

§ J–77 SUMMARY

Some 51 species of Polychaets were taken along shore in the Gulf during our brief survey. There were 30 ERRANTIA; the FAMILIES POLYNOIDAE with 5 species, AMPHINOMIDAE with 3, NEREIDAE with 5, and EUNICIDAE with 8 species, being particularly well represented. Among the Sedentaria, with a total of 20 species, only the SERPULIDAE was especially noticeable, with 4 species. The affiliations were strongly tropical—cosmopolitan, West Indian, or Indo-Pacific, in about that order. There were only slight northern (Californian) connections. Among the Polychaets, as with other Gulf invertebrates, the northern collector would find himself very much of a stranger, although the student familiar with tropical conditions elsewhere presumably would have little difficulty in orienting himself.

No distributional information, obviously, can be had on the undetermined and indeterminable species, and we have no zoogeographical data on *Eunice aphroditois,* on the *Eunice* cited as § J–54, and on *Spirobranchus incrassatus.* The information on *Eunice afra, Marphysa aenea,* and on *Aglaurides fulgida* may also be insufficient. However, the 37 species (including the 3 doubtful ones) on which we show geographic range are definitely tropical. The assemblage is less unique than we have found with other Panamic groups. Only three species (8⅓%) occur solely in the Gulf (in the Brachyura the uniqueness was about 15%, in anomurous forms, 50%). 21 extend south—always to Panama, often to the Galápagos. Only 3 extend north, all of them to San Diego. 9 extend both south and north, mostly to southern California only, but *Platynereis agassizi* is reported from British Columbia, *Stylarioides papillata* occurs at least to Puget Sound, *Travisia gigas* to San Francisco, and *Thelepus setosus* as *crispus* at least to southern British Columbia. 5 of these 9, incidentally, have been reported outside Pacific American waters.

Viewed in another light, only 13 of the 37 are restricted to Panamic waters. 23 occur elsewhere also. 6 are cosmopolitan, 4 are tropicopolitan, 5 are West Indian, 3 Indo-Pacific (including Hawaii), and one each are known elsewhere only from West Africa, from Japan, and from the Atlantic Coast of United States. Three others, of the forms limited to the Pacific American coast, extend northerly well out of Panamic limits, to San Francisco, Puget Sound, and to British Columbia.

However, although it derives from cosmopolitan rather than Panamic elements, the color of this assemblage of common shore species is decidedly tropical.

The (§ J–136) *Stylarioides* presents an interesting anomaly. It occurs at low tide not uncommonly along the Straits of Juan de Fuca and elsewhere in Puget Sound and British Columbia, in completely sheltered stations adjacent to oceanic waters. Along the California coast it occurs in deep water as a common inhabitant of crevices and interstices in rocks brought up by the fishermen from 60–110 fathoms. This would naturally be interpreted as a temperature situation, since the annual mean of Monterey Bay deep water is presumably not dissimilar to that of more northern tidal waters. But under tropical conditions in the Gulf, it occurs again along shore. The only common denominator would seem to be lack of wave shock.

The most spectacular species were the stinging worms, especially *Eurythoë complanata* which was incidentally the most abundant and ubiquitous annelid encountered; the introvert worm *Glycera dibranchiata;* the richly colored *Eulalia;* the handsome and large tube-dwelling *Polyondontes oculea* which occurred in great colonies at Agiabampo, and the tubed sabellids and serpulids. At Cape San Lucas, the serpulid *Spirobranchus incrassatus* literally covered the rocks with its coiled stony tubes from which the undisturbed animal extended brilliantly colored gills. Aside from the pelagic Syllids, and Nereids in their reproductive stages, and some of the species mentioned above, the commonest worms encountered were sand and mud-flat dwelling sessile Lumbrinerids, Ophelids, and Maldanids which we collected inadequately, unfortunately. Many of these worms are so nondescript to all

REFERENCES

except systematists that the average collector only with difficulty keeps his interest in them from flagging. The fact that the descriptions are usually remote, and popularly unrelated to the living animal, provides a further obstacle. Most of the annelid illustrations, furthermore, (except in Grube and Fauvel) are detailed drawings of appendages, often microscopic or at least highly magnified, and separated from the animal itself. However, Dr. Hartman remarks aptly, in a personal communication, that the remoteness of the Polychaet situation, on the Pacific coast at least, is a function of its "unknown-ness." She notes, furthermore, that under present conditions "the time necessary to field work on any group of our invertebrates is infinitely small compared to that involved in working up a collection—a fact not appreciated by non-systematists." Liaison writers are much needed here, but, pending the erection of a sound systematic foundation, are little likely to appear, owing to the inaccessibility and difficulties of the present subject matter.

Phylum Echinodermata

REFERENCES

§ K–1 Boone, Lee 1928. "Echinoderms from the Gulf of California and the Perlas Islands." Bull. Bingham Ocean. Coll., Vol. 2 (6): 14 pp., 9 pls.

A report on 142 specimens of 20 species, none new, all from shallow-water. 15 are well illustrated. 17 species, in 133 individuals, were taken in the Gulf. There are several oversights. Notably the photograph reproduced as Pl. 4 is not *Heliaster kubiniji* as labeled, but *Pycnopodia* of the temperate fauna. Deichmann 1936 (§ L–3) re-identifies as *Euapta godeffroyi* (Semper) the synaptid considered here to be *E. lappa*.

§ K–1a Boone, Lee 1933. "Coelenterata, Echinodermata, and Mollusca." Bull. Vanderbilt Mar. Mus., Vol. 4: 217 pp., 133 pls.

The Echinoderm section considers 12 Panamic species: *Nidorellia armata, Luidia columbia, Linckia columbiae, Heliaster multiradiatus, Amphiura diomedeae* (dredged only), *Ophiocoma aethiops, Ophioderma variegatum* (Punta Arenas, Costa Rica), *Eucidaris thouarsii, Strongylocentrotus gibbosus, Pelagothuria natatrix* (dredged only), *Holothuria impatiens,* and *H. kefersteinii,* mostly from the Galápagos littoral. Dredged material was from 50–300 fathoms. There are excellent toto-photographs of all the asteroids and echinoids, and of *H. impatiens.* However Dr. Fisher notes, in a personal communication, that while Pl. 39 and 40 are entitled *Linckia columbiae* per text subject on p. 79, the specimen photographed actually was *Pharia pyramidata.* Dr. Fisher remarks also that *Nidorellia,* p. 73, listed in the Astropectinidae, belongs actually in the Family Oreasteridae near *Oreaster,* p. 80.

§ K–2 Clark, A. H. 1939. "Echinoderms (other than holothurians) collected on the Presidential Cruise of 1938." Smith. Misc. Coll., Vol. 98 (11): 18 pp., 5 pls.

23 species from Galápagos, Clipperton, Cape San Lucas, and Magdalena Bay. Two new ophiurans (an *Ophiocomella,* new genus, and *Ophionereis roosevelti*), and a new cidarid urchin. Good toto-illustrations.

§ K–3 Clark, H. L. 1902. "Echinodermata." No. 12 in "Papers from the Hopkins Stanford Galápagos Expedition, 1898-1899." Proc. Wash. Acad. Sci., Vol. 4: 521-531.

146 individuals in 24 species from the Galápagos, none new, mostly from shore collecting. List to date of all littoral species known from the region. From Clipperton Island, 2 species of holothurians, one new.

§ K–4 .... 1910. "Echinoderms of Peru." Bull. Mus. Comp. Zool., Vol. 52 (17): 321-358, 14 pls.

REFERENCES

Among others, about 25 species of Panamic echinoderms are described, and in most cases excellently illustrated by lithography. About half of the Panamic forms range as far north as the Gulf. Good bibliography.

§ K–5 .... 1913 and 1923. "Echinoderms from Lower California, with descriptions of new species" and "Supplementary report." Bull. Am. Mus. Nat. Hist., Vol. 32: 185-236, 3 pls.; and 48: 147-163.

Albatross Expedition, devoted mostly to dredging. Coasts of California and Lower California, and up into the Gulf. Shallow-water collecting with boat dredge only in the Gulf. A statistical summary of both of the above reports indicates that a total of 118 species were taken, 9 of them new, with 2351 individuals. Of these, only 39 species in 482 individuals were from shallow water. 31 species in about 387 individuals were taken in the Gulf. The new species only are illustrated, and in the first report only. Per Dr. Clark's notes on p. 203 and 153, *Asterias forreri* becomes *Astrometis sertulifera* (Xantus).

§ K–6 .... 1935. "Some new echinoderms from California." Ann. Mag. Nat. Hist., (10) Vol. 15: 120-129.

One new variety of *Dendraster excentricus* and two new ophiurans, *Ophiacantha* and *Amphiodia*. From shallow dredgings in southern California. Possibly northerly extensions of Panamic forms into the Panamic-temperate overlap area.

§ K–7 .... 1940. "Notes on echinoderms from the west coast of Central America." No. 21 in: "Eastern Pacific Expeditions of the New York Zoological Society." Zoologica, Vol. 25 (Pt. 3, No. 22): 331-352, 2 pls., 4 text figs.

58 species, 822 specimens, comprising the echinoderm collections other than holothurians; of the 1937-38 *Zaca* Expedition. 52 of the species can be considered as Panamic, only 6 being northern or overlap forms. The majority were dredged in depths from a few fathoms to 61 fathoms; 15 however were taken along shore, or, lacking depth information, can probably be considered correctly as shore animals. There are 13 species of asteroids including one new variety, 29 ophiurans including 3 new species, and 16 echinoids including one new species.

§ K–8 Ives, J. E. 1889. "Catalogue of the Asteroidea and Ophiuroidea in the collection of the Academy of Natural Sciences of Philadelphia." Proc. Acad. Nat. Sci. Phil., 1889: 169-179.

14 Panamic asteroids are listed with locality, including undetermined *Asterias* and *Linckia* spp., such rare forms as *Acanthaster ellisii* and *Amphiaster insignis*, and several forms such as *Astropecten oerstedii*, *Luidia tesselatus* and *Archaster typicus* which have been lost track of since or have been synonymized. The *Heliaster multiradiata* deter-

REFERENCES

mination is probably an error for *H. kubiniji.* 12 ophiurans from the area under consideration.

§ K–9 Verrill, A. E. 1867. "Notes on the echinoderms of Panama and west coast of America, with descriptions of new species." Trans. Conn. Acad., Vol. 1: 251-322, Pl. 10 (part).

A check list of 67 species, 19 of them new (and most of these are still valid species), about 57 of which can be considered as Panamic, the balance Peruvian. Most of them are littoral, the result, evidently, of the most careful and discriminating collecting. A summary of what was known then of the Panamic echinoderms. Descriptions are superficial, there are no natural history notes, and only 2 illustrations.

§ K–10 .... 1867a. "On the geographical distribution of the echinoderms of the west coast of America." *Ibid.,* Vol. 1: 323-351.

Except for the Panamic Province, which could be defined fairly accurately even then, the lists are too fragmentary and poorly coordinated, and the deductions drawn from them are too loose to have any modern value.

§ K–11 .... 1868. "Notice of a collection of echinoderms from La Paz, Lower California, with descriptions of a new genus." *Ibid.,* Vol. 1: 371-376, Pl. 4 (part).

9 species with collecting notes. Lithograph of section of new sand dollar.

§ K–12 .... 1869. "On some new and imperfectly known echinoderms and corals." Proc. Bost. Soc. Nat. Hist., Vol. 12: 381-396.

3 asteroids (including *Gymnasterias spinosa*), 5 echinoids, and *Astrophyton panamense* ("adhering firmly to the branches of *Muricea*") mostly from La Paz.

§ K–13 .... 1870. "Descriptions of echinoderms and corals from the Gulf of California." Amer. Journ. Sci. (2), Vol. 49 (whole number, Vol. 99): 93-100.

Descriptions of an asteroid and 7 echinoids, with list of others.

§ K–14 .... 1871. "Additional observations on echinoderms, chiefly from the Pacific coast of America." Trans. Conn. Acad., Vol. 1: 568-593, Pls. 9 and 10 (part).

Includes notes on 28 species, some of them new, from La Paz.

§ K–15 .... 1871a. "The echinoderm fauna of the Gulf of California and Cape San Lucas." *Ibid.,* 1: 593-596.

50 species, mostly littoral, with localities. A modernization of the 1867 paper mentioned above.

REFERENCES

§ K–16 Ziesenhenne, Fred C. 1937. "Echinoderms from the west coast of Lower California, the Gulf of California, and Clarion Island." No. 10 in "The Templeton Crocker Expedition." Zoologica, Vol. 22: 209-239, 2 text figures.

65 species were taken in 2277 individuals of which the bulk were *Ophiura lütkenni.* 53 species can be considered as Panamic. Mostly dredged, in depths to 80 fathoms, but 21 were taken along shore or in water 3 fathoms deep or less. 41 species were taken in the Gulf. 2 of the ophiurans represent new species. Good bibliography.

CLASS ASTEROIDEA

§ K–100a Caso, Maria Elena 1941. "Contribución al conocimiento de los Astéridos de México. I. La existencia de *Linckia gouldingii* Gray en la costa pacífica." An. Inst. Biol., Vol. 12 (1): 155-160, 6 text figs.

A single specimen of this Atlantic and Indo-Pacific species was collected at Mazatlan. Differentiable from the similar and commoner *L. columbiae.*

§ K–101 Clark, A. H. 1916. "Six new starfish from the Gulf of California and adjacent waters." Proc. Biol. Soc. Wash., Vol. 29: 51-62.

Two from the gulf, dredged in 14 and 40 fathoms; an *Anthenea* possibly littoral from the west coast of Mexico; two from great depths off Clarion and Revillagigedo Islands; one from 31 fathoms off Lower California.

§ K–102 Clark, H. L. 1907. "The starfishes of the genus *Heliaster.*" Bull. Mus. Comp. Zool., Vol. 51 (2): 25-76, 8 pls.

Four are members of the Panamic fauna and one constitutes an inclusion from the Peruvian fauna to the south. Ample and excellent lithographic illustrations.

§ K–103 1920. "Asteroidea." (*Albatross,* 1904-1905.) Mem. Mus. Comp. Zool., Vol. 39 (3): 46 pp., 6 pls.

Four common littoral Panamic species are listed, and five dredged (300 to 1036 fathoms), one new.

§ K–103a Döderlein, L. 1917. "Die Asteriden der Siboga-Expedition; I: Die Gattung *Astropecten* und ihre Stammesgeschichte." Siboga-Expeditie, Vol. 46a: 1-191, 17 pls., 20 text figs.

103 species are recognized in this cosmopolitan genus, including the following Panamic forms: p. 165, *A. benthophilus* Ludwig 1905, 1408 M., Cocos Islands; p. 169, *A. brasiliensis peruvianus* Verrill 1867, 13–15 M., Payta; p. 169, *A. b. erinaceus* Gray 1840 (according to Fisher,

REFERENCES

equals *A. armatus* Gray 1840 which has page priority), Ballenas Bay (West Coast of Lower California) to Guayaquil; p. 170, *A. verrilli* de Loriol 1899 (Döderlein includes in this species, incorrectly according to Fisher, *A. californicus* Fisher 1911 from 10–244 M., San Francisco area to Cedros Island), Cedros and Guadelupe Islands and Mazatlan, 18–450 M.; p. 174, *A. regalis* Gray 1840, from Mazatlan and Panama; p. 175, *A. latiradiatus* Gray 1871 (Fisher, however, regards *Platasterias*, the original genus, as still correct).

§ K–103b .... 1920. "Die Asteriden der Siboga-Expedition; II: Die Gattung *Luidia* und ihre Stammesgeschichte." Siboga-Expeditie, Vol. 46b: 193-293, 3 pls., 5 text figs.

Another cosmopolitan genus, in which Döderlein recognizes 43 species, including the following within the Panamic area: p. 238, *L. brevispina* Lütken 1871, Mazatlan to Ecuador, doubtfully in Hawaii; p. 239, *L. columbia* Gray 1840, Magdalena Bay to northern Peru and Galápagos; p. 239, *L. ferruginea* Ludwig 1905, 280 M., Gulf of Panama; p. 239, *L. foliolata* (Grube) 1866, Alaska to California, doubtfully San Diego and Mazatlan, 18-146 (345) M.; p. 239, *L. marginata* Koehler 1911, Mazatlan; p. 242, *L. bellonae* Lütken 1865, Panama and Galápagos to the Straits of Magellan; p. 242, *L. b. lorioli* Meissner 1891 (1896), Mazatlan; p. 242, *L. armata* Ludwig 1905, 95 M., Cocos; p. 243, *L. phragma* H. L. Clark 1910, Magdalena Bay, Gulf of California and Peru; p. 245, *L. asthenosoma* Fisher 1906, Monterey Bay to Los Coronados (Lower California), 20–620 M.

§ K–104 Fisher, W. K. 1906. "The starfishes of the Hawaiian Islands." Bull. U.S. Fish Comm., Vol. 23 for 1903, Pt. 3: 987-1130, 49 pls.

Includes description and fine illustration of *Mithrodia bradleyi* and mention of 5 species assumed to be common to the Panamic and Hawaiian faunas. Bibliographic citations for Gray 1840 and 1866, Ives (1892), and Lamarck 1816, which contain reference to Panamic forms, and for Perrier 1875 which describes or otherwise treats 12 plus 1? Panamic forms, some of which are probably synonyms.

§ K–105 Fisher, W. K. 1911, 1928, and 1930. "Asteroidea of the north Pacific and adjacent waters." Bull. U.S.N.M., Vol. 76: 419 pp., 122 pls.; 245 pp., 81 pls.; and 356 pp., 93 pls.

Illustrations and descriptions of several Panamic species which extend into the southern California overlap area. Bibliography most comprehensive in Pt. I (listing the above mentioned references also). *Ceramaster patagonicus* (Sladen) of the Bering Sea and of the Straits of Magellan, at the opposite ends of the earth, is stated on p. 216 of Pt. I as occurring, depth unknown, in the Gulf of California also, but Dr. Fisher has verbally questioned the Gulf record. The north temperate *Patiria miniata* (as *Asterina*) is mentioned also on p 257, Pt. I, as having been dredged at La Paz in 33 fathoms.

REFERENCES

§ K–106 .... 1928. "Sea stars from the Arcturus Oceanographic Expedition." Zoologica, Vol. 8 (9): 487-493.

Six only Pacific species, Galápagos or no data, mostly dredged. *Mithrodia* and *Nidorellia* presumably taken along shore.

§ K–107 Leipoldt, F. 1895. "Asteroidea der *Vettor Pisani* Expedition (1882-1885)." Zeit. f. wiss. Zool., Vol. 59: 545-654, Pls. 31 and 32.

The following Panamic forms are considered: *Heliaster helianthus*, *H. cuminigii*, *H. multiradiatus*, *Echinaster panamensis* n. sp. in 7 examples from the Pearl Islands, *Luidia columbiae*, *Pharia pyramidata* (also reported from a different faunal zone at Valparaiso, far to the south), *Pentoceros occidentalis* (*Oreaster*), and *Nidorellia armata*. From an unknown station, a specimen of the north temperate *Pycnopodia helianthoides* strangely appeared in the collection.

§ K–107a de Loriol, P. 1891. "Notes pour servir à l'étude des Echinodermes." Mém. Soc. phys. Hist. Nat. Genève, Volume Supplémentaire, 1890, No. 8: 1-31, 3 pls.

p. 22, *Luidia bellonae* Lutken 1864, Mazatlan, with good partial toto as fig. 1, Pl. 3, now is recognized as *L. lorioli* Meissner 1891.

§ K–107b .... 1899. "Ibid., VII." Mém. Soc. phys. Hist. Nat. Genève, Vol. 33 (2, No. 1): 1-34, 3 pls.

p. 12, *Astropecten verrili* (apparently a typographical error for *verrilli*), n. sp., is described from Mazatlan, with a small toto-illustration as fig. 5 on Pl. 2. p. 16, *A. rubidus* n. sp. is described from "Mexique," station not stated, with good partial toto as fig. 1 on Pl. 2.

§ K–108 Ludwig, H. 1905. "Asteroidea," (*Albatross* 1891, 1899-1900). Mem. Mus. Comp. Zool., Vol. 32: 292 pp., 36 pls.

About 71 species, most of them new, but mostly abyssal forms dredged far from land. Within the Panamic area, 3 species are from depths shallower than 100 fathoms (the shallowest were from 45 fathoms), and two others are only slightly deeper. Good bibliography. Fine lithographic illustrations.

§ K–109 Verrill, A. E. 1914. "Monograph of the shallow-water starfishes of the north Pacific coast, from the Arctic Ocean to California." Harriman Alaska Series, Vol. 14 (1 and 2): 408 pp., 110 pls.

Obsolete and to be used with the greatest caution. Cited chiefly for the excellent lithographic illustrations of 3 Panamic species: *Mithrodia* and *Othilia* (as *Echinaster*) on Pl. 107, and *Amphiaster* on Pl. 98. Under "Geographic Distribution" there is a list of Panamic species revised or allegedly treated otherwise. The obsolescence of this costly work by Fisher's necessary revision (1928 and 1930) is to be regretted particularly because of the fine lithographs illustrating practically all the forms considered at that time.

REFERENCES

§ K–110 Xantus, John 1860. "Descriptions of three new species of starfishes from Cape San Lucas." Proc. Acad. Nat. Sci. Phil., 1860: p. 568.

Two species of *Heliaster* and what is known now as *Astrometis sertulifera* are characterized briefly without figures. Littoral.

§ K–111 In Stimpson's 1857 "Crustacea and Echinodermata . . ." § M–16a, there is only one echinoderm listed from the region under consideration, an asteroid from Mazatlan (p. 529), *Asterias helianthus* Lamarck. This is probably a *Helianthus,* perhaps *H. kubiniji* to be described three years later by Xantus.

§ K–112 There are also two Panamic asteroids described and illustrated in Verrill 1915 ("West Indian Starfish. . . . ," Bull. Labs. Nat. Hist. St. Univ. Iowa, Vol. 7 (1): 232 pp., 29 pls.) in connection with related Atlantic forms. These seem not to have turned up since the type specimens were found in the sixties. *Enoplopatiria siderea* Verrill (p. 63, Pl. 27) is listed from one dry specimen taken at Panama, and reference is made to p. 480 of a previous paper (1913 "Revision . . . Asteriniae," Amer. Journ. Sci., Vol. 35: 447-485) but no *siderea* could be found listed there. On p. 365, Verrill 1914 specifically implies that the specimen came from the Pacific side. *Asterinides modesta* (*Asterina modesta,* p. 277 of the 1867 paper [§ K–9]) is also redescribed on p. 61, and illustrated for the first time on Pl. 27, with a correct cross reference to p. 482 of the 1913 paper.

§ K–113 See also the asteroid section of the general echinoderm papers cited on our pp. 371-374.

§ K–114 There is no comprehensive account of the Panamic asteroids. Ziesenhenne's 1937 echinoderm account (§ K–16) comes nearest to serving as a key paper, the asteroid listing being fairly large, the treatment modern, and the bibliography ample, especially if used with the Clark 1940 (§ K–7) subsequent report.

§ K–115 ACKNOWLEDGMENT: The sea-star identifications are by Prof. W. K. Fisher, Director, Hopkins Marine Station, Pacific Grove, California, to whom thanks are due also for checking this portion of the phyletic index, for in-

REFERENCES

formation on the taxonomy and bibliography of this group, and for innumerable personal assistances and encouragements both subtle and patent.

LIST OF SPECIES TAKEN:

I: Many rayed-forms, with spiny or knobby skin.

§ K–116 *Heliaster kubiniji* Xantus 1860 PL. 24 FIG. 1

Complete description and fine illustration, p. 48, Pl. 4, H. L. Clark 1907 (§ K–102). General range: Gulf of California, shore to 5 fathoms, Ziesenhenne 1937 (§ K–16), p. 220.

The most common, obvious, and widely distributed shore starfish in the Gulf, often high in the intertidal. Taken or seen in quantity at every rocky shore collecting station without exception from Cape San Lucas to the northern tip of Angel de la Guardia Island and over into Sonora. Could have been collected by the hundreds. Unfortunately we may have confused this form in the field with the more compact, short rayed *H. microbrachius* Xantus, reported from Cape San Lucas, Magdalena Bay, La Paz, and south to Panama, since we failed to differentiate the two until after the trip was over, and no specimens of the latter form (Clark 1907 [§ K–102, p. 50, pl. 1]) were brought back, although this species was unquestionably seen in the field.

§ K–117 *Acanthaster ellisii* (Gray) 1840 PL. 37 FIG. 1

Never figured or completely described. General range: Gulf of California, Galápagos, Clarion Island, shore to 6 fathoms, Ziesenhenne 1937 (§ K–16) p. 220.

Taken only at Puerto Escondido, where several specimens were found on rocks in quiet water slightly below the low tide line. A thick, soft bodied, heavy animal with thin gray skin armed with erect thin spines. Ours seem to comprise one of the three or four extant lots of this form.

II: Typical "star-fish" with 5 fairly long rays, with skin smooth or at least lacking large spines.

§ K–118 *Leiaster teres* (Verrill) 1871

Never figured or completely described.

A single very large, brilliant purple-blue specimen was taken with *Acanthaster* at Puerto Escondido where others could have been taken had we supposed the form to be rare. Ours seems to be one of the three extant specimens. Verrill 1871 (§ K–14) p. 578, described it from a single La Paz specimen, and Ziesenhenne 1937 (§ K–16) p. 217 reports another single specimen from 20-30 fathoms on Gorda Bank in the Gulf.

REFERENCES

§ K–119 *Pharia pyramidata* (Gray) 1840 PL. 20 FIG. 1

There are good toto-photographs as Pls. 39 and 40, in Boone 1933 (§ K–1a), of a littoral Galápagos specimen described on p. 79 as *Linckia columbiae,* the illustration, according to Dr. Fisher, being titled incorrectly as it actually portrays *Pharia.*

Description and excellent illustration, H. L. Clark 1910 (§ K–4) fig. 2 on Pl. 5. General range: Gulf of California to Zorritos, Peru, and Galápagos, shore to 10 fathoms, Ziesenhenne 1937 (§ K–16) p. 217.

This large handsome form (up to 12" diameter) could have been taken by the dozen at all suitable reefs and rocky points, but only in the southern part of the Gulf. None, however, was seen at Cape San Lucas. None was taken and none surely recognized north of Puerto Escondido, but no attempt was made to differentiate this from related forms in the field. Large specimens, however, with their regular rows of lemon-colored plates, are quite noticeable. Madreporite large and irregular. Ground color dull purple-brown above; below dull lavender, with light, almost white, ambulacral plates and spines. There are rows of dermal papillae below the marginal arm plates.

§ K–120 *Phataria unifascialis* (Gray) 1840 PL. 15 FIG. 1

Exceedingly good lithographic illustration on Pl. 5, and description on p. 335, H. L. Clark 1910 (§ K–4). Boone 1928 (§ K–1) has a fair halftone as Pl. 3. General range: Gulf of California to Zorritos, Peru, shore to 10 fathoms, Ziesenhenne 1937 (§ K–16) p. 217.

In size between *Pharia* and *Linckia,* up to 8" in extreme diameter, this is the second most common, obvious, and widely distributed starfish in the Gulf. Taken or seen in quantity at all or almost all the rocky collecting places visited. The general body tone is dull ultramarine or dull brown, often a mixture of both; underneath tan, with reddish ambulacral spines; there are *no* rows of dermal papillae below the marginal arm plates. The areas of plates on the aboral surface are more regular and distinct than in *Linckia,* but much less so than in *Pharia.*

§ K–121 *Linckia columbiae* Gray 1840

Complete description and good illustration p. 242, Pl. 48, Fisher 1911 (§ K–105). Illustrations also in Boone 1928 (§ K–1) Pl. 3; and in Ricketts and Calvin 1939 (§ Y–3) Pl. 11. General range: Southern California to Magdalena Bay and into the Gulf to northern Peru and in the Galápagos, shore to 55 fathoms. Fisher 1911 (§ K–105) p. 242 and Ziesenhenne 1937 (§ K–16) p. 216. The photographs

REFERENCES

labeled *Linckia columbiae,* Pls. 39 and 40, Boone 1933 (§ K–1a) actually portray, according to Dr. Fisher, a specimen of *Pharia pyramidata.*

The smallest of the long armed forms, with irregular "bumps" over the entire aboral surface, and lacking the ventral rows of dermal papillae below the marginal arm plates. Very abundant especially in the northern part of the Gulf. *Linckia* is a notable feature of the intertidal fauna at Angeles Bay and at Puerto Refugio, although it tends to segregate in crevices and under rocks out of sight. We have no sure records from the southern part, having failed in the field to differentiate it from small *Phataria,* and no undoubted *Linckia* were collected there, but Verrill, Boone, H. L. Clark, etc., report it from La Paz, Espiritu Santo, San José Island, etc.

III: Asterias-like forms. Regular 5-rayed starfish with spiny skin.

§ K–122 *Mithrodia bradleyi* Verrill 1867 PL. 23 FIG. 3

Complete description and good illustration, p. 1094, Pl. 36, Fisher 1906 (§ K–104). Excellent litho of type, fig. 1, Pl. 107, Verrill 1914 (§ K–109). General range: Gulf, Panama, and the Galápagos. A common Hawaiian form. Not listed in Ziesenhenne 1937 (§ K–16).

A large to very large coral pink form with dark blotches, arms relatively longer than in the two following species. The largest specimen, about 15″ in diameter, was slightly subtidal at Puerto Escondido. Four or five smaller specimens were taken, usually far down in the intertidal, at Pulmo Reef, and at Pt. Lobos on Espiritu Santo. A small individual was found at Puerto Refugio. All were on rocks.

§ K–123 *Astrometis sertulifera* (Xantus) 1860

Complete description and good illustrations, Fisher 1928 (§ K–105) p. 119, Pls. 58 and 59. Also illustrated on Pl. 12, Ricketts and Calvin (§ Y–3). General range: Santa Barbara to Gulf of California, very low tide to 33 fathoms, Fisher 1928 (§ K–105) p. 119. Not listed in Ziesenhenne 1937 (§ K–16).

A distinctive spiny form, flexible and slimy in life. Skin soft, dark green or brown green, with red spines. Taken at Pt. Lobos on Espiritu Santo and at Marcial Point. The type locality also is in the south, but the animal seems to be commoner in the northern part of the Gulf, Coronado Island, Angeles Bay, Puerto Refugio, etc., where dozens of specimens could be taken at each rocky station.

REFERENCES

§ K–124 *Othilia tenuispina* (Verrill) 1871 as *Echinaster*
PL. 16 FIG. 3

Good lithographic illustration of type, fig 2, Pl. 107, Verrill 1914 (§ K–109). General range: Gulf, shore to 10 fathoms, Ziesenhenne 1937 (§ K–16) p. 219.

Ordinarily smaller than the above and with less obvious spines. Yellow-brown or tawny to brick red. Very abundant at all rocky shore collecting places in the north, Angeles Bay, Puerto Refugio, San Carlos Bay in Sonora, etc., where hundreds of specimens could be taken, but noted only once south of Puerto Escondido (where it occurred commonly): at Marcial Point. Extending high in the intertidal, the uppermost of the gulf asteroids.

IV: Thick armed, pentagonal and cushion stars with heavy plates and/or spines.

§ K–125 *Oreaster occidentalis* Verrill 1867 PL. 11 FIG. 1

Very fine lithograph as fig. 1, Pl. 4, H. L. Clark 1910 (§ K–4). Good illustration also on Pl. 1, Boone 1928 (§ K–1). General range: Gulf to northern Peru and Galápagos Islands, 3–40 fathoms, Ziesenhenne 1937 (§ K–16) p. 216.

A large and heavily spined form without the massive marginal plates of *Nidorellia*. Common at Cape San Lucas, La Paz, Pulmo Reef, Pt. Lobos on Espiritu Santo, and Puerto Escondido. Dozens of specimens could have been taken in the rocky subtidal and lower intertidal, at each suitable locality. The general tone is red to green-red; spines and marginal plates are bright red, spine tips lighter, against a background of green or grayish green. The hesionid annelid worm *Podarke pugettensis* Johnson (§ J–39) was taken commensally in the ambulacral groove of a Pulmo Reef specimen.

§ K–126 *Nidorellia armata* (Gray) 1840
PL. 10 FIG. 1 and PL. 11 FIG. 2

Fine lithograph as fig. 2, Pl. 4, H. L. Clark 1910 (§ K-4). Also illustrated on Pl. 1, Boone 1928 (§ K–1). General range: Guaymas to Zorritos, Peru, and Galápagos Islands, shore to 40 fathoms, Ziesenhenne 1937 (§ K–16) p. 216. There are also good photographs as Pl. 29, Boone 1933 (§ K–1a) from a littoral Galápagos specimen.

Fairly common in the southern part of the Gulf, where a dozen specimens were taken each at Pulmo Reef and at Puerto Escondido, rocky subtidal and lowest intertidal. The most spectacular Gulf asteroid, a spiny cushioned form with strongly armored disc, heavy marginal plates, and large and heavy aboral spines that break off

REFERENCES

at the slightest touch. General color lavender, points of spines almost black with lighter margins, center of disc gray.

§ K–127 *Astropecten armatus* Gray 1840
Equals *A. erinaceus,* Gray 1840, illustrated by very fine lithograph as fig. 1, Pl. 1, H. L. Clark 1910 (§ K–10) and as Pl. 2, Boone 1928 (§ K–1). Complete description and good illustration p. 56, Pl. 5, Fisher 1911 (§ K–105). Also illustrated on Pl. 36, Ricketts and Calvin 1939 (§ Y–3). General range: San Pedro to Ecuador, shore to 80 fathoms, Ziesenhenne 1937 (§ K–16) p. 211.

A sand-living form. One specimen only was taken at Angeles Bay, intertidal sand flat (although it must surely have been very common thereabout as it is in southern California, on suitable, well sheltered sand flats at very low tide). Gray-brown, sandy-colored.

§ K–128 *Luidia phragma* Clark 1910 PL. 15 FIG. 4
Originally described and illustrated from northern Peru, H. L. Clark 1910 (§ K–4) p. 329, fig. 1 on Pl. 2. Reported since only by Ziesenhenne 1937 (§ K–16) p. 214, from Magdalena Bay shore and from ½ to 13 fathoms at Santa Inez in the Gulf.

A black spotted form, very definite and contrasted and pretty. 8 or 10 were captured with the snails *Murex (Phyllonotus) bicolor* and the urchins *Astropyga pulvinata,* in crab nets put overboard at night, Concepcion Bay, 7 fathoms sand.

§ K–129 A minute *Patiria*-like form taken under a rock below the 0.0′ level at Cape San Lucas is an indeterminable juvenile, possibly the young of *Nidorellia* which Verrill at first described as a new species under the name *Goniodiscus stella* (§ K–9) 1867, p. 284.

SUMMARY

§ K–130 *Heliaster* is easily the commonest Gulf starfish, the most obvious of all Gulf animals and, next to the ophiurans and to *Holothuria lubrica* which are usually hidden, the commonest littoral invertebrate in the region. *Phataria* is the next most common asteroid. In the southern part of the Gulf, *Pharia* takes third place in abundance and occurrence, with *Oreaster* a poor fourth. *Othilia (Echinaster)* is easily third in the north, with *Linckia* and *Astrometis* a very poor fourth and fifth.

Hence the common shore asteroids of this region would seem to group themselves into three geographic categories:

REFERENCES

(1) Those distributed throughout the region: *Heliaster* (which, however, is surely larger and possibly a little more common in the south) and *Phataria;* (2) those more common in the northern part of the Gulf: *Othilia* (*Echinaster*), *Astrometis,* and *Linckia;* and (3) the more southerly types: *Pharia, Oreaster,* and *Nidorellia.* The few localities for *Mithrodia* were also chiefly in the south. *Luidia, Astropecten, Acanthaster,* and *Leiaster* were taken at only one locality each. Puerto Escondido would seem to be an overlap area since abundant representatives were taken there of both northerly and southerly ranging forms. This was, incidentally, possibly the richest asteroid collecting we have ever found on the entire Pacific coast, 11 out of the total of 13 species having been taken here.

The littoral asteroid fauna of the Gulf would seem to be composed entirely of tropical components; all the thirteen species collected were either Panamic, unique, or with Indo-Pacific affiliations. Only three, *Astrometis sertulifera, Astropecten armatus,* and *Linckia columbiae,* range as far north as southern California. None of the others has been recorded north of Magdalena Bay. *Mithrodia bradleyi* is common in Hawaii. *Mithrodia, Othilia* (which occurs also in the Atlantic), *Acanthaster,* and *Leiaster,* are genera of Indo-Pacific complexion. *Leiaster* is also Indo-Australian (with *Anthenea,* a new species of which was described from the Gulf by A. H. Clark). *Oreaster* is tropicopolitan. *Astropecten* and *Luidia* are cosmopolitan except in Arctic and Antarctic waters.

The completeness of the asteroid published reports, and Dr. Fisher's kindness in making accessible to us his personal library, readily permit a historical summary to be made of the work done to date on this group in the Gulf. Panamic species are mentioned by Lamarck 1816 and by Bell 1840 and 1866, and shell collectors during these early times sent in many echinoderms also, mostly dried, to the great museums and to the private cabinets then in vogue, but these were mostly from points to the south—Mazatlan, Acapulco, Panama, Ecuador. In the fifties and sixties two important collections were made in the Gulf and its environs. John Xantus, a tidal observer stationed by the United States at Cape San Lucas, was both discriminating and aggressive in zoological researches there. He himself described three starfish. But most of his echinoderms, and those collected by a Capt. Pedersen at La Paz, were described by Verrill in 1868-71, together with a great lot of more southern material forwarded from northern Peru and from Panama by Bradley who seems to have been employed by Yale University for that purpose. From various collections to that date, Verrill 1871 (§ K–15) had considered 16 Panamic asteroids, of which about 13, presumably littoral or from the slight depths reached by pearl divers, were from the Gulf. A few odd lots of Gulf asteroids have been reported upon between then and now, but otherwise it has been only during the past few years (although the *Albatross* Clark 1913 and 1923 papers reported 19-21 Panamic forms, only 11 of which were littoral and only 9 of which occurred in the Gulf) that collecting has been done again on a large scale in this region. The Bingham Expedition took 7 littoral Panamic asteroids, but only 4 of these were in the Gulf. The *Zaca's* take (Ziesenhenne 1937, H. L. Clark 1940) was 20-21 and 13 Panamic starfish, but none of the latter was in the Gulf, and of the 19 Gulf species reported by the former, only 7 were littoral. We seem to have taken all but one (*Amphiaster*) of the previously reported shore forms.

Finally, the asteroid reports from the Hancock expeditions are still forthcoming, seven years or more of combing the waters mostly of the Panamic Province, some of it in the Gulf. If this data could be presented co-ordinate to Fisher's 1911-1930 monograph of the forms north of San Diego, they could comprise a much needed revision and collation of what is known today of the Panamic asteroids, picking up just where the above-mentioned work leaves off.

CLASS OPHIUROIDEA

§ K–201 Clark, A. H. 1921. "A new ophiuran of the genus *Ophiopsila* from southern California." Proc. Biol. Soc. Wash., Vol. 34: 109-110.

O. californica, 30 fathoms; possibly Panamic.

§ K–202 Clark, H. L. 1915. "A remarkable new brittle star." Pomona Coll. Journ. Entom. Zool., Vol. 9: 64-66.

REFERENCES

Ophiocryptus maculosus from Laguna Beach kelp holdfasts, possibly identical with Ives (1889) *Ophioncus* and probably Panamic. Good toto-illustration in 1915a below.

§ K–203 .... 1915a. "Catalogue of recent ophiurans, based on the collections of the Museum of Comparative Zoology." Mem. Mus. Comp. Zool., Vol. 25 (4): pp. 252, 20 pls.

List of all species known to date. About 84 (+ 9?) are listed or mentioned as occurring within Panamic limits, but this includes many dredged forms, some of them abyssal. There are exceedingly good toto-lithographs of 9 Panamic species, most of them littoral. No bibliography.

§ K–204 .... 1917. "Ophiuroidea." (*Albatross* 1899-1900; 1904-1905.) Bull. Mus. Comp. Zool., Vol. 61: 429-453, 5 pls.

50 species, 7 of them new; mostly Panamic forms from deep and moderately deep water, with 5 common shore species.

§ K–205 Hill, A. 1940. "A new genus of brittle stars, *Amphicontus*." A. Hancock Pac. Exp., Vol. 8 (1): 1-4, 1 pl.

The small *A. minutus* from Galápagos 12 fathoms and from Peru.

§ K–206 Lütken, C. F., and Th. Mortensen. 1899. "The Ophiuridae." (*Albatross* 1891.) Mem. Mus. Comp. Zool., Vol. 23 (2): 116 pp., 23 pls.

Although most of the 64 spp. treated are abyssal and far from land, 5 were littoral, and 9 more were taken at depths shallower than 200 fathoms. Excellent lithographic illustrations of disc and arm details. Bibliography of 128 titles since 1882 only.

§ K–207 McClendon, J. F. 1909. "The ophiurans of the San Diego region." Univ. Calif. Publ. Zool., Vol. 6: 33-64, 6 pls.

8 Panamic shore forms, 9 dredged Panamic and northern forms including 4 new species, and 3 littoral species which extend north, all well illustrated by customary detail drawings. Several inaccuracies are corrected by Nielsen 1932.

§ K–208 Nielsen, Eigel 1932. "Ophiurans from the Gulf of Panama, California, and the Straits of Georgia." No. 59 in "Papers from Dr. Th. Mortensen's Pacific Expedition 1914-16." Vidensk. Medd. fra Dansk Naturh. Foren., Vol. 91: 241-346, 42 text figures.

A paper most excellently planned and executed from the viewpoint of the general biologist at least. Sums up the ophiuran work done to date at these three Pacific coast points, two of which come within Panamic limits. Descriptions of all littoral species heretofore recorded in the three areas considered, with keys to the genera and species.

REFERENCES

Consideration of the similarity between the West Indian and the Panamic fauna; some species are separable only by minute details. About 30 species from Panama, mostly along shore; geographic range, however, not given other than as taken by Dr. Mortensen. Excellent bibliography citing, among others, the following papers which consider Panamic forms but which we do not cite: Le Conte 1851, Lütken 1856, Lyman 1860, 1864, 1874, 1875, Verrill 1867, all of the Ives 1889 papers, and Campbell 1921. Note that Mortensen 1933, p. 16, volume 93 of the Vidensk. etc., makes a slight addendum and correction to this paper.

§ K-209 Ziesenhenne, Fred. 1935. "A new brittle star from the Galápagos Island." A. Hancock Pac. Exp., Vol. 2 (1): 1-4, 1 pl.

The littoral *Ophioplocus hancocki.*

§ K-210 .... 1940. "New ophiurans of the Allan Hancock Pacific Expeditions." A. Hancock Pac. Exp., Vol. 8 (2): 9-42, 8 pls.

19 new Panamic species, most of them from shallow dredgings, 2 to 60 fathoms. Illustrated by the customary detail drawings of the disc at an arm insertion.

§ K-211 See also the ophiuran sections of the general echinoderm papers cited on pp. 371-4 herewith.

§ K-212 The Nielsen 1932 report cited above comes nearest to being the key paper for this group in the Panamic area, bringing up to date all work done to the time of its publication.

§ K-213 ACKNOWLEDGMENT: The brittle star identifications are by Austin H. Clark, Curator, Division of Echinoderms, United States National Museum, Washington, D. C., to whom thanks are due also for checking this portion of the phyletic catalogue, for the gift of literature, and for much friendly co-operation.

LIST OF SPECIES TAKEN:

A. Ophiurans with arms of *short* or *medium* length in which the arm spines are perpendicular to the arm axis; an artificial group comprising such representatives of the FAMILIES OPHIOCOMIDAE, OPHIOTHRICHIDAE and OPHIACTIDAE as we took in the region under consideration.

REFERENCES

§ K–214 *Ophiocoma aethiops* Lütken 1859 PL. 13 FIG. 1
Excellent toto-illustrations as figs. 6 and 7, Pl. 13, H. L. Clark 1915a (§ K–203). Detailed account based on 56 specimens from Panama, 0–5 fathoms, p. 246, Nielsen 1932 (§ K–208). There is a description on p. 112, and detail drawings of the disc between arm insertions as Pl. 65, Boone 1933 (§ K–1a) based on 5 littoral specimens from the Galápagos. Ziesenhenne 1937, p. 226, (§ K–16) reports 9 specimens in Arena Bank coral (Gulf), and on shore at Clarion Island; recorded range, Lower California to Panama, Galápagos, Clarion. H. L. Clark 1940, p. 341, (§K–7) reports material from Mexico and Central America, and records a bathymetric range of 0–10 fathoms.

The most spectacular and the largest Gulf ophiuran. Coarse and bulky, with arm spines long, thick, and abundant. Uniform rich dark chocolate brown to purple brown. Very large specimens may have lighter mottlings on disc. Our largest specimen had disc 1¼″ in diameter, thick arms 5½″ long.

Very abundant at most rocky stations in the Gulf. Noted particularly at Pt. Lobos, Espiritu Santo Island, where any number up to several hundred could have been taken; east of La Paz; Coronado Island; Angeles Bay; Tiburon Island. 72 specimens from 5 stations. Prefers large boulders, where it may be found crawling about underneath, in the interstices, or in grottoes.

§ K–215 *Ophiocoma alexandri* Lyman 1860 PL. 9 FIG. 1
Excellent toto-illustrations as figs. 5 and 6, Pl. 16, H. L. Clark 1915a (§ K–203). Detailed account with one anatomical drawing, fig. 8, based on 62 specimens from Panama low tide region, p. 248, Nielsen 1932 (§ K–208). Ziesenhenne 1937, p. 227, (§ K–16) reports 1 low tide specimen from Clarion Island, and records a total range of Lower California to Galápagos and Clarion Islands, 0–5 fathoms. H. L. Clark 1940, p. 341, (§ K–7) reports material (with color notes) from Mexico, etc. and records a bathymetric range of 0–10 fathoms.

Another very large and very common under rock gulf ophiuran. Taken at Cape San Lucas; Pulmo Reef; Pt. Lobos on Espiritu Santo with *O. aethiops* where up to several hundred could have been collected; east of La Paz; Pt. Marcial; Coronado Island; Puerto Refugio on Angel de la Guardia. 96 specimens from 7 stations.

Delicate, proportionally longer and slimmer armed, and generally smaller than *aethiops*. The largest specimen had ⅞″ disc, with banded arms 6″ long. Arm spines long, thick, and abundant. Light seal brown; white dots at base of arm spines.

REFERENCES

§ K–216 *Ophiothrix spiculata* Le Conte 1851
Good toto-illustration as *O. dumosa* [since synonymized by May 1924, Proc. Calif. Acad. Sci. (4) Vol. 13 (18), p. 274], figs. 6 and 7, Pl. 12, H. L. Clark 1915a (§ K–203). There is also a description on p. 50, and detail figures 38 and 39 in McClendon 1909 (§ K–207). Many are recorded from 0–60 fathoms, La Jolla and Panama, p. 251, Nielsen 1932 (§ K–208). Ziesenhenne 1937, p. 225 (§ K–16) records 18 specimens from 1½ to 45 fathoms, Cedros Island, Arena Bank, and Santa Inez, and reports a total range of from Monterey Bay to Peru. H. L. Clark 1940, p. 340 (§ K–7) reports from various points in Mexico and records a bathymetric range of low water to 45 fathoms.

A highly- and vari-colored form which turns vividly blue after being stored in alcohol. This most common medium sized spiny form occurs along the outer coast of Lower California, is common also in southern California, and many be taken occasionally as far north at least as Pacific Grove. There is a fairly good toto-illustration in Ricketts and Calvin 1939 (§ Y–3) Pl. 18. *O. spiculata* can be differentiated from the related *rudis* only with hand lens; in the former the arm and disc spines are themselves heavily spiculated and even the radial shields, bare in *rudis*, have short spines.

A common gulf ophiuran. 30 specimens were taken at 7 stations; dozens more could have been taken at nearly every suitable point. Pulmo Reef (out of 16 *Ophiothrix* taken at this point, only 1 was *spiculata*); Pt. Lobos; El Mogote; Puerto Escondido; Concepcion Bay (common especially in the great red sponge *Tedania*, on *Pinna* and on submerged rocks): Puerto Refugio; Tiburon Island.

§ K–217 *Ophiothrix rudis* Lyman 1874
We have been unable to find any toto-illustration, but the superficial resemblance to *spiculata* is so great that the differentiation would be apparent only in a very clear or enlarged photo. There are detail illustrations as figs. 30 and 31 in McClendon 1909 (§ K–207) description on p. 51. Nielsen 1932, p. 251 (§ K–208) reports 27 specimens, 0–5 fathoms from the San Diego region. Not recorded in Ziesenhenne 1937 (§ K–16). In a personal communication, Dr. A. H. Clark reports the range as from San Pedro, California, to the entrance to the Gulf.

A common shore species only at Pulmo Reef, where 15 specimens were taken in the interstices of the coral *Pocillopora capitata* Verrill, with 1 only *O. spiculata*. For differentiation of these two similar species, see notes to *spiculata* above.

REFERENCES

§ K–218 *Ophiactis savignyi* (Müller and Troschel) 1842

PL. 15 FIG. 2

The most easily available toto-illustrations we have been able to find are on Pl. 65, figs. 5 and 6: R. Koehler 1922, "Ophiurans of the Philippine Seas and adjacent waters"; Vol. 5 of Bull. 100, U.S. Nat. Mus., pp. 486, Pls. 103. Nielsen 1932, p. 257 (§ K–208) describes numerous small specimens (disc up to 7 mm. diameter) usually with 6 arms, from 0–5 fathoms at Panama. Ziesenhenne 1937, p. 224 (§ K–16) records one only from Clarion Island, 20 fathoms, and reports a total range of San Pedro to Panama on the Pacific coast, and elsewhere cosmopolitan in warm waters. H. L. Clark 1940, p. 339 (§ K–7) records from Mexico and Nicaragua and gives the bathymetric range as low water to 27 fathoms.

Usually 6 armed. The color is green. Dr. A. H. Clark in a personal communication notes that *O. savignyi* may be distinguished from the excellent Nielsen 1932 (§ K–208) figs. 4a and 4b drawings of the similar *O. simplex* by the larger radial shields in contact along the mid-radial line, and the scattered disc spines, and the 2 mouth papillae.

We took 9 specimens at 4 stations in the Gulf: El Mogote; Coronado Island; Angeles Bay; and at Gabriel Bay on Espiritu Santo Island where they were very common among rocks at the base of the mangroves.

§ K–219 *Ophiactis simplex* (Le Conte) 1851 PL. 15 FIG. 2

Toto-illustrations as figs. 5 and 6, Pl. 10, Clark 1915a (§ K–203). There is a lengthy account in Nielsen 1932, p. 258 (§ K–208) with detail figures 4a and 4b, based on numerous specimens of the usual 5 armed and of the rarer 6 armed type, up to 5.5 mm. disc diameter, 0–5 fathoms, San Diego and Panama. Nielsen regards the description and detail drawings of O. *arenosa* Lütken in McClendon 1909, p. 42 (§ K–207) figs. 16a and 16b, based on 15 littoral specimens taken from San Diego sponges, as pertaining actually to the previously described *simplex*. Not recorded in Ziesenhenne 1937 (§ K–16). H. L. Clark 1940, p. 340 (§ K–7) reports material from (presumably) 25 fathoms off Cedros Island, reporting the geographic range as from La Jolla, California, south to Panama. 0–5 fathoms.

We collected *simplex* only three times; 5 specimens from Pulmo Reef, Coronado Island, and Tiburon Island.

REFERENCES

B. Long-armed brittle stars in which the arm spines are perpendicular to the arm axis. Including the common littoral gulf representatives of the FAMILY AMPHIURIDAE, and *Ophionereis* of the OPHIOCHITONIDAE.

§ K–220 *Amphipholis elevata* Nielsen 1932
Our specimen seems to comprise one of the two extant lots, this form having been known heretofore only from the 2 individuals (Panama, 15 fathoms) on which Nielsen erected the species, figs. 17a and 17b, p. 293.

A long-armed ophiuran, found burrowing in the sand at El Mogote. Probably not uncommon at this point, where further effort with a shovel at low tide might have uncovered additional specimens.

§ K–221 *Ophiophragmus marginatus* (Lütken) 1856
Since this was known heretofore only from the type specimen, we had to refer to a paper by the original describer for illustration citation. There are 3 good lithograph figures in a paper published in Copenhagen over 80 years ago, based on a single specimen from Punta Arenas, Costa Rica. Some details of these figures, however, Nielsen 1932, p. 297 (§ K–208) reports incorrect as a result of his study of the type specimen. The figures are 3a, b, and c in: Chr. Fr. Lütken, 1859, "Additamenta ad historiam Ophiuridarum," Danske Vidensk. Selskabs Skrifter, 5ts R., naturv. math. Afd., Vol. 5: 77-169, 5 pls. (An important paper, unfortunately in Danish, with a list on p. 79 of the species therein described or previously known from the west coast of Costa Rica, etc., from Panama, and from Peru.)

Another long-armed sand burrowing form, 1 only specimen of which was taken during the course of very superficial collecting at Estero de la Luna, Sonora.

§ K–222 *Ophiocnida hispida* Le Conte 1851
Good toto-illustrations as figs. 1 and 2, Pl. 9, H. L. Clark 1915a (§ K–203) based on 3 specimens from Panama. McClendon 1909 (§ K–207) has detail illustrations as figs. 34 and 35, and gives the total range on p. 46 as from Catalina to the west coast of Central America, no data as to depth. Not listed in Nielsen 1932 (§ K–208) nor in Ziesenhenne 1937 (§ K–16).

Only one specimen of this small, fairly bristly, long-armed form was saved from slightly subtidal rocks or sponge clusters at Concepcion Bay.

REFERENCES

§ K–223 *Ophionereis annulata* Le Conte 1851

The only easily available toto-illustration seems to be the fairly good one in Ricketts and Calvin 1939, Pl. 17, (§ Y–3). There are detail drawings in McClendon 1909 (§ K–207) figs. 36 and 37, with a description on p. 47: and in Nielsen 1932 (§ K–208) fig. 24, with a description on p. 309 based on many La Jolla and Panama specimens from 0–5 fathoms. Ziesenhenne 1937 (§ K–16), p. 226, reports 2 specimens from 20 fathoms, Clarion Island, and records the total range as San Diego to Panama, Galápagos, and Clarion. H. L. Clark 1940, p. 340 (§ K–7) reports from Central America.

11 specimens of this long and slim-armed, black and white form, plus several delicately variegated juveniles, were taken at 5 stations. Adults, not to be distinguished from specimens taken similarly but more abundantly in southern California, were collected at Pt. Lobos on Espiritu Santo, east of La Paz, at Concepcion Bay and at Angeles Bay. The comparatively short-armed juveniles were a feature of the mussel and alga beds of the upper intertidal, exposed to some wave shock, outside Puerto Refugio on Angel de la Guardia. Additional specimens of the adults undoubtedly could have been taken at several points, although this was one of the least abundant of the large ophiurans, despite its widespread distribution.

C. Serpent stars, in which the arm spines are short, scaly, and more or less *parallel* to the arm axis, so as to point toward the tip of the arm. FAMILY OPHIODERMATIDAE.

§ K–224 *Ophioderma teres* (Lyman) 1860

We have been unable to find any toto-illustration. There is a detail drawing as fig. 37 in Nielsen 1932 (§ K–208), and a description on p. 332 based on 17 specimens from Panama, under stones on shore. Not recorded in Ziesenhenne 1937 (§ K–16). A personal communication from Dr. A. H. Clark records the distribution as from Lower California to Panama and the Galápagos. H. L. Clark 1940, p. 342 (§ K–7) finds 2 varieties, a typical form in which the dark upper arm plates are light speckled, and an a-typical form uniformly dark (slightly banded in very young individuals only) possibly intergrading with *panamense*. *Teres* is ordinarily differentiated by its considerably fragmented upper arm plates.

A large, short-armed, slate-colored form, reminiscent of the Californian *Ophioplocus*, Ricketts and Calvin 1939, Pl. 7 (§ Y–3). The

REFERENCES

arm plates are broken up into sub-plates, whereas in *panamense* they are unbroken. We found this form to be physiologically unique in that practically 100% of the specimens curled up badly in preservation even when being given preliminary anesthetization with fresh water which extends most other ophiurans and prevents autotomy. Taken at Pt. Lobos on Espiritu Santo, east of La Paz, at Puerto Escondido, Coronado Island, Concepcion Bay, San Francisquito Bay, Angeles Bay, Puerto Refugio, and at San Carlos Bay in Sonora. 43 specimens from 9 stations. The commonest and most widely distributed large "smooth" ophiuran in the Gulf; dozens of additional specimens could have been collected had we wanted them, at practically any of the ecologically suitable stations.

§ K–225 *Ophioderma panamense* Lütken 1859 (often cited as "*panamensis*" per Lütken's original name.)

The rather good toto-photograph in Ricketts and Calvin 1939, Pl. 17 (§ Y–3) seems to be the only one readily available. There are detail drawings as fig. 1 in McClendon 1909 (§ K–207) with description on p. 35; and as figs. 35a, b, and c in Nielsen 1932 (§ K–208) with description on p. 327 based on 12 Panama and 1 La Jolla specimens. Nielsen p. 329 decides that the Ives 1889 geographic races are not substantiated. Ziesenhenne 1937 (§ K–16) reports one specimen from shore at Santa Inez Bay, and records a total range of San Pedro to Paita, Peru, and to the Galápagos, 0–10 fathoms. H. L. Clark 1940, p. 342 (§ K–7) notes that the banded arms differentiate it from *teres* since occasional *panamense* have the fragmented arm plates formerly considered specific to *teres*.

A large "smooth" ophiuran, with arms banded and longer and slimmer than in the above; arm plates undivided. In cross-section the arms are flatter than in *panamense*. Nielsen 1932, p. 332 (§ K–208) also gives under *teres* several other formulae for differentiating the very similar juveniles of these two species. Taken at one station only, 4 specimens at Pt. Lobos on Espiritu Santo—in contrast to southern California where it is the commonest large ophiuran.

SUMMARY

§ K–226 Several of the ophiurans are among the most abundant and widespread of all Gulf invertebrates. Of the large smooth species, the slate-brown *Ophioderma teres* could be found almost at will in some quantity, practically throughout the region, at low or even at half tide, under stones, especially on shelly or gravelly substratum. In more reefy areas, and below tide line, the commonest form was easily the bristly, highly colored, medium sized *Ophiothrix spiculata*. The large and banded *Ophiocoma alexandri* was widely distributed wherever there was a wide foreshore with boulders, especially in the south, and in such places the very large and dark *O. aethiops* was also abundant and noticeable.

Pt. Lobos on Espiritu Santo Island in the southern part of the Gulf about opposite La Paz was the finest ophiuran ground we collected over. 6 of the 12 species were taken there and *Ophiothrix rudis* and the two species of *Ophiactis* could probably have been turned out in the same place with a little extra effort, thus accounting for all the rocky shore ophiurans encountered in the entire trip. A collector concentrating on ophiurans at that point undoubtedly could pick up a thousand or more on one tide.

An analysis of the geographic distribution of the 12 species taken indicates the distinctiveness of this phase of the Panamic littoral as exemplified by the Gulf. The mean range of the 12 species would probably be from southern California or somewhere on the outer coast of Lower California, to Galápagos or northern Peru. None, except for the cosmopolitan *Ophiactis* and the northerly ranging *Ophiothrix spiculata,* occurs beyond Panamic limits, or occurs in any other area. Seven of them range to southern California (one clear to Pacific Grove, one cosmopolitan in warm waters): two are known only from a few specimens each, and the two *Ophiocoma* species and *Ophioderma teres* reach their northern limits somewhere along the Pacific coast of Lower California. A tabulated summary of several ophiuran reports is appended:

Name of trip and author	*Total ophiurids*	*Total taken in the Gulf*	
		Total	*Littoral*
Verrill 1871-b (his summary of all reports to 1871)	12, no depth data, some since synonymized.	12	probably 11
Albatross. Clark 1913, 1923	34. 17 only were Panamic, plus 4 common to both Panamic and north temperate, only 8 were littoral.	7	7

REFERENCES

Name of trip and author	Total ophiurids	Total taken in the Gulf	
		Total	Littoral
Bingham. Boone 1928	3: 2 Panamic, 1 cosmopolitan.	3	? 3
Zaca. Ziesenhenne 1937	21, of which 14 plus 2? were Panamic, 1 was cosmopolitan and 1 Indo-Pacific.	12	7 from 3 faths. or less
Present trip and report	12, 1 of which was cosmopolitan, others Panamic.	12	12

From the Hancock Expeditions, 21 new species have already been described within the area under consideration, and Dr. Ziesenhenne is at present monographing this group as represented in the collections of the Hancock Foundation.

CLASS ECHINOIDEA

§ K–301 Agassiz, Alexander 1872-74. "Revision of the Echini." (Ill. Cat. Mus. Comp. Zool., Vol. 7). Mem. Mus. Comp. Zool., Vol. 3. Pt. 1, text 796 pp. Pt. II, 94 pls.

List of all species known to date. Bibliography of several hundred titles. Geographic distribution starting on p. 205. His North Pacific or Boreal American Province extends from the Sea of Okhotsk to the Gulf of Georgia (Cape Flattery) "and some of the species even to San Diego"; the Panamic from Pt. Conception to Cape Blanco in northern Peru. His Californian, defined only inferentially, would seem to extend from Flattery to Conception or San Diego. Good descriptions of many, and toto-illustrations of some Panamic forms.

§ K–302 .... 1904. "The Panamic deep-sea Echini." (*Albatross* 1891.) Mem. Mus. Comp. Zool., Vol. 31: 243 pp., 112 pls.

Complete report to which the 1898 Bull. Mus. Comp. Zool., Vol. 32 (5) preliminary account is merely introductory. Mostly deep water material, but 3 species were taken in depths of 66 fathoms or less, one of which, *Moira clotho* was dredged at 7–14 fathoms from the Bay of Panama. Two others were taken inshore of the 200 fathom line. In the distribution table, 54 species are mentioned as Panamic, many of them shallow.

REFERENCES

§ K–303 Vols. 34 and 46 of the Memoirs of the Mus. Comp. Zool. are devoted to the Agassiz and Clark and to the Clark "Hawaiian and other Pacific Echini." Of these, Nos. 1, 2, and 3 of Vol. 34 treat only the pedicellariae of such Panamic echinoids as are mentioned, except for the deep water *Araeosoma leptaleum* (No. 3, p. 183). The following however are useful for Panamic shore and shallow-water forms:

§ K–304 Clark, H. L. 1912. "The Pedinidae . . . and Echinometridae." Vol. 34 (4): 180 pp., 32 pls.

Regular urchins. Descriptions and excellent toto-illustrations of 5 Panamic littoral forms and mention of two others.

§ K–305 .... 1914. "The Clypeastridae . . . and Scutellidae." Vol. 46 (1): 80 pp., 22 pls.

Cake urchins or sand dollars. Descriptions and excellent toto-illustrations of 3 Panamic species, with 6 others not illustrated.

§ K–306 .... 1917. "The Echinoneidae . . . and Spatangidae." Vol. 46 (2): 204 pp., 18 pls.

Heart urchins. Descriptions and excellent toto-illustrations of 5 Panamic forms; mention of many others. Index for the entire 6 sections.

§ K–307 .... 1939. "A remarkable new genus of Sea-urchin (Spatangidae)." A. Hancock Pac. Exp., Vol. 2 (11): 173-176, Pl. 17.

Idiobryssus coelus sp. nov. described from 2 small specimens dredged 40–70 fathoms at Tower Island, Galápagos. Good toto-drawings.

§ K–308 Grant, U. S. IV, and L. G. Hertlein. 1938. The west American Cenozoic Echinoidea. Publ. Univ. Calif. at Los Angeles in Math. and Phys. Sci., Vol. 2:vi 225 pp., incl. 30 pls. and 17 text figs.

Short diagnoses of orders, families, and genera; occasional species descriptions. Good synonymy and citations. Geographic ranges. 43 plus 2? recent species are listed as occurring in the Panamic area; a few, however, are deep and some are restricted to the Galápagos or to the area south of Ecuador. Many are illustrated by photoliths. Short account of morphology and classification with glossary. No bibliography except in citations. Key paper for this group for the entire Pacific coast of North and South America.

§ K–309 Lockington, W. N. (1875) 1876. "List of Echinidae now in the collection of the California Academy of Natural Sci-

REFERENCES

ences, May 1875." Proc. Calif. Acad. Sci., (1) Vol. 6: 152-159.

An annotated list, cited herewith because it seems generally to have been overlooked. 14 + 1? species from the Panamic area, with representatives usually from the Gulf. The following allegedly Panamic species, not readily allocatable, were commented upon by Dr. H. L. Clark per the following quotations from recent correspondence: *Heterocentrotus mammillatus* ("specimen surely from Hawaii"), *Strongylocentrotus mexicanus* A. Ag. ("probably *Heliocidaris stenopora*"), and *Toxopneustes pileolus* Ag. ("surely *T. roseus*") from the Gulf, and *Echinus margaritaceus* Lamarck (*"Strongylocentrotus fragilis* Jackson" or possibly *"Lytechinus anamesus"*) San Pedro, dredged at 40 fathoms.

§ K–310 Panamic species are treated in a number of other papers, notably in Agassiz 1863 (Bull. Mus. Comp. Zool., Vol. 1: 17-28), and in Agassiz and Pourtalès 1874 (Hassler Reports, Mem. Mus. Comp. Zool., Vol. 4 [1]), but the same species have been at least cited in the more recent accounts listed above.

§ K–311 See also the echinoid sections of the general echinoderm papers cited on our pp. 371-4.

§ K–312 ACKNOWLEDGMENT: The sea urchin identifications are by Dr. Hubert L. Clark of the Museum of Comparative Zoology, Harvard University, Cambridge, Mass., to whom thanks are due also for some additional assistance and for checking this portion of the phyletic catalogue. The quoted remarks are from correspondence with Dr. H. L. Clark.

LIST OF SPECIES TAKEN:

A. Regular Urchins

§ K–313 *Eucidaris thouarsii* (Valenciennes in L. Agassiz and Desor) 1846 PL. 22 FIG. 1

Good toto-illustrations as Pl. 5, Boone 1928 (§ K–1) and as fig. 2, Pl. 2, Grant and Hertlein 1928 (§ K–308) with citations to date, p. 7-8, and range reported as from Lower California (Cape San Lucas) and into the Gulf as far as San

REFERENCES

Felipe, to Panama and Galápagos, mostly along shore, but Ziesenhenne (§ K–16) reports from 50 fathoms (1937, p. 231). There is a description on p. 126, with good toto-photographs as Pls. 80 and 81, in Boone 1933 (§ K–1a) based on 6 littoral specimens from the Galápagos. H. L. Clark, 1940, p. 347 (§ K–7) reports from Mexico and Central America.

From 1" to 5" in diameter, with blunt primary and flat secondary spines, these were the commonest urchins in the Gulf, having been taken on solid rock, on reefs, and among boulders, often exposed to moderate surf, at practically every suitable collecting place. The spines of large specimens are almost invariably encrusted with white Bryozoan colonies. The general body tone is brown. The animal is very distinctive and abundant; several hundred specimens could have been taken.

§ K–314 *Echinometra vanbrunti* A. Agassiz 1863

PL. 17 FIG. 2

The only toto-illustrations are figs. 7 and 8, Pl. 5, Grant and Hertlein 1938 (§ K–308) where the range is reported, p. 41, as from outside Magdalena Bay (Santa Maria Bay) on the outer coast of Lower California, and in the Gulf, to Panama, Peru, and the Galápagos. Mentioned by Agassiz 1872-74, p. 706 (§ K–301) as occurring commonly at Panama with *Eucidaris thouarsii* in cavities in the solid rock of the reef. H. L. Clark 1940 (§ K–7), p. 349 records specimens from Mexico and Central America, noting a bathymetric range of low water to 10 fathoms.

Taken at Cape San Lucas, where they were the most common urchins —many dozens could have been collected: at Pulmo Reef, at Pt. Lobos on Espiritu Santo Island, and at Marcial Point, always associated with not less than moderate wave shock. Probably the second commonest urchin in the Gulf, especially in the south. Except that they are slightly elongate or oblong, they resemble the more northern *Strongylocentrotus*. The largest specimens had a diameter of 5" or 6" including spines which are short and stout, red purple to blue purple. In life, the animals are fairly inert, the spines can be moved only slowly. In preserved specimens they tend to point downward. A heavily-built powerful animal, comparable to the purple urchin of the North Pacific, and, like it, accustomed to holding on in the face of heavy surf.

§ K–315 *Arbacia incisa* A. Agassiz 1863. (Known previous to 1913 as *A. stellata*).

PL. 23 FIG. 2

Good toto-illustration as Pl. 6, Boone 1928 (§ K–1). The test only is illustrated as fig. 1, Pl. 4 in Grant and Hertlein

REFERENCES

1938 (§ K–308), where the range is reported as from Cedros Island (dredged), through the Gulf to San Felipe, to the Galápagos and to Zorritos in northern Peru; the depth being reported in Ziesenhenne 1937 (§ K–16) as from shore to 29 fathoms. Agassiz 1904 (§ K–302) p. 218 mentions this as occurring at Panama with *Eucidaris*. H. L. Clark 1940 (§ K–7), p. 348, records uniformly black specimens from 25 fathoms, Cedros Island.

Smaller and darker than *Echinometra*, with sharper, slimmer, and longer spines. They seem to be restricted also to more quiet waters. The spines are smooth and unicolored, never annulated. Taken at many points in the Gulf, at Pulmo Reef where the spines penetrated all but the heaviest part of the soles of our rubber boots, at Pt. Lobos on Espiritu Santo Island, on the rocks in Angeles Bay, at Puerto Refugio with juvenile *Centrostephanus*, etc. A large specimen had a diameter of 3½", of which the disc was only 1¼". In preserved specimens the spines bristle uniformly.

§ K–316 *Centrostephanus coronatus* (Verrill) 1867

Good toto-illustration as Pl. 5, Boone 1928 (§ K–1) and as fig. 2, Pl. 4, Grant and Hertlein 1938 (§ K–308) with range reported on p. 16 as Newport Bay to Panama via the Gulf, as far north as San Felipe. Ziesenhenne 1937 (§ K–16) p. 232 reports the depth as from shore to 34 fathoms.

This species represents still another step in the progression toward long and thin spines. There is considerable resemblance superficially to *Arbacia;* the general body tone in mature specimens is purple. The spines, however, are much longer, and in juvenile specimens, especially, are quite obviously banded light and dark; they are sharp and brittle, and the texture is noticeably annulated. We rate *Centrostephanus* as the fourth commonest shore urchin in the Gulf, having taken it at Marcial Point, Puerto Refugio, at the south end of Tiburon Island, and in several other places fairly sheltered from wave shock. In preserved specimens the spines are in great disorder, very much awry. In reply to an inquiry as to means of differentiating this from the very similar *Centrechinus* below, Dr. H. L. Clark writes under date of Oct. 23, 1940: "In *Centrostephanus* the peristomial plates around the mouth carry small spines as well as pedicellariae; in *Centrechinus* there are no spines." Thus he summarizes observations of Agassiz (§ K–301) Pt. 3, p. 409; and Grant and Hertlein (§ K–308) p. 154.

§ K–317 *Centrechinus mexicanus* (A. Agassiz) 1863 (as *Diadema mexicanus*).

There is no satisfactory published figure of this species, but Grant and Hertlein 1938 (§ K–308) reproduce the

REFERENCES

Lambert and Thiery 1910 figures. The range is given (p. 15) as from the Gulf to Panama and the Galápagos. The Puget Sound record mentioned by Ziesenhenne 1937 (§ K–16), p. 231, (who records the depth as from shore to 20 fathoms) has been stated by Clark 1913 (§ K–5) to be erroneous. Agassiz 1904, in the table on p. 228 (§ K–302) mentions this species also from the Indo-Pacific, "which is undoubtedly erroneous." (H. L. C.)

This is undoubtedly a very poisonous urchin, although we tend personally to avoid *Centrostephanus* equally, and to handle *Astropyga* particularly with sticks or wooden forceps only, just on general principles. The spines are very long, slim, and sharp; the animals as adults are quite large and active and they orient the spines very agilely toward the point of attack. Adults have black or blackish spines; in juveniles the banding is very obvious, and the annulation even in fully mature specimens is fairly apparent. The living animal is a veritable panoply, a diadem of armament; preserved, the very disordered and irregularly pointed spines tend to break readily. Ziesenhenne 1937 (§ K–16), p. 232, remarks, from confessedly unpleasant experiences, that the wound from even a single spine may be nauseating and numbing, and that multiple penetrations might be dangerous. We took this urchin only at Pt. Lobos on Espiritu Santo Island and at Marcial Point. In each place a considerable number were seen.

§ K–318 *Astropyga pulvinata* (Lamarck) 1816

PL. 19 FIG. 3

There are several toto-illustrations; figs. 1 and 2, Pl. 1, Agassiz and Pourtalès 1874, Mem. Mus. Comp. Zool., Vol. 4 (1); Boone 1928 (§ K–1), Pl. 6; and Grant and Hertlein 1938 (§ K–308); figs. 2 and 5, Pl. 28. The last account, p. 15, reports a range of from Cape San Lucas and the Gulf to Panama. Agassiz 1904 (§ K–301) also mentions this form from Hawaii "but *radiata* is the species found there." H. L. Clark 1940 (§ K–7) p. 348 records 3 specimens with color notes, a small specimen from Colombia extending the range to South America, low water to 36 fathoms.

This is the most active and most highly colored urchin we have ever seen; the animals are fiercely and menacingly spined, although we have no evidence that the spines are poisonous as they are known to be in *Centrechinus*. The ground color is light lavender; the aboral surface is quite fleshy, with meridianal rows of bright ultramarine dots, and five vividly colored triangles. The spines are smooth and brittle, longer than those of any urchin we took except possibly *Centrechinus*, and banded toward the base, probably a juvenile character. Probably subtidal only, since the only time this species was taken was in crab nets cast overboard at night, at an anchorage in

REFERENCES

Concepcion Bay, sand bottom at several fathoms, with the asteroid *Luidia phragma* and the snail *Phyllonotus bicolor;* 8 or 10 specimens were taken within a few minutes; the animals invaded the net as soon as we lowered it to the bottom, apparently attracted by the fish bait.

§ K–319 *Tripneustes depressus* A. Agassiz 1863

Never illustrated, but so similar to the analogous West Indies species, according to Dr. H. L. Clark, that an illustration of one will serve for the other. Cited recently by Ziesenhenne 1937 (§ K–16) from Clarion Island, 1 tide pool specimen only, with bathymetrical range recorded from shore to 40 fathoms. Grant and Hertlein 1938 (§ K–308), p. 27, records a range from the west coast of Lower California and Clarion and Socorro Island, from the Gulf, and from the Galápagos. A very large, close cropped urchin, with light, almost white spines. We took this only once, at Pt. Lobos on Espiritu Santo Island. In a personal communication, Dr. H. L. Clark rates this as a rare form, stating that the British Museum has one specimen only, and Museum of Comparative Zoology only seven, none in good condition.

B. Cake urchins or sand dollars.

§ K–320 *Encope californica* Verrill 1870

Good toto-illustration as Pl. 7, Boone 1928 (§ K–1); the test only of a Pleistocene specimen is figured on Pl. 11 (fig. 4), Grant and Hertlein 1938 (§ K–308) and on p. 96 the range is given as Gulf of California only.

This is the large, nearly circular *Encope* with one small "keyhole" and 5 other lunules, adults of which were seen literally by the thousand on the sand flats at Concepcion Bay and at Angeles Bay. Juveniles were the stouter, darker, and unicolored (with "keyhole" round and noticeable) of the two species of juveniles taken at Estero de la Luna in Sonora, the other being the more delicate and variegated *Mellita.* These and/or the following had very frequent peacrab commensals, *Dissodactylus nitidus* Smith and *D. xantusi* Glassell (§ R–62 and 63):

§ K–321 *Encope grandis* L. Agassiz 1841

There are two good toto-illustrations: figs. 5 and 6, Pl. 13d, Agassiz 1872-74 (§ K–301) and Pl. 8, Boone 1928 (§ K–1). The range is reported on p. 97 of Grant and Hertlein (§ K–308) as restricted to the Gulf, but Agassiz 1904

REFERENCES

(§ K-302) records it also from Panama, in the p. 228-234 table, "this is almost certainly an error." (H. L. C.)

This is the grotesquely shaped large keyhole urchin with enormous keyhole and five marginal indentations. A few were brought in by small boys at La Paz; thousands, literally, were seen at Concepcion Bay where they were mixed with *E. californica* above, and many at Angeles Bay, in both cases on the sand flats. These and/or the above (they were both preserved in the same trays, where the crabs became detached), had frequent commensals in the form of the above-mentioned pea-crabs.

§ K-322 *Clypeaster rotundus* (A. Agassiz) 1863

PL. 17 FIG. 1

Toto-illustrations as Pl. 32, H. L. Clark 1914 (§ K-305) from a single giant specimen dredged at 33 fathoms, Panama, p. 38; and as fig. 10, Pl. 21, Grant and Hertlein 1938 (§ K-308), who doubt the Agassiz 1904 San Diego report ("no doubt this is an error"), record it from Cape San Lucas and Acapulco to Panama and the Galápagos. To these must be added our Gulf record.

A very large (5½" x 5") cake urchin without perforations, and with a very obvious "starfish" pattern etched in the center of the aboral surface. 5 or 6 only of these were taken with the 2 species of *Encope* at Concepcion Bay sand flats.

§ K-323 *Mellita longifissa* Michelin 1858 PL. 24 FIG. 3

Grant and Hertlein 1938 (§ K-308) reproduce as figs. 1 and 3, Pl. 21, the original toto-illustrations of Michelin, and report the range on p. 101-2 as from the Gulf, as high as San Felipe, to Panama.

Taken only once, at Estero de la Luna, Sonora, where the delicate and color-variegated (light and dark green in alcohol) juveniles were very abundant in the sandy substratum, with juvenile *E. californica*. The keyhole is a mere slit in our specimens.

C. Heart Urchins.

§ K-324 *Meoma grandis* Gray 1852 PL. 19 FIG. 2

Excellent toto-illustration as Pl. 8, Boone 1928 § K-1). Grant and Hertlein 1938 (§ K-308) p. 130 report it from the Gulf to Acapulco, and Ziesenhenne 1937 (§ K-16) reports the bathymetrical range as from 20 to 60 fathoms, p. 236. Reported from Mexico by H. L. Clark 1940 (§ K-7) p. 352.

Slightly subtidal, and below the belt of *Encope* in Concepcion Bay, we found these giant heart urchins to be very abundant, where some

REFERENCES

2 dozen were taken with a hand net from a skiff. A typical specimen was 5″ long by 4½″ wide by 2½″ thick, oval-shaped, with closely cropped spines.

§ K–325 *Agassizia scrobiculata* Valenciennes 1846

There is a toto-illustration in Agassiz 1872-74 (§ K–301) figs. 1, 2, and 3, Pl. 19b, and illustrations of the test as figs. 2 and 3, Pl. 29, Grant and Hertlein 1938 (§ K–308), where on p. 115 the range is recorded from the Gulf (as far as San Felipe), from Magdalena Bay and on to Panama, Peru, Juan Fernandez Island, although this last is somewhat doubtful.

Small, ovate, close cropped. Several dozen specimens were taken in all, at El Mogote, San Carlos Bay (Lower California Gulf shore), and at Estero de la Luna in Sonora, in each case burrowing in the sand. The largest was 1¼″ long by about 1″ in diameter, the smallest about ⅜″ long.

§ K–326 *Metalia spatagus* (Linnaeus) 1758

There is a good toto-illustration in Grant and Hertlein 1938 (§ K–308) figs. 4 and 5, Pl. 8, where on p. 128 the range is recorded on the Pacific coast from the Gulf as far north as San Felipe to Panama, and elsewhere throughout the Indo-Pacific region.

Close cropped and broadly ovate, very similar to the above, this larger form was taken only once, in several representatives, buried in the sandy mud substratum of tide pools at Puerto Refugio. Larger, flatter, and more elongate than *Agassizia*, with shorter, coarser spines. The largest specimen was 2¼″ long by 1¾″ wide, by 1¼″ high, the smallest was less than an inch long.

§ K–327 *Lovenia cordiformis* A. Agassiz 1872

Good toto-illustrations as figs. 8-12, Pl. 161, H. L. Clark 1917 (§ K–306). There is also a toto-photo reproduced as fig. 2, Pl. 35, in Ricketts and Calvin 1939 (§ Y–3). There are illustrations of the test only on Pls. 11 and 13, Grant and Hertlein 1938 (§ K–308), where on p. 136 the range is reported as from Santa Barbara to the Gulf, to Panama, Galápagos, and probably to northern Peru. The bathymetric range is reported by Ziesenhenne 1937, p. 236 (§ K–16) as from 8 to 54 fathoms, who also reports this form from Hawaii, (specimens in Museum of Comparative Zoology). "These specimens from Hawaii are of recent

REFERENCES

collecting and there seems to be no room for doubts about them." Reported from 25 fathoms, Cedros, by H. L. Clark, 1940, p. 352 (§ K–7). Occasional specimens may be found on quiet water sand flats in Southern California, as at Corona del Mar.

With its long, sharp, and highly erectile spines, this "sea porcupine," with its habit of burrowing just under the surface on sandy bottoms, is a menace to bathers and to diggers of the clam *Chione*. Either the spines are slightly poisonous, or wounds resulting from them are prone to secondary infections, as will be attested to by one of the sailors aboard our ship, who, searching for submerged clams, raked his fingers through the sand more vigorously than wisely. We took this heart urchin only at Gabriel Bay on Espiritu Santo Island, where possibly a dozen specimens were found just below the tide line.

SUMMARY

§ K–328 All of the 15 species taken along shore during this Gulf trip were definitely Panamic. Except for the records of *Centrostephanus* from Catalina and Newport Bay, of *Arbacia incisa* from Cedros, and of *Lovenia* which definitely occurs even in the littoral in southern California, none ranges north of Magdalena Bay, and none goes south of northern Peru. Two spatangids, however, occur outside Panamic limits; *Lovenia* occurs in Hawaii, while *Metalia* is fairly widespread in tropical waters. One species, *Encope californica,* is restricted to the Gulf. *Eucidaris* is easily the most common and widely distributed of the Gulf urchins being with *Holothuria lubrica* and *Heliaster kubiniji* one of the most ubiquitous of all Gulf invertebrates. The blue purple *Echinometra* is common in rocky areas subject to wave shock. *Arbacia* occurs abundantly in rather more quiet waters often with *Centrostephanus* and/or *Centrechinus* as scarcer associates. Quiet water sand flats seem to be heavily populated by *Encope grandis* and *E. californica,* with *Meoma grandis* common in slightly deeper water. That urchins are among the most common of the invertebrates in the Gulf intertidal will be seen from the fact that 15 species were taken in many hundred individuals, and from the fact that no station had fewer than 2 or 3 species.

CLASS HOLOTHURIOIDEA

REFERENCES:

§ L-1 Clark, H. L. 1920. "Holothurioidea." *Albatross* 1899-1900 and *Albatross* 1904-1905. Mem. Mus. Comp. Zool., Vol. 39 (4): 40 pp., 4 pls.

Mostly abyssal forms far from land, but 6, including one new species, were littoral. On p. 152 there is a summary of the total Agassiz Pacific echinoderm collections.

§ L-2 Deichmann, E. 1922. "On some cases of multiplication by fission and of coalescence in holothurians; with notes on the synonymy of *Actinopyga parvula* (Sel.)." Vidensk. Medd. fra Dansk naturh. Foren., Vol. 73; 199-214, 10 text figs.

Ecological data on a common Panamic shore form.

§ L-3 Deichmann, E. 1936. "Notes on the Pennatulacea and Holothurioidea collected by the first and second Bingham Oceanographic Expositions, 1925-26." Bull. Bingham Ocean. Coll., Vol. 5 (3): 11 pp., text figs.

Reidentifies Boone's 1928 *Euapta lappa* as *E. godeffroyi;* records *Holothuria lubrica* Selenka from San Francisquito Bay; other notes.

§ L-4 .... 1936a. "A new species of *Thyone* from the west coast of Mexico." Proc. New Eng. Zool. Club, Vol. 15: 63-66, text fig.

T. glasselli burrowing in the intertidal sand of the Gulf Sonora shore. Key to other Mexican *Thyone* spp. By now obsolete.

§ L-5 .... 1937. "Holothurians from the Gulf of California, the west coast of Lower California, and Clarion Island." No. 9 in "The Templeton Crocker Expedition." Zoologica, Vol. 22: 161-176, 3 text figs.

14 species taken, 4 of them new. 8 or 9 are Panamic. The Cedros Island records probably constitute the extreme southerly limits of the northern forms. Bibliography, including Clark 1902 (Galápagos), Ludwig 1875 and 1898, and Panning, all of which mention Panamic forms.

§ L-6 .... 1938. "Holothurians from the western coasts of Lower California and Central America, and from the Galápagos Islands." No. 16 in "Eastern Pacific Expeditions of the New York Zoological Society." Zoologica, Vol. 23: 361-387, 15 text figs.

23 species, littoral or from shallow dredgings, of which 4 are new, 16 are typically Panamic, and 4 are Indo-Pacific or tropicopolitan.

REFERENCES

Good bibliography, including H. L. Clark 1907-8 (*apodous*), 1922 (*Stichopus*), 1924 (*Synaptinae*), and 1935 (*Caudina*), Ludwig 1874 (same as 1875 in bibliography next above), and Panning, '29-35 (Holothuria), which revise groups in which Panamic forms occur.

§ L–7 .... 1938a. "New holothurians from the western coast of North America, and some remarks on the genus *Caudina*." Proc. New Eng. Zool. Club, Vol. 16: 103-115, 4 text figs.

Caudina arenicola (Stimpson), *Pentamera pseudopopulifera* new species, from Southern California. Possibly Panamic, etc.

§ L–8 .... 1938b. "New records of *Paracaudina chilensis* (J. Müller) from the west coast of Central America and Mexico." Proc. New Eng. Zool. Club, Vol. 17: 23-25, text figs.

Records from Mexico and Guatemala, plus Elkhorn Slough (Monterey), and Aleutians, establish this as a cosmopolitan Pacific form.

§ L–9 .... 1939. "A new holothurian of the genus *Thyone* collected on the Presidential Cruise of 1938." Smith. Misc. Coll., Vol. 98 (12): 7 pp.

T. lugubris from shallow dredging (10–15 fathoms) Magdalena Bay, Lower California.

§ L–10 .... 1941. "The Holothurioidea collected by the *Velero III* during the years 1932 to 1938. Part I, Dendrochirota." A. Hancock Pac. Exp., Vol. 8 (3): 61-195 including 21 pls.

The Dendrochirote zoogeographical affiliations are with the West Indies and to some extent with California rather than with Indo-Pacific waters. On p. 61 there is a modern definition of the Panamic region, as extending from Cerros (or Cedros) Island southward, and including Peru. 43 valid species are treated, of which three are extra-territorial (2 Californian, 1 Antarctic) and 3 others occur only from Peru southward. Bibliography of 52 titles. Detail drawings of calcareous plates only.

§ L–11 Heding, S. G. 1928. "Synaptidae." No. 46 in "Papers from Dr. Th. Mortensen's Pacific Expedition, 1914-16." Vidensk. Medd. fra Dansk naturh. Foren., Vol. 85: 105-323, 2 pls. (colored), 68 text figs.

3 littoral species from Panama, 2 of them new. *Euapta godeffroyi* is described from 5 Hawaiian specimens. Keys. No bibliography.

§ L–12 .... 1938-39. "The holothurians collected during the cruises of the M.S. *Monsunen* in the tropical Pacific

REFERENCES

in 1934." Vidensk. Medd. fra Dansk naturh. Foren., Vol. 102: 213-223.

Holothuria impatiens (Forskål) from Galápagos, presumably littoral.

§ L–13 Ludwig, Hubert 1886-1887. "Die von G. Chierchia auf der Fahrt der Kgl. Ital. Corvette *Vettor Pisani* gesammelten Holothurien." Zool. Jahr., Abt. für Syst., Vol. 2: 1-36, 2 pls.

8 spp. from Panama and the Galápagos, 3 of them new, all presumably littoral.

§ L–14 .... 1894. "The Holothurioidea." *Albatross* 1891. Mem. Mus. Comp. Zool., Vol. 17 (3): 183 pp., 19 pls.

Mostly abyssal forms far from land, but 8 species are from depths less than 200 fathoms and several of these, including one new species, are from shore. All the colored illustrations are of deep water forms. No bibliography. The 1893 Bull. Mus. Comp. Zool., Vol. 24 (4) paper serves merely as a preliminary report.

§ L–15 Selenka, E. 1867. "Beiträge zur anatomie und systematik der Holothurien." Zeit. f. wiss. Zool., Vol. 17: 29-374, Pls. 17-20.

A summary of what was known scientifically then with reference to holothurians. Panamic forms are described or mentioned on pp. 316, 318, 328, 329, 335, 338, and 356, often with good toto-illustrations, some lifesize from well preserved specimens. Florida and Acapulco material, however, seems to have been confused, possibly in the original labeling. California species, some of them probably overlap forms, are described or mentioned. An important paper.

§ L–16 Semper, C. 1867-68. "Holothurien." Reis. Arch. Phil. Semper, Bd. 1, Thiel 2: 288 pp., 40 pls. (22 by color lithography), Jena.

Some of these Philippine holothurians are known already to occur in the Panamic Province, and the affinities between the Gulf-Panamic fauna and that of the Indo-Pacific area are known to be close. This account, with that of Selenka 1867 above, covers what was known scientifically about the holothurians at that time. Admirable color lithographs from Mrs. Semper's paintings of the living animals.

§ L–17 ACKNOWLEDGMENT: The sea cucumber identifications are by Dr. Elisabeth Deichmann of the Museum of Comparative Zoology, Cambridge, Massachusetts, to whom thanks are also due for checking this portion of the phyletic catalogue, for literature, and for much friendly co-operation. The quoted passages which follow are mostly from personal correspondence with Dr. Deichmann, and from her papers.

LIST OF SPECIES TAKEN:

ORDER ASPIDOCHIROTA

REFERENCES

§ L–18 Usually large and slug-shaped holothurians with numerous tube feet. Tentacles not completely retractile, and not greatly extendable, in no case forming the great and spectacular tentacle nets characteristic of some of the Dendrochirota. Adapted to a creeping habitat, whereby the stubby tentacles brush along the rock surface so as to pick up and ingest the detritus film, or to a burrowing habitat in which the sand is ingested for its contained nutriment.

§ L–19 *Holothuria arenicola* Semper 1868
p. 364 Deichmann 1938 (§ L–6), short description, range reported as almost circumtropical: Philippines, Hawaii, West Indies in shallow water often buried in the sand, Galápagos, Central America, and Mexico. There is a fine colored lithograph from the living animal as Pl. XX in Semper 1867-8 (§ L–16).

The following descriptive statements are quoted from Dr. Deichmann's personal communication: "cigar-shaped, finely speckled or mottled gray, with either two rows of dark spots on the dorsum or irregularly spotted . . . mouth almost terminal with 20 small tentacles, anus terminal."

A fairly common form taken at Pt. Lobos on Espiritu Santo Island, at Puerto Escondido, at Puerto Refugio on Angel de la Guardia Island (note on label: "sand-eating"), and at the south end of Tiburon Island.

§ L–20 *H. difficilis* Semper 1867
p. 164 Deichmann 1937 (§ L–5), short description, range reported "widespread in western tropical Pacific . . . Samoa . . . common in Hawaii . . . Clipperton Island . . . Clarion Island. . . . In tide pools, often under rocks."

Small, yellow, and soft skinned "with small inconspicuous papillae on the dorsal side and numerous soft feet on the ventral side . . . related to *H. parvula* of the West Indies . . . frequently multiplies (like *parvula*) through transverse fission." (Personal communication from Deichmann) (see § L–2).

2 specimens only were taken at Pt. Lobos on Espiritu Santo Island, among rocks.

§ L–21 *H. impatiens* (Forskål) 1775 PL. 10 FIG. 2
p. 365 Deichmann 1938 (§ L–6), short description, range

REFERENCES

reported as almost circumtropical, described originally from the Red Sea, common in the West Indies and Hawaii, shallow water to several fathoms; Galápagos, Mexico, Lower California. There is a fine colored lithograph natural size from life (as *H. botellus* = *impatiens*), Pl. XXII, in Semper 1867-8 (§ S–16), and a photo as Pl. 99, Boone 1933 (§ K–1a) with reference on p. 155 to two specimens from Galápagos.

A knobby, mottled form, medium-sized "slender, often distinctly bottle-shaped with long narrow neck. Tentacles 20, small, mouth and anus both terminal. Feet cylindrical, slightly more papilliform on the dorsal side, scattered over entire body without any apparent order, often placed distinctly on wart . . . Color mottled, gray, skin decidedly sandy to the touch." In a personal communication, Deichmann also remarks the resemblance between this and *H. languens* (§ S–24) but notes that *impatiens* is "more sandy, more knobby, and has (among the calcareous plates buried in the skin) buttons and tables with well developed discs, while the other has tall tables with a Maltese cross on top and usually no disc; also it lacks buttons. . . ."

The second most common holothurian encountered. Found at Pt. Lobos on Espiritu Santo Island, east of La Paz, at El Mogote, at Puerto Escondido, and at Puerto Refugio on Angel de la Guardia.

§ L–22 *H. inhabilis* Selenka 1867

p. 164 Deichmann 1937 (§ L–5), short description, distribution reported as: Hawaii, Cocos, Clarion Island, Lower California, 1 to 50 fathoms. p. 365 Deichmann 1938 (§ L–6) reports from Nicaragua. The Selenka 1867 (§ L–15) illustrations are detail figures only.

A very large, flat, brown, creeping form, resembles *S. fuscus* (§L–28) but with smaller warts dorsally; small and not very obvious tube feet scattered at random ventrally, and less distinction than in *fuscus* between the upper and lower (creeping) surface. Mouth ventral and anus terminal as in *S. fuscus*. About 20 small tentacles. Bears a considerable superficial resemblance to *Stichopus variegatus* Semper 1867-68 illustrated from life in color on Pls. 16 and 17 (§ L–16).

Very abundant slightly subtidally (reached by handnet from skiff), creeping about on the alga covered sandy shores at Concepcion Bay.

§ L–23 *Holothuria kefersteini* (Selenka) 1867

p. 318 Selenka 1867 (§ L–15), from Acapulco, good life-sized toto-illustration, as fig. 37, Pl. 18. In a personal communication from Deichmann, Semper's 1868 (§ L–6) *H.*

REFERENCES

inornata, p. 252, detail. Figure only, from Mazatlan, is stated to be synonymous.

A large black, reddish-gray, or greenish form, apparently not reported since the above two citations. One specimen was taken on the rocks at Pt. Lobos on Espiritu Santo, and another, described as a "sand living form" was taken in the La Paz area, possibly at El Mogote.

§ L–24 *H. languens* Selenka 1867

p. 367, Deichmann 1938 (§ L–6) short description, range reported as entire Panamic and Galápagos. The Selenka 1867 illustrations are detail figures only.

Small and slender, somewhat resembling *H. impatiens* (see § L–21 for distinctions). Mouth and anus terminal. Tube feet on both surfaces; those above are papilliform, below they are cylindrical. Skin slightly sandy to the touch, but not so much as in *impatiens.* Color often a dull purple. Resembles *H. surinamensis* from the West Indies. Abundant specimens from the rocky tide pools at Pt. Lobos on Espiritu Santo were of two types: one was yellowish, the other mottled; both were elongate. A chunkier specimen was taken at El Mogote.

§ L–25 *H. lubrica* Selenka 1867. Sulphur cucumber

PL. 15 FIG. 3

p. 165, Deichmann 1937 (§ L–5), short description, with range reported as Panamic south at least to Panama, north to Gulf, Hawaii; Malay record doubtful. p. 368 Deichmann 1938 (§ L–6) records some distribution and taxonomic notes in listing Costa Rican and Nicaraguan specimens. The Selenka 1867 illustrations are detail figures only.

"Soft-skinned forms with large tentacles, dorsally soft papillae, ventrally large tube feet . . . to withstand effects of surf." Medium sized, yellow-green to mustard-yellow beneath; very dark yellow-green to purple above.

Observed practically everywhere in the Gulf in suitable stations: rock pools, boulders, gravel, and ledges, from half tide or higher, down almost to the low tide line. The commonest cucumber and one of the commonest marine invertebrates in the Gulf. Cape San Lucas, Pt. Lobos, Puerto Escondido, Coronado Island, Angeles Bay, Puerto Refugio, San Carlos Cove in Sonora and Gabriel Bay at Espiritu Santo Island among the mangrove roots.

§ L–26 *H. paraprinceps* Deichmann 1937 PL. 19 FIG. 1

p. 166 Deichmann 1937 (§ L–5) original description, from Arena Bank (Gulf) and from Panama, both at 35 fathoms, detail figures only.

REFERENCES

A large, stout "pepper-and-salt" form. "Color almost black with a whitish ring around the base of the papillae . . . mouth ventral, anus terminal . . . dorsal side with small conical papillae . . . ventral side with relatively small feet often completely retracted . . . related to Philippine and Hawaiian forms."

Many specimens were seen burrowing in the sand at El Mogote.

§ L–27 *H. rigida* Selenka 1867

Selenka 1867 (§ L–15). The illustrations are detail figures only.

El Mogote, sand flats, March 22, 1940, and Puerto Escondido, March 26, 1940. Dr. Deichmann reports in a personal communication: "10 cm. long, dirty gray with two rows of brownish spots on the dorsal side; somewhat flattened with papillae on the dorsum and small feet on the ventrum; skin thin but rigid, packed with spicules. The latter are robust tables with knobbed edge and low spire with numerous blunt teeth, and an inner layer of knobbed buttons. It was reported from Society Islands and apparently ranges all over the Indo-Pacific; it is a burrowing form. *H. rigida* from Florida is now called *fossor* and is very closely related to it."

§ L–28 *Stichopus fuscus* Ludwig 1875

p. 163 Deichmann 1937 (§ L–5), short description; range: Gulf to Ecuador, 0–20 fathoms. Deichmann, p. 363, 1938 (§ L–6) reports from Galápagos.

A large flat, brown, creeping form with thickened flanks. Upper surface with blunt warts. Ventral surface bearing mouth, and tube feet in crowded bands. Anus terminal.

One specimen taken on the rocks at Pt. Lobos, Espiritu Santo Island. Very abundant within the lagoon at Puerto Escondido where fifty or more could have been collected.

Order Dendrochirota

§ L–29 Usually sessile forms—as contrasted with the creeping Aspidochirota. Tentacles dendritic and (in the undisturbed living animal) often very large and spectacular, although completely contractile so that preserved or contracted specimens may not be separable from similar Aspidochirota in this feature. Morphology associated with a detritus-feeding habitat whereby the more or less permanently ensconced animal extends its huge net of tentacles, and allows itself to be fed by the rain of detritus from above and by the particles circulated by the current around it.

REFERENCES

§ L–30 *Cucumaria californica* Semper 1868
p. 79, Deichmann 1941 (§ L–10), short description, "west coast of Mexico and Central America, possibly to Peru." The Semper 1868 illustration is detail figure only.

"Medium-sized forms (about 10 cm.) with 10 bushy tentacles of equal size. Skin soft, smooth; feet large, soft, completely retractile, arranged in 5 bands, not scattered in the interambulacra . . . color varying from almost black to slate-colored or almost white, with dark tentacles, and anterior end which always seems to be blackish." (Personal communication from Dr. Deichmann.)

Small specimens, a few cm. long, were very abundant under rocks at the half tide level at Angeles Bay, April 1.

§ L–31 *Euthyonidium ovulum* (Selenka) 1867
p. 124 Deichmann 1941 (§ L–10), short description, Lower California to Peru. The Selenka 1867 illustration is detail figure only.

Medium-sized, soft-skinned form. Reddish brown color which is extracted in alcohol. 10 large external tentacles arranged in pairs; five inner pairs which are smaller and completely retractile into small pockets. Tube feet are numerous and distributed throughout the surface.

Taken only at Pulmo Reef. Presumably fairly common in the Gulf however, since the Hancock Expeditions took 40 specimens on shore at Puerto Escondido.

§ L–32 *Euthyonidium veleronis* Deichmann 1941
p. 126, Deichmann 1941 (§ L–10) original description of a single specimen from Costa Rica shore, detail figures only.

A mottled grayish medium-sized form, "body tapering toward both ends with numerous soft feet scattered over the interambulacra," feet arranged, however, in definite bands toward the ends.

Several specimens were taken east of La Paz.

§ L–33 *Neothyone gibbosa* Deichmann 1941
p. 113, Deichmann 1941 (§ L–10), original description of a number of specimens from the upper end of the Gulf to Lobos de Afuera Islands, Peru, all from the shore, detail figures only.

A small to medium-sized form with feet all over the surface. Color, reddish to cinnamon brown, typically with darker tentacles. Skin rigid with spicules.

REFERENCES

Small examples common in dead coral heads east of La Paz; at Puerto Escondido; and under rock with *Cucumaria californica* at Angeles Bay.

§ L–34 *Pentamera chierchia* (Ludwig) 1887
p. 86, Deichmann 1941 (§ L–10), short description from numerous specimens ranging from Ecuador and the Galápagos to Lower California, shore to 40 fathoms; detail illustrations only. The Ludwig 1887 illustrations are detail figures only. There seems never to have been a toto-illustration of this form published.

Small 3-6 cm., soft-skinned with numerous cylindrical feet in 5 bands, ventral tentacles small. Color dark-brown, almost black.
April 1, under rocks in Angeles Bay, with *C. californica.*

§ L–35 *Thyone parafusus* Deichmann 1941
p. 106, Deichmann 1941 (§ L–10), original description and detail figures from 2 specimens, Tenacatita Bay, Mexico, 25–35 fathoms.

This was a white, slender, doubled-over, medium-sized, burrowing form taken in the sandy mud-flats at El Mogote, with *Paracaudina* (§ L–37). ". . . fairly stout feet, more or less distinctly arranged in bands and scattered in the interambulacra."

Order Molpadonia

§ L–36 No tentacles. No tube feet (except anal papillae). Sipunculid-like forms with skin smooth and slippery, apparently adapted to a burrowing and sand-eating existence.

§ L–37 *Paracaudina chilensis* (J. Müller) 1850
p. 383, Deichmann 1938 (§ L–6), two specimens from 35–40 fathoms, Costa Rica. p. 23-24, Deichmann 1938b (§ L–8) records the distribution: Straits of Magellan through the Pacific west coast to the Aleutians, clear around to Japan, China, North Australia, and New Zealand; related form known from the West Indies; 0–40 fathoms.

A white, smooth, burrowing, elongate form; looks like a large slim sipunculid. Single specimen taken with white *Thyone parafusus* (above) in the sand flats at El Mogote.

Order Apoda

§ L–38 Tentacles pinnate. No tube feet. Skin roughly unpleasant to the touch, from the supporting calcareous plates which tend to be rubbed off by contact.

REFERENCES

§ L–39 *Chiridota aponocrita* H. L. Clark 1920

p. 125, Clark 1920 (§ L–1), with detail figures, 3 fragmentary specimens from Panama.

A small synaptid with white polka dots and 12 tentacles. A single specimen was taken, presumably under rocks, at Pt. Lobos on Espiritu Santo Island, March 20.

§ L–40 *Euapta godeffroyi* (Semper) 1867-8

p. 9, Deichmann 1936 (§ L–3), Puerto Escondido, "widespread Indo-Pacific form which ranges from Mauritius to Hawaii." Detailed description in Heding 1928 (§ L–11) based on 5 specimens from Hawaii.

Giant worm-like synaptid, up to 3 or 4 feet in length; active, crawls about on the bottom with slow undulatory movements; very obvious. Probably the most spectacular animal observed. Skin very rough, clings unpleasantly to the hand. Constricts and fragments when preserved, rupturing the thin body wall so that the viscera protrude. Head surmounted by crown of tentacles. A species of *Euapta* almost identical in superficial appearance is illustrated as Pl. 1 in Semper 1867-68 (§ L–16).

Puerto Escondido, very abundant in coves, just below low tide. Gabriel Bay on Espiritu Santo Island.

§ L–41 *Leptosynapta* sp., undetermined

May be *Epitomapta tabogae* Heding, reported from Panama, p. 233 Heding 1928 (§ L–11), up to 9 cm. long, alcoholic color bright yellow.

Taken under rocks at Angeles Bay.

SUMMARY

§ L–42 That holothurians are among the most varied and abundant of the Gulf littoral invertebrates will be seen from the fact that 20 different forms were taken, and that one, *Holothuria lubrica,* was possibly the commonest shore animal seen there. An analysis of the zoogeography of the 20 species indicates that 13 are strictly Panamic, none occurring north of the Gulf nor south of northern Peru, nor anywhere outside this area which includes also the Galápagos. Seven are Indo-Pacific (including Hawaii), two of these being reported also from the West Indies. 9 of the total were species of the tropical genus *Holothuria,* which is especially well developed in the Indo-Pacific. Of the 10 Aspidochirotes, 4 only are strictly Panamic. The others are

REFERENCES

Indo-Pacific, including Hawaiian, 2 of them occur also in the West Indies. All of the 6 Dendrochirote species are strictly Panamic, although affiliated with West Indies and to some extent with California relatives, as noted by Deichmann, 1941 (§ L–10), p. 62. Of the total, our experience shows the following 10 species to be widely distributed along Gulf shores, or common, or both: *H. arenicola, impatiens, inhabilis, languens, lubrica,* and *paraprinceps, Stichopus fuscus, Cucumaria californica, Neothyone gibbosa, Euapta godeffroyi:* 5 are Indo-Pacific, 5 Panamic only. Of these, the two most frequently encountered are Indo-Pacific forms; *Holothuria lubrica,* the commonest of all is widely distributed in the Gulf, north and south; *impatiens,* although possibly commoner in the south, was taken also at Puerto Refugio, the most northerly point examined.

Phylum Arthropoda

CLASS CRUSTACEA

REFERENCES TO THE GROUP AS A WHOLE:

§ M-1 Boone, Lee 1930a. "Crustacea; Stomatopoda and Brachyura. Scientific results of cruises of the yachts *Eagle* and *Ara,* 1921-1928, William K. Vanderbilt, commanding." Bull. Vanderbilt Mar. Mus., Vol. 2: 228 pp., 74 pls.

Descriptions and toto-illustrations of 15 species from the Panamic area, usually littoral. Should be used in connection with Glassell's 1934 "Corrections" (§ M-5).

§ M-2 .... 1930b. "Crustacea: Anomura, Macrura, Schizopoda, Isopoda, Amphipoda, Mysidacea, Cirripedia, and Copepoda. Scientific results . . . etc." Bull. Vanderbilt Mar. Mus., Vol. 3: 231 pp., 83 pls.

5 Anomura, 9 Macrura (2 from fresh water), 1 Schizopod, 3 Isopods, 1 Amphipod, and 2 Copepods from the Panamic area; descriptions and good toto-illustrations. Note Schmitt's 1935 (§ O-13), p. 214 corrections to her *Emerita analoga* item on p. 67. Probably should be used with caution throughout.

§ M-3 .... 1934. "Crustacea: Stomatopoda and Brachyura." Bull. Vanderbilt Mar. Mus., Vol. 5: 210 pp., 109 pls.

Grapsus grapsus only of the Panamic area, from Galápagos, p. 178, excellent toto-photograph as Pl. 90. Also good toto-drawings of *Lysiosquilla maculata* as Pl. 5; and of *Trapezia ferruginea,* as Pl. 88; which occur also on the tropical west American coast.

§ M-4 Cano, G. 1888. "Crostacei raccolti dalla R. Corvetta *Caracciolo* nel viaggio intorno al globo duranti gli anni 1881-82-83-84." Boll. Soc. Nat. Napoli, Vol. 2: 161-206.

Collections were made at Payta, Guayaquil, Panama, and Acapulco, where 9 species were taken, 3 barnacles and 6 Brachyura. See § O-4 remarks as to the untrustworthiness of these reports.

§ M-5 Glassell, S. 1934. "Some corrections needed in recent carcinological literature." Trans. San Diego Soc. Nat. Hist., Vol. 7: 453-454.

Applicable to the 1927-30 papers of Lee Boone on Pacific stomatopods and decapods.

§ M-6 Faxon, W. 1895. "The Stalk-Eyed Crustacea." *Albatross* 1891. Mem. Mus. Comp. Zool., Vol. 18: 292 pp., 67 plates of which 11 are colored, 1 map.

A complete report, mostly of dredged material often from very great depths, to which the Faxon 1893 new species diagnoses were merely

REFERENCES

preliminary. However, some 47 shallow-water species (on shore or in depths less than 100 fathoms) are also reported—all Decapods—frequently with complete descriptions and excellent toto-illustrations. The considered Schizopods are all from deep water, and the single Stomatopod, *Squilla biformis,* is from 85–259 fathoms in the Gulf.

Treatment is by families; no tribes, etc., are indicated. 25 of the littoral species are Brachyura, and as such will have been treated to date in the Rathbun monographs, but 14 are Anomura, and 8 are Macrura. Of the 8 last, however, only one was intertidal; *Gnathophyllum panamense,* described, with fine color drawing, from the low tide reef at Panama. 4 of the others were taken pelagically at the surface on the high sea; 1 was from fresh water; *Alpheus panamensis* was lacking a data label; and *Sicyonia affinis,* with fine toto-illustrations as fig. 1, Pl. 46, occurred in depths of 52–112 fathoms.

Coenobita compressus, Calcinus obscurus, and *Petrolisthes occidentalis* are merely listed from Galápagos, Panama, and/or Acapulco. *Petrolisthes agassizii* and *Pachycheles panamensis* are described, with good toto-figures, from the reef at Panama.

An important paper. The illustrations are little short of magnificent; the color drawings were made from the living animals in a most vivid and convincing manner. We cannot pass on its specialist accuracy, but everything else about this collection, from its procurement through its preservation, sorting, and handling, into the final write-up seems to have been done on a perfectionist and non-expedient scale, as were most of the Agassiz projects, so that the result is a joy to work with.

§ M–7 Harford, W. G. W. 1877. "Description of three new species of sessile-eyed crustacea, with remarks on *Ligia occidentalis.*" Proc. Calif. Acad. Sci. (1), Vol. 7: 116-117.

The amphipod, *Dexamine scitulus,* and the isopods *Ligia* and *Idothea marmorata,* were from Lower California, the two former from Magdalena Bay.

§ M–8 Holmes, S. J. 1894. "Notes on west American crustacea." Proc. Calif. Acad. Sci. (2), Vol. 4: 563-588.

3 pinnotherids from the Panamic area. A mysid and a phyllopod, the only crustacea considered other than decapods, are from upper California.

§ M–9 .... 1900. "Synopsis of the California Stalk-Eyed Crustacea. . . ." Occasional Papers, Calif. Acad. Sci., Vol. 7: 262 pp., including 4 pls.

Although concerned primarily with more northern species, the Panamic and overlap forms then known to enter Californian waters are considered. Crustaceans occurring along the Pacific coast of Lower California, in the Gulf, at Panama, etc., are mentioned in the Hippidea, p. 103-104-105; in the Galatheidea, p. 107 and 112; in the Thalassinidea, p. 157 and 161; in the Loricata (spiny lobster) p. 168;

REFERENCES

in the Caridea, p. 182, 196, and 199; and in the Stomatopoda on p. 220. Southern California schizopods are considered on pp. 221 ff. The apropos Brachyura will have been dealt with in the Rathbun monograph. Splendid bibliography.

§ M–10 .... 1904. "On some new and imperfectly known species of west American crustacea." Proc. Calif. Acad. Sci. (3), Vol. 3: 307-330, Pls. 31-37.

4 Panamic species, a *Uca,* a new Calianassid, a new *Crangon,* and an *Amphithoë* (amphipod) with mostly detail drawings.

§ M–11 Lockington, W. N. (1876) 1877. "Descriptions of 17 new species of crustacea." Proc. Calif. Acad. Sci. (1), Vol. 7: 41-48.

Amphipods, isopods, and decapods, but only 5 plus 2? of the last, and 1? isopod are Panamic.

§ M–12 Nobili, Giuseppe 1897. "Decapodi e Stomatopodi raccolti dal Dr. Enrico Festa nel Darien, a Curaçâo, La Guayra, Porto Cabello, Colón, Panama, ecc." Bol. Mus. Zool. Anat. comp. R. Univ. Torino, Vol. 12 (280): 8 pp., 1 text fig.

1 Stomatopod, 6 Macrura (including *Panulirus martensii* sp. nov. per 1901 synonymy below, and some fresh-water forms), 2 Anomura, plus some 13 Brachyura which will have been considered in the Rathbun monograph, are Panamic, apparently littoral.

§ M–13 .... 1901. "Viaggio del Dr. Enrico Festa nella Repubblica dell Ecuador e regioni vicine." Boll. Mus. Zool. Anat. comp. R. Univ. Torino, Vol. 16 (415): 56 pp.

84 species of which 13 are new. 1 Stomatopod. 6 Natantia. 2 lobsters with synonymy of *Panulirus inflatus* (see *P. martensii* above), 14 porcellanids, 8 hermit crabs, 1 Hippa. The balance are Brachyura and as such will have been considered in the Rathbun monograph.

§ M–14 Rathbun, W. M. 1910. "The stalk-eyed crustacea of Peru and the adjacent coast." Proc. U. S. Nat. Mus., Vol. 38: 531-620, 20 pls., 3 text figs.

A monographic treatment with many toto-illustrations of all species known to inhabit Peru (the northern part of which is well within the Panamic Province), with good bibliography.

§ M–15 Saussure, M. H. de 1853. "Description de quelques Crustacés nouveaux de la côte occidentale de Mexique." Rev. et mag. de Zool. (2), Vol. 5: 354-368, Pls. 12, 13.

Porcellana edwardsii and *Albuminea lucasia,* new species, both with good toto-illustrations, with notes on *Hippa emerita* Fabr., and *Squilla* (*scabricauda?*) The balance are Brachyura and as such will have been dealt with in the Rathbun monograph. All from Mazatlan.

REFERENCES

§ M–16 Schmitt, W. L. 1939. "Decapod and other crustacea collected on the Presidential Cruise of 1938 (with introduction and station data.)" Smithson. Misc. Coll., Vol. 98 (6): 29 pp., 3 pls.

Some 100 species of decapods, with one new species (*Thalamita roosevelti*), and 2 new subspecies (*Crangon hawaiiensis clippertoni,* and *Callianidea laevicauda occidentalis*). 7 of the 18 species taken at Clipperton have strong Indo-Pacific affiliations. *Nebalia bipes* was taken at Magdalena Bay, a barnacle from Cocos Island (p. 27, *Conchoderma* sp.) and 8 species of copepods. Most of this material was taken on the Pacific coast, only one station (Old Providence Island, off Colombia, having been occupied in the Caribbean.

§ M–16a Stimpson, William 1857. "On the Crustacea and Echinodermata of the Pacific shores of North America." Bos. Journ. Nat. Hist., Vol. 6: 444-532, 6 pls.

An interesting, newsy article with new species and catalogue of those already known. 13 Decapods from the west coast of Mexico, 5 of which are Macrura and Anomura. One mantis shrimp.

§ M–17 Streets, T. H. 1877. "Contributions to the natural history of the Hawaiian and Fanning Islands and Lower California." Bull. U.S.N.M., Vol. 7: 172 pp., no illustrations.

Crustacea start on p. 103. In addition to the Brachyura, there are 2 Anomura, 4 Macrura, 2 Euphasids (1 new), 1 mysid, 11 amphipods (6 of them new) and 6 copepods, etc., from Lower California, the pelagic forms having been taken in the open ocean at various points from Lower California to Hawaii.

ORDER AMPHIPODA

§ M–201 Holmes, S. J. 1908. "The Amphipoda collected by the U. S. Bureau of Fisheries Steamer *Albatross* off the west coast of North America. . . ." Proc. U.S.N.M., Vol. 36: 489-543, 46 text figs.

19 species, many of them new, taken only at southern California stations, 7 of which were in water shallower than 100 fathoms. Others from southern California *and* more northern stations, and from localities solely in the north.

§ M–201a Schellenberg 1938. "Littorale Amphipoden des Tropischen Pazifiks." Kongl. Sven. Vet. Akad. Handl. (3), Vol. 16 (6):

Not accessible when this section was being constructed. Information supplied through the kindness of Mr. Shoemaker of the United States National Museum.

REFERENCES

§ M–202 Shoemaker, C. R. 1916. "Description of three new species of amphipods from southern California." Proc. Biol. Soc. Wash., Vol. 29: 157-160.

Aruga macromerus, Ampelisca venetiensis, and *Podoceropsis concava* from Venice, Southern California. No data as to depth.

§ M–203 .... 1925. "The Amphipoda collected by the U.S. Fisheries Steamer *Albatross* in 1911, chiefly in the Gulf of California." Bull. Am. Mus. Nat. Hist., Vol. 52: 21-61, 26 text figs.

16 species, 8 of which are new. Mostly captured by electric light at ship anchorages in the Gulf. (Strangely, only one of these appears in our findings.) Mostly detail drawings. Bibliography.

§ M–204 .... 1926. "Amphipods of the FAMILY BATEIDAE in the collection of the United States National Museum." Proc. U.S.N.M., Vol. 68 (25): 26 pp., 16 text figs.

More detailed description of previously described Gulf of California *Batea rectangulata* 1925. *B. transversa,* n. sp., p. 13, from southern California.

§ M–204a .... 1932. "Notes on *Talorchestia fritzi* Stebbing." Journ. Wash. Acad. Sci., Vol. 22 (7): 184-187, text figs.

Costa Rican material, including large mature males necessitating emendations to the original Stebbing 1903 description.

§ M–205 .... 1934. "Two new species of *Corophium* from the west coast of America." Journ. Wash. Acad. Sci., Vol. 24: 356-60.

One is widespread in shallow water throughout the Panamic area.

§ M–205a .... 1935. "A new species of amphipod of the genus *Grandidierella* and a new record for *Melita nitida* from Sinaloa, Mexico." Journ. Wash. Acad. Sci., Vol. 25 (2): 65-71, text figs.

From Mazatlan shrimp fisheries, brackish water; the *Melita* had been recorded previously from the Atlantic coast of United States, New England, and Louisiana. Good toto- and near toto-drawings.

§ M–206 .... 1938. "Three new species of the amphipod genus *Ampithoe* from the west coast of America." Journ. Wash. Acad. Sci., Vol. 28 (1): 15-25, 4 text figs.

Including one Panamic species, *A. plumulosa.*

§ M–207 Stebbing, T. R. R. 1903. "Amphipoda from Costa Rica." Proc. U.S.N.M., Vol. 26: 925-931, 2 pls.

Talorchestia fritzi, n. sp. from Cocos; the other is terrestrial.

REFERENCES

§ M–208 .... 1908. "A new amphipod crustacean, *Orchestoidea biolleyi,* from Costa Rica." Proc. U.S.N.M., Vol. 34: 241-44, 2 text figs., 1 pl.

In sand under trees at Punta Arenas on the west coast.

§ M–209 Streets, T. H. 1878. "Pelagic Amphipoda." Proc. Acad. Nat. Sci. Phil., 1878: 276-290, 1 pl.

Mostly new species in the FAMILY OXYCEPHALIDAE of the HYPERIIDEA, based on four years' collecting over the entire Pacific Ocean both north and south of the equator, but excluding high latitudes. Mostly taken at the surface by tow net at night, being absent there during the daylight or on moonlit nights.

§ M–210 Woltereck, R. 1909. "Amphipoda." *Albatross.* Bull. Mus. Comp. Zool., Vol. 52 (9): 145-168, 8 pls. Devoted to the Hyperiidea only, and hence to entirely pelagic forms largely outside our scope.

§ M–210a The Walker 1910 Peru paper (Proc. U.S.N.M., Vol. 38: 621-22) treats only one species, and that extra-limital. See also amphipod items listed under Crustacea, pp. 415-8, especially Holmes 1904 (§ M–10).

Five papers on Laguna Beach amphipods are cited on p. 280 in Ricketts and Calvin 1939 (§ Y–3), some species from which presumably may be encountered in the Panamic area. One of them, Stout, Vinnie R. 1913, "Studies in Laguna Amphipods II," Zool. Jahr., Abt. f. Syst., Vol. 34: 633-659, with 3 text figs., describes a species we encountered, and gives an interesting picture of the amphipod situation at Laguna Beach.

§ M–211 ACKNOWLEDGMENT: For determining the amphipods listed below, for co-operation in the matter of the literature and of data on geographic distribution, and for reading over this portion of the phyletic catalogue, we are grateful to Mr. C. R. Shoemaker of the Division of Marine Invertebrates, United States National Museum.

LIST OF SPECIES TAKEN:

§ M–212 *Ampithoe plumulosa* Shoemaker 1938
Shoemaker 1938 (§ M–206), p. 16. Southern California to Ecuador, including the Gulf. One record from British Columbia.

REFERENCES

Taken at four stations, hence the most, or one of the two most widely distributed species we found in the Gulf. El Mogote; Coronado Island; Concepcion Bay; San Lucas Cove. The species was littoral, in pools or around rocks, but at the last station it was netted pelagically at night, by electric light hung overside at the anchorage.

§ M–213 *Ampithoe ramondi* (Audouin)
Schellenberg 1938 (§ M–201a), p. 87. Cosmopolitan.

Taken only once, on the drowned coral flats east of La Paz, in a cluster of *Plumularia* (§ B–9).

§ M–214 *Ampithoe* sp.

Also taken only once, 1 female, from Coronado Island.

§ M–215 *Aruga* sp.
The genus diagnosis is on p. 504, Holmes 1908 (§ M–201).

7 specimens were taken at Puerto Refugio.

§ M–215a Aruga dissimilis (Stout)
p. 638, Stout 1913, § M–210a, as *Nannonyx dissimilis* n. sp., very numerous in subtidal holdfasts of kelp at Laguna Beach.

15 plus specimens were taken at Coronado Island, from bottom at 7 fathoms anchorage.

§ M–216 *Bemlos macromanus* Shoemaker 1925
Shoemaker 1925 (§ M–203), p. 36, "Lower California" presumably from the Gulf, well-figured with several semi-toto-drawings in Fig. 10.

Taken at Coronado Island, and in the estuary at Gabriel Bay on Espiritu Santo Island.

§ M–217 *Elasmopus pocillimanus* (Bate)
Schellenberg 1938 (§ M–201a), p. 56. Cosmopolitan.

High on the rocks at Amortajada Bay, at half tide or above.

§ M–218 *Elasmopus* sp.

Taken only at San Carlos Bay, Baja California, high in the sand, at half tide or above.

§ M–219 *Hyale hawaiensis* (Dana)
Schellenberg 1938 (§ M–201a), p. 66. Tropicopolitan.

Taken with *Elasmopus pocillimanus* high on the rocks at Amortajada Bay.

§ M–220 *Pontharpinia* sp. or spp.

Taken at night by electric lights set overside at anchorages. Entirely pelagic. Concepcion Bay; San Lucas Cove, very abundant (100 plus);

REFERENCES

San Carlos Bay in Baja California, several; Port San Carlos, Sonora, several.

SUMMARY

§ M–221 Of the 10 or 10 plus species taken, 4 or 4 plus represent material at present unallocated to species. Of the remaining 6, 3 are cosmopolitan or tropicopolitan. One of the others was described from southern California. The other two are strictly Panamic, *Bemlos* being known only from the Gulf, and *Ampithoe plumulosa* occurring widely from southern California to Ecuador, with one record from British Columbian waters. A previous reconnaissance of Gulf amphipods had been made during the *Albatross* 1911 expedition, but strangely, we took in our 10 plus only one of the 16 species they reported. Amphipods, however, are unquestionably numerically very important in the Gulf, and the collector devoting himself to minute and subtle material, instead—as we did—to the obvious and large items, should be able to procure up to 20 or 30 species fairly easily.

§ M–222 *Addenda:* 4 additional species were reported upon while this account was in press. Under date of September 30, 1941, Clarence R. Shoemaker, Assistant Curator, Division of Marine Invertebrates, United States National Museum, informs us that the following were determined from material taken from clusters of *Obelia plicata* (§ B–11) attached to a gorgonian skeleton at Estero de la Luna, Sonora: *Pontogeneia* sp. and *Caprella aequilibra* Say were both very abundant (the *Obelia* appeared to be alive with the skeleton shrimps), with 6 specimens of *Ericthonius brasiliensis* (Dana), and 10 of *Parajassa* sp.

Order Isopoda

§ M–300a Abbott, C. H. 1940. "Shore Isopods: niches occupied, and degrees of transition toward land life with special reference to the Family Ligydidae." Proc., 6th Pacific Science Congress, Vol. 3: 505-511.

Stressing *Ligyda occidentalis* at La Jolla. An ecological study.

§ M–301 Boone, Pearl L. 1918. "Descriptions of ten new isopods." Proc. U.S.N.M., Vol. 54: 591-604, 4 pls.

REFERENCES

Including *Braga occidentalis*—an ovigerous female taken off the California coast in 1866, and *Astacilla californica* from southern California.

§ M-302 Richardson, Harriet 1905. "Monograph on the isopods of North America." Bull. U.S.N.M., Vol. 54: pp. LIII plus 727, 740 text figs.

Including a 16 page bibliography which cites the Richardson 1901 Galápagos paper. A comprehensive and functional monograph cited for its orientation value, and for its consideration of Panamic and potentially Panamic species.

§ M-303 .... 1905a. "Description of a new species of *Livoneca* from the Coast of Panama." Proc. U.S.N.M., Vol. 29: 445-446, 2 text figs.

Presumably from the west coast. *Livoneca convexa* with good totofigure.

§ M-304 .... 1910. "Report on isopods from Peru collected by Dr. R. E. Coker." Proc. U.S.N.M., Vol. 38: 79-85, 6 text figs.

The collection included one previously known species extending from Mazatlan to Chile, and 2 new species. List of 2 additional species previously known from the region, only the northern part of which is not extra-limital to our work.

§ M-305 Searle, Harriet Richardson 1914. "Isopoda." (*Albatross* 1899-1900 and 1904-1905.) Bull. Mus. Comp. Zool., Vol. 58 (8): 14 pp., etc. Not consulted.

§ M-306 Van Name, Willard G. 1924. "Isopods from the Williams Galápagos Expedition." Zoologica, Vol. 5 (18): 181-210, Pls. 8-19.

2 only were marine. List (p. 183) of 16 species recorded from the islands and from neighboring waters.

§ M-306a Two papers by Stafford on Laguna Beach isopods, some of which may turn up in the south, are cited on p. 281, Ricketts and Calvin 1939 (§ Y-3).

§ M-307 ACKNOWLEDGMENT: For determination of the following material and for checking this portion of our phyletic catalogue, we are indebted to Mr. J. O. Maloney of the Division of Marine Invertebrates, United States National Museum.

LIST OF SPECIES TAKEN:

REFERENCES

§ M–308 *Cirolana harfordi* (Lockington) 1877
p. 109, Richardson 1905 (§ M–302), British Columbia to Lower California. p. 28, Ricketts and Calvin 1939 (§ Y–3), with toto-fig. 14.

Abundant under rocks at the south end of Tiburon Island.

§ M–309 *Dynamella* sp.

One of the very abundant small animals taken in a region generally barren, from high in the rocky intertidal of a rocky islet off Amortajada Bay.

§ M–310 *Eurydice caudata* Richardson 1899
p. 124, Richardson 1905 (§ M–302), known at that time only from Catalina Island in southern California.

This was one of the most abundant small animals swarming in the aggregations attracted by lights hung overside at night anchorages. At Pt. San Marcial, 13 fathoms anchorage, 10 p.m.; at Concepcion Bay; at San Lucas Cove; San Carlos Bay, Baja California; and at Puerto San Carlos, Sonora.

§ M–311 *Exocorallana tricornis* (Hansen) 1890
p. 139, with toto-fig. 121-t, Richardson 1905 (§ M–302); Caribbean, Florida to Honduras; west coast of Nicaragua. 24–27 fathoms.

A single specimen of this very common Caribbean species turned up in a tray in which material from Concepcion Bay had been preserved. It was probably brought in on the clusters of encrusting animals attached to shells of the hatchet clam *Pinna*.

§ M–312 *Exosphaeroma yucatanum* (Richardson) 1901
p. 291, Richardson 1905 (§ M–302), 24 faths., Yucatan.

A single specimen appeared in trays filled with the results of our day's collecting at Pulmo Reef. Known also previously from the Atlantic.

§ M–313 *Ligyda exotica* (Roux) 1828
p. 676, with toto-fig. 716, Richardson 1905 (§ M–302), tropicopolitan.

From intertidal rocks at Puerto Escondido, and from the highest intertidal at San Carlos Bay. An active, extremely prevalent tropical species, on wharves, rocks, etc.

REFERENCES

§ M-314 *L. occidentalis* (Dana) 1853
p. 681, Richardson 1905 (§ M-302), San Francisco to the Gulf.
p. 15, with toto-fig. 5, Ricketts and Calvin 1939 (§ Y-2), "specimens rarely occur singly, there will be hundreds or none."

Angeles Bay, high on rocks. Called "cucaracha" by the Mexicans in the northern part of the Gulf—"cucaracha del mar" more properly, since to generalize the name too widely would be to leave Mexico's most abundant household concomitant without specific designation.

§ M-315 *Mesanthura* sp.

Another of the unintentionally collected animals which turned up in our preserving trays. From Concepcion Bay, at which point we took the hatchet clam and three types of sand dollars, among other items.

§ M-316 *Paracerceis gilliana* (Richardson) 1899
p. 313, with toto-fig. 341, Richardson 1905 (§ M-302), northern California and Catalina, 30–40 fathoms.

Taken with amphipods on the *Ulva*-like algae growing on rocks in the estuary at San Lucas Cove.

§ M-317 *Paracerceis* sp.

Taken incidentally with other material at El Mogote, sandy mud-flats.

§ M-318 *Paranthura* sp.

Taken at the south end of Tiburon Island, April 3, with larval stages of *Squilla* in pelagic material attracted to the boat by a light hung overside at the night anchorage.

§ M-319 *Rocinella aries* Schioedte and Meinert 1879-80.
p. 210, Richardson 1905 (§ M-302) with toto-figs. 213 and 214, Lower California, Gulf to Panama.

San Francisquito Bay rocky shore, presumably at mid-tide level.

SUMMARY

§ M-320 12 species of isopods seem to have resulted from our survey of the Gulf littoral. On 4 of these, possibly representing new species, we have no data other than the name of the genus. One of the other eight, *Ligyda exotica,* is a very common tropicopolitan species, and two others are West Indian. The other 5 range from the Gulf northward, one

REFERENCES

to "Lower California," one to southern California, two to central and northern California, and one, *Cirolana harfordi,* possibly the commonest isopod on the coast of California, ranges into British Columbia.

ORDER SCHIZOPODA, an obsolete unit containing two unrelated groups, the MYSIDACEA and the EUPHAUSIACEA, but still a convenient collective term for these very similar appearing crustacea.

§ M-401 Esterly, C. O. 1914. "The SCHIZOPODA of the San Diego region." Univ. Calif. Pub. Zool., Vol. 13: 1-20, 2 pls.

Classification, family diagnoses. 8 EUPHAUSIACEA and 3 MYSIDS. Bibliography.

§ M-402 Colosi, G. 1924. "EUPHAUSIACEA e MYSIDACEA raccolti dalla R. Nave *Vettor Pisani.*" Nel 1882-1885. Annuario del Museo Zoologico della R. Universitá Napoli (N.S.), Vol. 5 (7): 1-7, 9 text figs.

Euphausia mutica Hansen, *E. diomedeae,* and *Siriella thompsonii,* p. 2 and 3, are cited from Panama, Galápagos, and Payta (Peru).

§ M-403 Hansen, H. J. 1912. "The Schizopoda." *Albatross* 1899-1900 and 1904-5. Mem. Mus. Comp. Zool., Vol. 35 (4): 175-296, 12 pls.

63 species, 23 MYSIDACEA and 40 EUPHAUSIACEA; all the latter are oceanic and pelagic. All the mysids taken in waters off the Panamic coasts were far from shore, pelagic.

§ M-404 .... 1915. "The Crustacea EUPHAUSIACEA of the United States National Museum." Proc. U.S.N.M., Vol. 48: 59-114.

Nyctiphanes simplex p. 70 is recorded in enormous quantities from San Luis Gonzales Bay in the Gulf, and elsewhere between L. 5° 57′ South and 35° 31′ North. Errors and mis-identifications applying to Panamic species in the Ortman 1894 (§ M-405) paper are recorded in pp. 70, 75, 76, 78, 85, 86, 107, 109, 110, and 112.

§ M-405 Ortmann, A. 1894. "The pelagic Schizopoda." (*Albatross* 1891.) Bull. Mus. Comp. Zool., Vol. 25 (8): 99-110, 1 pl.

Pelagic and shallow-water forms. Faxon (§ M-6) 1895 considers the deep water material. 14 species of EUPHAUSIACEA, 4 of MYSIDACEA, mostly in Panamic waters. Synopsis of the known species of *Boreomysis.* Observations on vertical distribution. Should be used only in connection with Hansen 1915 (§ M-404) which makes important corrections.

REFERENCES

§ M–406 Tattersall, W. M. 1932. "Contribution to a knowledge of the MYSIDACEA of California. I. On a collection of Mysidae from La Jolla, Calif." Proc. Univ. Calif. Publ. Zool., Vol. 37 (13): 301-413, 38 text figs.

4 species, one new, from surface tow-nettings taken from the pier. Mysid situation brought up to date for the region.

§ M–407 See also Mysid items in general crustacean papers on pp. 415-8. The Hansen 1913 Univ. Calif. Publ. Zool., Vol. 11 paper treats mostly North Pacific species.

§ M–408 ACKNOWLEDGMENT: For determining the following species, we have to thank Dr. Waldo L. Schmitt of the United States National Museum, to whom thanks are due also for reading this portion of the phyletic catalogue.

LIST OF SPECIES TAKEN:

EUPHAUSIACEA

§ M–409 *Nyctiphanes simplex* Hansen 1911
Esterly 1914 (§ M–401), p. 9; Tropical and north temperate regions of the East Pacific.
Detail figs. Hansen 1915 (§ M–404), p. 70, Tropical East Pacific.

Abundant specimens were taken inshore at Pt. Marcial at 5 p.m., the water at this point being soupy with them.

MYSIDACEA

§ M–410 *Archeomysis* sp., near to *maculata* Holmes
The genus is considered on p. 303, and Holmes's *maculata* on p. 304, Tattersall 1932 (§ M–406).

Taken with the above *Nyctiphanes* at Pt. Marcial at 5 p.m., close inshore, occasional specimens of the larger Mysid occurring with the smaller Euphausiaceans.

§ M–411 *Mysidopsis* sp.
Two species of this genus are well described on p. 307 and 310, Tattersall 1932 (§ M–406).

Pelagic at surface off Pt. Marcial, 10 p.m. at boat anchorage, a few hundred feet off shore in 13 fathoms. Attracted to the side of the boat in great numbers by light hung overside.

ORDER STOMATOPODA

REFERENCES

§ M–601 Bigelow, R. P. 1894. "Report upon the crustacea of the ORDER STOMATOPODA collected by the Steamer *Albatross* between 1885 and 1891, and on other specimens in the United States National Museum." Proc. U.S.N.M., Vol. 17: 489-550, text figs.

A monograph on American stomatopods complete to its date. 7 Panamic species.

§ M–602 Lunz, G. Robert 1937. "STOMATOPODA of the Bingham Oceanographic Collection." Bull. Bingham Ocean. Coll., Vol. 6 (50): 17 pp., text figs.

Gonodactylus oerstedii Hansen from Puerto Refugio and from San José Island, both in the Gulf (and from the Bahamas). *Chloridella aculeata* (Bigelow) Panama, and *C. panamensis* var. A (Bigelow) from Angeles Bay.

§ M–603 Schmitt, W. L. 1940. "The Stomatopods of the west coast of America." A. Hancock Pac. Exp., Vol. 5 (4): 129-225, 33 text figs.

A monographic account, with excellent illustrations. Key paper for this group. 28 Panamic species are treated, 8 of them new. Only 7 occur outside Panamic waters. Only 3 of the drawings are of the entire animal, but the others are sufficiently comprehensive so as to serve diagnostically for identifications by laymen. No bibliography other than in the species and synonymy citations.

§ M–604 See also stomatopod items in the general crustacea papers cited on our pp. 415-8 and Gravier's annelid paper which treats and illustrates a Stomatopod commensal with *Balanoglossus* in the Gulf (§ J–8).

§ M–605 ACKNOWLEDGMENT: For determinations of the following species, and for innumerable assistances and personal kindnesses, here and elsewhere, we are indebted to Dr. Waldo L. Schmitt of the United States National Museum.

LIST OF SPECIES TAKEN:

§ M–606 *Gonodactylus oerstedii* Hansen 1895
Schmitt (§ M–603), p. 211. Range recorded as from Carolina to Brazil in the Atlantic, including Bermuda and the West Indies; from Fernando de Noronha in the Pacific exclusive of America; and from the Gulf to Ecuador and the Galápagos in the Panamic Province.

One female was taken east of La Paz on the drowned coral flats, and 1 small male under rocks at Pt. Lobos on Espiritu Santo Island.

§ M–607 *Gonodactylus stanschi* Schmitt 1940
Schmitt (§ M–603), p. 215, from Gulf (upper end; Angel de la Guardia) to Tangola-Tangola Bay and Tres Marias Islands, Mexico.

A single male was taken under rocks at Pt. Lobos with the *G. oerstedii* male above.

§ M–608 *Pseudosquilla lessonii* (Guerin) 1830
Schmitt (§ M–603), p. 175, southern California to Chile.

The first littoral stage was taken pelagically at San Carlos Cove, Baja California, by light hung overside at night anchorage.

§ M–609 Pelagic larvae of some STOMATOPODA, genus and species indeterminable, were netted pelagically with the above *Pseudosquilla,* attracted by light hung overside at night anchorage.

CIRRIPEDIA

§ N–1 Broch, Hjalmar 1922. "Studies on Pacific Cirripedia." No. 10 in: "Papers from Dr. Th. Mortensen's Pacific Expedition 1914-16." Vidensk. Medd. fra Dansk naturh. Foren., Vol. 73: 215-358, 77 text figures.

Lepas pectinata Spengler, *Catophragmus pilsbryi* sp. nov., *Balanus tintinabulum* Linn. f. *coccopoma* Darwin, and *Tetraclita squamosa* (Brugière) f. *viridis* Darwin, from Panama littoral. Remarks on Pacific zoogeography on p. 351-3. *Mitella polymerus* is mentioned as being characteristic of the Californian region; actually this occurs in suitable ecological niches from outside Sitka clear down into Lower California. Good bibliography.

§ N–2 Cornwall, I. E. 1941. "A new genus and species of barnacle from Ecuador." A. Hancock Pac. Exp., Vol. 5 (5): 227-230, Pl. 27.

Tetrabalanus polygenus on shore stones at Puna Island.

§ N–3 Pilsbry, H. A. 1907. "The barnacles (CIRRIPEDIA) contained in the collections of the U. S. National Museum." Bull. U.S.N.M., Vol. 60: 122 pp., 36 text figs., 11 pls.

Pedunculate forms. *Mitella polymerus* (San Quentin Bay and Rosario), *Scalpellum soror* and *S. galapaganum* (deep water off Galá-

REFERENCES

pagos), *Lepas anatifera* (Panama), and *L. anserifera* (Gulf of California) are mentioned as occurring in the region under consideration. On p. 80 and 103, *Lepas hilli californiensis* and *Heteralepas quadrata* are mentioned as having been described or mentioned from Lower California and from the Gulf respectively by Gruvel and Aurivillius. There are no further references, and nothing important or pertinent in the bibliography. A cursory search of the more important papers on these two was unfruitful.

§ N–4 .... 1909. "Report on the barnacles of Peru, collected by Dr. R. E. Coker and others." Proc. U.S.N.M., Vol. 37: 63-74, Pls. 16-19.

Tetraclita porosa (Gmelin), *Balanus peruvianus* sp. nov., *Mitella elegans* (Lesson), and *Conchoderma auritum* (Linn.) on *Coronula* from whale, reported from area (northern Peru) in which Panamic forms occur.

§ N–5 .... 1916. "The sessile barnacles (CIRRIPEDIA) contained in the collections of the U. S. National Museum." Bull. U.S.N.M., Vol. 93: 366 pp., 76 pls., 99 text figs.

In addition to some of those we list, specimens from the Panamic Province are recorded as follows: Balanus *tintinabulum coccopoma* (Panama to Guaymas), *B. regalis* (Abrojos Pt.), *B. orcutti* (San Ysidro), *Tetraclita squamosa panamensis* (Panama to Peru), *T. s. milleporosa* (Galápagos), *Chthamalus panamensis,* and *C. imperatrix,* in addition to four whale and turtle barnacles.

§ N–6 See also cirriped items treated in the general crustacean papers cited on our pp. 415-8, especially § M–4 (Cano 1888 which reports *Balanus tintinabulum* from Taboga Island, Panama, *B. nigrescens* from Acapulco, and *Tetraclita porosa* from Payta, Peru), and in the general mollusk paper § S–42 (Hertlein reporting *Tetraclita squamosa milleporosa* from Malpelo and Cocos Island).

§ N–7 ACKNOWLEDGMENT: The barnacle identifications are by Ira E. Cornwall of the Hopkins Marine Station, Pacific Grove, California, to whom thanks are due also for information as to the geographic distribution of these representatives of this difficult group, and for access to cirriped literature in his personal library.

LIST OF SPECIES TAKEN:

§ N–8 *Tetraclita squamosa stalactifera* (Lamarck)
p. 254, pl. 39, Pilsbry 1916 (§ N–5).

REFERENCES

Typically a medium to large form, eroded, symmetrical, fairly flat, or low. Previously known from Florida to Brazil, West Indies, west coast of Mexico including the Gulf. Eleven lots of specimens comprising several hundred individuals, were taken at eight stations as follows, generally high in the littoral: Cape San Lucas; Pulmo Reef; Pt. Lobos, Espiritu Santo Island; Amortajada Bay, San José Island; Puerto Escondido; San Carlos Bay; San Francisquito Bay; Angeles Bay. The most common, ubiquitous, and obvious barnacle in the Gulf.

§ N–9 *Chthamalus anisopoma* Pilsbry
Pilsbry 1916, p. 317, Pl. 74 (§ N–5).

An ivory white barnacle, small to very small, typically delicate and symmetrical, erect and uniform, with starry base. Previously reported only from San Luis Gonzales Bay, Gulf of California. Eight lots of specimens were taken from the following six localities, high in the littoral: Cape San Lucas; Pt. Lobos, Espiritu Santo Island; San Gabriel Bay, Espiritu Santo Island; Marcial Pt., Angeles Bay; Puerto Refugio, Angel de la Guardia Island. The second most common and ubiquitous barnacle in the Gulf.

§ N–10 *Balanus tintinabulum californicus* Pilsbry
Pilsbry 1916, p. 65, Pls. 14 and 15 (§ N–5).

Typically a medium to moderately large barnacle, clean cut, with white base and vertically radiating definite red lines. Recorded by Pilsbry from Santa Barbara to San Diego, but known to be common at least as far north as Monterey. Three lots of specimens from half tide or below: Cape San Lucas; Concepcion Bay (submerged); Estero de la Luna, Sonora.

§ N–11 *B. t. peninsularis* Pilsbry
Pilsbry 1916, p. 66, Pl. 15 (§ N–5).

Clear grayish red, heavily thatched. In old eroded specimens the thatching may be worn off. Previously reported from Cape San Lucas. Two lots of specimens were taken, at Cape San Lucas at the 0.0 level, and at Pulmo Reef growing on *Tetraclita* at half tide or above.

§ N–12 *Balanus trigonus* Darwin
Pilsbry 1916, p. 111, Pl. 26 (§ N–5).

Formerly reported from various warm waters, cosmopolitan. Two lots were taken, both submerged at low tide, minute and small specimens at Cape San Lucas and at Estero de la Luna, Sonora.

Of the following there were taken only one lot each:

§ N–13 *B. amphitrite inexpectatus* Pilsbry
Pilsbry 1916, p. 97, Pl. 20 (§ N–5).

Large barnacles growing on waterlogged timbers on the Mogote sand flats, La Paz Bay. Previously reported from the Gulf, where they grew on oyster shells.

REFERENCES

§ N-14 *B. concavus* Bronn
Pilsbry 1916, p. 100, Pl. 21 (§ N-5).

Growing with *B. t. californicus* on a submerged calcareous tube at Estero de la Luna, Sonora. Previously reported from California to Peru, and from the Philippines.

§ N-15 *B. improvisus* Darwin
Pilsbry 1916, p. 84, Pl. 24 (§ N-5).

In vial with no data, but probably taken from submerged *Pinna* shells at Concepcion Bay. Previously reported from Scotland, France, Nova Scotia, Patagonia, and Columbia.

§ N-16 *Chthamalus fissus* Darwin
Pilsbry 1916, p. 317, Pl. 74 (§ N-5).

A small barnacle growing on waterlogged timbers on the Mogote mud-flats. Previously reported from southern California.

§ N-17 *Tetraclita squamosa*
Probably *T. s. rubescens Darwin*. Pilsbry 1916, p. 257, Pl. 61 (§ N-5).

Growing on rocks at the south end of Tiburon Island. Previously reported as ranging from the Farallone Islands, California, to Cape San Lucas.

SUMMARY

§ N-18 Of the ten species taken, several are herein reported for the first time from the Gulf, and two represent considerable extensions of range, *B. t. californicus* apparently not having been reported previously south of San Diego, and *Chthamalus fissus* not having been reported south of southern California, although Ricketts and Calvin 1939 (§ Y-3), p. 18, find it common in northern Lower California. It would appear that the barnacle fauna of the Gulf, both as to species and numbers of individuals, derives from tropical or unique species with an infiltration of northern (Californian) and cosmopolitan forms.

§ N-19 *Addenda:* At the suggestion of the United States National Museum, we sent some of the barnacles which turned up subsequent to our original sorting to Dr. Dora Priaulx Henry, Research Associate of the University of Washington Oceanographic Laboratories, who wished to examine Panamic species. Her report, which arrived when this account was in press, adds two species to the above list,

REFERENCES

bringing the total to 12, and increases the number of stations for 3 species already reported:

A new subspecies of *Chelonibia patula,* to be named and described by Dr. Henry, is recorded from Agiabampo Estuary, Sonora, where it occurred on the great claw of *Callinectes bellicosus,* the swimming crab (§ R–42), and on the carapace and great claw of another individual of the same species, illustrated here as Pl. 14, fig. 2.

Tetraclita squamosa stalactifera, f. *confinis,* is the barnacle on *Acmaea dalliana* (§ S–397), photographed on Pl. 34, fig. 6.

Smaller barnacles on another specimen of *A. dalliana,* are *Chthamalus anisopoma* (§ N–9), which also occur on the specimen of *Acmaea atrata* (§ S–396) from Cape San Lucas, photographed as Pl. 34, fig. 5. Additional *Ch. anisopoma* were recorded from specimens of *Ostrea mexicana* (§ S–232, per photo on Pl. 37, fig. 2) taken on the mangrove roots at Gabriel Bay on Espiritu Santo Island; and on the turban shell *Tegula rugosa* (§ S–407) at Port San Carlos, Sonora.

Balanus trigonus (§ N–12) occurred with the above mentioned *Chelonibia,* on the great claws of swimming crabs at Agiabampo Estuary, Sonora. The same barnacle was reported from Concepcion Bay on *Navicula pacifica* (§ S–224).

Specimens of the conch *Strombus gracilior* (§ S–372) from Estero de la Luna and Agiabampo Estuary, both in Sonora, were commensalized by *Balanus amphitrite inexpectatus* (§ N–13), in one case heavily.

Unidentifiable barnacles were reported from the same conches, and from the paper-shell clam *Isognomon anomioides* (§ S–227).

Dr. Henry mentions also two of her forthcoming papers which, if published, would have been cited in our bibliography, both presumably 1941. The first, "Notes on some sessile barnacles from Lower California and the west coast of Mexico" is to appear in the *Proceedings of the New England Zoological Club.* "Studies on the Sessile CIRRIPEDIA of the Pacific Coast of North America" is scheduled for the University of Washington Publications in Oceanography.

Order Decapoda

GENERAL REFERENCES TO THE ORDER AS A WHOLE:

§ O-1 Boone, Lee 1931. "Anomuran, Macruran Crustacea from Panama and the Canal Zone." Bull. Amer. Mus. Nat. Hist., Vol. 63: 137-189, 23 text figs.

21 species, 4 of them new. Fair to good toto-drawings or photos of all the species involved. No bibliography except as citations in synonymy. Schmitt notes (1935, § P-18) several incorrect identifications and mislabeled shrimp drawings in this work, and Glassell states that the p. 152, fig. 7, porcellanid is *not Petrolisthes edwardsii* as labeled, that *P. eriomerus*, p. 154, and fig. 8 also is incorrectly determined and that *Pontonia margarita,* p. 180, fig. 20 is *not* commensal in *Pinna,* and is *not* a synonym of *P. pinnae* Lockington. Probably this paper should be used also otherwise with caution.

§ O-2 Bouvier, E. L. 1895. "Sur une collection de Crustacés décapodes recueillis en Basse-Californie par M. Diguet." Bull. Mus. Hist. Nat. Paris, Vol. 1: 6-8.

Very concise descriptions of 7 new species, 6 of them Macrura and Anomura. No collecting data, no illustrations. Notes on additional previously known species taken at the same time.

§ O-3 .... 1898. "Sur quelques crustacés anomures et brachyures recueillis par M. Diguet en Basse-Californie." Bull. Mus. Hist. Nat. Paris, Vol. 4: 371-384.

13 species, 7 (all Anomura) of them new. Notes on synonymy. No illustrations. No natural history data.

§ O-4 Cano, Gavino 1889. "Crostacei Brachyuri ed Anomuri raccolti nel viaggio della *Vettor Pisani* intorno al globo." Boll. Soc. Nat. Napoli, (1), Vol. 3: 79-105, and 169-268.

Anomura are treated on pp. 95-96. Lists of species taken, at Payta, p. 101, in Ecuador, p. 101, in the Gulf of Panama, p. 102. *Hippa emerita* on p. 266. Otherwise mostly Brachyura. Good bibliography. Noted by Glassell as an untrustworthy paper, see Rathbun (§ M-14), p. 609.

§ O-5 Glassell 1935a. "New or little known crabs from the Pacific coast of northern Mexico." Trans. San Diego Soc. Nat. Hist., Vol. 8: 91-106, 8 pls.

9 species are considered, all probably Panamic, 7 of them new. One is an Anomuran (a porcellanid), and the others are Brachyura. Extension of range and new locality records for 14 additional species. Superb and ample illustrations including toto-drawings.

§ O-6 .... 1936. "New porcellanids and pinnotherids from tropical North American waters." Trans. San Diego Soc. Nat. Hist., Vol. 8: 277-304, 1 pl.

REFERENCES

16 new species, all probably Panamic. 4 only figures, one of which, *Orthochela pumila,* is a toto-drawing. *Pachycheles sonorensis,* p. 291, is stated in a personal communication from Glassell, March 1941, to be a synonym of *P. panamensis* Faxon.

§ O–7 .... 1938. "New and obscure decapod crustacea from the west American coast." Trans. San Diego Soc. Nat. Hist., Vol. 8: 411-454, 10 pls.

2 Natantia, 11 Anomura, 4 Brachyura, all probably Panamic, of which 9 are new. Key to the species of the new porcellanid genus *Pisonella,* and to the west coast species of *Petrolisthes* and *Pachycheles* with *sonorensis* which is a synonym of *panamensis* Faxon. The usual superb illustrations, with 12 species in toto-drawings.

§ O–7a Hult, J. 1938. "Crustacea Decapoda from the Galápagos Islands collected by Mr. Rolf Blomberg." Ark. f. Zool. K. Svenska Vetens., Vol. 30a (5): 18 pp., 1 pl., 4 text figs.

10 Macrura Natantia, 1 Macrura Reptantia, 2 Anomura, and 12 Brachyura, of which 5, including *Callianassa hartmeyeri* (formerly known only from the West Indies, with good toto-illustrations), are new to the Galápagos.

§ O–8 Lockington, W. N. 1873-74. "On the Crustacea of California." Proc. Calif. Acad. Sci. (1), Vol. 5: 380-84.

Mentions also 2 species of *Porcellana* from Mazatlan (in addition to a Mexican *Callinectes,* then called *Amphitrite* which will have been considered subsequently in the Rathbun Brachyura monograph).

§ O–9 Rathbun, M. J. 1902a. "Brachyura and Macrura." No. 8 in: "Papers from the Hopkins Stanford Galápagos Expedition, 1898-99." Proc. Wash. Acad. Sci., Vol. 4: 275-292, 4 text figs., 1 pl.

Nine new species illustrated by lithographs of drawings. Mostly Brachyura, but including one in the FAMILY PENAEIDAE and one in the new FAMILY DISCIDAE.

§ O–10 Schmitt, W. L. 1921. "The marine decapod crustacea of California." Univ. Calif. Publ. Zool., Vol. 23: 470 pp., 50 pls., etc.

Monographic, complete, well illustrated, and with good bibliography. Transcending the *Albatross*-San Francisco Bay subject matter on which it was based. In addition to some others cited herein directly, the bibliography lists papers by Baker 1912, and Hilton 1916 and 1918, which treat Panamic and overlap decapods found at Laguna Beach. We use this as key paper for the decapods.

REFERENCES

§ O–11 .... 1924. "Crustacea (Macrura and Anomura)." No. 36 in: "Expedition of the California Academy of Sciences to the Gulf of California in 1921." Proc. Calif. Acad. Sci. (4), Vol. 13 (24): 381-388.

17 previously known species, plus a *Sicyonia* and a *Penaeopsis* to be described—these presumably have been attended to by Burkenroad (§ P–1). No illustrations or bibliography.

§ O–12 .... 1924a. "The Macrura and Anomura collected by the Williams Galápagos Expedition, 1923." Zoologica, Vol. 5 (15): 161-171.

10 species, 2 of them new in the genera *Hippolyte* and *Lysmata.*

§ O–13 Schmitt, W. L. 1935. "Crustacea, Macrura and Anomura of Porto Rico and the Virgin Islands." N.Y. Acad. Sci., Sci. Survey of Porto Rico and the Virgin Islands, Vol. 15 (2): 125-227, 80 text figs., 4 pls.

A Pacific item is described on pp. 214, 215, and 216, figs. 77a and b. *Emerita rathbunae* sp. nov. Key to the genus *Emerita.* There are descriptions and toto-illustrations of several forms whicl. occur also in the Panamic area. Extensive bibliography.

§ O–14 Smith, S. I. 1869. "Descriptions of a new genus and two new species of Scyllaridae and a new species of *Aethra* from North America." Amer. Jour. Sci. (2), Vol. 48: 118-121.

Evibacus princeps, one of the Palinura, and the brachyuran *A. scutata* from La Paz.

§ O–15 Stimpson, Wm. 1859. "Notes on North American Crustacea, No. I." Ann. Lyc. Nat. Hist. N.Y., Vol. 7: 49-93, 1 pl.

Decapods oñly. 17 plus 2? Panamic forms, mostly new species. 8 only belong to the Anomura. The balance are Brachyura and will have been considered in the Rathbun monograph.

§ O–16 .... 1860. "Notes on North American Crustacea, in the Museum of the Smithsonian Institution, No. II." Ann. Lyc. Nat. Hist. N.Y., Vol. 7: 176-246, 2 pls.

Almost entirely the results of the collecting activities of John Xantus, U. S. tidal observer at Cape San Lucas. 57 species of decapods, mostly new. 6 are Anomura. The balance are Brachyura and as such will have been considered in the Rathbun monograph.

REFERENCES

§ O–17 .... 1871. "Notes on North American Crustacea, in the Museum of the Smithsonian Institution, No. III." Ann. Lyc. Nat. Hist. N.Y., Vol. 10: 92-136.

Largely the results of the collecting activities of Xantus at Manzanillo, and of Capt. Dow in Central America. 33 Panamic and overlap decapod species, many of them new. All are Brachyura except for the Natantians *Hippolysmata californica* Stimpson from San Diego, and *Penaeus stylirostris* sp. nov. from Panama.

§ O–18 Streets, T. Hale 1871. "Description of five new species of Crustacea from Mexico." Proc. Acad. Nat. Sci. Phil., Vol. 23: 225-227.

New species of *Pachycheles, Panulirus,* and *Palaemon* from the Gulf of Tehuantepec.

§ O–19 .... 1871a. "Catalogue of Crustacea from the Isthmus of Panama, collected by J. A. McNeil." Proc. Acad. Nat. Sci. Phil., Vol. 23: 238-243.

In addition to the Brachyura, 12 Anomura and Macrura are listed, with 3 new hermits, a Thallassinid, and 2 shrimps. No ecological or distributional data, no illustrations or bibliography.

§ O–20 See also Decapod items in papers listed under "Crustacea."

§ O–21 Schmitt, W. L. 1921 (§ O–11) is used as key paper for the decapods, especially as augmented by Schmitt 1935 (§ O–13) with reference to more characteristically tropical families and genera.

SCHEME OF CLASSIFICATION ADOPTED HEREIN

§ O–22 In the confusion of an assemblage so large and varied as that provided by the Gulf decapods, we were able to proceed efficiently, only by constructing a large taxonomic picture along the lines of the following conspectus. Into this we were able to fit the new information on tropical decapods, relating it to previous similar data from temperate waters. In the thought that a graphic presentation of this sort might have supra-personal value —other minds also being presumably susceptible to confusion from a mass of detail so intricate—we are reproducing these tabulated working notes as a general orientation picture for whoever may have similar problems.

MACRUROUS FORMS

§ P numbers

COMPRISING ALL THE NATANTIA AND ALL THE MACRUROUS REPTANTIA EXCEPT THE GALATHEIDAE

SUB-ORDER NATANTIA (§ O–10, p. 18)

TRIBE STENOPIDES

FAMILY STENOPIDAE (Stenopus) (§ O–13, p. 170)

TRIBE PENEIDES (§O–10, p. 19)

FAMILY SERGESTIDAE (Sergestidae, Lucifer) (§ P–3)

" PENEIDAE (§ P–1, 2, 4, 18)

TRIBE CARIDES (§ P–5, p. 26 and § O–10)

FAMILY PASIPHAEIDAE (Leptochela and the deep Pasiphaea)

FAMILY ACANTHEPHYRIDAE OR OPHLOPHORIDAE (Hymenodora etc., deep)

FAMILY PALAEMONIDAE (Palaemon, Palaemonetes, the fresh-water Macrobrachium, etc. Chace includes Pontonia)

FAMILY ATYIDAE (fresh-water forms)
" PONTONIIDAE (Pontonia, etc.)
" PANDALIDAE (Pandalus, Plesionika, etc.)
" HIPPOLYTIDAE (Hippolyte, Hippolysmata, Spirontocaris, etc.)

FAMILY CRANGONIDAE (or Alpheidae) (§ P–6, P–11)
(Snapping shrimps, Crangon, Synalpheus, Betaeus, Automate, etc.)

FAMILY GNATHOPHYLLIDAE (Gnathophyllum, etc.)

FAMILY LYSMATIDAE OR PROCESSIDAE (Processa)
" CRAGONIDAE (formerly CRANGONIDAE) (Crago) (§ P–13)

FAMILY GLYPHOCRANGONIDAE (Glyphocrangon, deep)

SUB-ORDER REPTANTIA, all the rest of the Decapods (§ O–10, p. 104)

TRIBE PALINURA (Scyllaridea, formerly called also Loricata) (§ O–10, p. 105)

FAMILY ERYONTIDAE (Eryoniscus, Polycheles, deep)

FAMILY PALINURIDAE (Sea crayfish, spiny lobsters; Panulirus)

FAMILY SCYLLARIDAE (Stone lobsters, Scyllarus, Evibacus, etc.)

TRIBE ASTACURA (Nephropsidea)

FAMILY ASTACIDEA, fresh-water crayfish
" HOMARIDAE (Nephropsidae) Atlantic lobsters

TRIBE ANOMURA (§ O–10, p. 109)

SUPERFAMILY THALASSINIDEA (ghost shrimps, etc.)

FAMILY AXIIDEA (Loamediidae). (Calastacus, Axius, Axiopsis, etc.)

FAMILY CALLIANASSIDAE (Upogebidae). (Callianidea, Callianassa, Upogebia, etc.)

ANOMUROUS FORMS
§ *Q numbers*

ALL THE ANOMURA EXCEPT THE THALLASINIDEA

SUPERFAMILY PAGURIDEA

FAMILY PAGURIDAE. The hermit crabs
" COENOBITIDAE. Robber crabs, land crabs

FAMILY LITHODIDAE. The stone crabs (Hapalogaster, Cryptolithodes, etc.)

SUPERFAMILY GALATHEIDAE

FAMILY GALATHEIDAE, macrurous forms, but treated here with the Anomura for taxonomic consistency

FAMILY PORCELLANIDAE. Porcelain crabs

SUPERFAMILY HIPPIDEA. Sand bugs, sand crabs

FAMILY ALBUNEIDAE (Albunea and Lepidopa)
" HIPPIDAE (Hippa, Emerita)
(§ O–10, p. 182 and
§ R–15, 18, 19, and 22)

***TRIBE BRACHYURA* — the true crabs**
§R numbers

Div. One Oxystomatous Crabs. Vol. 4 of Rathbun Monograph, § R-22, 1937

SUBTRIBE GYMNOPLEURA
- FAMILY RANINIDAE

SUBTRIBE DROMIACEA
- FAMILY DROMIIDAE
- " DYNOMENIDAE
- " HOMOLODROMIIDAE
- " THELIOPEIDAE
- " LATREILLIIDAE

SUBTRIBE OXYSTOMATA
- FAMILY DORIPPIDAE
- " LEUCOSIIDAE
- " CALAPPIDAE

SUBTRIBE HAPALOCARCINIDEA
- FAMILY HAPALOCARCINIDAE

First three volumes of Rathbun Monograph, 1918, 1925, 1930

SUBTRIBE BRACHYGNATHA

Div. Two Spider Crabs § R-18

- SUPERFAMILIES OXYRHYNCHA
  - FAMILY MAJIDAE
    - SUBFAMILY INACHINAE, ACANTHONYCHINAE, PISINAE, AND MAJINAE
  - FAMILIES PARTHENOPIDAE
    - SUBFAMILY PARTHENOPIDAE AND EUMEDONINAE
  - FAMILY HYMENOSOMIDAE

Div. Three Cancroid Crabs § R-19

- SUPERFAMILY BRACHYRHYNCHA
  - FAMILY EURYALIDAE
  - " PORTUNIDAE
  - " ATELECYCLIDAE
  - " CANCRIDAE
  - " XANTHIDAE

Div. Four Grapsoid Crabs § R-15

- FAMILY GONEPLACIDAE
- " PINNOTHERIDAE
- " CYMOPOLIIDAE
- " GRAPSIDAE
- " GECARCINIDAE
- " OCYPODIDAE

REFERENCES

§ O–23 ACKNOWLEDGMENT: For determinations of the decapods as a whole, we are indebted to Dr. Waldo L. Schmitt of the U. S. National Museum, and to Mr. Steve Glassell of Beverly Hills, California. Detailed acknowledgments will be found under the separate sections.

MACRUROUS FORMS

(NATANTIA PLUS ALL THE MACRUROUS REPTANTIA)

§ P–1 Burkenroad, M. D. 1934. "Littoral PENAEIDAE chiefly from the Bingham Oceanographic collection. . . ." Bull. Bingham Ocean. Coll., Vol. 4 (7): 109 pp.

15 Panamic species and subspecies, 11 of them new, in the genera *Penaeopsis Eusicyonia,* etc. Structural drawings. Bibliography.

§ P–2 .... 1936. "The ARISTAEINAE, SOLENOCERINAE and pelagic PENAEINAE of the Bingham Oceanographic Collection." Bull. Bingham Ocean. Coll., Vol. 5 (2): 151 pp.

A few Panamic forms, *Solenocera* and *Funchalia* spp., in this consideration chiefly of tropical Atlantic forms, all pelagic. Data on geographic distribution.

§ P–3 .... 1937. "SERGESTIDAE (Crustacea Decapoda) from the Lower California region, with descriptions of two new species, and some remarks on the organs of Pesta in *Sergestes.*" Zoologica, Vol. 22: 315-329, text figs.

Pelagic, usually in very deep water (the shallowest was 145 fathoms). 6 species, of which 2 are new. List of 6 additional species previously reported from the region.

§ P–4 .... 1938. "PENAEIDAE from the region of Lower California and Clarion Island, with descriptions of four new species." No. 13 in: "The Templeton Crocker Expedition." Zoologica, Vol. 23: 55-91, 34 text figs.

18 species, 4 of them new, shore to 117 fathoms, Cedros to Galápagos. It is remarked that the *littoral* penaeid fauna of the Panamic area is most closely related to that of the Caribbean, but that the respective *deep water* faunas show little similarity. Short bibliography. *Penaeus californiensis* Holmes is resurrected and differentiated from *P. brevirostris* Kingsley.

§ P–5 Chace, Fenner A., Jr. 1937. "Caridean decapod Crustacea from the Gulf of California and the west coast of Lower California." No. 7 in: "The Templeton Crocker Expedition." Zoologica, Vol. 22: 109-138, 9 text figs.

29 species, 9 new, 21 new to the region, shore to several hundred fathoms. An important paper, covering the snapping shrimps, hippolytids, etc., with 6 good toto-drawings, but lacking a bibliography.

REFERENCES

§ P–6 Coutière, Henri 1909. "The American species of snapping shrimps of the genus *Synalpheus*." Proc. U.S.N.M., Vol. 36: 1-93, etc.

8 species from Lower California, mostly the result of M. Diguet's collecting. 7 are new. *S. lockingtoni* is proposed as a new name for Lockington's *Alpheus leviusculus*. Structural drawings only.

§ P–7 Garcia, Antonio G. 1939. "La pesca de camarón en la costa del Pacifico." Rev. Soc. Mex. Hist. Nat., Vol. 1: 45 to (?)52, 2 pls.

A consideration of the co-operatively operated stationary net fisheries in Nayarit and Sinaloa. No data on the trawler operations off Sonora (p. 247 herein) which are not even mentioned. Some information on the amount of production and on the hydrobiology of the region, but the only copy of this paper to which we had access was incomplete in this feature due to error in binding.

§ P–8 Kingsley, J. S. 1878. "Notes on the North American Caridea in the Museum of the Peabody Academy of Science at Salem, Mass." Proc. Acad. Nat. Sci. Phil., Vol. 30: 89-98.

Panamic new species in the genera *Atyoida, Alpheus,* and *Peneus* on p. 93 and 98, plus *Atya occidentalis* with no locality stated.

§ P–9 .... (1878) 1879. "List of the North American Crustacea belonging to the SUB-ORDER CARIDEA." Bull. Essex Inst., Vol. 10: 53-71.

10-11 marine species from the west coasts of Mexico, Nicaragua, and Panama, and from Peru.

§ P–10 Lockington, Wm. N. 1878. "Remarks upon the Thalassinidea and Astacidea of the Pacific coast of North America." Ann. Mag. Nat. Hist. (5), Vol. 2: 299-304.

7 Panamic ana overlap species are mentioned in the genera *Gebia* (now *Upogebia*), *Callianidea,* and *Panulirus,* from La Paz, Puerto Escondido, Central America, etc., with *Callianassa longimana* from the north Pacific coast of Lower California.

§ P–11 .... 1878a. "Remarks on some new Alphei, with a synopsis of the North American species." Ann. Mag. Nat. Hist., (5), Vol. 1: 465-480.

Snapping shrimps. 15 species are Panamic or overlap. 4 new species. Many are from Puerto Escondido, Mulege, etc., in the Gulf.

§ P–12 .... (1878) 1879. "Notes on Pacific coast Crustacea." Bull. Essex Inst., Vol. 10: 159-165.

7 Natantia from Lower California, mostly from the Gulf, of which 3 are new: *Palaemon longipes, Pontonia pinnae,* and *Sicyonia pencillata.*

REFERENCES

§ P–13 Ortmann, A. E. 1895. "A study of the systematic and geographic distribution of the decapod FAMILY CRAGONIDAE Bate." Proc. Acad. Nat. Sci. Phil., Vol. 47: 173-197.

Now known as the CRAGONIDAE. 5 Panamic species, all except one from very deep water.

§ P–14 Rathbun, M. J. 1902. "Descriptions of new decapod crustaceans from the west coast of North America." Proc. U.S.N.M., Vol. 24: 885-905.

A study of the *Albatross* shrimps from Bering Sea to San Diego, but including 3 Gulf of California Carideans, one of which, *Pasiphaea pacifica,* seems not to be recorded in the only recent caridean report from the region (Chace 1937 [§ P–5]).

§ P–15 Rioja, E. 1940. "Descripción de un órgano setiforme en el tercer maxilípedo de algunos PENAEIDAE." No. 3 in: "Estudios carcinológicos." An. Inst. Biol. Mex., Vol. 11: 261-266.

Morphological differentiations in the organ involved as applied to the taxonomy of Pacific species of *Penaeus.*

§ P–16 .... 1940a. "Observaciones sobre las anténulas de algunas especies del género *Penaeus.*" No. 4 in: "Estudios carcinológicos." Ibid. 267-273.

Concerning the Pacific *P. occidentalis,* etc.

§ P–17 .... 1940b. "La morfología de las cerdas de las piezas bucales de los peneides (Crust. Decap.) y su valor diagnóstico." Ciencia, Vol. 1 (3): 116-117.

A general statement of an exact taxonomic method for differentiating Penaeids, and its application to Mexican shrimps, including Pacific coast forms.

§ P–17a Rioja, E. 1941a. "Estudios carcinológicos. VIII. Morfología e interpretación del petasma de los PENAEIDAE." An. Inst. Biol., Vol. 12 (1): 199-221, 23 text figs.

Including considerations and drawings of these organs in 7 species including *P. stylirostris, P. vannamei,* etc., involving implications significant to any evaluation of the taxonomy, morphology, and phylogeny of this group.

§ P–17b .... 1941b. "Estudios carcinológicos. IX. El macho maduro de *Penaeus vannamei* de las costas del Pacífica de México." An. Inst. Biol., Vol. 12 (1): 223-229, 5 text figs.

Description of 3 sexually mature males from Guaymas. The largest, with incomplete rostrum, measured 165 mm. indicating a greatest stature of 180-182 mm. or more in this species.

REFERENCES

§ P–18 Schmitt, Waldo L. 1935a. "The west American species of shrimps of the genus *Penaeus*." Proc. Biol. Soc. Wash., Vol. 48: 15-24, 2 pls.

With the exception of one species mentioned by Burkenroad 1938 and of the Panama forms considered in Boone 1930, this is the only modern paper treating the distribution and natural history of the several species of commercial shrimps so abundant on the Mexican west coast. Five species, into one of which, *P. brasiliensis* (according to a personal communication from Dr. Schmitt), too much material was concentrated, per corrections in Burkenroad 1938, p. 66 (§ P–4).

§ P–19 See also Macrura items in the general Decapoda bibliography, especially the Boone 1931 (§ O–1) and the Schmitt 1935 (§ O–13) papers, and occasionally in the general Crustacea citations. The fact that there is no key paper for the shrimps of this region, nor even any adequate local NATANTIA bibliography, has necessitated a more comprehensive listing in this group than we ordinarily should have employed.

§ P–19a ACKNOWLEDGMENT: For determining the majority of the following Macrura, for gifts of and help with the literature, but most of all for checking through this portion of the phyletic catalogue, we have to thank Dr. Waldo L. Schmitt of the United States National Museum, who reports, however, that credit for the *Penaeus* identifications should go to Mr. Milton J. Lindner, in charge of the U. S. Fish and Wildlife shrimp identifications at New Orleans, La. For identifying a few macrurous forms, we have to thank Steve Glassell of Beverly Hills, California.

LIST OF SPECIES TAKEN:

SUB-ORDER NATANTIA

TRIBE PENEIDES

FAMILY PENEIDAE

§ P–20 *Eusicyonia penicillata* (Lockington) 1878 (1879)
Burkenroad 1938 (§ P–4), p. 83. Range: Lower California, both coasts, to Cedros Island; beach to 40 fathoms. Toto-illustration as Pl. 36, Boone 1930b (§ M–2), description on p. 115.

REFERENCES

Apparently very common in the Lower California region, since it has been listed in the reports of most of the Gulf collections since Schmitt 1924. This shrimp was netted free-swimming, by light at anchorage at night, at Port San Carlos, Sonora, and was taken again, with the commercial *Penaeus stylirostris*, in sorting the catch of the Japanese shrimp trawler south of Guaymas.

§ P–21 *Penaeus californiensis* Holmes 1900

Burkenroad 1938 (Zoologica, § P–4), p. 67. San Francisco (two records only), Santa Monica. Gulf, west coast of Mexico to "Esquinapa." Formerly merged with *P. brevirostris* Kingsley.

One specimen found dead on sand flat at El Mogote.

§ P–22 *Penaeus stylirostris* Stimpson 1871 PL. 39 FIG. 1

Schmitt 1935a (§ P–18), p. 19. Range: Guaymas, Sonora, to Tumbes, Gulf of Guayaquil (South America). This seems not to have been illustrated, the Boone 1931 (§ O–1) Fig. 15 being a drawing of *Xiphopeneus riveti* Bouvier, according to Schmitt.

This is the common commercial shrimp at Guaymas, all the market specimens we examined belonging apparently to this one species. Several small boats of the Mexicans, and twelve large boats (up to 175 feet long) operated out of Tokyo, Japan, were engaged in trawling for this form south of Guaymas in the spring of 1940; the daily catch must have been many hundreds of tons.

TRIBE CARIDES

FAMILY PALAEMONIDAE

§ P–23 *Macrobrachium jamaicense* (Herbst) 1792

Rathbun 1910 (§ M–14), p. 561, with good toto-fig. 1, pl. 51. Fresh waters of the Pacific slope from Lower California to Peru; Atlantic coast from Texas to Brazil, including the West Indies.

2 specimens identified with this *fresh-water* species were given us at Guaymas by Capt. Corona of the Mexican shrimping fleet. At the same time we personally selected 2 or 3 contrasting small shrimps of unknown identity out of great piles of *Penaeus stylirostris* heaped on the deck of one of the Japanese shrimp dredgers working out of Guaymas. The label on the latter container was confused or turned out to be missing except for date and locality somewhere along the line. The United States National Museum subsequently determined three of our Guaymas shrimps about which there was little informa-

REFERENCES

tion as *M. jamaicense,* and the obvious inference is that these were the same animals we ourselves took aboard the shrimp dredger which was operating presumably in entirely salt water. However there are reasons, other than incongruity, for assuming that the drag boat material must have been some other species of *Penaeus,* rather than the different appearing *Macrobrachium.* Which leaves us with two specimens of *M. jamaicense* of known history (donations from Capt. Corona), and three others from an unknown source with ecological niche unknown also. Undoubtedly, *Macrobrachium* may occur in brackish water, but that it should be taken at a depth of several fathoms in the ocean, even off the mouth of a river (where the *Penaeus* were being fished) seems most unlikely.

§ P–24 *Palaemon ritteri* Holmes 1895. Broken-back shrimp.

Rathbun 1910 (§ M–14), p. 561, 1 specimen near Tumbes, toto-fig. 1, Pl. 53.

Schmitt 1921 (§ O–10), p. 35, San Diego to Gulf to Ecuador; detail figure.

Schmitt 1924 (§ O–11), p. 386, Carmen Island (Gulf), 8 fathoms, 1 specimen.

Vividly colored, abundant and active *Spirontocaris*-like forms observed in quantity at Puerto Refugio and at the south end of Tiburon Island in the tide pools.

FAMILY PONTONIIDAE

§ P–25 *Pontonia pinnae* Lockington (1878) 1879 PL. 21 FIG. 5

Kemp, Stanley 1922. "Notes on Crustacea Decapoda in the Indian Museum. XV. PONTONIIDAE." Rec. Ind. Mus., Vol. 24 (2): p. 287-88. From the Gulf of California, collected by M. Diguet.

We took this very common commensal shrimp, which seems to be restricted to the Gulf of California, in the hatchet clam, *Pinna,* at Concepcion Bay, and at Gabriel Bay on Espiritu Santo Island, and (presumably free-living) at Pt. Lobos on Espiritu Santo Island.

FAMILY CRANGONIDAE (ALPHEIDAE)

§ P–26 *Crangon malleator* (Dana) 1852

Rathbun 1910 (§ M–14), p. 607, recorded from Ecuador, Galápagos, Brazil (?), Cape Verde Islands; with citation of Dana, p. 557 "Crustacea." U. S. Explor. Exped. Pt. 1, 1852; fig. 9, Pl. 31, Atlas 1855.

Pulmo Reef, 2 males, 1 ovigerous female.

REFERENCES

§ P–27 *Crangon ventrosus* (H. Milne-Edwards) 1837
Chace 1937 (§ P–5), p. 118. 10 specimens off Arena Bank. Range recorded as entire Indo-Pacific, Hawaiian Islands, and Gulf.
Pulmo Reef. 3 small.

§ P–28 *Crangon* sp. No. 1
Pt. Lobos, Espiritu Santo Island; Pt. Marcial Reef; Gabriel Bay, Espiritu Santo Island, from rocks at base of mangrove roots.

§ P–29 *Crangon* sp. No. 2
Pt. Lobos, Espiritu Santo Island; Gabriel Bay, Espiritu Santo Island, from rocks at base of mangroves.

§ P–30 *Crangon* sp. No. 3
Pt. Lobos, Espiritu Santo Island.

§ P–31 Specimens of *Crangon* which are indeterminable as to species because the chelae are lacking were taken at Gabriel Bay and at El Mogote.

§ P–32 ? *Synalpheus opiocerus sanjosei* Coutière
Coutière 1909 (§ P–6), p. 29; San José Island (Gulf), detail figures.
Taken at Pulmo Reef; El Mogote; and at Concepcion Bay, mostly under sponge and tunicate clusters on "hachas."

§ P–33 *Synalpheus digueti* Coutière
Chace 1937 (§ P–5), p. 123. Range: Lower California (probably La Paz—these were collected by Diguet); Arena Bank. In a forma *ecuadorensis,* from St. Helena, Ecuador, taken by M. Festa.
Taken at Pulmo Reef; Concepcion Bay; mostly under sponges and tunicates on "hachas."

§ P–34 *Synalpheus sanlucasi* Coutière
Chace 1937 (§ P–5), p. 123. General range: Cape San Lucas taken by Xantus, and Arena Bank only.
Taken at Pulmo Reef (many) and at Pt. Lobos on Espiritu Santo Island.

§ P–35 *Synalpheus townsendi mexicanus* Coutière
Chace 1937 (§ P–5), p. 123. General range: Ceralbo Island and Arena Bank (both in southern part of Gulf). Coutière

REFERENCES

1909 (§ P–6), p. 34; southern part of Gulf, 9½ fathoms. *S. townsendi* and its formae extend from the West Indies to Hawaii.

Concepcion Bay, mostly under sponges and tunicates on "hachas."

§ P–36 Specimens of *Synalpheus* which are indeterminable as to species because the chelae are lacking, were taken also at Pulmo Reef.

Sub-order Reptantia

TRIBE PALINURA

§ P–101 Allen, B. J. 1916. "Notes on the spiny lobster." Univ. Calif. Publ. Zool., Vol. 16: 139-152, 2 text figs.

Natural history data, the result of a California Fish and Game Commission Survey, on this commercially important species.

§ P–102 See also Macrura and Palinura items mentioned in papers (especially in Kingsley and in Lockington) cited in the general Crustacea and in the general Decapod bibliographies, pages 415-8 and 434-7.

LIST OF SPECIES TAKEN:

Family Palinuridae (spiny lobsters, sea crayfish)

§ P–103 *Panulirus inflatus* Bouvier

This and/or similar forms from the Pacific Coast have been described and listed variously as *P. inflatus* and *P. ornatus*, etc., p. 163, Boone 1931 (§ O–1). Toto-fig. 12, color notes. Gulf to Galápagos.

Only a single specimen of this large and brilliantly colored form was taken, at night, in the reefs and tide pools at Pt. Marcial. Not common in the La Paz area, where it is eagerly sought for food. Well illustrated as fig. 12, Boone 1931 (§ O–2), from Panama material described on p. 162, where the range is recorded from Lower California to Peru and Galápagos and doubtfully from the Sandwich Islands; and (*P. ornatus*) as fig. 1, Pl. 52, Rathbun 1910 (§ M–14) from Payta, Peru, where the range is also recorded, p. 560, from the Indo-Pacific. Nobili already in 1901 gives several synonyms (§ M–13). There may or may not be several species involved. No one seems to know. In order to untangle this, it would probably be necessary for a competent specialist (all of whom already have more demands put on their time than they can hope to fulfill) to devote a great deal of time and energy to a great collection of representative specimens

REFERENCES

from all the areas involved, a collection which probably doesn't exist. In the meantime, the informed general public and the fishermen and fish dealers in the area involved must remain in ignorance concerning the taxonomy and listing of this commercially important food animal. From the Bouvier plate, it would appear that two or more species actually *are* involved, and that Boone was incorrect in synonymizing *inflatus* and *ornatus*. The specimen we took was similar to the *inflatus* illustration, and definitely *not* similar to the illustration of *ornatus*.

§ P–104 *P. interruptus* (Randall) 1839
Schmitt 1921 (§ O–10), p. 108, fig. 73; abundant from Santa Barbara southward at least to Cedros Island. There is a poor toto-illustration in Ricketts and Calvin 1939 (§ Y–3) as Pl. 24. Range given in Schmitt as San Luis Obispo to Rosalia Bay, Lower California. Apparently a true overlap form as we define the term.

Carapaces, recently cleared by the amphipods, were abundant at several bays, notably San Francisquito, in the northern part of the Gulf, but there was no hint of it in the south, where *P. inflatus* is known to occur though not abundantly.

Family Scyllaridae

§ P–105 *Evibacus princeps* Smith 1869 PL. 21 FIG. 1
Described without illustration, p. 119, Smith 1869 (§ O–15). Range given in Rathbun 1910 (§ M–14), p. 603, as La Paz, Panama, and Peru. However, in a personal communication, Glassell remarks that there seem to be two Mexican west coast sand lobsters, to either one of which (according to the brief and inadequate original descriptions) any of several names could be applied. There is a small, dorso-ventrally flattened form such as we took and per our illustration. There is also a large, more laterally flattened type, up to several feet in length, such as the specimen mounted in the Museo Nacional de Historia Natural in Mexico. The former, however, may be a juvenile of the latter.

Two specimens of this grotesque "langusta" were taken aboard the Japanese shrimper April 9, 1940, in trawl nets on the shrimping grounds outside Guaymas. They are called by the Mexicans "langusta de Arena" or "Zapatera."

TRIBE ANOMURA, SUPERFAMILY THALASSINIDEA

FAMILY AXIIDAE

REFERENCES

§ P–201 *Axius (Neaxius) vivesi* (Bouvier 1895) PL. 36 FIG. 4
J. G. de Man 1925 "Sur deux espèces encore imparfaitement connues du genre *Axius* Leach." pp. 56-61. Bulletin de la Société Zoologique de France, Vol. 50: 50-61, 2 text figs.

Dr. Schmitt writes (No. 160366, U.S. Nat. Mus.): "The unique type of this species was briefly characterized by Bouvier (1895, § O–2) p. 7, more fully described by de Man (above) fig 2. (one chela and telson only) and keyed out in de Man, Siboga Expedition, Monograph 39A5, p. 14, 1925, but nothing is contributed by either author regarding habits or occurrence other than that Bouvier states that the collection of which the specimen was a part was made by M. Diguet in Lower California (presumably in the La Paz area)."

Taken abundantly east of La Paz, in burrows in the gravelly mud flats, under and around the (mostly) dead heads of the coral *Porites*. One of the commonest animals of the region, presumably very important in the littoral economy, since it is both large and active. Difficult to collect, wary, and fast. We were successful in procuring many specimens by enlisting the aid of the small boys of La Paz who routed them out alive by manipulating flexible harpoons in the burrows and who would have kept an indefinite supply coming in had we continued to accept the material.

FAMILY CALLIANASSIDAE

§ P–202 *Callianassa* sp.
Near if not *C. uncinata* H. Milne-Edwards 1837, but probably new species.

Dr. Schmitt of the U. S. National Museum remarks (No. 160366, August 9, 1941): "Except for minor differences could be *C. uncinata* H. Milne-Edwards, Hist. Nat. Crust., vol. 2, p. 310, Pl. 25 bis, fig. 1-3, 1837. Otherwise it might be *C. rochei* Bouvier from Lower California, but of this de Man saw the incomplete type and in at least two particulars—shape of front and armature of the merus of the small cheliped, it cannot be that species either."

Dug out of sandy mud at Estero de la Luna, Sonora.

§ P–203 *Upogebia* sp.
"Resembles, but on close inspection reveals several distinct differences from, *U. pugettensis*. Very different from Lockington's *rugosa,* which I wish you had found." (Communication from Dr. Schmitt, U.S.N.M. No. 160366.)

REFERENCES

Gabriel Bay on Espiritu Santo Island. No further data. We collected here on coral, on sand flat, and on rocks at base of mangroves.

§ P–204 Curiously, we turned out none of the Smith, Street, or Lockington species listed from the region by Lockington 1878 p. 300 (§ P–10).

§ P–205 SUMMARY

21 determinable macrurous species were taken not including *Pleuroncodes* which is treated with the Anomura, (§ Q–21).

In this group the differentiation of two Gulf faunas, south and north, is very marked. 6 species of snapping shrimps and the ornate *Panulirus inflatus* were found in the extreme south only: Pulmo reef and Espiritu Santo Island; and 3 other snapping shrimps plus *Pontonia pinnae* occurred to Concepcion Bay but no farther. In the north, the lobster was *P. interruptus,* the same that occurs in California, and there were hippolytid shrimps *Palaemon ritteri* (described originally from San Diego) very abundantly in the tide pools.

5 represent new or little-known species for which no geographical information is available. Of the 21 total, 7 have been reported (on the Pacific coast) from the Gulf only, but one of these is Indo-Pacific also, and one, *S. digueti,* extends in a subspecies to Ecuador. 5 extend from the Gulf southwardly generally to Panama and Galápagos. 4 extend northwardly, one, *Penaeus californiensis,* apparently to San Francisco, although that northerly extreme has never been substantiated since the original report. *Eusicyonia* extends to Cedros, the others to southern California. Of the affiliations outside Pacific North America, 2 are West Indian and one is Indo-Pacific. The snapping shrimps would seem to represent a group more or less restricted to the Gulf and occurring only in its southern part. The penaeid fauna shows both north and south elements pretty well scattered. In general, the Macruran population of the southern part of the Gulf is strongly Panamic or unique, of the northern part, strongly overlap, there is little intermingling and there is a gap wherein (possibly due to personal or to circumstantial collecting factors) no Macrura were found.

ANOMUROUS FORMS

TRIBE ANOMURA (Less the THALASSINIDEA)

REFERENCES

§ Q–1 Benedict, James E. 1892. "Preliminary descriptions of thirty-seven new species of hermit crabs of the genus *Eupagarus* in the United States National Museum." Proc. U.S.N.M., Vol. 15: 1-26.

11 new species from the Gulf of California. No illustrations or collecting data.

§ Q–2 .... 1902. "Description of a new genus and 46 new species of crustaceans of the FAMILY GALATHEIDAE, with a list of the known marine species." Proc. U.S.N.M., Vol. 26: 243-334, 47 text figs.

Several Gulf of California and other Panamic forms, usually from deep water, with good line drawings of the entire animals.

§ Q–3 .... 1903. "Revision of the crustacea of the genus *Lepidopa*." Proc. U.S.N.M., Vol. 26: 889-985.

Two new Panamic species, some natural history notes.

§ Q–4 .... 1904. "A new genus and two new species of Crustaceans of the FAMILY ALBUNEIDAE from the Pacific Ocean. . . ." Proc. U.S.N.M., Vol. 27: 621-625, 5 text figs.

Lophomastix diomedeae, n.g., n.sp., from 26 fathoms off Cartes Bank, possibly a Panamic or overlap form. The other is from Samoa.

§ Q–5 Boone, Lee 1931. "The littoral crustacean fauna of the Galápagos Islands, Part II. Anomura." Zoologica, Vol. 14: (1): 1-62, figs. 1-19.

14 species of which 4 were new. Entirely littoral. Good illustrations of all the species taken. Some data on natural history, etc. No bibliography.

§ Q–6 Glassell, Steve A. 1937a. "Porcellanid crabs from the Gulf of California." No. IV in: "The Templeton Crocker Expedition." Zoologica, Vol. 22: 79-88, 1 pl.

11 species (2 new, 1 redescribed, 1 ms. name) in 4 genera, mostly from shallow dredgings of coral on Arena Bank. Synonymy, general and local range, and various remarks on habitat, etc.

§ Q–7 .... 1937b. "Hermit crabs from the Gulf of California and the west coast of Lower California." No. 11 in: "The Templeton Crocker Expedition." Zoologica, Vol. 22: 241-263. No illustrations.

26 species of which 8 are new. Data per above.

REFERENCES

§ Q–8 .... 1938. "Three new anomuran crabs from the Gulf of California." A. Hancock Pac. Exp., Vol. 5 (1): 1-6.

Dredged 15-65 fathoms.

§ Q–9 Lockington, W.N. 1878. "Remarks upon the Porcellanidea of the west coast of North America." Ann. Mag. Nat. Hist. (5) Vol. 2: 394-406.

Eleven species from Lower California, mostly in the Gulf, 9 of them new.

§ Q–10 See also Anomura items in general papers on decapods, pp. 434-7 and on crustacea, pp. 415-8.

§ Q–11 ACKNOWLEDGMENT: For determining our ANOMURA material, for help with the literature and for other assistances, and for checking over this portion of the phyletic catalogue we have to thank Steve A. Glassell of Beverly Hills, California.

LIST OF SPECIES TAKEN:

SUPERFAMILY PAGURIDEA, FAMILY PAGURIDAE
(The Hermit Crabs)

§ Q–12 *Calcinus californiensis* Bouvier 1898
Glassell 1937b, p. 252, (§ Q–7), 13 specimens from Arena Bank, coral, 2½ fathoms, largest was 25 mm.; common in lower half of intertidal in various small shells, Gulf to Acapulco.

The common small red hermit crab, at Cape San Lucas, at Pulmo Reef, at Pt. Lobos on Espiritu Santo Island, east of La Paz (with *Clibanarius digueti*), and at San Carlos Bay (Baja California), hence mostly in the southern part of the Gulf. Alcoholic specimens are warm ivory in color. The two pig-clawed hands are unequal in size, broad, and naked.

§ Q–13 *Clibanarius digueti* Bouvier 1898
Bouvier 1898, p. 379 (§ O–3) 12 specimens from La Paz. Also listed without range, recent citations or illustration on p. 382, Schmitt 1924 (§ O–11) from San Carlos Bay, Sonora.

Another small red form, the commonest small hermit in the Gulf, especially in the upper part. Amortajada Bay (high up), Puerto Escondido, Pt. Marcial Reef, Coronado Island, Concepcion Bay, San Lucas Cove, San Francisquito Bay, Angeles Bay, south end Tiburon Island, Port San Carlos in Sonora, and at Gabriel Bay on Espiritu

REFERENCES

Santo Island. Alcoholic specimens are brick red in color. The two big-clawed hands are equal in size, elongate, and hairy.

§ Q–14 *Clibanarius panamensis* Stimpson (1859) 1862

PL. 16 FIG. 1

Toto-illustration as fig. 4, Pl. 47, Rathbun 1910 (§ M–14); Lower California to Peru, where it is used for food. Listed on p. 382, from Puerto Escondido by Schmitt 1924 (§ O–11).

Medium-sized hermit with longitudinal stripes on dactyls. From mangroves at El Mogote in Estero de la Luna, and at entrance to Agiabampo Estuary.

§ Q–15 *Dardanus sinistripes* (Stimpson) 1859.

Toto-illustration as fig. 2, Pl. 49, Rathbun 1910 (§ M–14), largest was 80 mm. Glassell 1937b. p. 251 (§ Q–7), Gulf to Peru; largest of the 45 *Zaca* specimens was 110 mm. from 10–60 fathoms.

A large species taken at San Lucas Cove, at Port San Carlos (or from the Jap shrimp trawlers at Guaymas, there may have been confusion of labeling here), and at Estero de la Luna.

§ Q–16 *Petrochirus californiensis* Bouvier 1895. PL. 12 FIG. 1

Glassell 1937b p. 251 (§ Q–7), Gulf to Ecuador; the Gulf material was mostly in *Phyllonotus nigritus,* 98 to 128 mm.

A giant, thick-clawed, chunky, red hermit, in shells of *Strombus galeatus, Fasciolaria,* etc. Taken at El Mogote, at Concepcion Bay, and at the entrance to Agiabampo Estuary.

§ Q–17 *Paguristes digueti* Bouvier 1892-93.

Glassell 1937b, p. 243 (§ Q–7), 5 specimens in shells of *Strombus,* etc. 7–33 fathoms, largest was 80 mm. Gulf only.

A large hermit taken in crab nets set at night in 7 fathoms at Concepcion Bay anchorage, along shore in Concepcion Bay, and at San Lucas Cove.

§ Q–18 *Pagurus albus* (Benedict) 1892.

Glassell 1937b, p. 258 (§ Q–7). The *Zaca* took one specimen from 3 fathoms. From previous collecting, Glassell notes it as a fast traveler, conspicuous in color, intertidal, but nowhere very common. Gulf only.

Our specimens were taken at San Lucas Cove, and at Estero de la Luna, as slender, yellow, medium-sized hermits in *Polinices* shells.

REFERENCES

§ Q–19 *Pagurus benedicti* Bouvier 1892.
Toto-illustration as fig. 1, Pl. 48, Rathbun 1910 (§ M–14), specimens up to 33 mm. recorded, but 15 mm. was the largest Peruvian form. P. 382, Schmitt 1924 (§ O–11) records 1 specimen from San José Island, 2 fathoms. P. 143, Boone 1931 (§ O–1) with detail fig. 2, records the range: Gulf, Panama, Peru, and Galápagos.

Taken as small hermits with *Clibanarius digueti* at San Lucas Cove.

§ Q–20 *Pagurus lepidus* (Bouvier) 1898.
Glassell 1937b, p. 256 (§ Q–7), 5 at Santa Inez shore, length of largest, 13.5 mm.

Ours were taken as small specimens at El Mogote. In a personal communication, Glassell notes, "the juveniles of this beast inhabit tiny *Dentalium* shells, thus starting out in life with an uncoiled abdomen . . ."

SUPERFAMILY GALATHEIDEA, FAMILY GALATHEIDAE (treated here, rather than with the Macrura which they resemble, for taxonomic consistency).

§ Q–21 *Pleuroncodes planipes* Stimpson 1860.
Good toto-illustration as fig. 2, Pl. 31, and description on p. 163, Schmitt 1921 (§ O–10). Distribution: open ocean, San Francisco to Cape San Lucas.

A brilliantly red, pelagic, lobster-like form. Found in great schools, with ovigerous females, about 70 miles N.W. of Pt. Lazaro in the Magdalena Bay region, its brilliant red color very striking against the blue of "tuna water." Carapaces found washed ashore in hordes in the Gulf, at Pt. Lobos on Espiritu Santo Island, and again near La Paz, March 1940. The turtles in the open ocean, and tuna which we examined in the Gulf were found to be feeding upon them exclusively.

FAMILY PORCELLANIDAE (porcelain crabs)

§ Q–22 *Pachycheles biocellatus* (Lockington) 1878.
Glassell 1937a p. 84, (§ Q–6), no illustration. The *Zaca* specimens were from 2½ fathoms, Arena Bank. General range: Lower California.

On Pulmo Reef, in coral interstices, with *Pachycheles panamensis*.

REFERENCES

§ Q–23 *Pachycheles panamensis* Faxon 1895. PL. 29 FIG. 2
Equals (according to personal communication from Glassell) *P. sonorensis* Glassell 1936 (§ O–6) p. 291, from Sonora; listed also in Glassell 1937a (§ Q–6) p. 86 from 2½ fathoms coral on Arena Bank. Faxon 1895, p. 71 (§ M–6) with toto-figures 2 and 2a, Pl. 15, 1 ovigerous female from Panama. P. 601 Rathbun 1910 (§ M–14), recorded from Panama, Bahamas, and Ecuador. Total range compiled from above: Gulf of California to Ecuador, Bahamas.

A very common form on Pulmo Reef, in interstices of coral.

§ Q–24 *Pachycheles setimanus* (Lockington) 1878.
Lockington 1878, p. 204 (§ Q–9), no illustration. Mulege and San José Island in the Gulf.

On Pulmo Reef in coral interstices, with *P. panamensis*. Taken also on or about Pinna at Concepcion Bay.

§ Q–25 *Pachycheles* sp. undetermined.

Minute porcellanids from Pulmo Reef, associated with the red coral crabs *Trapezia*.

§ Q–26 *Petrolisthes armatus* (Gibbes) 1850.
Boone 1931, p. 151 (§ O–1) with good toto-drawing as fig. 6, range: Florida to Brazil in the Atlantic; Lower California to Galápagos and Peru on the Pacific, and in the Indo-Pacific.

A "large-elbowed" porcelain crab taken at El Mogote and at Concepcion Bay.

§ Q–27 *Petrolisthes edwardsii* (Saussure) 1853. PL. 22 FIG. 2
The Boone 1931 (§ Q–27) item, p. 152, with toto-photograph as fig. 7, was incorrectly determined according to a personal communication from Glassell. Range: Gulf to Ecuador and Galápagos.

The big-clawed, very flat crab taken at Cape San Lucas.

§ Q–28 *Petrolisthes gracilis* Stimpson 1859 (1862). PL. 29 FIG. 4
Schmitt 1921, p. 181 (§ O–10) with toto-fig. 4, Pl. 32, recorded from Monterey, Catalina Island, and Guaymas.

REFERENCES

Schmitt 1924 (§ O–11), p. 385, 5 specimens from Tepoca and San Carlos Bay (Gulf).

San Carlos Bay (Baja California): Angeles Bay, Port San Carlos (Sonora) with *Petrolisthes* sp. (§ Q–32). With *hirtipes* among rocks at foot of mangroves in Gabriel Bay on Espiritu Santo Island.

§ Q–29 *Petrolisthes hirtipes* Lockington 1878. PL. 29 FIG. 3
Schmitt 1924, p. 383 (§ O–11), one female, Tepoca Bay.

Pt. Marcial Reef, San Carlos Bay (Baja California), Angeles Bay, Puerto Refugio on Angel de la Guardia Island. With *gracilis* among rocks at base of mangroves in Gabriel Bay on Espiritu Santo Island.

§ Q–30 *Petrolisthes hirtispinosus* Lockington 1878.
Glassell 1937a (§ Q–6) p. 80, no illustration. Gulf only.

Another big-clawed very flat crab, similar to *edwardsii*. Taken at Pt. Lobos on Espiritu Santo Island, and at Coronado Island.

§ Q–31 *Petrolisthes nigrunguiculatus* Glassell 1936. PL. 29 FIG. 5
Glassell 1936 (§ O–6) p. 382, from Santa Catalina Island in the Gulf.

Pt. Lobos on Espiritu Santo Island, Amortajada Bay fairly high up, Puerto Escondido, San Carlos Bay (Baja California), San Francisquito Bay.

§ Q–32 *Petrolisthes* sp.
Glassell MSS.

With *gracilis* at Port San Carlos (Sonora).

§ Q–33 *Pisonella tuberculipes* (Lockington) 1878.
Glassell 1938 (§ O–7) p. 440 with toto-fig. 1, Pl. 34; material examined was from upper part of Gulf. Range: Gulf to Ecuador.

Taken twice only; Coronado Island, a minute *Pachycheles*-like form removed from sponge mass; and at the south end of Tiburon Island.

§ Q–34 *Pisonella sinuimanus* (Lockington) 1878.
Glassell 1938 (§ O–7) p. 437, with toto-fig. 2, Pl. 34; material examined was from Puerto Escondido. Range: Gulf to Ecuador.

Pt. Lobos on Espiritu Santo Island. Puerto Escondido. San Gabriel Bay on Espiritu Santo Island (one of the vividly white crabs taken on the coral sand which is also vividly white at this point).

REFERENCES

§ Q–35 *Pisosoma lewisi* Glassell 1936.
Glassell 1936 (§ O–6), p. 287, no illustration, from the states of Jalisco and Guerrero on the Mexican west coast.

Taken once only, at Pt. Lobos on Espiritu Santo Island.

§ Q–36 *Pisosoma flagraciliata* Glassell 1937.
Glassell, 1937a (§ Q–6), p. 82, with toto-fig. 2, Pl. 1, from 2½ fathoms coral at Arena Bank.

Minute under-rock *Pachycheles*-like specimens at Cape San Lucas. Also attached to the sharp-spined urchin probably (§ K–316) *Centrostephanus* on Pulmo Reef.

§ Q–37 *Polyonyx quadriungulatus* Glassell 1935.
Glassell 1935a (§ O–5), p. 93, with toto fig. 1, Pl. 9, commensal in tubes of the worm *Chaetopterus* in estuary south of Ensenada, northwestern Lower California.

A big-clawed flat crab taken apparently free-living at El Mogote.

§ Q–38 *Porcellana cancrisocialis* Glassell 1936.
Glassell 1937a (§Q–5), p. 86, no illustration; commensal with hermit crabs, Magdalena Bay to Gulf.

Found in container of hermit crabs from Estero de la Luna (where *Dardanus sinistripes* (§ Q–15) and *Pagurus albus* (§ Q–18) were taken). Was described originally from the upper part of the Gulf, where it was commensal with the large *Petrochirus californiensis* (§ Q–16).

§ Q–39 *Porcellana paguriconviva* Glassell 1936.
Glassell 1936 (§ O–6), p. 293, no illustration, commensal with large hermit crab *Petrochirus* from upper part of Gulf.

Taken only once, at Concepcion Bay, where it appeared in the trays wherein we preserved the large hermit crabs *Paguristes digueti* (§ Q–17) taken in crab nets set at night at the 7 fathom anchorage.

§ Q–40 An indeterminable porcellanid juvenile was taken in a worm tube at Cape San Lucas.

§ Q–41 We took also an unnamed porcellanid with *Mithrax*, etc. at Pulmo Reef.

SUPERFAMILY HIPPIDEA, FAMILY ALBUNEIDAE

REFERENCES

§ Q–42 *Albunea lucasii* (Saussure) 1853.
Saussure 1853 (§ M–15), p. 367, as *Albuminea lucasia,* with good toto-illustration as fig. 4, Pl. 12, from Mazatlan.
Burrowing in sand of intertidal beach, San Lucas Cove, B. C.

FAMILY HIPPIDAE

§ Q–43 *Emerita rathbunae* Schmitt 1935. PL. 32 FIG. 5
Schmitt 1935 (§ O–13) p. 214, with good toto-figs. 77a and b, from Pacific coast of Panama, ranging from La Paz to Capon, Peru, and formerly called *E. emerita.*
Juvenile and adult typical "sand bugs" burrowing in the beach sand of San Francisquito Bay, fairly high up.

SUMMARY of the ANOMURA

§ Q–44 32 species of littoral Anomura, including juveniles and indeterminable forms, but excluding the Thalassinids, were collected in the Gulf. 9 were hermit-crabs, 1 was a pelagic rock lobster, 20 were porcelain crabs, and 2 were sand bugs. Of this total, the geographic range of 16 species is restricted to the Gulf only; 11 range from there in a southerly extension, 4 to the north, and 1 both north and south, forming a closely knit group geographically.

Considered by taxonomic divisions, 4 of the hermits are restricted to the Gulf, and 5 extend south, 1 to Acapulco, the remainder clear on to Ecuador or northern Peru. The Galatheid ranges north to San Francisco (but rarely; there have been only one or two inshore reports of these north of Lower California for nearly a hundred years). The porcellanid fauna, well developed on Pacific American shores generally, is extraordinarily well developed in the Gulf, both as to species and numbers. Porcelain crabs are literally everywhere, but mostly in small species, several of which are commensal with worms, hermit-crabs, urchins, etc. There are 4 species of *Pachycheles* and 7 of *Petrolisthes* which are common enough to have been turned out in our short survey of the intertidal. Of the total 20, 11 are restricted to the Gulf (or are known so inadequately as to preclude the compilation of their range), 5 (?-1) extend southwardly, 1 (?+1) both north and south,

REFERENCES

and 3 north of the Gulf, one (*Petrolisthes gracilis*) ranging clear to Monterey, the others limited by Estero de Punta Banda (south of Ensenada), and Magdalena Bay respectively. 2 occur elsewhere also, *Pachycheles panamensis* being reported from the Bahamas, and *Petrolisthes armatus* from Indo-Pacific and West Indian waters. Both the sand crabs are strictly Panamic, one being restricted to the Gulf, the other ranging to northern Peru.

Some of the Anomura are among the most common and characteristic Gulf inhabitants. Of the 7 species rated as very common or abundant, *Calcinus californiensis* and *Clibanarius digueti* (small hermits, the latter one of the commonest invertebrates in the Gulf), the pelagic *Pleuroncodes,* the porcelain crabs, *Petrolisthes gracilis, P. hirtipes* and *P. nigrunguiculatus,* and the sand bug *Emerita rathbunae,* 2 are primarily northerly or north ranging, 2 are south ranging, and 3 are restricted to the Gulf.

From the above, but particularly from the zoogeographic analysis of all the 32 species taken, it will be seen that the Gulf Anomura assemblage is rather strictly Panamic, and fairly unique. Of the 32 total, half are restricted to the Gulf, and only 4 occur outside Panamic waters as defined in this work, the pelagic form ranging rarely to San Francisco, one of the porcellanids ranging north to Monterey, one occurring in the West Indies and in the Indo-Pacific, so that it can be considered tropicopolitan, and the third having been reported from the Bahamas.

TRIBE BRACHYURA

§ R–1 The Rathbun monograph in 4 parts, issued separately in 1918 (§ R–15), 1925 (§ R–18), 1930 (§ R–19), and 1937 (§ R–22), will be considered as the key paper, although it has no bibliography other than in the synonymy citations. Other papers referring to the Brachyura are cited only when they consider species described more recently than these four parts. Of previous work, the A. Milne-Edwards 1875-1881 "Études sur les Xiphosures et les Crustacés podophthalmaires," Mission Scientifique au Mexique . . . (5) Vol. 1, perhaps should be specifically mentioned for the magnificent color plates which figure,

REFERENCES

usually lifesize, most of the Brachyura known at that time from Panamic waters. All the data, however, will have been abstracted into the Rathbun monographs.

§ R–2 Boone, Lee 1926. "The littoral crustacean fauna of the Galápagos Islands. Pt. I, Brachyura." Zoologica, Vol. 8 (4): 127-288, figs. 34-102c incl.

46 species of which 5 are new. Brachyuran situation in the Galápagos brought to date by the inclusion of descriptions and toto-illustrations of all 69 species known to inhabit the region from shore to 100 fathoms. Data on habits, color, diagnostic characters, local and general distribution. Bibliography includes Rathbun 1924 report on the rather unimportant Williams Galápagos collection. See also corrections in Glassell 1934 (§ M–5).

§ R–3 .... 1929. "A collection of Brachyuran Crustacea from the Bay of Panama . . ." Bull. Am. Mus. Nat. Hist., Vol. 58: 561-583, 18 text figs.

17 marine and 1 fresh-water species, all well illustrated by toto-photographs. Data on color, local and general distribution, with some collecting notes. See also corrections in Glassell 1934 (§ M–5).

§ R–4 Contreras, Francisco 1930. "Contribución al conocimiento de las jaibas de Mexico." An. Inst. Biol. Univ. Mex., Vol. 1: 227-241, 11 figs.

4 Pacific coast species including the new *Callinectes ochoterenai* from La Paz and Guaymas. Dimensions, anatomical descriptions, and distribution notes. Toto-illustrations of each species.

§ R–5 Crane, Jocelyn 1937. "Brachygnathous crabs from the Gulf of California and the west coast of Lower California." No. 3 in "The Templeton Crocker Expedition." Zoologica, Vol. 22: 47-78, 8 pls.

73 species, including 1 new species described herein and 6 described by Glassell 1936a (§ R–14). Notes on general and local distribution, color, food, breeding, etc. Toto-illustrations of 14 species.

§ R–6 .... 1937a. "Oxystomatous and dromiaceus crabs from the Gulf of California and the west coast of Lower California." No. 6 in "The Templeton Crocker Expedition." Zoologica, Vol. 22: 97-108, 2 pls.

17 species with notes on distribution, color, habits, breeding, food, etc.

§ R–7 .... 1940. "On the post-embryonic development of brachyuran crabs of the genus *Ocypode*." No. 18 in "Eastern Pacific Expeditions of the N.Y. Zool. Soc." Zoologica, Vol. 25: 65-82.

REFERENCES

Descriptions of the zoea and megalops of *O. gaudichaudii*, and of the megalops of *O. occidentalis*. Consideration of the relation between ecology and structure in these larvae.

§ R–8 Finnegan, Susan 1931. "Report on the Brachyura collected in Central America, the Gorgona and Galápagos Islands by Dr. Crossland on the *St. George* Expedition to the Pacific, 1924-25." Journ. Linn. Soc. London, Zool., Vol. 37: 607-673, 2 charts, 6 text figs.

65 species, 8 new, with 1 new variety of which only 7 were collected on the Atlantic side of the Isthmus; some ecological information, depth, etc. Zoogeographical considerations start on p. 653. She rates the Panama region as extending from Panama to Cape San Lucas; 3 northern areas: from Cape San Lucas to San Diego; San Diego to Monterey Bay, and Monterey Bay to Puget Sound; and 3 southern areas: from Panama to Guayaquil, Guayaquil to Chinchas Island, and Chinchas Island to Valparaiso. Fine bibliography.

§ R–9a Garth, John S. 1939. "New brachyuran crabs from the Galápagos Islands." A Hancock Pac. Exp., Vol. 5 (2): 9-48 incl. 10 pls.

3 new species in the MAJIDAE, 4 in the XANTHIDAE, 1 in the PINNOTHERIDAE and 1 in the CYMOPOLIIDAE; all with good toto-illustrations. Descriptions and figures for differentiating the several confused *Glyptoxanthus* species. Bibliography.

§ R–9b .... 1940. "Some new species of brachyuran crabs from Mexico and the Central and South American mainland." A. Hancock Pac. Exp., Vol. 5 (3): 53-126 incl. 16 pls.

New species as follows: 1 Leucosid, 1 Calappid, 4 Majids, 2 Parthenopids (1 described by Glassell), 4 Xanthids, and 3 Goneplacids (1 described by Ziesenhenne). Redescription of Stimpson's 1871 *Portunus acuminatus* which is removed from synonymy. Good toto-figures of all 16 species, by Anker Petersen, who so ably illustrated many of Glassell's papers. Descriptions and detail figures to illustrate subtle or heretofore unrecognized differentials between closely related species of *Osachila, Mithrax,* and *Macrocoeloma*. All except three or four occurred only in dredge hauls from 3 to 165 fathoms. Bibliography.

§ R–10 Glassell, Steve A. 1933. "Notes on *Parapinnixa affinis* Holmes and its allies." Trans. San Diego Soc. Nat. Hist., Vol. 7 (27): 319-330, 2 pls., 2 text figs.

Holmes original description reprinted. A pea-crab from southern California—which may represent its northern limit as a possible Panamic form—commensal in *Amphitrite* tube. A redescription also of Lockington's *P. nitida,* from specimens found at San Felipe in the Gulf, and at Magdalena Bay. Good toto-drawings.

REFERENCES

§R–11 .... 1933a. "Description of five new species of Brachyura collected on the west coast of Mexico." Trans. San Diego Soc. Nat. Hist., Vol. 7: 331-344, 5 pls.

Littoral to 15 fathoms, Magdalena Bay and Gulf of California. One each, Leucosid, Grapsid, Majid, Xanthid, and Pinnotherid. Fine toto-drawings of each species.

§R–12 .... 1934. "Affinities of the brachyuran fauna of the Gulf of California." Journ. Wash. Acad. Sci., Vol. 24 (7): 296-302.

Geographic distribution lists of 197 spp. of Brachyura (excluding oxystomatous and dromiaceus crabs) known from the Gulf; 48% of which are intrusions from waters south of a line drawn from Cape San Lucas to Mazatlan; 40% are indigenous; and 12% are intrusions from waters north of Magdalena Bay.

§R–13 .... 1935. "Three new species of *Pinnixa* from the Gulf of California." Trans. San Diego Soc. Nat. Hist., Vol. 8: 13-14.

Shore at San Felipe. Preliminary descriptions only; no figures.

§R–14 .... 1936a. "Six new brachyuran crabs from the Gulf of California." No. 1 in "The Templeton Crocker Expedition." Zoologica, Vol. 21: 213-218.

Dredged, 33 to 50 fathoms, 2 each FAMILIES MAJIDAE and XANTHIDAE, 1 each GONEPLACIDAE and CYMOPOLIIDAE. No illustrations.

§R–15 Rathbun, Mary J. 1918. "The grapsoid crabs of America." Bull. U.S.N.M., Vol. 97: 461 pp., 161 pls., text figs.

Part 1 of a four part monograph on American Brachyura. FAMILIES GONEPLACIDAE, PINNOTHERIDAE, CYMOPOLIIDAE, GRAPSIDAE, and OCYPODIDAE. About 76 spp. can be rated as Panamic forms, usually with good toto-photographs. Transcript on p. 228, of Catesby's remarks on *Grapsus grapsus*, published originally some 200 years ago.

§R–16 .... 1923. "The brachyuran crabs collected by the U. S. Fisheries Steamer *Albatross* in 1911, chiefly on the west coast of Mexico." Bull. Am. Mus. Nat. Hist., Vol. 48 (20): 619-637, 11 pls., text figs.

56 species, including the Cancroid *Pilumnus townsendi* n. sp., and 2 new Pinnotherids. Good toto-photographs of these and of 4 other forms previously not well known. Toto-drawings of several larvae.

§R–17 .... 1924a. "Crustacea (Brachyura)." No. 35 in "Expedition of the California Academy of Sciences to the Gulf of California in 1921." Proc. Calif. Acad. Sci. (4), Vol. 12 (23): 373-379.

REFERENCES

26 species with station data. One, No. 12, is to be described elsewhere. No illustrations, usually no descriptions.

§ R–18 .... 1925. "The spider crabs of North America." Bull. U.S.N.M., Vol. 129: 613 pp., 283 pls., text figs.

Pt. 2 of a four-part monograph on American Brachyura. FAMILIES MAJIDAE, PARTHENOPIDAE, and HYMENOSOMENIDAE. About 91 species plus 5? can be rated as Panamic forms. Good toto-photographs of nearly all species treated.

§ R–19 .... 1930. "The Cancroid crabs of America." Bull. U.S.N.M., Vol. 152: 609 pp., 239 pls., text figs.

Pt. 3 of a four-part monograph on American Brachyura. FAMILIES EURYALIDAE, PORTUNIDAE, ATELECYCLIDAE, CANCRIDAE, and XANTHIDAE. About 99 plus 5? species can be considered as Panamic.

§ R–20 .... 1933. "Description of new species of crabs from the Gulf of California." Proc. Biol. Soc. Wash., Vol. 46: 147-150.

5 new species of XANTHIDAE and one of DROMIIDAE collected along shore at San Felipe by H. N. Lowe. No illustrations.

§ R–21 .... 1935. "Preliminary descriptions of six new species of crabs from the Pacific coast of America." Proc. Biol. Soc. Wash., Vol. 48: 49-52.

A Xanthid, a Goneplacid, a Pinnotherid, and 3 spp. of *Uca* from northern Peru, Ecuador, and Panama.

§ R–22 .... 1937. "The Oxystomatous and allied crabs of America." Bull. U.S.N.M., Vol. 166: 278 pp., 86 pls., text figs.

Pt. 4 of the four-part monograph on American Brachyura. 11 families including 1 additional species in the FAMILY GONEPLACIDAE (see Pt. 1, Rathbun 1918). About 52 species can be considered as Panamic forms, 29 known from the Gulf, or whose total range includes the Gulf area.

§ R–23 See also Brachyura items in papers listed under Crustacea, pp. 415-8 and under Decapoda, pp. 434-7.

LIST OF SPECIES TAKEN:

DIVISION I: Oxystomatous and related crabs (per Rathbun 1937, § R–22).

SUBTRIBE DROMIACEA

§ R–24 *Dromidia larraburei* Rathbun 1910.
Rathbun 1937 (§ R–22), good totos as figs. 4 and 5, Pl. 7:

REFERENCES

Monterey Bay to Peru and Galápagos, low tide to 60 fathoms, but reported only twice north of U. S.-Mexico boundary.

Pelagic larvae at night via light overside at anchorage, San Lucas Cove, March 29. Small specimens in tide pools, Puerto Refugio on Angel de la Guardia, with occasional adults.

§ R–25 *Hypoconcha panamensis* Smith in Verrill 1869.

PL. 21 FIG. 4

Rathbun 1937 (§ R–22), p. 47, good totos as figs. 6 and 7, Pl. 9, Mexico to Peru, 3 to 60 fathoms. None of Rathbun's records are north of the Gulf.

Collected only once, in crab net on bottom at night, at 7 fathoms anchorage in Concepcion Bay, where half a dozen specimens of this curious reddish, furry crab were taken in half shells of the clam *Chione*.

SUBTRIBE OXYSTOMATA

§ R–26 *Hepatus kossmanni* Neumann 1878.

Rathbun 1937 (§ R–22), p. 239, good toto-illustrations as figs. 3 and 4, Pl. 72, west coast of Mexico to Ecuador, 2 to 25 fathoms. There is only one record cited by Rathbun north of Sinaloa, 1 male at Abrojos Pt., Pacific coast of Lower California.

This fairly large species, one of the few large crabs excepting *Callinectes* found in the entire area, superficially bears a striking resemblance to a *Cancer*. Only a single specimen was procured; this was a donation from Capt. Corona of Guaymas, who operates a fleet of shrimp dredgers in the shallow water off the mouth of the River Mayo.

DIVISION II: *SUBTRIBE BRACHYGNATHA*, SUPERFAMILY OXYRHYNCHA, not listed alphabetically but in the order in which they appear in Rathbun, 1925, "The Spider Crabs of America." (§ R–18)

FAMILY MAJIDAE, SUBFAMILY INACHINAE

§ R–27 *Stenorynchus debilis* (Smith) 1871.

Rathbun 1925, p. 18, toto-illustrations on Pls. 4 and 5. Magdalena Bay through Gulf to Panama, Galápagos, and Chile. Shore to 31 fathoms.

The greatly attenuated spider crab taken on the rocks at the south end of Tiburon Island, just below extreme low tide level.

REFERENCES

§ R–28 *Podochela latimanus* (Rathbun) 1893.
Rathbun 1925, p. 56, toto-illustrations on Pl. 21. Gulf, to 11 fathoms.

El Mogote, Puerto Escondido. From Sponge cluster of *Tedania* on slightly subtidal rock at Concepcion. Another very attenuated form.

§ R–29 *Eucinetops lucasii* Stimpson 1860.
Rathbun 1925, p. 86, no illustration. From Cape San Lucas only.

Concepcion Bay, associated with the hatchet clam *Pinna*. Puerto Refugio. A masking crab.

§ R–30 *Eucinetops panamensis* Rathbun 1923.
Rathbun 1925, p. 87, toto-figures 3 and 4, Pl. 23. Gulf to Panama.

San Carlos Bay, Baja California.

Subfamily Acanthonychinae

§ R–31 *Epialthus minimus* Lockington 1876 (1877).
Rathbun 1925, p. 155, toto-fig. 1, Pl. 47. Gulf only, shore to 4½ fathoms.

Coronado Island. Associated with the hatchet clam *Pinna* at Concepcion.

Subfamily Majinae

§ R–32 *Thoë sulcata* Stimpson 1860. PL. 24 FIG. 2
Rathbun 1925, p. 349, toto-figs. 3 and 4, Pl. 125. Gulf to Oaxaca.

One of the small coral living crabs taken at Pulmo Reef. Also occurring at Pt. Lobos on Espiritu Santo, in the coral heads east of La Paz, at El Mogote, and at Cape San Lucas on sponge mass below 0.0′ level.

§ R–33 *Pitho picteti* (Saussure) 1853.
Rathbun 1925, p. 359, toto-figs. 2 and 3, Pl. 130; and 1, Pl. 252. West coast of Mexico and Central America.

Taken only at El Mogote.

§ R–34 *Pitho sexdentata* Bell 1835 (1836).
Rathbun 1925, p. 367, toto-fig. 1, Pl. 130. Cape San Lucas and Galápagos.

REFERENCES

Pt. Marcial Reef, San Gabriel Bay on Espiritu Santo (one of the vividly white crabs taken at this point on the vividly white coral sand), and at Coronado Island as a decorator crab.

§ R–35 *Anaptychus cornutus* Stimpson (1860) 1862.
Rathbun 1925, p. 378, toto-figs. 4 and 5, Pl. 134, and fig. 1, Pl. 254. Gulf to Panama.

Pt. Marcial Reef. Associated with the hatchet clam *Pinna* at Concepcion.

§ R–36 *Mithrax areolatus* (Lockington) 1876 (1877). PL. 29 FIG. 1
Rathbun 1925, p. 433, toto-fig. 1, Pl. 154. San Diego (doubtful), Gulf to Panama.

The commonest small decorator or spider crab encountered in the Gulf littoral. A curiously etched form. This was the sluggish, tightly-gripping, small spider crab found abundantly in interstices of coral at Pulmo Reef under rocks at Pt. Lobos on Espiritu Santo, in the dead coral heads east of La Paz, at El Mogote, and at Coronado Island.

§ R–37 *Mithrax sonorensis* Rathbun in Glassell 1933.
Glassell 1933a, p. 338 (§ R–11) good toto-drawing as fig. 1 on Pl. 24, San Pedro and Miramar, Sonora.

A decorator crab taken only at one locality, Coronado Island.

§ R–38 *Teleophrys cristulipes* Stimpson 1860.
Rathbun 1925, p. 441, toto-figs. 1, 2, and 7, Pl. 159; and fig. 7, Pl. 262. Cape San Lucas to Panama.

Taken only once, at El Mogote.

§ R–39 *Microphrys platysoma* (Stimpson) 1860. PL. 32 FIG. 6
Rathbun 1925, p. 497, toto-figs. 1 and 2, Pl. 176. Lower California (probably not known north of the Gulf) to Panama.

Under rock at Pt. Lobos, Espiritu Santo. East of La Paz, El Mogote, Coronado Island, Concepcion Bay, Puerto Refugio. Among rocks at foot of mangroves at San Gabriel Bay, Espiritu Santo Island. At San Gabriel Bay this also was one of the vividly white crabs on the dazzling white coral sand. Another of the small spider crabs very common throughout the Gulf.

REFERENCES

DIVISION III: (*SUBTRIBE BRACHYGNATHA* continued). SUPERFAMILY BRACHYRHYNCHA, Section "A," the first five families, known as Cancroid or Cyclometopous crabs, not listed alphabetically, but in the order in which they appear in Rathbun 1930, "The Cancroid Crabs of America," Bull. 152, U.S.N.M. (§ R–19).

FAMILY PORTUNIDAE

§ R–40 *Portunus minimus* Rathbun 1898. PL. 18 FIG. 3
Rathbun 1930, p. 76, toto-figures on Pl. 36. Gulf to Tres Marias.

Taken only once, at night, but then in considerable numbers, free swimming at Port San Carlos, Sonora, attracted to a light hung overboard at the anchorage.

§ R–41 *Portunus pichilinquei* Rathbun 1930.
Rathbun 1930, p. 78, toto-figures on Pl. 37. Magdalena Bay to head of Gulf.

Taken at Concepcion Bay during the night, via crab net cast overboard at 7 fathom anchorage. Also taken during the night, by light hung overboard at Puerto Refugio, free swimming at the surface.

§ R–42 *Callinectes bellicosus* Stimpson 1859. PL. 14 FIG. 2
Rathbun 1930, p. 112, toto-figures on Pl. 49. Pt. Loma, California (only record north of San Bartolomé Bay) into Gulf.

The commonest (or possibly next to Sally Lightfoot, the commonest) large crab in the Gulf. Found at practically every suitable station, especially in bays and estuaries, and on sand flats, from La Paz on the south to Port San Carlos on the north Sonora shore, thence as far south on the mainland as we collected (Agiabampo). Taken only once with other species of swimming crabs with which it might be confused: at Concepcion Bay during the night via crab net cast overboard in 7 fathoms, with *Portunus pichilinquei.* An aggressive and rapidly swimming animal rarely or never found except submerged. Pinching claws brilliant ultramarine. Large specimens may be encrusted with barnacles and hydroids. Ordinarily very abundant along the sandy shore in a few inches of water, juveniles attracted only once to light hung overboard at night. Used as food at Guaymas. Specimens from the Sonoran estuaries carried a new subspecies of the barnacle *chelonibia patula* (§ N–19), per illustration, and, in one case, *Balanus trigonus* (§ N–12).

Family Cancridae

REFERENCES

§ R–43 *Platypodia rotundata* (Stimpson) 1860.
Rathbun 1930, p. 248, toto-figs. 1-3, Pl. 102. La Paz to Ecuador.

A minute furry crab, very "un-*Cancer*-looking," taken at Pulmo Reef, at Pt. Lobos on Espiritu Santo, at Puerto Escondido associated with Zoanthidean anemone clusters, and under rock at Angeles Bay.

Family Xanthidae

§ R–44 *Glyptoxanthus meandricus* (Lockington) 1876.
Removed from synonymy with *labyrinthicus* (Stimpson) 1860 (where it was placed by Rathbun 1930, p. 266, toto-figs. 1-3 on Pl. 108, west coast of Mexico to Galápagos), by Garth 1939 (§ R–9a), p. 17, with comparative detail drawings on Pl. 4 and 5. Glassell also notes, in personal correspondence that *G. felipensis* Rathbun is identical to *meandricus* of Lockington.

A very spectacularly sculptured small *Lophopanopeus*-like form taken only once, at San Carlos Bay, Baja California.

§ R–45 *Daira americana* Stimpson 1860. PL. 18 FIG. 5
Rathbun 1930, p. 268, toto-figs. 1 and 2, Pl. 110. Lower California to Ecuador.

A black, knobby crab common at Pulmo Reef and at El Mogote.

§ R–46 *Leptodius occidentalis* (Stimpson) 1871.
Rathbun 1930, p. 301, toto-figures 3 and 4, Pl. 137, and fig. 2, Pl. 138. Magdalena Bay to Gulf to Panama to Galápagos.

A very common medium-sized brown to fawn-colored *Lophopanopeus*-like crab with black tipped big claws. East of La Paz, El Mogote, San Carlos Bay, Puerto Refugio, Port San Carlos. Among rocks at base of mangroves, Gabriel Bay, Espiritu Santo Island.

§ R–47 *Leptodius cooksoni* Miers 1877.
Rathbun 1930, p. 310, toto-figures on Pl. 142. Clarion Island to Galápagos and Chile.

Pt. Lobos on Espiritu Santo Island. Puerto Escondido. San Carlos Bay, Baja California.

REFERENCES

§ R–48 *Xanthodius hebes* Stimpson 1860.
Rathbun 1930, p. 313, toto-figures on Pl. 147. Magdalena Bay and Gulf to Maria Madre Island.

Another black-clawed *Lophopanopeus*-like and slow-moving crab. High up at Amortajada Bay. Pt. Marcial Reef. Among rocks at foot of mangroves, Gabriel Bay, Espiritu Santo Island.

§ R–49 *Panopeus bermudensis* Benedict and Rathbun 1891.
Rathbun 1930, p. 360, toto-figures on Pl. 165. Florida to Bermuda and Brazil in the Atlantic. On the Pacific, Magdalena Bay to Peru.

El Mogote; Puerto Escondido.

§ R–50 *Eurypanopeus planissimus* (Stimpson) 1860.
Rathbun 1930, p. 421, toto-figs. 1 and 2, Pl. 175. Gulf to Maria Madre Island.

Angeles Bay. Puerto Refugio. South end of Tiburon Island.

§ R–51 *Micropanope areolata* Rathbun 1898.
Rathbun 1930, p. 450, toto-figs. 1 and 2, Pl. 182. Southern California and Gulf.

Coronado Island.

§ R–52 *Pilumnus gonzalensis* Rathbun 1893. PL. 26 FIG. 1
Rathbun 1930, p. 505, toto-figs. 3 and 4, Pl. 204. Gulf only.

Coronado Island. San Carlos Bay, Baja California. Angeles Bay. Puerto Refugio. South end of Tiburon Island.

§ R–53 *P. pygmaeus* Boone 1927.
Rathbun 1930, p. 515, toto-figs. 4 and 5, Pl. 207. Galápagos.

A small fuzzy blue-gray crab removed from sea-fan at Pulmo Reef.

§ R–54 *Pilumnus townsendi* Rathbun 1923. PL. 26 FIG. 2
Rathbun 1930, p. 504, toto-fig. 1, Pl. 202, figs. 1 and 2, Pl. 204. Magdalena Bay to Manzanillo via Gulf. Galápagos?

A fuzzy or hairy black-clawed crab taken East of La Paz, at El Mogote, at Coronado Island, and associated with the hatchet clam *Pinna* at Concepcion Bay.

REFERENCES

§ R–55 *Heteractaea lunata* (Milne-Edwards and Lucas) 1843. Rathbun 1930, p. 532, toto-fig. 1, Pl. 212, and on Pl. 214. San Diego to Valparaiso. Chile.

A medium-sized black-clawed knobby crab taken only once, in the coral heads east of La Paz.

§ R–56 *Pilumnoides.* sp. indeterminable, juvenile. Genus on p. 534, Rathbun, 1930.

Cape San Lucas, under rocks.

§ R–57 *Ozius tenuidactylos* (Lockington) 1876 (1877). Glassell 1935a (§ O–5), p. 104, corrected name. Equals *O. agassizii* Milne-Edwards 1880 on p. 554, Rathbun 1930, toto-figs. 3 and 4, Pl. 221. Costa Rica to Ecuador.

A smooth, yellow-brown, sluggish, *Lophopanopeus*-like crab taken twice, from Pt. Lobos on Espiritu Santo Island, and abundantly among rocks at roots of mangroves on the opposite side of the island, at Gabriel Bay.

§ R–58 *Eriphia squamata* Stimpson 1859. PL. 18 FIG. 4
Rathbun 1930, p. 550, toto-figs. on Pl. 223, and fig. 1, Pl. 224. Magdalena Bay to Chile.

Pulmo Reef. Pt. Lobos, Espiritu Santo Island. San Carlos Bay, Baja California. Angeles Bay. Among rocks at foot of mangroves, Gabriel Bay, Espiritu Santo Island.

§ R–59 *Eriphides hispida* (Stimpson) 1860. Rathbun 1930, p. 552, toto-figs. on Pls. 225 and 226, west coast of Costa Rica and Panama. Galápagos.

Pulmo Reef in interstices of coral.

§ R–60 *Trapezia cymodoce ferruginea* Latreille 1825. PL. 18 FIG. 6
Rathbun 1930, p. 557, toto-figs. 2 and 3, Pl. 228. Clarion Island and Acapulco to Panama and Galápagos. Red Sea to Indo-Pacific.

One of the cherry-colored coral crabs taken at Pulmo.

§ R–61 *Trapezia digitalis* Latreille 1825. Rathbun 1930, p. 559, toto-figs. 5 and 6, Pl. 228. Cape San Lucas to Panama. Red Sea to Indo-Pacific.

Taken with the above at Pulmo, very common among the coral interstices.

REFERENCES

DIVISION IV: Grapsoid (or Catometopous) crabs.

(*SUBTRIBE BRACHYGNATHA* continued). SUPERFAMILY BRACHYRHYNCHA, section "B," the last 6 families. Not listed alphabetically but in the order in which they appear in Rathbun 1918 (§ R–15), "The Grapsoid Crabs of America." Bull. 96, U.S.N.M.

FAMILY PINNOTHERIDAE

§ R–62 *Dissodactylus nitidus* Smith 1870.
Rathbun 1918, p. 116; with toto-illustrations as figs. 6 and 7, Pl. 26; from Santa Maria Bay, Lower California to Peru, to 5½ fathoms.

We took our representatives at Concepcion Bay, on the sand dollars *Encope grandis* and *E. californica*, and, as a minute Grapsoid, found in miscellaneous Concepcion Bay material.

§ R–63 *Dissodactylus xantusi* Glassell 1936. PL. 29 FIG. 6
Glassell 1936 (§ O–6) p. 299; no toto-illustration; from many points in the Gulf, commensal on sand dollars in association with *D. nitidus*.

As above, Concepcion Bay on sand dollars.

§ R–64 *Parapinnixa nitida* (Lockington) 1876 (1877).
Rathbun 1918, p. 107, text fig. 58. Glassell 1933, p. 324, (§ R–10) toto-figure 1a. Magdalena Bay plus three stations in the Gulf.

We took this form only once, free swimming at night, attracted to anchored boat by light hung overside, Port San Carlos, Sonora.

§ R–65 *Pinnixa transversalis* (Milne-Edwards and Lucas) 1843.
Rathbun 1918, p. 131, toto-figs. 1-3, Pl. 29. Panama to Patagonia, in *Chaetopterus*-like tubes.

Commensal with sandy-tubed worm on sand flat, Angeles Bay.

§ R–66 *Pinnotheres* sp. Glassell MSS.

94 pea-crabs taken at night by light overboard at anchorage, Concepcion Bay.

FAMILY GRAPSIDAE

REFERENCES

§ R–67 *Grapsus grapsus* (Linnaeus) 1758. Sally Lightfoot.
Rathbun 1918, p. 227, excellent toto-photographs as Pls. 53 and 54 (equals Stimpson's 1860 *G. altifrons* from Cape San Lucas, and Milne-Edwards's 1853 *G. ornatus* from Chile). This, with a subspecies, is cosmopolitan in the tropics; and occurs on the Pacific coast from San Benito Island, Lower California, to Chile. Many times illustrated elsewhere.

Abundant at all suitable rocky points visited in the Gulf; commonest well above the water line on cliffy and bouldery shores especially where there is wave shock. A tremendously agile crab; very difficult to capture despite its great abundance; the commonest large crab in the Gulf, and one of the commonest invertebrates in the region.

§ R–68 *Geograpsus lividus* (Milne-Edwards) 1837. PL. 18 FIG. 2
Rathbun 1918, p. 232, excellent toto-photograph as Pl. 55 (equals Stimpson's 1860 *G. occidentalis* from Cape San Lucas). Atlantic tropical America; Cape Verde Islands; Hawaii; on the Pacific coast from Lower California (La Paz) to Chile.

This is the very plentiful, active, and belligerent high-intertidal crab found among rocks associated with *Goniopsis pulchra* at the base of mangrove roots at Gabriel Bay on Espiritu Santo Island; and again on the high rocks at San Francisquito Bay with *Pachygrapsus crassipes.*

§ R–69 *Goniopsis pulchra* (Lockington) 1876 (1877).
Rathbun 1918, p. 239, excellent toto-photograph as Pl. 59. Magdalena Bay to Peru.

Similar to *Geograpsus* (but with slimmer big claws), with which it was taken among rocks at the base of mangrove roots, Gabriel Bay on Espiritu Santo Island.

§ R–70 *Pachygrapsus crassipes* Randall 1839 (1840).
Rathbun 1918, p. 241, toto-photographs on Pl. 59. Range: Oregon to Gulf; Galápagos, Chile; Japan and Korea. The commonest shore crab along the central Californian coast.

Taken only once, at San Francisquito Bay in the northern part of the Gulf, with *Geograpsus lividus.*

REFERENCES

§ R–71 *Pachygrapsus transversus* (Gibbes) 1850.
Rathbun 1918, p. 244; toto-photographs as figs. 2 and 3, Pl. 61 (equals Stimpson's 1871 *P. socius* from several stations in the Peru to Cape San Lucas stretch). Cosmopolitan in the tropics; on the Pacific from California (only 1 doubtful report) and Pichilinque Bay near La Paz, to Peru.

A small grapsoid taken under rock at Cape San Lucas, and, as a black grapsoid, from rocks at the base of mangroves, at Gabriel Bay on Espiritu Santo Island, and with miscellaneous small crabs at Pulmo Reef.

§ R–72 *Planes minutus* (Linnaeus) 1758.
Turtle Crab. Gulf Weed Crab.
Rathbun 1918, p. 253, excellent toto-photographs as Pl. 63; cosmopolitan in the tropics.

Male and female taken from flipper of sea turtle south of Pt. Abrojos, west coast of Lower California.

§ R–73 *Sesarma sulcatum* Smith 1870.
Rathbun 1918, p. 289; fig. 3, Pl. 78 is toto-photograph. La Paz to Panama.

A hairy-legged, dark, *Pachygrapsus*-like form occurring with *Uca* along the banks of a bitter-water pond a few rods inland at Concepcion Bay. Very alert and hard to catch.

§ R–74 *Percnon gibbesi* (Milne-Edwards) 1853. PL. 21 FIG. 6
Rathbun 1918, p. 337; Pl. 105 is toto-photograph. Tropical Atlantic. Pacific Coast from Cape San Lucas to Chile.

A very characteristic, highly flattened, rock crab, very common in the middle intertidal at Cape San Lucas.

FAMILY OCYPODIDAE

§ R–75 *Ocypode occidentalis* Stimpson 1860. PL. 14 FIG. 1
Rathbun 1918, p. 372; Pl. 129, figs. 2 and 3. From Turtle Bay to Peru.

The very stalk-eyed, light-colored ghost crabs found running about on the shore high up, and burrowing in the sand around the entrance of Agiabampo Estuary, Sonora. Clear cut, having a characteristic sandy texture more apparent to the eye than to the touch.

§ R–76 *Uca crenulata* (Lockington) 1876 (1877).
Rathbun 1918, p. 409; toto-photographs as Pl. 146. San Diego to Mazatlan and Gulf. Common in southern Cali-

REFERENCES

fornia; there is a fair photo as fig. 1, Pl. XL, Ricketts and Calvin 1939 (§ Y–3). In a personal communication, Glassell notes the northern limit as Magu slough, north of Malibu Beach.

The very common fiddler crab, and the only one we encountered in the Gulf. Plentiful at La Paz; at Amortajada Bay on San José Island; sand flats a few rods inland on the peninsula side of Concepcion Bay, and on the sandy mud-flats inland of San Lucas Cove below Santa Rosalia.

In a personal communication, Glassell differentiates from the similar *U. musica:* "*crenulata* has red cheliped at base underneath, is highest up the Gulf shores and *does* go down into *musica's* zone. *Musica* is *not* in California (as reported in the literature) but is common in Gulf in sand mid-tidal to upper third, and has no permanent burrow (whereas *crenulata* has); the male is purple, and smaller than *crenulata* and does *not* come up in *crenulata's* zone."

§ R–77 *Geotice americanus.*

Rathbun 1923 (§ R–16), p. 629, toto-photo Pl. 31. Range, San Bartolomé Bay into Gulf. Formerly considered to be *Hemigrapsus oregonensis* per Gulf records, p. 273. Rathbun 1918.

Angeles Bay under rocks; Port San Carlos, Sonora.

§ R–78 SUMMARY

In our rapid and cursory survey of the Gulf littoral, we collected 54 species of true crabs. 197 species (excluding oxystomatous and dromiaceus forms) were listed from this region in 1934 by Glassell (§ R–12), marking it as possibly the richest area on record with regard to this group. From the well-explored and proverbially rich California coast, Schmitt in 1921 (§ O–10) cited 181 total decapods, and only 74 of these were Brachyura (less exclusions as above). However, except for 2 species, Sally Lightfoot and the swimming crab *Callinectes bellicosus,* none of the tropical species occurs in the great hordes that characterize many of the more northern forms.

Of the separate groups, there were 3 Oxystomatous and related crabs, none common. Of the 13 spider crabs, the masking *Mithrax areolatus* and *Microphrys platysoma* were common, together with the coral crab *Thoë sulcata.*

22 were Cancroids, with 3 Portunids (including the very common *Callinectes bellicosus*), 1 Cancrid (the fairly

REFERENCES

common *Platypodia*), and 18 Xanthids among which are included some of the commonest crabs encountered: *Leptodius occidentalis, Pilumnus gonzalensis* and *P. townsendi, Eriphia squamata,* and the cherry-colored coral crabs *Trapezia.*

There were 16 grapsoids, with 5 Pinnotherids (usually small or commensal forms), 8 grapsids including the ubiquitous *Grapsus grapsus,* the common *Geograpsus lividus* and the flat *Percnon* which occurred only at Cape San Lucas but which was common and characteristic there; and 3 ocypodids including the fiddler *Uca crenulata* which occurred in great colonies. Remarkably, no land crabs (FAMILY GECARCINIDAE) were encountered.

Of the total 54, 8 seem to be limited to the Gulf, 34 species range from there southward (including one species reported only once slightly out of range to the north, at Pt. Abrojos; and 1 which extends clear to Chile), 5 range northwardly (including the familiar California *Pachygrapsus crassipes* which ranges north to Oregon but which we took only once, at San Francisquito Bay), and 5 both north and south (1 to Monterey Bay, 1 to Chile). Only 12 of the 54 occur outside Panamic limits as here defined: 1 extends to Monterey Bay, 1 to Oregon and is reported from Japan, 2 are also Indo-Pacific, 2 range to Chile, another clear to Patagonia, 2 occur in the tropical Atlantic (West Indies, etc.), and 3 are tropicopolitan.

The 14 commonest or most abundant forms mentioned by name in the above analysis are all restricted, on Pacific-American shores, to Panamic waters, but 4 are also more or less tropicopolitan. None ranges north of Pt. Conception, and only 3 occur north of Magdalena Bay: *Callinectes* to San Bartolomé Bay (there is a single old record from the San Diego area); *Grapsus* to San Benito Island in the Cedros Island region; and *Uca* to San Diego—actually above San Pedro. Only 1, *Pilumnus gonzalensis,* is limited to the Gulf.

Hence the relationships would seem to be strongly southern, a typical range reading "Gulf to Galápagos." The collector from southern California, who in other groups would find occasional familiar species in the Gulf, would see utterly strange Brachyura almost exclusively,

REFERENCES

only one of the 14 common species—*Uca*—being found in both areas. However, a Panama collector would be at home in the Gulf with this group, and students acquainted with the crab fauna so far south as Ecuador and the Galápagos would find here mostly familiar forms.

Phylum Mollusca

REFERENCES APPLYING TO THE ENTIRE GROUP:

§ S–1 Conchological literature of the Pacific coast is so extensive, and has already been reviewed so expertly, that there is little need for a complete résumé. However, some of the very best accounts, being old and difficult to procure, have been popularly lost sight of. Furthermore, Panamic papers generally have not been cited in recent bibliographies, and the treatment of the literature situation in conchological papers is less extensive than the standard common to most zoological writing. Hence it has been thought wise to include herewith, citations and abstracts of:

(1) All the preponderantly important accounts relating to this area since 1852 which have come to our attention, however obsolete or unavailable, and even though they may have been listed in literature summaries which are also cited:

(2) Pacific coast molluscan bibliographies, or papers which summarize the literature, even though chiefly to the north of the Panamic region. There are six of these. Four are fundamentally important: Carpenter's 1856 "Report" and his 1863 "Supplementary Report," Dall 1909, and Keen 1937. Two: Cooper 1895 for the Gulf, and Strong and Hertlein 1939 for Panama, are of subsidiary bibliographic importance only: and

(3) Papers which, because they are recent, or extralimital, or so obscure as to have been overlooked, have not been cited in these literature summaries.

Accounts essential to the study of Panamic littoral mollusks, in addition to the six above mentioned, are those of Adams 1852 (320 pp.), Bartsch and Rehder 1938 (18 pp.), Carpenter 1855-57 (552 pp.), Dall 1908 (282 pp.), Dall 1921 (217 pp., includes only those species whose range includes points at or north of the San Diego region), Grant and Gale 1931 (1036 pp., same limitations), Oldroyd 1924-27 (4 pts. totaling 1187 pp., some limitations), Pilsbry and Lowe 1932 (111 pp.), Rogers 1902 (485 pp., popular), Stearns 1894 (65 pp.), Tomlin 1927-28 (28 pp.), and apropos parts of Tryon and Pilsbry's "Manual of Conchology."

REFERENCES

§ S–2 Adams, C. B. 1852. "Catalogue of shells collected at Panama, with notes on their synonymy, station and habitat." Ann. Lyc. Nat. Hist., Vol. 5: 229-549.

An important paper which should be used, however, only in connection with Carpenter's "Review" of same, cited herein as "B" in § S–8. Noteworthy introductory remarks especially as regards zoogeography. Lists of species taken which occur also in the Galápagos. Consideration of methods of dispersal, problems of ecological distribution, history of conchological collecting in the Panamic Province. Narrative of the trip, with topographic description of the region collected over. P. 245: "all the way from the low water mark, up to the ledges of rocks where some species of *Littorina* live out of the reach of the highest tides, species are found, most of which are limited to a very narrow vertical zone." Noting the presence at low water of certain species which were obtained by Cuming at 15 fathoms, Adams remarks, p. 246: "The difference between such stations is obviously of less importance than that between this extreme low water mark, where the air will but slightly reach the animal every fortnight, and that of a few feet higher, whence the water wholly recedes twice every day." 41,830 specimens of 516 species of mollusks (including one brachiopod) were collected. There is an index of place names on p. 250. The annotated catalogue extends from p. 253 to p. 527; the Latin descriptions of new species from p. 528 to p. 548.

§ S–3 Bartsch, Paul and Harald A. Rehder 1938. "Mollusks collected on the Presidential Cruise of 1938." Smith. Misc. Coll., Vol. 98 (10): 18 pp., 5 pls.

Descriptions with toto-illustrations, of 14 new species taken at Galápagos, Central America, Cocos, and Clipperton Islands, and at four points on the Lower California Pacific coast from San José del Cabo to Cedros Island, by shore collecting, dredging, etc.; in addition to the Caribbean material. Faunal lists of the various regions. "Indo-Pacific relationship of the marine mollusks of Clipperton Island . . . suggests a drift fauna." Clipperton is a French coral atoll in the latitude of Costa Rica (about 10° N.), but some 1500 miles west, hence in the longitude (about 108°) of the Gulf of California.

§ S–4 Boone, Lee 1928. "Mollusks from the Gulf of California and the Perlas Islands." Bull. Bingham Ocean. Coll., Vol. 2 (5): 17 pp., 3 pls., 1 map.

An Argonaut, 2 Octopi, 2 squid, 23 Pelecypods including *Tellina barbarae* sp. nov., 33 Gastropods and 1 chiton. 6 species are well illustrated by toto-photographs. Except for the *Tellina,* no color notes appear in the annotations, despite the p. 3 remarks. Probably this, like others of the Boone papers, should be used with caution.

§ S–4a Boone, Lee 1933. "Coelenterata, Echinodermata and Mollusca." Bull. Vanderbilt Mar. Mus., Vol. 4: 217 pp., 133 pls.

REFERENCES

The Mollusk section considers 9 Panamic species, and illustrates in addition, as fig. A, Pl. 125, the Panamic *Ischnochiton (Stenoplax) limaciformis,* one specimen of which was dredged in 5 fathoms off Florida. *Onychoteuthis banksii, Dosidicus gigas, Pyrgopsis schneehageni, Loligo diomedeae, Argonauta argo, Octopus bimaculatus, Chiton latus, C. goodallii,* and *C. sulcatus,* all with good toto-drawings or photographs, are reported and described, mostly from the Galápagos littoral, with a few dredged forms off Central America. Probably should be used with caution.

§ S–5 Carpenter, Philip P. 1855-57. Catalogue of the Collection of Mazatlan Shells, in the British Museum, collected by Frederick Reigen . . . pp. xii plus 552. British Museum.

Known as the Mazatlan Catalogue. See also the author's abstract in par. 51, pp. 265-81 of the 1856 Report, and the corrections, emendations, and remarks in the 1863 Supplementary Report, par. 51, p. 542. Annotated catalogue of 8873 specimens, with data on synonymy, geographic range, abundance, and ecological station. Bryozoa, p. 1; Palliobranchiata (the modern Brachiopoda), p. 7; Lamellibranchiata, p. 8; Gastropoda, p. 170, with Opisthobranchs on p. 170, Pulmonates on p. 174, and Prosobranchs on p. 185. Considered then as belonging to the Gastropoda were the Lateribranchiata (the present Scaphopoda) p. 188, and the chitons (then included in the Scutibranchiata) on p. 189. The Gadinidae, p. 211, at that time were grouped with the Prosobranchs also. There is an appendix of species out of order, and a bibliography.

§ S–6 Carpenter, Philip P. (1856) 1857. "Report on the present state of our knowledge with regard to the Mollusca of the West Coast of North America." Rep. Brit. Assoc. Adv. Sci. 1856 (published in 1857): 159-368, 4 pls.

Known as the "1856 Report." Should be used in connection with the "1863 Supplementary Report" cited below. The most important single paper encountered in these investigations; the product, obviously, of a disciplined, humble, and competent mind. The author, although "a learner who came fresh to the subject without previous aquaintanceship with books and naturalists," expertly summarizes and evaluates all the information then available on mollusks of the American Pacific from arctic to equatorial waters. In addition to its factual and conceptual value as a source book, the account has literary charm and reflects a personality both warm and significant. Like most of Darwin's writings, and like the monographs of Fisher, Rathbun, and Schmitt today, this transcends its time and subject matter and achieves a quality of universalness. Pacific coast faunal provinces are provisionally defined, fairly accurately in the case of the Panamic area; synonymy is considered; there are annotations and geographical lists; and suggestions are made which are pertinent today. Previous literature is analyzed critically by reference to the explorations and shell collections which serve as its bases.

REFERENCES

The account proceeds by means of numbered paragraphs. Since its aims are as applicable now as they were 85 years ago, par. 1 is quoted herewith:

"Before results can be obtained of permanent value . . . it is necessary that the foundations shall be laid by patient and accurate examination of every minute point in our enquiries: else, as the wrong measurement of a degree of gravity nearly prevented Newton's elimination of the great law of gravitation, so the deficiency or hasty examination of details respecting particular species or their abodes, may lead the great master minds of science to erroneous conclusions, which, through their well earned influence, retard rather than stimulate the progress of future research. It is proposed therefore, (1) to state the physical conditions and the cautions to be observed in the enquiry, (2) to present the different sources of information in historical order, and (3) after tabulating these geographically and zoologically, to draw such inferences as the present state of knowledge may warrant."

Many workers today will echo the opening remark of his par. 17: "As human life is so short, and those who have the inclination for scientific pursuits have generally so little leisure, it is a serious evil when so large a proportion of that little has to be devoted to the labour of making out the errors of predecessors." Significant concepts, writing, or information, will be found also in paragraphs 18, 19, 20, 21, 51, 69, 70, and 71, and in the final paragraph 92.

Hertlein (1937, p. 310) also, in commenting on the relation between Indo-Pacific mollusks and those of the islands off the west coast of America, has quoted the following from Carpenter, p. 346:

". . . having, when examining the shells of the Marquesas in the center of the Pacific, found several conspicuous and well-known forms of the Asiatic Seas, in spite of (in parts) the profound depth of ocean that lies between; he will naturally expect, as he reaches the American shores, to find also not a little in common with the opposite shores. He crosses the vast unbroken expanse of the West Pacific; one flank of the hemisphere of waters, which of itself almost rivals the Atlantic in extent. He pauses at the solitary Archipelago of the Galápagos, in the very longitude of the Gulf of Mexico, guarding (as it were) the great bay of Central America, and within 600 miles of its shores. Even here his eye rests with pleasure on a few well-known Cones and other forms, which have crossed the fathomless depths and come to claim kindred with their molluscan brotherhood of the New World. But here they stop. They could traverse half a world of waters. The human spirit that gives them understanding and a voice, beholds them on the very threshold of the promised continent, in whose bays and harbours, protected by

REFERENCES

the chain of everlasting mountains, they shall find the goal of their long pilgrimage. But the Word of the unknown Power has gone forth; and the last narrow channel they attempt to cross in vain."

Among the conclusions reached with reference to the Pacific coast fauna, Carpenter reports: "there does not exist on the surface of the earth a more separate, independent assemblage of mollusks than is to be found, under three great typical divisions, from Oregon to Chile," the divisions of which he finds fairly distinct also.

§ S-7 Carpenter, Philip P. (1863) 1864. "Supplementary Report on the Present State of our Knowledge with regard to the Mollusca of the West Coast of North America." Rep. Brit. Assoc. Adv. Sci. 1863 (publ. 1864): 517-686.

This, with twelve other Carpenter papers on west coast mollusks, was published as:

§ S-8 Carpenter, Philip P. 1872. "The Mollusks of Western North America." Smith. Misc. Coll., Vol. 10 (publ. in 1873) (252): 446 pp. Also published as a separate, No. 252, in 1872.

Another fundamental account, wisely undertaken and well written, again reflecting internationalism in scientific co-operation. In the pertinent introduction to the 1872 edition, the author remarks: "In the First Report, I was a novice in the scientific world, and rarely ventured on criticisms; in the second, I allowed myself with more confidence to state my own conclusions, because I found that others had not enjoyed the remarkable facilities of comparing types which fell to my lot, and which (in many instances) cannot be renewed." In this paper also, numbered paragraphs are used, which, in the corrections and emendations, correspond with those in the First Report to which they refer, the paragraph subjects being stated in the opening sentences. At the end there is a table of contents with parallel columns for pagination in the first and second reports, a great convenience, since Carpenter's visit to this country disclosed that some of the species described as new then and in the Mazatlan Catalogue, actually were synonyms of earlier American species. There are noteworthy quotations from early works not easily accessible even then. The account reflects also an interesting flavor of the times, of the difficulties in the way of seekers after true things, then as now. Behind the kind and ponderous wording, a picture emerges of work difficultly accomplished in bringing a degree of order into things which very obviously had little of it previously, and against the usual odds of indifference, carelessness, and stupidity, highlighted occasionally by a Darwin or a Gould or an Adams, or by the self-effacing Mexican War naturalists, as contrasted with the petty politics and limitations of the average European museum management. The Reigen Mazatlan collections, par. 51, are treated again on p.

REFERENCES

542, the par. 54 Adams Panama catalogue on p. 549. The new material starts on p. 577 with par. 93. Par. 105 reports the very significant Xantus collections from Cape San Lucas. The northern Mexico list is in par. 113; a small listing from Acapulco in par. 114. Par. 122 brings to date the zoogeographical conclusions, without much revision. In par. 123 the then new Darwinian theory is mentioned favorably, with a slight fence-straddling tendency, but with an admonition to the reactionary biologist "to guard against seeing that only which accords with his previous belief." In par. 126 the similarities between the west coast fauna and that of Great Britain are mentioned. Par. 128, p. 683: "The Vancouver and California districts have so many characteristic species in common (111 out of 492), that they must be regarded as constituting one fauna . . ." Par. 129, p. 684: "Of the blending of the temperate and tropical faunas on the peninsula of L. California we are still in ignorance. All we know is, that at Margarita Bay [on the open coast outside Magdalena Bay] the shells are still tropical, and that at Cedros Island they are strangely intermixed."

Of the twelve additional papers reprinted in the 1872 edition, "B" in which Prof. Adams Panamic Catalogue is reviewed, "C" with Xantus new species, "G," "H," and "I" on Mexican and Panamic forms, are most important to our purpose. The descriptions of new species are in Latin. There is a 108 page index to the genera and species in *all* Carpenter's papers published outside America which relate to American forms.

Despite the charm of Carpenter's writing, his obviously lofty aims, and the comprehensiveness of his program, at least one modern systematist evaluates his scientific results adversely. Dr. S. Stillman Berry, foremost chiton authority, regards many of Carpenter's new species as inadequately and too briefly described, founded on insufficient evidence, and erected from incomplete specimens. However, this would seem to be a fault which, with too great dependency on type specimens, he shared with C. B. Adams and others of his time.

§ S–9 Cooper, J. G. 1895. "Catalogue of marine shells collected chiefly on the eastern shore of Lower California for the California Academy of Sciences." Proc. Calif. Acad. Sci. (2), Vol. 5: 34-48.

An important paper. Reviews the work done to date in the Gulf, which is noted as the richest field for mollusks on the entire west American coast, about 700-800 indigenous species having been recorded by that time. Referring to the northern part of the Gulf, Cooper remarks (p. 37): "it appears that the species found there are more largely of the temperate fauna, many of them being identical with those from the same latitude on the west coast of the peninsula. This seems to indicate that the dividing ridge, now 3000 feet or more in altitude, was crossed by one or more channels within geologically recent times." [However, in this connection see our p. 178.] Collecting localities and methods are stated; collectors were Bryant, Brandegee, and Eisen, not primarily interested in marine

REFERENCES

invertebrates. Annotated list of 110 gastropods and 90 pelecypods. Some natural history notes. Record of one immature *Haliotis fulgens* from San José del Cabo.

§ S–9a Dall, W. H. 1899. "Preliminary Report on the collection of Mollusca and Brachiopoda obtained in 1887-88." *Albatross*. Proc. U.S.N.M., Vol. 12: 219-362, 10 pls.

Classification and conditions of life of mollusks, especially of abyssal forms. Within the Panamic area, 7 mostly new species are described from shallow water (5 to 40 fathoms), and 27 from water over 100 fathoms.

§ S–10 Dall, W. H. 1902. "Illustrations of new, unfigured, or imperfectly known shells, chiefly American, in the United States National Museum." Proc. U.S.N.M., Vol. 24: 499-566, Pls. 27-40.

The west American species start on p. 511. 14 are Panamic, 3 probably overlap: an Argonaut, a chiton, the remainder snails and pelecypods, some very deep water, some from the Gulf. Index to the genera.

§ S–11 Dall, W. H. 1908. "The Mollusks and the Brachiopoda." (*Albatross* 1891, and 1904-05.) Bull. Mus. Comp. Zool., Vol. 43 (6): 205-487, Pls. 1-22.

Despite its emphasis on deep water forms, this comprehensive paper comes near to being a handbook of Panamic marine mollusks. Materials are listed from several other *Albatross* cruises in the eastern tropical Pacific, and from miscellaneous sources. In addition to the usual shelled forms, Cephalopods, Pteropods, Tectibranchs, Chitons and Brachiopods are treated; only the Nudibranchs (see Bergh, § T–1) are missing. Many shallow water and shore forms are considered, some from the Gulf, with p. 435 notes on littoral species. 63 and 14 species of shore mollusks are listed from Panama Bay and from Cocos Island. There is a list of *Albatross* stations, a complete alphabetical index, and many toto-figures, maps, and charts.

§ S–12 Dall, W. H. 1909. "Material towards a bibliography of publications on the post-Eocene marine mollusks of the northwest coast of America, 1865-1908."

Appendix XIII of: "Contributions to the Tertiary paleontology of the Pacific Coast. I. The Miocene of Astoria and Coos Bay, Oregon." U. S. Geol. Surv. Prof. Paper., Vol. 59: 195-216.

REFERENCES

"The present bibliography is intended to take the subject up where it was dropped by Dr. Philip P. Carpenter in 1864. . . . Some titles of papers on the tropical west coast fauna, important for comparative purposes, have been included, though I have not attempted to make the bibliography complete for Panaman or South American faunas."

§ S–13 Dall, W. H. 1909a. "Report on a collection of shells from Peru, with a summary of the littoral marine Mollusca of the Peruvian Zoological Province." Proc. U.S.N.M., Vol. 37: 147-294, Pls. 20-28.

Although the Peruvian Province itself is out of our range, this considers all species occurring within the boundaries of Peru proper, by listing some 869 mollusks reported to date from the Bay of Guayaquil southward, with citations and reference to figures, following summaries and zoogeographical considerations starting on p. 185. The paper opens with a list of the mollusks collected by R. E. Coker, incorporating his field notes on ecology and economic importance, with geographic range. Much of the collecting was done in the extreme north, where the Panamic animals occur en masse.

§ S–14 Dall, W. H. 1921. "Summary of the marine shell-bearing mollusks of the northwest coast of America, from San Diego, California, to the Polar Sea . . . with illustrations of hitherto unfigured species." Bull. U.S.N.M., Vol. 112: 217 pp., 22 pls.

Known as the Dall Summary. See also: "Additions and emendations . . ." by the same author, Proc. U.S.N.M., Vol. 63 (10): 1-4. Should be used with the Keen § S–19 check list. "To the preparation of this summary the author has brought the results of more than 50 years' study of the molluscan fauna of the northwest coast." (p. 1.)

The Introduction largely considers zoogeographical data. A *Temperate Division* is stated to extend from the Bering Sea to Point Conception, a *Tropic Division* to Pt. Aguja on the coast of Peru (L. 6° S.). The first is subdivided into an Aleutian with 291 species, an Oregonian with 371, and a Californian with 996 species. The second is subdivided into a Gulf of Californian, a Panamic (of which 131 species are recorded in the scope of his work), and an Ecuadorian. No limits are stated for the subdivisions, which are defined in greater detail but unfortunately not conformably, in an earlier paper by the same writer (1912, Journ. Acad. Nat. Sci. Phil.). He remarks on p. 4, "In endeavoring to assort according to their respective faunas the species in the following discussion I have met with the difficulty of properly placing a number of species which overlap the limits of one or more faunal districts in their geographic range. In such cases I have tried to assign them to the fauna which seemed to be their natural center of distribution." Still we have no clue as

REFERENCES

to his criterion of assorting. This may have been on the basis of intuitive apperception based on his profound knowledge of the material, which may be a perfectly valid method, but which is hard to check.

Exclusive of nudibranchs and cephalopods, 2122 species are catalogued, with citation and geographic range. Many are overlap or Panamic forms. The treatment is systematic, by family, genus, and sub-genus. The sub-generic name is used in the binomial, an occasional source of confusion. The Index is generic only, there is an alphabetical list of genus synonyms, and a short bibliography.

§ S–14a Dall, W. H. 1925. "Illustrations of unfigured types of shells in the collection of the United States National Museum." Proc. U.S.N.M., Vol. 66 (17): 41 pp., 36 pls.

Applicable to previous Dall papers, to his 1921 Summary, etc., with figures of many Gulf and other Panamic pelecypods and gastropods. Over 200 species in all are illustrated, usually by toto-photographs or drawings.

§ S–15 Gould, Augustus A. 1855-57. "Descriptions of shells from the Gulf of California and the Pacific coasts of Mexico and California." Bost. Journ. Nat. Hist., Vol. 6 (26): 374-408, 3 pls.

Based on a collection made from Panama to San Francisco, by officers in the Mexican War. Some were characterized briefly in the Proc. Bost. Soc. Nat. Hist. for 1851. About 22 new species from the region under consideration, mostly illustrated by small but very good lithos. Other new species are from the upper California coast. For some of the items, no locality is stated. This should be used with the paper of the same title, Pt. II, by Gould and Carpenter 1856, Proc. Zool. Soc. London, Vol. 26: 198-208, wherein Gould's and Carpenter's sometimes synonymous names are mutually allocated.

§ S–16 Grant, U. S. IV, and H. R. Gale 1931. "Catalogue of the marine Pliocene and Pleistocene Mollusca of California." Mem. San Diego Soc. Nat. Hist., Vol. 1: 1036 pp., incl. diagrams, text figs., tables, and 32 pls.

The catalogue covers shelled forms only, and excludes the chitons and the Cephalopoda. The treatment is by families, 16 of which are revised. Although the approach is paleontological, the detailed treatment, references, and illustrations, mark this as a source book for California mollusks, incidentally including a good many overlap and truly Panamic forms. Geographic ranges are given for all recent species. Key paper for shelled mollusks.

§ S–17 Hertlein, L. G. 1938. "A note on some species of marine mollusks occurring in both Polynesia and the western Americas." Proc. Amer. Phil. Soc., Vol. 78 (2): 303-312, 1 map and 1 pl.

26 Gastropods and 1 Pelecypod from the Pacific mainland or offshore islands, which are identical or analogous to Polynesian or Indo-Pacific forms. Largely in the FAMILIES CONIDAE, CYMATIDAE, and CYPRAEIDAE, 9 of these are reported from the Gulf, and 2 others from Mazatlan and Cape San Lucas. There are 23 rather good toto-figures in which specimens from the Pacific coast are photographed alongside of comparable material from the Indo-Pacific.

§ S–18 Jordan, E. K. 1924. "Quaternary and Recent molluscan faunas of the west coast of Lower California." Bull. South. Calif. Acad. Sci., Vol. 23 (5): 145-56.

"Two distinct faunas exist on the west coast of Lower California. The Southern Californian now ranges southward from Pt. Conception to Cedros Island . . . probably extends a little farther. . . . The fauna of the Gulf of California ranges to the north on the west coast of the peninsula approximately to Scammon's Lagoon, which is a little farther up than Cedros Island." p. 146. Present geographic ranges are given for 124 species collected in lower quaternary beds at Magdalena Bay, all of which occur living today, but rarely so far south. "It . . . appears that when these quaternary beds were laid down there was a southward displacement of the isotherms sufficient to carry the conditions today prevailing at Cedros down as far as the latitude of Magdalena Bay." p. 148. There are similar lists of 32 and of 24 such species from the upper quaternary of San Ignacio Lagoon and of Scammon's Lagoon, at which time conditions similar to those of today are thought to have obtained. Starting on p. 153 is a list of about 150 living species which have their range extended to the north or the south by records from the collections made by Hemphill along this coast.

§ S–19 Keen, A. Myra 1937. An abridged check list and bibliography of West North American marine Mollusca. Stanford University Press. Photolith. 88 pp.

Brings to date the information gathered in Dall's 1921 Summary, § S–14, and in his 1909 Bibliography, § S–12. The treatment is alphabetical. The ranges of all species of shell-bearing mollusks (excluding chitons) recorded north of the Coronado Islands, just south of San Diego, are expressed in terms of latitude, with mid-range stated. There are remarks on statistical methods in conchology and on zoogeography. The numbered titles are alphabetical; there is a subject index digest. Information and bibliography on Panamic forms is incident to their occurrence in the area being considered.

§ S–20 Lowe, H. N. 1913 to 1934. A group of articles on shell collecting along the Mexican west coast, as follows, in Nautilus:

1913. Vol. 27 (3): 25-29. Record of a four weeks' cruise from Punta Banda to Cedros Island and including San Gerónimo Island [hence involving only

REFERENCES

the overlap area as herein defined], with locality lists.

1930. Vol. 43 (4): 135-38 (in "Correspondence"). Description of conditions and collecting at Guaymas, at Topolobampo, and at Mazatlan.

1933. Vol. 46 (3): 73-76, and Vol. 46 (4): 109-115, with map. Record of a ten weeks' cruise principally involving Magdalena Bay and points in the Gulf.

1933. Vol. 47 (2): 45-47. Conditions and collecting at San Felipe Bay, 75 miles below the head of the Gulf, on the Lower California side.

1934. Vol. 48 (1): 1-4, and Vol. 48 (2): 43-46. Conditions and collecting at Punta Peñasco, near the head of the Gulf on the Sonora side.

§ S–21 Lowe, Herbert N. 1935. "New marine Mollusca from west Mexico, together with a list of shells collected at Punta Peñasco, Sonora, Mexico." Trans. San Diego Soc. Nat. Hist., Vol. 8 (6): 15-34, Pls. 1-4.

Descriptions, with fine toto-illustrations, collecting notes and localities, of 30 new species. Annotated list of 285 species, with collecting notes, from Punta Peñasco on the northern Sonora coast, many of them littoral and on the rocky reef. Scientific results of collecting trips popularly reported in Nautilus by Lowe, § S–20.

§ S–22 Mabille, J. F. 1895. "Mollusques de la Basse Californie recueillis par M. Diguet." Bull. Soc. Phil. Paris (8), Vol. 7: 54-76.

Annotated list of 159 species, 26 of them new. Most of the new species, described in Latin, are land or fresh-water forms. There are 2 octopi, 73 Prosobranchs (4 new), 4 Opisthobranchs, and 54 Pelecypods (4 new), in addition to the 26 Pulmonates.

§ S–23 Oldroyd, Ida Shepard 1924 and 1927. "The Marine shells of the West Coast of North America." Stanf. Univ. Publ., Univ. Ser., Geol. Sci., Vol. I: 247 pp., 57 pls. (Pelecypoda and Brachiopoda); Vol. II (1, 2, and 3): 297 pp., 29 pls., 304 pp., Pls. 30-72, 339 pp., Pls. 73-108 (Scaphopoda, Gastropoda, and chitons).

Original description, translation if not in English, type locality, depository of type, geographic range, and, often, illustration, of all the marine shells then known to occur from the Arctic Sea to San Diego. Many overlap and Panamic species are therefore treated as they occur also within this range. There are lists which classify the species as circumboreal, extending to the Gulf of California, to Lower California, to Central America and Panama, etc. The treatment is by families, genera, subgenera, etc. There are short and rather inadequate bibliographies. Some of the generally observed taxonomic conventions (involving methods of citation, the use of parentheses, etc.) seem not always to have been regarded, minimizing the value of this otherwise comprehensive catalogue as a reference work. Should be used with the corrections noted by Keen 1937 § S–19 pp. 82-84.

§ S–24 Olsson, A. A. 1924. "Notes on marine mollusks from Peru and Ecuador." Nautilus, Vol. 37 (4): 120-130.

The results of fairly comprehensive shell collecting, mostly of Panamic species, Sechura Bay having been the most southerly point visited. Consideration of the boundary line between the Panamic and the Peruvian provinces, set here at Punta Parinas, the most westerly point of South America, "although many of the Panaman species extend beyond into the warm, protected bays of Paita and Sechura. . . . Going north up the coast from Parinas, we can note a progressive warming of the waters with shells becoming continually more abundant and varied." The annotated check list of almost 200 species, Gastropods and Pelecypods only, includes many extensions of range.

§ S–25 Pilsbry, H. A. 1931. "The Miocene and Recent Mollusca of Panama Bay." Proc. Acad. Nat. Sci. Phil., Vol. 83: 427-440, Pl. 41, 5 text figs.

Describes, with good toto-illustrations, 7 new or slightly known living species dredged and from shore, 2 Pelecypods and 5 Gastropods, with mention of 10 other recent forms, correcting a paper of Li 1930.

§ S–26 Pilsbry, H. A. and H. N. Lowe 1932. "West Mexican and Central American mollusks collected by H. N. Lowe, 1929-31." Proc. Acad. Nat. Sci. Phil., Vol. 84: 33-144, Pls. 1-17. 7 text figs. (the last, p. 102, curiously numbered).

Narrative, with two photos, includes descriptions of collecting grounds. A Nicaraguan muricid is found to be not separable from *Coraliophila madreporarum* Sowerby of the South Seas, p. 37. 133 species and sub-species are described, all but twelve of them new, all illustrated by toto-photographs. Some of the genera and subgenera are new. There is an annotated list, by family, of the 763 species taken, with collecting notes. 3 are Cephalopods (shells of *Argonauta* spp.), 492 are Gastropods, 18 are chitons, 8 are tooth shells, and 242 are Pelecypods.

§ S–27 Pilsbry, H. A. and A. A. Olsson 1935. "New mollusks from the Panamic Province." Nautilus, Vol. 48 (4): 116-121, Pl. 6, and Vol. 49 (1): 16-19, Pl. 1.

An anastomosing vermetid, and three new pelecypods, with description of a fourth, previously known, but herein placed in a new subgenus; all from Panama, Ecuador, and northern Peru; all with toto-illustrations. 3 new species of *Modiolus,* a fossil *Chione* (Pleistocene), and a Lyonsid, all from northern Peru or Ecuador; all with toto-illustrations.

§ S–28 Pilsbry, H. A. and E. G. Vanatta 1902. "Marine Mollusca." No. XIII in: "Papers from the Hopkins Stanford Galápagos Expedition, 1898-1899." Proc. Wash. Acad. Sci., Vol. 4: 549-560, Pl. 25.

Reference to the Stearns § S–36 paper. There are 6 new species in the present list of 101, of which about 25 were not listed in the aforementioned Stearns 1893 catalogue. 19 species collected at Cocos.

§ S–29 Rochebrune, A. T. de 1895. "Sur les propriétés toxiques du *Spondylus americanus,* Lamarck." Bull. Mus. Hist. Nat. Paris, Vol. 1: 151-156, 6 text figs.

A spiny or rock oyster collected in Lower California by M. Diguet, who reports that the flesh emits an unpleasant odor of sulphuretted hydrogen and is reputed to be poisonous. A pharmacological examination affirms the toxicity, said to be due to an alkaloid related to the ptomaines and leucomaines and similar to muscarine (of poisonous fungi), which is elaborated by the living animal as a normal product. A modern investigator might consider whether this toxicity may not be due to metabolic waste products arising from the ingestion by the bivalve of such planktonic organisms as are involved in mussel poisoning along the California coast during the summer months.

§ S–30 Rochebrune, A. T. de 1895a. "Diagnoses de Mollusques nouveaux, provenant du voyage de M. Diguet en Basse-Californie." Bull. Mus. Hist. Nat. Paris, Vol. 1: 239-243.

Pertaining to the collection mentioned on p. 36 of the same Bulletin, which was worked up by Mabille (1895, § S–22). *Pleurobranchus digueti* and the nudibranch *Doris umbrella,* with 4 species of *Ostrea,* 2 of *Chama,* and *Avicula vivesi.*

§ S–31 Stearns, R. E. C. 1893. "On rare or little known mollusks from the west coast of North and South America, with descriptions of new species." Proc. U.S.N.M., Vol. 16: 341-352, Pl. 50.

"The number of forms heretofore associated in the monographs and by the principal authors with an Indo-Pacific habitat, will attract attention." There are several Gulf species, including a sea-hare, *Dolabella californica* Sterns 1878, and the shell-less marine pulmonate *Onchidella binneyi* Stearns 1878 (from San Francisquito, Las Animas and Angeles Bays, detailed description with toto-figure). There is a reference to the 1878 paper, where the aplysid was originally described, with other Gulf species, and where the pulmonate was identified as *O. carpenteri* (Binney) 1860.

REFERENCES

§ S–32 Stearns, R. E. C. 1893a. "Report on the Mollusk Fauna of the Galápagos Islands, with descriptions of new species." Proc. U.S.N.M., Vol. 16: 353-450, 2 pls.

Based on collections made by the *Albatross*. Littoral, shallow water, and land forms. Data on climate, currents, origin of the fauna, natural history of the land snails; previous lists of local mollusks. The annotated catalogue, complete to date for the Galápagos, starts on p. 419. There are lists of species which occur also elsewhere, in Lower California, etc. Included among the new species are the marine shell-less pulmonates *Onchidium lesliei* and *Onchidella steindachneri*, both with illustrations on Pl. 51.

§ S–33 Stearns, R. E. C. 1894. "The shells of the Tres Marias and other localities along the shores of Lower California and the Gulf of California." Proc. U.S.N.M., Vol. 17: 139-204.

Results of collections made by W. J. Fisher, who chartered a vessel for this work in 1876. An annotated catalogue, with some ecological notes, of a total of 294 species of mollusks, including shell-less forms and chitons. 89 species are recorded from the Tres Marias. ". . . the detection of so many familiar forms, heretofore associated in our minds with Indo-Pacific or rather Polynesian waters, is almost a revelation and of exceeding interest." (p. 139) Stearns regards the exotics as having arrived in coral rock ballast from the South Sea trading vessels that loaded dyewood in the Tres Marias and at Altata. Bibliography. No illustrations.

§ S–34 Strong, A. M. 1937. "Marine Mollusca of San Martin Island, Mexico." Proc. Calif. Acad. Sci. (4), Vol. 23 (12): 191-194.

San Martin is off San Quentin, Lower California, about latitude 30° N. "The fauna . . . is of particular interest, in that it marks the southern known limit of the range of a considerable number of California species." List of 199 mollusks known to date from this island, including a revision by consultation with the original author, of the Baker 1903 list in Nautilus, Vol. 16: 40.

§ S–35 Strong, A. M. 1938. "New species of west American shells." Proc. Calif. Acad. Sci. (4), Vol. 23 (14): 203-216, Pls. 15 and 16.

17 new species of Gastropods and Pelecypods from Guadalupe and the Tres Marias, all with good toto-figures.

§ S–36 Strong, A. M. and G. D. Hanna 1930. "Marine Mollusca of Guadalupe Island, Mexico." "Marine Mollusca of the Revillagigedo Islands, Mexico." "Marine Mollusca of the Tres Marias Islands, Mexico." Proc. Cal. Acad. Sci. (4) 19 (1, 2, and 3): 1-22.

Lists, with geographical ranges, of 87, 61, and 211 species respectively. Zoogeographical summaries. The fauna of the entirely vol-

REFERENCES

canic Guadalupe Island, about 29° N., resembles that of Catalina or Coronado Islands, with a majority of overlap and northern forms, 10 of which range clear to Puget Sound, indicating "a current at sometime running southerly along the southern California coast and turning out to sea at an angle which carried it past Guadalupe Island. The almost complete absence of species which are characteristic of the Gulf of California or the coast of the southern portion of Lower California, indicates that there has been no similar current from the south since the time when the shores of Guadalupe Island were first able to support marine life. It is well recognized that the fauna of the coast of southern California and the neighboring islands, containing over 1000 species, is distinctly southern in its affiliations with many species at the extreme northern end of their geographic range." There is no evidence, however, of the effects of this northern current in the Revillagigedos, about 18° north and far to the west, the character of the mollusc fauna being strongly southern. 2 species only were Indo-Pacific, 18 had been known previously from southern California or its islands, "the remainder, 41 species, would seem more properly to belong with the Galápagos Islands or Panamanian faunas than with that of the Gulf of California." For the Tres Marias, 21° N., the belief is expressed that Stearn's 1894 list (§ S–33) was based on collections made on the western shores as well as the eastern, where the faunas are thought to differ. Otherwise: "The fauna is found to be similar to that long known from the Gulf of California; however, an unsuspected number of migrants frcm northern waters appeared."

§ S–37 Strong, A. M., G. D. Hanna, and L. G. Hertlein 1933. "Marine Mollusca from Acapulco, Mexico, with notes on other species." Proc. Calif. Acad. Sci. (4), Vol. 21 (10): 117-130, Pls. 5 and 6.

List of 118 species dredged in Acapulco Bay. Description of 3 new species, with consideration of 6 additional species mostly slightly known; all 9 are illustrated.

§ S–38 Strong, A. M. and L. G. Hertlein 1937. "New species of recent mollusks from the coast of western North America." Proc. Calif. Acad. Sci. (4), Vol. 12 (6): 159-178, Pls. 34-35.

Descriptions, with toto-illustrations, of 21 mostly new species of Gastropods and Pelecypods, from southern California to southern Mexico, including the Gulf. The Philippine *Cymatium amictum* is reported from 35 fathoms off the state of Chiapis: "several other species of this group have been generally recognized as occurring on both sides of the Pacific."

§ S–39 Strong, A. M. and L. G. Hertlein 1939. "Marine Mollusks from Panama . . ." A. Hancock Pac. Exp., Vol. 2 (12): 177-245, incl. Pls. 18-23.

Pacific coast forms only. Review of the more important papers on Panama mollusks. Remarks as to geological and zoogeographical

REFERENCES

conditions. The area is said still to have about 50 living species in common with the Caribbean, presumably attributable to the open seaway which connected the two regions in Miocene times, p. 180. Alphabetical list, with localities, of "336 species of mollusks. The assemblage consists of the following: 1 brachiopod, 91 pelecypods, 2 scaphopods, 242 gastropods." Description of 57 mostly new species from shallow dredging and shore collecting, with collotype toto-illustrations.

§ S–40 Tomlin, J. R. le B. 1927-28. "The Mollusca of the St. George Expedition, I" and Ibid. "II." Journ. Conchol., Vol. 18 (6): 153-70, and (7): 187-198.

Deals with mollusks of the Panamic area only, collected by Dr. Hornell in Panama Bay, off the Colombian coast, in the Galápagos, and at Cocos Island. Annotated list, by families, of some 263 species, with data on locality, station, abundance, and geographic range. There are 4 chitons, 3 Pulmonates, 2 Opisthobranchs, 162 Prosobranchs, 1 Scaphopod, and 85 Pelecypods.

§ S–41 There are several general orientation accounts, important as such, and which consider Panamic forms incidentally.

There is the catalogue of the shell dealer, Walter F. Webb, published by its compiler at Rochester, N. Y., in 1936, called *Handbook for Shell Collectors,* also titled *Ocean and Land Shells,* 291 pp., 2200 cuts. Although so impaired by misspellings and errors as to militate against its serious use, it nevertheless figures—not always clearly—some 62 Gastropods and 14 Pelecypods of the Panamic area.

There is also the popular but more serious account published as *The Shell Book* by Julia Rogers, 1902, Doubleday, Page and Co., 485 pp., profusely illustrated with drawings, photographs, and colored plates, and reissued in 1908 as one of the volumes of "The Nature Library." Many Gulf of California and other Panamic shells are described and illustrated, and there are helpful characterizations of the families.

Most useful of all, and significant by any standards, however strict, is the two volume handbook issued in 1931 and 1935 by Gustav Fischer, Jena, Germany, under the title *Handbuch der systematischen Weichtierkunde* by Johannes Thiele. The first part, with 778 pp. and 783 text figures, covers the chitons and the Gastropods including the shell-less forms. The second, with 114 figures, carries the text through p. 1154 with a consideration of the Scaphopods, bivalves, and Cephalopods. The treatment is

REFERENCES

by orders, families, and genera; characteristic species are mentioned occasionally. There are remarks on geographic distribution, on taxonomy, phylogeny, paleontology, and on the literature, with separate alphabetical indexes to the genera in each volume.

§ S–42 The following short or obsolete papers seem not to have been cited, due to their extra-limital nature or to their recentness, by Dall 1909 or by Keen 1937 (§ S–12 and S–19). All were examined for pertinent data:

Bales 1938. "Marine collecting on the west coast of Mexico." Nautilus, Vol. 52: 46-48, list of Acapulco species.

Biolley 1907. "Mollusques de l'Isla del Coco," Mus. Nacional de Costa Rica, 30 pp., 2 maps. In French, lists only 30 marine species.

Eyerdam 1939. Nautilus, Vol. 53: 108. Extends the range of a group of north temperate and Panamic Pelecypods to Arica, Chile.

Hertlein 1932. "Mollusks and barnacles from Malpelo and Cocos Islands," Nautilus, Vol. 46: 44-45. 2 Gastropods and *Tetraclita squamosa milleporosa* from Malpelo; 30 species of mollusks and the aforementioned barnacle from Cocos.

Lamy 1907. Bull. Mus. Hist. Nat. Paris, Vol. 13: 530-39. List from Costa Rica, Panama, and Chile.

Lamy 1909. *Ibid.,* Vol. 15: 264-270. List of 105 Gulf species, fruit of M. Diguet's energy.

Pilsbry 1929. "Notes from the Pinchot South Sea Expedition," Nautilus, Vol. 43 (2): 37-38. Covers chiefly Galápagos land snails but mentions also some rocky marine collecting there.

Zetek 1918. "Los Moluscos de la Republica de Panama," La Revista Nueva (Panama), Vol. 5 (2): 511-575. (This is cited by Strong and Hertlein 1939, § S–39, p. 179, as Nos. 1 and 2, July and Aug., 69 pp., but the only copy to which we had access, a separate, was titled as above.) In Spanish. List (p. 516) of the 92 eurythermal species which occur from southern California to Peru or Chile. Unannotated catalogue. Bibliography of mostly Pacific coast species.

REFERENCES

§ S–43 The following less important papers are listed by Dall 1909 (§ S–12), and the citations need not be repeated here:

Dall 1871. Amer. Journ. Conchology described 60 new species from the Pacific, including a squid (from the Gulf) and 7 Gastropods from the Panamic region.

Dall 1891. Proc. U.S.N.M., *Albatross* dredgings.

Dall 1908. Proc. U.S.N.M., misc. Pacific coast species, mostly new.

Kelsey 1907. List of San Diego species.

Orcutt 1885. With notes by Dall; San Diego and Todos Santos Bay.

Stearns 1891. Proc. U.S.N.M., list of 211 species from the west coast of South and Central America and Mexico.

Verrill 1870. Gulf of California shells.

§ S–44 Also the following short or partially apropos papers are cited, among others, by Keen 1937 (§ S–19):

Dall 1913. Miscellaneous new species, some from the Gulf (as No. 76).

Dall 1918. Magdalena Bay (as No. 97).

Orcutt 1899. Lagoon Head, Lower California (cited on p. 71).

§ S–45 In the early reports of the California Academy of Sciences, and in Nautilus, there are dozens of articles, short or unimportant or both, comprising notes, lists, comments, and reports on species occurring within overlap or Panamic limits as here defined. Most of them, the more northern items especially, have been picked up by Dall or Keen and are cited in the 1909 and 1937 bibliographies.

§ S–46 There is at present no summary, nor even any co-ordinated data from which a summary could be drawn, of the considerable but very scattered information available on mollusks south of the United States border. As preparation for his 1921 North Pacific check list, Dall reviewed a great many families in which southern species occur, attending to any situations he was forced to clean up in order to provide clear sailing in the summary, and in this way incidentally straightening out portions of the Panamic fauna. But groups which were entirely or even principally

REFERENCES

tropical escaped attention at that time, and seem not to have been treated since.

A summary which shall accomplish for Panamic mollusks what Dall 1921 and Keen 1936 have done for temperate species, or, better, what Keen and Frizzell 1939 have done for the Pelecypod genera of the U. S. west coast, is very badly needed. It is too much for workers to hope for an illustrated handbook. Not even the better known American shores support such luxuries. But *something* should be done, if for no purpose other than to make comparisons possible with better known areas elsewhere in the world.

§ S-47 It is a matter for regret that in the mollusks we were not able to provide that severe and consummate synopsis which, we dare to hope, will have been accomplished for other groups. We had planned a source book, a summary of information to date, which should be equally thorough and fundamental in all its parts. But human effort is limited—as many a worker will have bewailed long before this, in the world's devious history. Irretrievable time passes. We spend energy not wholly replaceable. And the funds which sustain such a task prove equally limited and ephemeral.

To achieve a whole picture for each of our 165 or more species of mollusks, by examining, comparing, and considering all available examples, by following each name through its circuitous history, and by allocating each taxonomic position by understanding the larger unit to which it keys in—an ideal not impossible of achievement in less prolific groups—would have entailed delays outreaching our schedule. It seemed wiser merely to map out the more obvious canyons and tributaries, deliberately neglecting the detailed topography, rather than to jeopardize the entire undertaking by too strict adherence to perfectionist aims.

For the moment, such a project can be at best merely a scant trail over hills previously accessible only to explorers. Clearing and grading and widening, possibly even paving, can come later if they may. In the present maze we need most of all a path straight through the wilderness—a path traversable in its whole length not only by relays of voyageurs with special techniques and local knowledge,

REFERENCES

but by average travelers whose interests or businesses justify the expenditure of no great or special effort.

CLASS SCAPHOPODA

§ S–101 Baker, Fred 1925. "A new species of mollusk (*Dentalium hannai*) from Lower California, with notes on other forms." Proc. Cal. Acad. Sci. (4), Vol. 14 (4): 83-87, Pl. 10.

Dredged material from southern California, possibly representing overlap forms.

See also Scaphopod items in general mollusk papers cited on our pp. 471-496.

§ S–102 ACKNOWLEDGMENT: The scaphopod identifications under U.S.N.M. No. 159097 are by Dr. Paul Bartsch, Curator, Division of Mollusks, United States National Museum, to whom thanks are due also for data on citations and on geographic ranges.

LIST OF SPECIES TAKEN:

§ S–103 *Dentalium* (*Graptacme*) *semipolitum* Broderip and Sowerby 1829. PL. 13 FIG. 2

Description and reference to Manual of Conchology illustration: p. 10, Oldroyd 1927 (§ S–23). Mentioned on p. 15, Strong and Hanna 1930 (§ S–36) as having been taken in abundance at the Tres Marias Islands; and on p. 130, Pilsbry and Lowe 1932 (§ S–26) as having been dredged at San Juan del Sur, Nicaragua. In a personal communication from Dr. Bartsch, the range is given as extending from San Diego and the Gulf of California to Costa Rica.

This was the commoner of two species taken very abundantly living in the sand of the gradually sloping beach at low tide at El Mogote, opposite La Paz. The shell has longitudinal striations at least at the apical end.

§ S–104 *Dentalium* (*Graptacme*) *splendidum* Sowerby 1832 (Proc. Zool. Soc. London, 1832, p. 29). PL. 13 FIG. 2

Mentioned by Pilsbry and Lowe 1932 (§ S–26), p. 130, as having been dredged at Acapulco, Mexico, and at San Juan del Sur, Nicaragua, and found living on the sand bar at La Paz. Also mentioned by Lowe 1935 (§ S–21), p. 29, as having been dredged at 10 fathoms off Punta

REFERENCES

Peñasco, Sonora, in the Gulf of California. In a personal communication from Dr. Bartsch, the range is given as extending from Cerros (Cedros) Island, Lower California to Colombia.

This was the less common of two species taken in the sand of the gradually sloping beach at El Mogote, opposite La Paz, at low tide. The shell is smooth throughout.

CLASS PELECYPODA

The Bivalves—Clams, Oysters, Mussels, Cockles, Scallops, etc.

§ S–201 Contreras, F. 1932. "Datos para el estudio de los ostiones mexicanos." An. Inst. Biol., Vol. 3: 193-213, plates.

11 Pacific species are recognized, with good toto-illustration of each, usually reproductions of the original figures. The Gulf (Guaymas) commercial oyster is considered to be *Ostrea chilensis* Philippi, Sonora to Chile, "distinguished easily by the white or pale green border of its valves."

§ S–202 Dall, W. H. 1899. "Synopsis of the Recent and Tertiary Leptonacea of North America and the West Indies." Proc. U.S.N.M., Vol. 21: 873-897, 2 pls.

In the synopsis of the west coast species, pp. 879-882, 15 Panamic species are listed, plus several North Temperate species that range to Mazatlan, and several overlap forms. New Panamic and overlap species are described on pp. 885, 889, and 893, with good toto-illustrations.

§ S–203 .... 1899a. "Synopsis of the Solenidae of North America and the Antilles." Proc. U.S.N.M., Vol. 22: 107-112.

In the synopsis of west coast species, there are 3 Panamic *Solen* and 1 *Ensis* with one new species of each, *mexicanus* and *californicus* described respectively from the Gulf of Tehuantepec and the Gulf of California.

§ S–204 .... 1900. "Synopsis of the Family Tellinidae and of the North American species." Proc. U.S.N.M., Vol. 23: 285-326, 3 pls.

Exclusive of Central American forms. Although the enumeration of Panamic forms was deliberately incomplete, 11 new species of *Tellina* and 1 of *Macoma* were described with toto-illustrations from Panama, the Gulf, Cedros and Guadalupe Islands, and many other previously known Panamic species were listed, for details of which the reader is referred to more complete data in the Transactions of the Wagner Institute of Science.

REFERENCES

§ S-205 .... 1901. "Synopsis of the FAMILY CARDIIDAE and of the North American species." Proc. U.S.N.M., Vol. 23: 381-392, no illustrations.

18 Panamic species are listed in the genera *Cardium* and *Protocardia.*

§ S-206 .... 1901a. "Synopsis of the LUCINACEA and of the North American species." Proc. U.S.N.M., Vol. 23: 779-833, 4 pls.

Several Panamic species listed in the FAMILIES THYASIRIDAE, DIPLODONTIDAE, and LUCINIDAE. New Panamic species, with toto-illustrations, described on pp. 818, 821, 823, 827, and 828, in the genera *Thyasira, Phacoides,* and *Codakia.*

§ S-207 .... 1902. "Synopsis of the FAMILY VENERIDAE and of the North American recent species." Proc. U.S.N.M., Vol. 26: 335-412, 5 pls.

With bibliography and alphabetical index, a great convenience in a paper so large and involved. Many Panamic species, including many new species from the Gulf, Magdalena Bay, Panama, etc., with good toto-illustrations.

§ S-208 .... 1903. "Synopsis of the FAMILY ASTARTIDAE with a review of the American species." Proc. U.S.N.M., Vol. 26: 933-955, 2 pls.

Although Panamic or overlap forms are described, some of the *figures* apply to Panamic species treated, or to be treated, elsewhere.

§ S-209 .... 1915. "A review of some bivalve shells of the group Anatinacea from the west coast of America." Proc. U.S.N.M., Vol. 49: 441-456, no illustrations.

25 Panamic and overlap species, mostly new, in the FAMILIES THRACIIDAE, PERIPLOMATIDAE, PANDORIDAE, and LYONSIIDAE.

§ S-210 .... 1916. "Diagnoses of new species of marine bivalve mollusks from the northwest coast of America in the collection of the United States National Museum." Proc. U.S.N.M., Vol. 52: 393-417, no illustrations.

40 species, mostly dredged, 24 to 1100 fathoms off southern California. Several, however, are littoral and others may be. One was collected many years ago at Santa Barbara by Major Rich, one of the Mexican War naturalists. 6 new strictly Panamic species are described on pp. 399 403, 412, and 415.

§ S-211 Gilbert, Chas. H. 1891. "Reports upon certain investigations relating to the planting of oysters in southern California." Bull. U.S. Fish Comm., Vol. 9 for 1889: 95-98, maps and pl.

REFERENCES

Topographical description of natural beds south of Guaymas. Shells of dead and living oysters form huge hummocks in lagoons at the mouth of rivers, etc., where the winter and early spring temperature of the water is 70° F.

§ S–212 Hertlein, L. G. 1925. "The recent PECTINIDAE." Proc. Calif. Acad. Sci. (4), Vol. 21: 301-328, 2 pls.

12 plus 1? Panamic species, and 1 overlap species, including 2 new species and 4 new names, mostly from material collected on the Crocker Expedition. Good toto-illustrations.

§ S–213 Keen, A. M. and D. L. Frizzell 1939. "Illustrated Key to West North American Pelecypod Genera." Stanford University Press, photolith, 28 pp., many illustrations.

Although concerned primarily with extra-limital forms, there are diagrammatic drawings of a number of genera encountered in Panamic waters.

§ S–214 Soot-Ryen, T. 1932. "Pelecypods from Floreana (Sancta Maria), Galápagos Islands." Medd. Zool. Mus. Olso., Nr. 27: 313-324, 2 pls.

19 species, of which 8 are illustrated. *Semele floreanensis* is new, and *Diplodonta subquadrata* is new to the Galápagos. Review of the literature.

§ S–215 ACKNOWLEDGMENT: For determining the following material, for help with the literature, and for information as to geographic distribution, we have to thank Dr. Bartsch and Dr. Rehder of the Division of Mollusks, United States National Museum, Washington, D. C. To Dr. Rehder we owe a special debt of gratitude for his checking of this section in manuscript.

§ S–216 It will be noted that the material collected consists only of specimens living at the time of capture. No dead shells were collected. Hence there can be no confusion with empty shells cast up from deeper water or collected away from the environment normal for the living animal.

If the species involved is cited in Grant and Gale 1931, no citations previous to that are listed, and only Pilsbry and Lowe 1932 after that (chiefly for collecting notes, since in this case the report was written by the person who did the field work), and the latter reference only for Gulf stations, extensions of range, etc.

LIST OF SPECIES TAKEN:

ORDER PRIONODESMACEA

FAMILY ARCIDAE

Family diagnosis on p. 133 Grant and Gale 1931.

REFERENCES

§ S–217 *Arca multicostata* Sowerby 1833. PL. 33 FIG. 3
The cockle-like Arca.
Grant and Gale 1931, p. 139. Balboa (Southern California) to Gulf. Good toto-figures of Pliocene subspecies as fig. 5, Pl. 2.
Pilsbry and Lowe 1932, p. 141. Newport Beach (Southern California) to Tres Marias, rare on sand bars. La Paz.
A *Cadmium*-like form taken at El Mogote in sandy mud-flats, and on quiet sand flats at Concepcion Bay.

§ S–218 *Arca tuberculosa* Sowerby 1833.
G. and G. 1931, p. 141. Cedros to Peru, mangrove swamps and muddy places.
P. and L. 1932, p. 141. Very abundant in the soft mud of mangroves, used for food throughout Central America.
A *Cardium*-like form taken with *A. multicostata* on the sandy mud-flats at El Mogote.

§ S–219 *Barbatia reeveana* d'Orbigny 1846. PL. 32 FIG. 2
The bristly mussel-like Arca.
G. and G., p. 143, as *Arca (B) reeveana*. Lower California to Peru and Galápagos. Possibly Balboa, in southern California.
P. and L. 1932, p. 141. Under rocks at extreme tide, Guaymas, La Paz, etc.
A very large and hairy ribbed mussel-like (elongate) form taken on the drowned coral flats east of La Paz, juveniles and adults at Puerto Escondido, Concepcion Bay. Resembles *Volsella* superficially.

§ S–220 *Fossularca solida* Sowerby 1833. PL. 39 FIG. 4
Garbanzo clam.
Dall 1921 (§ S–14) p. 16, as *Barbita solida* Broderip and Sowerby 1833; San Diego to Panama.
P. and L. 1932, p. 141, as *Arca solida* Broderip and Sowerby, common under rocks, Guaymas, Cape San Lucas.
This is one of the most abundant Gulf invertebrates, and one of the two most common clams, reported in the notes as "Garbanzo clam." Occasional as a small under-rock form at Pt. Lobos, Espiritu Santo Island; more abundant at Puerto Escondido and Coronado

REFERENCES

Island; exceedingly common in the north at San Carlos Bay, Angeles Bay, Puerto Refugio, at the south end of Tiburon Island, and at Port San Carlos, Sonora.

§ S–221 *Fugleria illota* Sowerby 1833.
P. and L. 1932, p. 140 as *Arca solida,* rare, under rocks at extreme tides, Cape San Lucas.

Reported in our notes as "the triangular *Arca.*" Angeles Bay, Puerto Refugio, south end of Tiburon Island, Port San Carlos, Sonora—hence commoner in the northern part of the Gulf.

§ S–222 *Glycymeris giganteus* Reeve.
Rogers (§ S–41) 1908, as *Pectunculus giganteus.* By inference: tropical America and Gulf.

Superficially similar to a gigantic Venerid.

§ S–223 *Navicula mutabilis* Reeve.
P. and L. 1932, p. 141, as *Arca mutabilis* Sowerby, abundant under rocks, La Paz and Cape San Lucas (south to Panama).

Pulmo Reef. As "a minute *Zirfaea*-like form" at Pt. Lobos, Espiritu Santo Island; reef at Marcial Point. Hence largely in the southern part of the Gulf.

§ S–224 *Navicula pacifica* Sowerby 1833. PL. 37 FIG. 5
The elongate, irregular Arca.
G. and G. 1931, p. 143, as *Arca (N.) pacifica.* Scammon's Lagoon to Tumbes, north Peru.
P. and L. 1932, p. 141, as *Arca pacifica.* Rare, under rocks, La Paz, Guaymas.

Remarked in our collecting notes as "common large irregular elongate form," at Concepcion Bay.

FAMILY PINNIDAE

Family diagnosis on p. 144 G. and G. 1932.

§ S–225 *Atrina tuberculosa* Sowerby. Hatchet clam.
P. and L. 1932, p. 140, as *Pinna tuberculosa.* Mud-flats, La Paz.

The wider, more convex, and less common of the two "hachas" or hatchet clams. Taken at La Paz, and probably at Concepcion Bay, possibly elsewhere—we failed to differentiate the two species in the field, and specimens were not tagged individually.

§ S–226 *Pinna rugosa* Sowerby. Hatchet clam.
P. and L. 1932, p. 140. Mud bars. Guaymas and La Paz

REFERENCES

(to Montijo Bay, Central America, 200 miles north of Panama City).

The "hacha" of the Gulf. Very common, the largest Pelecypod encountered, occasional specimens were nearly 2 feet long; an important food product. Pt. Lobos on Espiritu Santo Island; Puerto Escondido; Concepcion Bay, very plentiful. In mud, muddy sand, or in sand. Most of our specimens however, and the largest ones, came from flats of nearly pure sand, as at Concepcion Bay.

FAMILY ISOGNOMIDAE MELINIDAE in P. & L. 1932. Entirely tropical, not treated in G. and G. 1931.

§ S–227 *Isognomon anomioides* Reeve. PL. 33 FIG. 5

Paper-shell clam.

P. and L. 1932, p. 140, as *Melina (Perna) anomioides*. Under rocks between tides, La Paz, Cape San Lucas, Guaymas (to Tres Marias).

A small, flat, very delicate, and opalescent, *Hinnites*-like form attached usually to the underside of rocks. One of the three commonest Pelecypods in the Gulf, occurring at all suitable stations from Cape San Lucas, clear up to the north and down the Sonora side to Guaymas.

Cape San Lucas; Pulmo Reef (juvenile, determination not certain); and as a *Pectin*-like adult. Pt. Lobos on Espiritu Santo Island. The larger (with axial flutings) of two species, (the other being *I. chemnitziana* below) at Puerto Escondido. As "an oyster-like form" from mangrove roots, Gabriel Bay on Espiritu Santo Island. *Isognomon* sp., not differentiated in the field but probably *anomioides*, was also taken east of La Paz, at Puerto Escondido, Coronado Island, Concepcion Bay (probably on shells of *Pinna*), Angeles Bay, Puerto Refugio, south end of Tiburon Island, and Port San Carlos, Sonora.

§ S–228 *Isognomon chemnitziana* d'Orbigny.

P. and L. 1932, p. 140, as *Melina* (*Perna*) *chemnitziana*, with the preceding at La Paz and Cape San Lucas (to Tres Marias).

Taken with the above at Puerto Escondido (the smaller with black coloring) and at San Carlos Bay, Baja California.

FAMILY PTERIIDAE: THE PEARL OYSTERS
Family diagnosis on p. 147, G. and G. 1931.

§ S–229 *Pinctada fimbriata* Dunker 1852.

G. and G. 1931, p. 148. Gulf only.

A small flat dark form similar to the following, taken, attached, once only at Puerto Escondido.

REFERENCES

§ S–230 *Pinctada mazatlanica* (Hanley) 1855. PL. 32 FIG. 3
The Gulf pearl oyster.
G. and G. 1931, p. 148. Gulf to Panama.
P. and L. 1932, p. 140, as *Margaritiphora* (*Pinctada*) *mazatlanica.* La Paz, Cape San Lucas.

Formerly the source of an important industry, La Paz having been the second pearling port of the world for many years. In our experience, specimens of this sort are exceedingly rare, at least in the littoral. We took only one specimen, slightly sub-littoral, at Puerto Escondido. However, Indians still fish for pearls in the Gulf, so there must be some representatives of this formerly abundant species still available. So closely related to the Oriental *P. margaritifera* that many authorities still regard it merely as a subspecies.

Family Ostreidae

Family diagnosis on p. 149, G. and G. 1931.

§ S–231 *Ostrea cumingiana* Dunker 1846.
G. and G. 1931, p. 150, Gulf and west coast of Mexico.

Taken only once, at Cape San Lucas, where it occurred attached to the rocks with *Chama echinata* and *O. mexicana.*

§ S–232 *Ostrea mexicana* Sowerby 1871. PL. 37 FIG. 2
The mangrove oyster.
G. and G. 1931, p. 150, mention only as possible synonym of *O. cumingiana.*
P. and L. 1932, p. 139. "On rocks and mangrove roots in protected locations . . . Guaymas, La Paz. . . ." (To Montijo Bay, Central America).

With *O. cumingiana* and *Chama echinata* at Cape San Lucas. At Angeles Bay noted as "rock oysters" in the collecting report. From mangrove roots at Gabriel Bay on Espiritu Santo Island. The Gabriel Bay specimens were heavily encrusted with barnacles, *Chthamalus anisopoma* (§ N–19), per photo.

§ S–233 There is a large oyster industry in Guaymas. Unfortunately we neglected to bring back any of the material. Pilsbry and Lowe 1932, p. 139, report the species involved as *O. chilensis* Philippi, and it is under this name that the commercial west coast form is known in the museums at Mexico City.

"On mud bars, especially near river mouths. This is the large edible form of the Gulf, often reaching a length of ten inches or more." Farther south, from Manzanillo to Panama, the edible oyster is said to be *O. iridescens* Gray, P. and L. 1932, p. 139.

§ S–233a *Ostrea* sp.

Rock oyster from Pulmo Reef, noted in our collecting reports as "a *Chama*-like form."

FAMILY SPONDYLIDAE: Entirely tropical, not treated in G. and G. 1931.

REFERENCES

§ S–234 *Spondlyus* sp. probably *limbatus* Sowerby. PL. 28 FIG. 2
Thorny oyster.
Illustrated in color on Pl. 76, p. 264, Smithsonian Series, Vol. 10, as *Spondylus princeps* Broderip.

P. and L. 1932, p. 139, as *Spondylus crassisquama* Lamarck (for the one with the long spines) and *S. limbatus* Sowerby (for the one with reduced spines and internal purple margin). They were not differentiated in our field notes, although we recall the purple margin in material being prepared for food at Puerto Escondido. The latter is recorded by Pilsbry and Lowe from the Gulf only, the former from La Paz to Panama.

Taken only once, but in considerable numbers, slightly subtidally, at Puerto Escondido, where they were got for food. Eaten by the natives, who call them "abalon" and who bring them up from the bottom at low tide by harpoons or fish spears. Among the most spectacularly shaped and colored of the large bivalves. The lower valve is attached to the rock.

FAMILY ANOMIIDAE

Family diagnosis on p. 240, G. and G. 1931.

§ S–235 *Anomia peruviana* d'Orbigny 1846.
G. and G. 1931, p. 240, with fair toto-illustration to show inside of valve as fig. 5, Pl. 12. San Pedro to Peru and the Galápagos.

Taken only once, at Puerto Escondido, where it appears in our notes merely as a rock oyster.

FAMILY MYTILIDAE

Family diagnosis on p. 244, G. and G. 1931.

§ S–236 *Botulina opifex* Say.
P. and L. 1932, p. 138, as *Modiolus opifex*. Under rocks, rare. La Paz (to Montijo Bay).

A chunky boring clam common, with *Lithophaga aristata*, at Pulmo Reef.

§ S–237 *Brachidontes multiformis* Carpenter. PL. 32 FIG. 1
Small shore mussel.
P. and L. 1932, p. 138, as *Mytilus multiformis*, on rocks, Mazatlan.

REFERENCES

The commonest very small Pelecypod in the Gulf. Resembling small specimens of the California and edible mussels and like them attached by byssal hairs to rocks exposed to sun and wave shock.

Pulmo Reef. High in the intertidal at Amortajada Bay, Coronado Island. Attached to *Isognomon chemnitziana* at San Carlos Bay, Baja California. High up at Puerto Refugio. On mangrove roots or on stones at foot of mangroves, Gabriel Bay on Espiritu Santo Island. The tropical analogue of the northern surf-swept *M. californianus,* occurring high in the intertidal and exposed to very unmarine conditions, but able to exist apparently in quiet water communities, as on the mangroves at Gabriel Bay.

§ S–238 *Lithophaga (Myoforceps) aristata* Dillwyn.

P. and L. 1932, p. 138, as *Lithophagus aristatus,* boring in rocks at La Paz.

The more elongate of the two boring forms at Pulmo Reef. In dead coral heads east of La Paz. Concepcion Bay (no data, possibly boring in some of the larger shells; the collecting grounds were sand flats). As minute boring clams under coral at Puerto Refugio.

§ S–239 *Volsella capax* (Conrad) 1837. PL. 33 FIG. 4

The horse mussel.

G. and G. 1931, p. 249, Santa Barbara to Payta, Peru.

P. and L. 1932, p. 138, as *Modiolus capax.* Attached to stones on mud-flats, La Paz, Guaymas.

Very common at La Paz as a "*Mytilus*-like form." El Mogote. Puerto Escondido.

ORDER TELEODESMACEA

FAMILY CARDITIDAE

Family diagnosis on p. 272, G. and G. 1931.

§ S–240 *Carditamera affinis californica* Deshayes 1852. PL. 33 FIG. 2

Ruffled clam.

G. and G. 1931, p. 278, as *Glans affinis* (Sowerby). Margarita Bay, Gulf to Panama.

P. and L. 1932, p. 137, in var. *californica,* as *Cardita affinis.* Mud-flats. "This is the large heavy form from the Gulf." Guaymas, La Paz, Mazatlan.

The commonest bivalve in the Gulf, especially to the north where it is larger and heavier. Under rocks and in tide pools; we failed to find it on uncomplicated mud-flats or sand flats. Pt. Lobos on Espiritu Santo Island. La Paz area (largest was 1¾"); San Francisquito Bay;

REFERENCES

San Carlos Bay, Baja California; Angeles Bay; Puerto Refugio; south end of Tiburon Island; Port San Carlos, Sonora. A brilliantly colored form was taken from stones about the mangrove roots in Gabriel Bay, Espiritu Santo Island. El Mogote (small); Puerto Escondido, small. A large specimen from Angeles Bay was 4″ long.

FAMILY CHAMIDAE

Family diagnosis on p. 279, G. and G. 1931.

§ S–241 *Chama echinata* Broderip.
P. and L. 1932, p. 137. "Attached to wave beaten rocks, next to impossible to obtain lower valve." Cape San Lucas, La Paz (to Panama).

Cape San Lucas with *Ostrea cumingiana* and *O. mexicana.* Also recorded as an "oyster-like form (*Spondylus*?)" from the reef at Pulmo, an important form physiographically, since it, with the coral *Pocillopora,* makes up the reef and therefore must be regarded as exceedingly abundant locally.

§ S–242 *Chama mexicana* Carpenter. PL. 39 FIG. 2
The Mexican rock oyster.
Range reported by Dr. Rehder, in a personal communication, as La Paz to Acapulco.

Common at Concepcion Bay, where we considered it to be a species of *Spondylus.*

§ S–243 *Chama squamuligera* Pilsbry and Lowe 1932.
p. 103, with toto-fig. 10, Pl. 14, P. and L. 1932. Cape San Lucas to Tres Marias and Nicaragua.

Pt. Lobos on Espiritu Santo Island; reef at Pt. Marcial; from stones about mangrove roots, Gabriel Bay, Espiritu Santo Island.

FAMILY LUCINIDAE (CODAKIIDAE)

Family diagnosis on p. 283, G. and G. 1931.

§ S–244 *Divaricella eburnea* Reeve.
Dall 1901 (§ S–205), p. 815. Cape San Lucas to Panama.

A small pearly-white form with diagonal sculpture. From El Mogote.

FAMILY UNGULINIDAE (DIPLODONTIDAE)

Family diagnosis on p. 292, G. and G. 1913.

§ S–245 *Felaniella sericata* Reeve 1850. PL. 39 FIG. 3
The Satin Diplodon.
G. and G. 1931, p. 295, as *Taras parilis* (Conrad) var. *sericatus* (Reeve), San Diego to Panama.

REFERENCES

P. and L. 1932, p. 137, as *Diplodonta (F) sericata.* Mud-flats, Guaymas, to Corinto.

Small clams very abundant on the sand bar almost at high tide, San Lucas Cove. As a flat smooth light form, Angeles sand flats, and again as small smooth clams at Estero de la Luna. Dall 1901 (§ S–205) gave the range (p. 796) as Cape San Lucas to Guayaquil and it is likely uncommon at its more recently reported northern limit.

§ S–246 *Taras subquadrata* Carpenter 1855.
P. and L. 1932, p. 137, as *Diplodonta subquadrata.* Mud-flats, La Paz.
Dall 1901 (§ S–205), p. 795, as *Diplodonta subquadrata,* Catalina to Panama, 16-36 fathoms. And Soot-Ryen 1932 (§ S–214) to Galápagos, but recent California synopses fail to list it as a United States species.

South end of Tiburon Island, with the triangular Arca *Fugleria illota.*

FAMILY CARDIIDAE

Family diagnosis on p. 302, G. and G. 1931.

§ S–247 *Trachycardium (Mexicardia) procerum* Sowerby 1833.
G. and G. 1931, p. 305, as *Laevicardium (M) procerum* (Sowerby), Scammon's Lagoon to Peru.
P. and L. 1932, p. 136, as *Cardium (Bingicardium,* typographical error for *Ringicardium) procerum* Broderip and Sowerby. "A common mud-flat species." La Paz, Guaymas, and south.

Recorded from the dead coral flats east of La Paz as a typical cockle.

FAMILY VENERIDAE

Family diagnosis on p. 315, G. and G. 1931.

§ S–247a *Anomalocardia subrugosa* Sowerby.
p. 134 P. and L. "A mud-flat species living between tides, not common." Guaymas to Central America.

A brilliantly colored *Area cardium*-like form taken at Estero de la Luna.

REFERENCES

§ S–248 *Chione succincta* (Valenciennes) 1821. PL. 33 FIG. 1
Hard-shell cockle.
G. and G. 1931, p. 321, with good toto-illustrations of fossil specimen, figs. 1-4, Pl. 16. As *Venus (Chione) succincta* Val. San Pedro to northern South America.
P. and L. 1932, p. 134. "Common on mud-flats." Guaymas, La Paz.

On sandy mud-flats at El Mogote. Very abundant in sandy beach at San Lucas Cove. Very abundant on sand flats at Angeles Bay, where several hundred were taken in an hour or so for food. Very plentiful at Agiabampo Estuary. Abundant (by the washtub full) at Gabriel Bay on Espiritu Santo Island. An important food item throughout the Gulf. Typically a sand flat form.

§ S–249 *Dosinia dunkeri* (Philippi) 1844.
G. and G. 1931, p. 354, Gulf of California to Tumbes (north Peru) and Galápagos.
P. and L. 1932, p. 135. "A common form on mud-flats from Gulf of California to Panama." Guaymas, La Paz, etc.

A large white or pearly white clam on the drowned coral flats east of La Paz, and on the sandy mud-flats of El Mogote, and as a smooth white clam at Agiabampo Estuary.

§ S–250 *Macrocallista* (*Paradione*) *aurantiaca* Sowerby 1841.
G. and G. 1931, p. 347, as *Pitar* (*Megapitaria*) *aurantiacus* (Sowerby), Gulf to Guayaquil.
P. and L. 1931, p. 135, as *Macrocallista aurantiaca* Hanley "rather an uncommon form at or below extreme low water." La Paz, etc., to Panama.

Two specimens marked "Large *Tivela*-like form" taken at Puerto Escondido; the larger was 125 mm.

§ S–251 *Macrocallista (Paradione) squalida* (Sowerby) 1835.
The Gulf Pismo clam. PL. 33 FIG. 6
G. and G. 1931, p. 347, as *Pitar* (*Megapitaria*) *squalidus* (Sowerby), Cedros Island and Scammon's Lagoon to Peru.
P. and L. 1932, p. 134, "common throughout west Mexico and Central America on sandy mud-flats." Guaymas and La Paz, to Panama.

A *Tivela*-like form from El Mogote (sandy mud-flats), San Lucas Cove sandy beach. Fairly abundant on the sand flats at Angeles Bay. Agiabampo Estuary, Sonora. Reminiscent of the California Pismo clam, but more vividly marked; the commonest smooth and glistening clam of this type in the Gulf.

REFERENCES

§ S-251a *Periglypta multicostata* Sowerby 1835.
G. and G. 1931, p. 316, as *Venus (Antigona) multicostata,* Gulf to Panama. P. and L. 1932, p. 135, as *Cytherea multicostata,* sand under or between rocks, La Paz to Panama.

Noted in our collecting reports from Puerto Escondido as "*Chione*-like form, 90 mm."

§ S-252 *Protothaca grata* (Sowerby). COLOR PL. 4 FIG. 1
G. and G. 1931, p. 328, as *Venerupis (Protothaca) grata* (Say) 1831, Turtle Bay (outer coast of Lower California just below Cedros) to Chile ("has been reported from San Pedro," p. 329).
P. and L. 1932, p. 134, as *Tapes (Paphia) grata* Say. "A common species of the mud-flats which in Central America blooms out in several exotic color patterns, a number of which have been given names." Guaymas and La Paz to Panama.

Puerto Escondido (2 only). Several at Angeles Bay from sandy shore. Several hundred brilliantly colored specimens were dug for food at half tide along the gravel shore under a rock ledge at the Pajaro Island anchorage just outside Guaymas.

FAMILY TELLINIDAE

Family diagnosis on p. 357, G. and G. 1931.

§ S-252a *Macoma (Rexithaerus)* sp.
A small smooth clam, very common in the sand flats of Angeles Bay. ". . . apparently a new species . . . close to *M. secta* and *M. indentata* which I hope you will allow me to name *rickettsi* . . ." Personal communication from Dr. Rehder, U. S. National Museum.

FAMILY SEMELIDAE

Family diagnosis on p. 375, G. and G. 1931.

§ S-253 *Semele corrugata* Sowerby.
Range reported from the literature by Dr. Rehder, in a personal communication, as "Lower California."

Remarked in the collecting notes from Puerto Escondido as "unribbed *Paphia*-like form," a single specimen only was taken.

FAMILY SANGUINOLARIIDAE (PSAMMOBIIDAE)
Family diagnosis on p. 381, G. and G. 1931.

REFERENCES

§ S–254 *Tagelus affinis* C. B. Adams.
P. and L. 1932, p. 131. "Tide flats, not common." (Acapulco to Montijo Bay, not taken in the Gulf.)
A single specimen was taken at San Lucas Cove, as a typical razor clam.

§ S–255 A small *Paphia*-like form from Angeles Bay sand flats remains still undetermined.

§ S–256 SUMMARY

42 species of Pelecypods are recorded above, all from Gulf shores, intertidal or slightly subtidal. Geographic information is lacking on the above, on *Ostrea* sp., *Macoma* sp., and on *Semele corrugata* (§ S–253). On the following, reported by Pilsbry and Lowe from the area between the Gulf and Panama, we have only their records with no other information: *Fugleria illota* Sowerby and *Navicula mutabilis* Reeve of the ARCIDAE, *Pinna rugosa* Sowerby and *Atrina tuberculosa,* of the PINNIDAE, *Isognomon* 2 spp. from the ISOGNOMONIDAE, *Botulina opifex* and *Lithophaga aristata* of the MYTILIDAE, *Chama echinata* and *squamuligera* of the CHAMIDAE, and *Tagelus affinis* C. B. Adams of the SANGUINOLARIIDAE (PSAMMOBIIDAE). However, on the latter group, we can assume that the material is Panamic only, since if these species occurred in the overlap or North Temperate provinces, they would have been treated in Grant and Gale 1931, in Dall 1921, Oldroyd 1924, etc.

From the geographical information available on the 38 species, it appears that only 6 are restricted to the Gulf, 1 (*Arca multicostata*) extends only to the North (to Newport Bay in southern California), 18 extend from the Gulf (in one case including Magdalena Bay) to the south, and 13 extend both north and south. Of the 13 extending both north and south, none overreaches the overlap stretches of the Panamic in the north; 6 reach into southern California. But at least two of these, *Taras subquadrata* and *Felaniella* must be very uncommon in the southern California region, which represents the most northerly extension of their range. *Taras* is not included in any of the present California lists, and *Felaniella* was recorded north of Cape San Lucas only since 1901. The

REFERENCES

southern California records, furthermore, are so complete, and the conchological search there is so intense, that strays and extreme rarities are likely to be listed as members of the fauna.

1 extends into "Lower California," and 5 are limited by the Cedros Island-Scammon's Lagoon-Turtle Bay region. Of the southward reaching species, 12 extend to northern South America (Guayaquil, Galápagos, or northern Peru), 1 reaches Chile. Many are recorded no farther south than Panama.

The commonest rocky shore forms, the garbanzo clam (*Fossularca solida,* San Diego to Panama), the delicate attached *Isognomon anomioides* (Gulf to Tres Marias), *Ostrea mexicana* (Gulf to Central America), *Brachidontes multiformis* (the Gulf *"Mytilus"* apparently restricted to this area), and *Carditamera affinis*—the commonest of them all (Magdalena Bay to Panama), are all strongly Panamic. Soft bottom species, *Pinna rugosa*—the "hacha" (Gulf to Panama), the small smooth *Felaniella* of upper sand flats (San Diego to Panama), the exceedingly common edible clams *Chione succincta* (San Pedro to northern South America), *Macrocallista squalida* (Cedros Island to Peru), and *Protothaca grata* (Turtle Bay to Chile), have more northern tendencies, but only *Chione succincta* occurs commonly in southern California, and even here the related *C. undatella* seems to be more common. Hence the tropical affiliations, even though not so marked as in other groups, are nevertheless fairly predominant. Of the Families, the tropical ARCIDAE is best represented, with 8 species, and the tropical PINNIDAE, ISOGNOMIDAE, PTERIDAE, and SPONDYLIDAE, sparsely or not at all represented on Pacific United States shores, are important features of the Gulf fauna.

CLASS GASTROPODA

SHELLED GASTROPODS

§ S–301 Baker, Fred 1926. "Mollusca of the family TRIPHORIDAE." Proc. Calif. Acad. Sci. (4), Vol. 15 (6): 223-239, Pl. 24.

From the Academy's 1921 Gulf Expedition. 11 species and subspecies, of which 10 are new, all small (largest was 8 mm.). Good toto-illustrations of the new species.

REFERENCES

§ S–302 Baker, Fred and V. D. P. Spicer 1935. "New species of mollusks of the genus *Triphora.*" Trans. San Diego Soc. Nat. Hist., Vol. 8 (7): 35-46, Pl. 5.

The two Panamic species were taken in the Gulf by Capt. Porter who, with his companion, "was ambushed and killed in 1896 on Tiburon Island in the Gulf of California by the Seri Indians. Their boat, which was owned by the late Miss J. M. Cooke of San Diego, was looted and burned." (§ S–303.) Minute specimens with good toto-illustrations.

§ S–303 .... and G. Dallas Hanna, and A. M. Strong 1928. "Some PYRAMIDELLIDAE from the Gulf of California." Proc. Calif. Acad. Sci. (4), Vol. 17 (7): 205-246, Pls. 11 and 12.

53 species, usually small, many of them new, from the Academy's 1921 Gulf Expedition, and from the ill-fated activities of Capt. Porter.

§ S–304 .... 1930. "Some Rissoid Mollusca. . . ." and "Some Mollusca of the FAMILY EPITONIIDAE from the Gulf of California." Proc. Calif. Acad. Sci. (4), Vol. 19 (4 and 5): 23-56, Pls. 1-3, 4 text figs.

27 and 13 species, all small, many of them new, resulting from the activities of Capt. Porter, aforementioned, and of the Academy.

§ S–305 .... 1938. "Some Mollusca of the FAMILIES CERITHIOPSIDAE, CERITHIIDAE, and CYCLOSTREMATIDAE from the Gulf of California and adjacent waters." Proc. Calif. Acad. Sci. (4), Vol. 23 (15): 217-244, Pls. 17-23.

41 species, many of them new, small to medium, resulting from various expeditions of the Academy, from Capt. Porter's collections, etc. Excellent collotype illustrations from photographs.

§ S–306 .... 1938a. "COLUMBELLIDAE from western Mexico." Proc. Calif. Acad. Sci. (4), Vol. 23 (16): 245-254, Pl. 24.

24 species, many new, small to 26 mm. with excellent collotype illustrations from photographs of 11 species.

§ S–307 Bartsch, Paul 1907. "New mollusks of the FAMILY VITRINELLIDAE from the west coast of North America." Proc. U.S.N.M., Vol. 32: 167-176, 11 text figs.

3 small new species (1 or 2 mm. diameter) are from Ecuador, Cape San Lucas, and Cedros Island. Others from Pt. Loma to Pt. Abrojos.

§ S–308 .... 1917. "Descriptions of new west American marine mollusks and notes on previously described forms." Proc. U.S.N.M., Vol. 52: 637-681, Pls. 42-47.

REFERENCES

Among extra-limital and fossil forms, some 43 new species, mostly small, are described from the southern California to Gulf area, in the genera *Pyramidella, Turbonilla, Odostomia, Cerithopsis, Bittium, Alvania,* and *Rissoina,* with toto-illustrations.

§ S-308a .... 1917a. "A monograph of west American Melanellid Mollusks." Proc. U.S.N.M., Vol. 53: 295-356, Pls. 34-59.

About 47 Panamic and 18 overlap species, many of them new, with good toto-illustrations of the new species. All small. Mostly from shore, but several were dredged, some from 1772 fathoms.

§ S-309 .... 1924. "New mollusks from Santa Elena Bay, Ecuador." Proc. U.S.N.M., Vol. 66 (14): 9 pp., 2 pls.

Small Gastropods in the FAMILIES PYRAMIDELLIDAE and MELANELLIDAE collected by Olsson in the area made famous by the activities of Hugh Cuming.

§ S-310 .... 1926. "Additional new mollusks from Santa Elena Bay, Ecuador." Proc. U.S.N.M., Vol. 69 (20): 20 pp., 3 pls.

As before. 14 new species from a teaspoonful of shells.

§ S-311 .... 1927. "New west American marine mollusks." Proc. U.S.N.M., Vol. 70 (11): 36 pp., 6 pls.

10 species from southern California and northern Lower California, 3 from Magdalena Bay, all minute Gastropods.

§ S-312 .... 1931. "Descriptions of new marine mollusks from Panama, with a figure of the genotype of *Engina*." Proc. U.S.N.M., Vol. 79 (15): 10 pp., 1 pl.

7 new species, all under 20 mm., in the genera *Anachis, Eudaphne, Mitra, Rissoina,* and *Engina,* all with good toto-illustrations.

§ S-313 Dall, W. H. and Paul Bartsch 1909. "A monograph of west American pyramidellid mollusks." Bull. U.S.N.M., Vol. 68: 258 pp., 30 pls.

Small and minute forms, many of which occur in the Panamic area. Well illustrated.

§ S-314 Dall, W. H. 1910. "Summary of the shells of the genus *Conus* from the Pacific coast of America, in the United States National Museum." Proc. U.S.N.M., Vol. 38: 217-228.

Most of which occur in the overlap or Panamic area. No illustrations.

§ S-315 .... 1915a. "West American species of *Nucella*." Proc. U.S.N.M., Vol. 49: 557-572, 2 pls.

REFERENCES

Mostly from the north Pacific, but one form is reported from Mazatlan.

§ S–316 .... 1917. "Summary of the mollusks of the FAMILY ALECTRIONIDAE of the west coast of America." Proc. U.S.N.M., Vol. 51: 575-579, no illustrations.

Catalogue, with geographic range, of all species then known from the area; many Panamic and overlap forms. Descriptions of 6 new species.

§ S–316a .... 1917a. "Notes on the shells of the genus *Epitonium* and its allies of the Pacific coast of North America." Proc. U.S.N.M., Vol. 53: 471-488.

About 47 Panamic and 9 overlap species of these small shells, usually from along shore, the deepest in our region being 62 fathoms.

§ S–317 .... 1919. "Descriptions of new species of mollusks of the FAMILY TURRITIDAE from the west coast of America and adjacent regions." Proc. U.S.N.M., Vol. 56: 1-86, 24 pls.

A consideration of over 200 species of these generally small snails—the largest shell illustrated was 52 mm., the smallest 3.5 mm. 181 new species are described and usually illustrated by a toto-photograph. Over half are Panamic or overlap.

§ S–317a .... 1919a. "Descriptions of new species of Mollusca from the north Pacific Ocean in the collection of the United States National Museum." Proc. U.S.N.M., Vol. 56: 293-371, no illustrations.

Among the 222 new species of Gastropods here described, including shelled Tectibranchs and Heteropods, 37 plus 1? are Panamic (including the Galápagos) but 5 are from depths of more than 100 fathoms. 61 plus 1? have ranges which rate them as overlap forms, but 17 are from deep water.

§ S–318 Strong, A. M. 1928. "West American Mollusca of the genus *Phasianella*." Proc. Calif. Acad. Sci. (4), Vol. 17 (6): 187-203, Pl. 10.

11 species, one new, with one new name. All small, the largest was 9.2 mm. high. All but one illustrated. Mostly from Cape San Lucas.

§ S–319 There are many shorter or less important items on Panamic and southern California shells in Nautilus, especially in Vols. 52 and 53, 1938-9. Most of the previous accounts, especially as they relate to California species, have been picked up in the general mollusk bibliographies of Dall 1909 and Keen 1937. See also Gastropod items in the general mollusk papers cited under § S–1 ff. No attempt was

REFERENCES

made to abstract the Gastropod literature as it related to Panamic species, but the main highlights presumably will have been picked up in the papers which relate specifically to collections made in the Panamic area, most of which are thought to have been examined.

§ S-320 ACKNOWLEDGMENT: For determining the following shelled Gastropods, for gifts of and help with the literature, and for reading over this section of the phyletic catalogue, we are grateful to Dr. Harald A. Rehder, Assistant Curator, Division of Mollusks, United States National Museum.

LIST OF SPECIES TAKEN:

§ S-321 As with the Pelecypods, it will be noted that the material consists only of specimens living at the time of capture, hence all the animals will be characteristic of the environment from which we report them. It should be noted also that we made no attempt to procure very small or minute specimens, so that representatives of several FAMILIES, very common in the Gulf, the TURRIDAE, COLUMBELLIDAE, EPITONIIDAE, the MELANELLIDAE, and PYRAMIDELLIDAE, the TURRITELLIDAE, and RISSOIDAE, will be sparse or entirely missing from our list. The key paper, as with the Pelecypods, is Grant and Gale 1931 (§ S-16).

ORDER PULMONATA

FAMILY ELLOBIIDAE (p. 461 G. and G. 1931.)

§ S-322 *Melampus olivaceous* Carpenter 1857 (also as *olivaceus*).
p. 461 G. and G. 1931, range: Salinas River (Monterey Bay) to Mazatlan.
p. 54 Oldroyd 1927 (§ S-23), Pt. I, Vol. II, toto-fig. 16, Pl. I.

Taken only once, in the lagoon off Amortajada Bay.

FAMILY SIPHONARIIDAE (p. 463 G. and G. 1931.)

§ S-323 *Siphonaria aequilirata* Reeve. PL. 28 FIG. 1
Digitate Pulmonate Limpet.

A limpet-like form taken at 3 stations: this (or the following) on the starfish *Heliaster* at Cape San Lucas, as "a sand flat limpet" at

REFERENCES

Angeles Bay, clustered on occasional rocks buried in sand, and from Gabriel Bay on Espiritu Santo Island, ecological niche unknown.

§ S–324 *S. maura* Sowerby.

Taken only once, as a limpet (possibly this was the species taken on a starfish) at Cape San Lucas.

§ S–325 *S. pica* Sowerby. PL. 26 FIG. 3
Starry Pulmonate Limpet.
p. 108 Pilsbry and Lowe 1932 (§ S–26), San Juan del Sur, Central America.

This was recorded in our collecting notes as "a stellate *Acmaea*" (although widely separated phyletically from the Prosobranchs) at Pulmo Reef.

ORDER CTENOBRANCHIATA (PROSOBRANCHIATA)

SUPERFAMILY TOXOGLOSSA

FAMILY TEREBRIDAE (p. 464 G. and G. 1931.)

§ S–326 *Terebra (Myurella) variegata* Gray 1834.
COLOR PL. 5 FIG. 1
p. 466 G. and G. 1931, as *T. (Strioterebrum) variegata.* Guaymas and Scammon's Lagoon to Galápagos.
p. 109 P. and L. 1932, "living specimens taken on tide flats, La Paz."

A turreted shell very common on the sand flats at Angeles Bay.

FAMILY CONIDAE (p. 471 G. and G. 1931.)

§ S–327 *Conus brunneus* Mawe. Royal Cone.
p. 109 P. and L. 1932, "in rock ledges at extreme tide." La Paz to Panama.

With *C. princeps* and *C. nux* at Pt. Lobos on Espiritu Santo Island.

§ S–328 *C. nux* Broderip.
p. 109 P. and L. 1932, "nestling in crevices of rocks exposed to the surf." Cape San Lucas to Panama.

Cape San Lucas. At Pt. Lobos on Espiritu Santo Island, with the other cones.

§ S–329 *Conus princeps* Linné 1758. PL. 34 FIG. 1
p. 475 G. and G. 1931, Cape San Lucas to Payta, Peru.
p. 109 P. and L. 1932, "on coral reefs or rocks at extreme tide." La Paz and Cape San Lucas to Central America.

REFERENCES

At Pt. Lobos on Espiritu Santo Island, with the other cones. On the reef at Marcial Pt. At Port San Carlos, Sonora. The largest cone. Often encrusted.

FAMILY TURRIDAE (p. 477 G. and G. 1931.)

§ S–330 *Monilispira monilifera* Carpenter.
p. 111 P. and L. 1932, as *Crassipira monilifera,* "between tides and under rocks." Central America.
A small snail from under rock at Pt. Lobos on Espiritu Santo Island.

SUPERFAMILY RACHIGLOSSA
FAMILY OLIVIDAE (p. 623 G. and G. 1931.)

§ S–331 *Oliva venulata* Lamarck.
COLOR PL. 6 FIG. 2 AND COLOR PL. 7 FIG. 1
Very abundant on the sand flats at El Mogote. Concepcion Bay (no data, collecting here was mainly on slightly subtidal sandy shore). San Carlos Bay, Baja California. In a color form, from the dazzling white coral sands at Gabriel Bay on Espiritu Santo Island.

§ S–332 *Olivella dama* Mawe 1883.
p. 626 G. and G. 1931, as *Olivella dama* (Wood) 1828. Range: Gulf to Mazatlan.
p. 112 P. and L. 1932, as *O. dama* Gray. "Beach shells at Guaymas; La Paz; Cape San Lucas."
At San Lucas Cove, sand flats at half tide.

FAMILY VOLUTIDAE (p. 633 G. and G. 1931.)

§ S–333 *Enaeta cumingii* Broderip.
p. 113 P. and L. 1932, as *Lyria* (*Enaeta*) *cumingii.* "Three live shells taken on the reef at La Paz."
From rocks on sandy beach at Concepcion Bay.

FAMILY MITRIDAE (p. 634 G. and G. 1931.)

§ S–334 *Mitra tristis* Broderip.
p. 113 P. and L. 1932, "under stones at low tide, rare." La Paz and Cape San Lucas to Panama.
Taken only once, under rock at Pt. Lobos on Espiritu Santo Island.

§ S–335 *Strigatella dolorosa* Dall.
Several taken on the rocky shore at Angeles Bay.

FAMILY FASCIOLARIIDAE (p. 637 G. and G. 1931.)

REFERENCES

§ S–336 *Fasciolaria princeps* Sowerby. PL. 40 FIG. 3
p. 637 G. and G. 1931, Magdalena Bay to Gulf to Peru and Galápagos.
p. 114 P. and L. 1932, "brought in occasionally by fishermen or on beaches after storms. Mazatlan; La Paz."
Several specimens were taken at Puerto Escondido slightly subtidally by net or spear from skiff. A shell only was brought back from Concepcion Bay as the largest Gastropod occurring in the region.

§ S–337 *Fusinus dupetit-thouarsi* (Kierner) also as *dupetit-thouarsii.*
p. 638 G. and G. 1931, Lower California and west coast of Mexico to Galápagos.
p. 114 (in the BUCCINIDAE) P. and L. 1932, "on mud-flats at extreme tide. La Paz; Acapulco; San Juan del Sur."
In our collecting notes as "large snail . . . *Fasciolaria*" from San Lucas Cove sand flats. Port San Carlos, Sonora.

§ S–338 *Melongena patula* Broderip and Sowerby. COLOR PL. 8 FIG. 2
p. 12 Boone 1928 (§ S–4) as *Melongena patula,* seven specimens from Angeles Bay.
Another of the large Fasciolarids. . . . Fairly common at Estero de la Luna.

FAMILY BUCCINIDAE (p. 667 G. and G. 1931.)

§ S–339 *Engina ferruginosa* Reeve. Rusty Engina. PL. 38 FIG. 2
p. 435 Dall 1908 (§ S–11), listed from shore, Panama Bay.
Common east of La Paz. Puerto Escondido. In the upper tide pools at Puerto Refugio. Very abundant at Port San Carlos, Sonora, where juveniles were also taken.

FAMILY NASSARIIDAE (p. 670 G. and G. 1931 [also as Alectrionidae].)

§ S–340 *Nassarius ioaedes* Dall. COLOR PL. 5 FIG. 2
p. 30 Lowe 1935 (§ S–21) as *Nassa iodes* Dall, "many living in sand flats" at Punta Peñasco, Sonora.
As small sand flat snails at San Lucas Cove. As minute snails on the sand flats at Estero de la Luna.

§ S–341 *Nassarius luteostoma* (Broderip and Sowerby). COLOR PL. 5 FIG. 2
p. 671 G. and G. 1931, mention only, Panamic (Gulf species is stated to be *Arcularia tiarula* Kierner).

REFERENCES

p. 115 P. and L. 1932 as *Nassa*. "Mud-flats, common." Guaymas and La Paz to Central America.

Common on the sand flats at Estero de la Luna in Sonora.

§ S–342 *Nassarius tegula* (Reeve).
p. 671 G. and G. 1931, good toto-fig. 43, Pl. 26, San Francisco to Lower California.
p. 267 Oldroyd 1927 (§ S–23), Pt. I, Vol. II, with toto-fig. 10, Pl. 27 as *Alectrion tegulus*.

San Lucas Cove, sand flats; Angeles Bay, ecological niche not certain, probably sand flat; Estero de la Luna, Sonora, sand flats.

FAMILY COLUMBELLIDAE (PYRENIDAE) (p. 679 G. and G. 1931 [see § S–306].)

§ S–343 *Anachis coronata* (Sowerby) 1832.
p. 685 G. and G. 1931, Magdalena Bay to Ecuador.
p. 116 P. and L. 1932, "under rocks between tides, rare," La Paz and Guaymas to Central America.
p. 249 Baker, Hanna, and Strong 1938a (§ S–306), 8 stations in Gulf, toto-fig. 5, Pl. 24.

Angeles Bay, ecological niche unrecorded. At this point we collected on rocky beach and on sand flats.

§ S–344 *Anachis milium* (Dall) 1916.
p. 688 G. and G. 1931, Gulf to Ecuador.
p. 117 P. and L. 1932, "crab specimens only, rare. Mazatlan; Manzanillo."
p. 249 Baker, Hanna, and Strong 1938a (§ S–306), one specimen collected on Carmen Island. No illustration.

Port San Carlos, Sonora.

§ S–345 *Anachis reevei* Carpenter 1864.
p. 689 G. and G. 1931, as *A. santa-barbarensis* Carpenter 1856, Gulf to Acapulco.
p. 116 P. and L. 1932, as *Mitrella santabarbarensis,* "under rocks, exceedingly rare; only 5 specimens taken in three years collecting." La Paz to Central America.

Cape San Lucas.

§ S–346 *Columbella fuscata* Sowerby 1832.
p. 681 G. and G. 1931, as *Pyrene fuscata* (Sowerby), Gulf to Paita and Galápagos; (Orcutt reported from southern California).

REFERENCES

p. 116 P. and L. 1932, "under rocks between tides, common," Guaymas, etc. to Panama.
p. 247 Baker, Hanna, and Strong 1938a as *Pyrene fuscata* (Sowerby); widespread in Gulf; toto-fig. 2, Pl. 24.

At Cape San Lucas, rocky shore. In interstices of coral, east of La Paz. On bouldery shore at Puerto Refugio on Angel de la Guardia Island.

§ S–347 *Nitidella densilineata* Carpenter.

Taken only once, at Cape San Lucas.

§ S–348 *Nitidella guttata* (Sowerby) 1832. Variegated Dove-shell.
p. 696 G. and G. 1931, as *Mitrella ocellata* (Gmelin) 1791, Magdalena Bay to Ecuador and Galápagos; West Indies.
A subspecies of *guttata, baileyi,* is described with good toto-fig. 6, Pl. 2, in Bartsch and Rehder 1939 (§ S–3).

A most common small variegated snail. Cape San Lucas; Pt. Lobos on Espiritu Santo Island; Amortajada Bay; Marcial Point; Puerto Escondido; San Carlos Bay; Angeles Bay; Gabriel Bay on Espiritu Santo Island. Could have been taken by the hundred; the commonest obvious small Gastropod encountered, occurring especially where there were sand or gravel filled crevices between rocks and boulders.

§ S–349 *Parametaria coniformis* Sowerby.
p. 117 P. and L. 1932. "Among small rocks in muddy bays. Guaymas."

Port San Carlos, Somora; two color phases, one was of an orange persimmon color, very striking; about 1″ long; rocks on gravel flats.

§ S–350 *Strombina maculosa* (Sowerby) 1832.
p. 699 G. and G. 1931, Cape San Lucas to Central America, common in the Gulf.
p. 118 P. and L. 1932, "dredged at 20 fathoms. Also on mud-flats. La Paz; Guaymas."
p. 251 Baker, Hanna, and Strong 1938a (§ S–306). Collected at 5 stations in the Gulf. No illustration.

Taken on sandy mud flat of estuary at San Lucas Cove.

FAMILY MURICIDAE (p. 704 G. and G. 1931.)

§ S–351 *Acanthina lugubris* (Sowerby) 1821. PL. 38 FIG. 1
Gulf Unicorn-shell.
p. 719 G. and G. 1931, San Diego to Magdalena Bay and the Galápagos.

A small Thais-like snail very common in the northern half of the Gulf. San Carlos Bay; San Francisquito Bay, in the high rocky

REFERENCES

littoral; on the rocky beach at Angeles Bay; Puerto Refugio where the juveniles were very common in the uppermost tide pools, and the adults occurred farther down; Port San Carlos, Sonora.

§ S–352 *Coralliophila californica* A. Adams.

Taken only once, at Cape San Lucas.

§ S–353 *C. costata* Blainville (equals *C. nux* Reeve).
p. 120 P. and L. 1932, "under rocks at low tide," La Paz to Central America.
p. 50 Oldroyd 1927 (§ S–23), Vol. II, Pt. II; San Miguel Island, California, to Panama.

Pt. Lobos on Espiritu Santo Island; Amortajada Bay (white, as were most of the animals found on these white rocks which were imbedded in dazzling white coral sand): Puerto Escondido.

§ S–354 *Galeropsis madreporarum* Sowerby.
p. 120 P. and L. 1932, as *Coralliophila madreporarum,* "beach shells taken at Tres Marias and Acapulco; living specimens on coral reefs at San Juan del Sur and Tagoguilla Island."

Pulmo Reef; Pt. Lobos on Espiritu Santo Island, sufficiently variable so that we assumed that representatives at this point belonged to two species.

§ S–355 *Murex rectirostris* Sowerby. COLOR PL. 2 FIG. 1
p. 20 Strong and Hanna 1930 (§ S–36), one specimen from the Tres Marias.

Living specimens were taken only once, and then fairly abundantly, on the sand of San Lucas Cove, but the number of beach shells brought in by the crew suggests that this may be a very common species subtidally.

§ S–356 *Muricopsis squamulata* Carpenter.
p. 119 P. and L. 1932, "among rocks on mud flats at extreme tide. La Paz; Guaymas."

East of La Paz; and at Port San Carlos in Sonora.

§ S–357 *Purpura patula* Lamarck. PL. 34 FIG. 2
Flattened Purple-shell.
p. 119 P. and L. 1932, "common on rocks in exposed locations. This species on being disturbed gives off several drops of a milky liquid which on white cloth first turns to

REFERENCES

a metallic green color and later to a permanent royal purple." Cape San Lucas to Panama.

One of the common snails on the rocks on Cape San Lucas, exposed to considerable wave shock. Occurs in colonies in the upper intertidal where it must suffer intense sun and drying. The counterpart in these waters of California species of *Thais* and *Acanthina*.

§ S–358 *Phyllonotus bicolor* (Valenciennes) 1832. COLOR PL. 1
p. 730 G. and G., as *Chicoreus* (*Phyllonotus*) *hippocastatum* (Philippi) 1845, Guaymas to Paita.
p. 118 P. and L. 1932. "Tide flats at extreme tide." La Paz and Guaymas to Acapulco.

The commonest large snail in the Gulf, with shell brilliantly colored, and with animal active and voracious. The specimens determined as such by the United States National Museum were from east of La Paz, but unquestionably similar material was taken at Puerto Escondido, Concepcion Bay, Port San Carlos in Sonora, and elsewhere, chiefly in the southern half of the Gulf. The animals appeared to be most abundant just below low tide level; at Concepcion Bay they immediately attacked the bait in crab nets put overside at night in about 7 fathoms of water on a bottom of hard-packed sand.

§ S–359 *P. brassica* Lamarck.
p. 118 P. and L. 1932, "tide flats, quite rare." Mazatlan and Nicaragua.

A single small specimen was taken at La Paz.

§ S–360 *P. nigritus* Philippi.
p. 31 Lowe 1935 (§ S–21), as *M. negritus* Meusch. "Abundant on reefs feeding on *Cerithium stercus-muscarum*." Punta Peñasco, Sonora.

Possibly confused in our field notes with *P. radix* for which we show no records, and with *P. Princeps*. Murices of this type were large and heavy, and one group only (of three specimens) was sent to the United States National Museum for determination per the above. Other similar Murex were taken at Pulmo Reef, Pt. Lobos on Espiritu Santo Island, Marcial Point; Puerto Escondido; Concepcion Bay; and Estero de la Luna in Sonora. More than one species may be involved. Pilsbry and Lowe 1932 fail to report this species from a collecting experience of several years, but a large Murex from La Paz and Guaymas is reported on p. 118 as *Phyllonotus nigrescens* Sowerby, also taken in southern Mexico.

§ S–361 *P. princeps* (Broderip) 1832. COLOR PL. 3 FIG. 2
p. 730 G. and G. 1931, west coast of Mexico and Central America.

REFERENCES

p. 118 P. and L. 1932, "on rocks and lowest tides," Cape San Lucas and La Paz to Panama.

Several specimens taken on the boulders at Pt. Lobos, Espiritu Santo Island, and at Pulmo Reef, whence the specimen photographed in the color plate was taken.

§ S-362 *Thais biserialis* Blainville.
p. 119 P. and L. 1932, as *Purpura biserialis,* "on rocks between tides, not rare," Guaymas, La Paz, and Cape San Lucas to Central America.

Cape San Lucas; south end of Tiburon Island; Port San Carlos, Sonora; always on rocks.

§ S-363 *T. centriquadrata* Duclos. PL. 37 FIG. 4
Four-pronged Rock-shell.
Fairly common on the rocks at Cape San Lucas. Among boulders at Pt. Lobos on Espiritu Santo Island. At the south end of Tiburon Island. On the rocks at Port San Carlos in Sonora. Gabriel Bay on Espiritu Santo Island, exact ecological niche not certain, but probably on rocks among the roots of mangroves. One of the most widespread of the typical *Thais,* but nowhere common.

§ S-364 *Thais (Tribulus) planospira* Lamarck.
Masked flat snail. PL. 38 FIGS. 3 AND 4
p. 119 P. and L. 1932, "on rocks at extreme tide, very rare. Cape San Lucas."

A masked flat snail. Very common on the reef at Pulmo, but hard to see. They look like sea-weed encrusted stones. In the photographed specimen, an annelid worm will be noted, *Eunice antennata* (§ J-50).

§ S-365 *T. triangularis* Blainville.
p. 119 P. and L. 1932, as *Purpura triangularis,* "on rocks between tides, rather rare," Cape San Lucas to Central America.

Two specimens only taken, Pt. Lobos on Espiritu Santo Island. White. Confused in our notes with *T. centriquadrata.*

§ S-366 *T. triserialis* Blainville.
p. 119 P. and L. 1932, as *Purpura triserialis,* "on rocks between tides, common." Cape San Lucas to Panama.

Common only at Pt. Lobos on Espiritu Santo Island.

§ S-367 *T. (Neorapana) tuberculata* Gray. COLOR PL. 3 FIG. 1
The commonest small murex-like form in the Gulf. East of La Paz; Puerto Escondido; a light color phase at Coronado Island; large at

REFERENCES

San Lucas Cove; Angeles Bay; Port San Carlos, Sonora. Brown, small to large, but never as large as average *Phyllonotus bicolor.* Markedly larger and more abundant in the north.

SUPERFAMILY TAENIGLOSSA

FAMILY AMPHIPERASIDAE (OVULIDAE)

§ S–368 *Simnia variabilis* C. B. Adams 1850.
p. 235 Oldroyd 1927 (§ S–23), Vol. II, Pt. II; Monterey to Panama.

A vividly purple small snail, of a group associated ordinarily with gorgonians. Taken abundantly at Cape San Lucas and at Pulmo Reef, at the latter point noted as "vividly red-bodied."

FAMILY CYPRAEIDAE (p. 752 G. and G. 1931.)

§ S–369 *Cypraea albuginosa* Gray. Violet cowry.
p. 122 P. and L. 1932, "beach specimens common at Tres Marias and La Paz; a few live shells taken at Cape San Lucas."

Pulmo Reef. Taken with *C. annettae* at Pt. Lobos on Espiritu Santo Island. The latter has a brown base; *albuginosa,* a violet base.

§ S–370 *C. annettae* Dall 1909. Brown cowry. COLOR PL. 4 FIG. 2
p. 752 G. and G. 1931, Gulf of California to northern Peru.
p. 122 P. and L. 1932, "under rocks between tides, living examples not common. Guaymas; La Paz; Cape San Lucas."

Taken with *C. albuginosa* at Pt. Lobos on Espiritu Santo Island; east of La Paz; El Mogote; Puerto Escondido; fairly common at Coronado Island.

FAMILY STROMBIDAE (p. 755 G. and G. 1931.)

§ S–371 *Strombus galeatus* Swainson 1823. PL. 36 FIG. 5
Giant Conch.
p. 756 G. and G. 1931, Gulf to Panama.
p. 122 P. and L. 1932, "living just below the low tide line, not common." La Paz to Central America.

The heaviest and one of the largest Gastropods taken in the Gulf. Several specimens were taken at Puerto Escondido, where the natives capture them for food. Concepcion Bay, subtidal.

REFERENCES

§ S–372 *S. gracilior* Sowerby 1825. COLOR PL. 8 FIG. 1
p. 755 G. and G. 1931, La Paz and Gulf to Ecuador.
The commonest mangrove large snail, also common on sand flats. At the base of sandy shore mangroves at El Mogote. On the sand flats at San Lucas Cove. Along the sandy shore of the channel entrance, Agiabampo Estuary in Sonora. Very active, eyes at the end of fairly long and movable stalks. The Sonora specimens bore the barnacles *Balanus amphitrite inexpectatus* (§N–19), in one case rather heavily.

FAMILY CERITHIIDAE (Clavidae of G. and G. 1931, p. 756. See also Baker et al 1938 § S–305).

§ S–373 *Cerithidea mazatlanica* Carpenter 1857. PL. 25 FIG. 1
p. 123 P. and L. 1932, mud-flats at Mazatlan. Mazatlan Horn-shell.
p. 227 Baker et al 1938 (§ S–305), as *C. albonodosa mazatlanica* Carpenter.
Very common in the lagoon off Amortajada Bay.

§ S–374 *Cerithium maculosum* Kierner. COLOR PL. 2 FIG. 2
Spotted Cerithium.
p. 225 Baker et al 1938 (§ S–305), with good toto-fig. 2, Pl. 17, abundant in the Gulf.
p. 123 P. and L. 1932, "Beach shells at Tres Marias; Mazatlan; Cape San Lucas; some fine living specimens taken on the reef at La Paz."
Very abundant and noticeable at Port San Carlos, Sonora.

§ S–375 *Cerithium sculptum* (Sowerby) 1855. PL. 40 FIG. 4
Sculptured Cerithium.
p. 757 G. and G. 1931, as *Thericium incisum* (Sowerby), Lower California and Guaymas.
p. 123 P. and L. 1932, as *Cerithium incisum,* "alive under rocks between tides, not common. La Paz; Cape San Lucas."
p. 227 Baker et al 1938 (§ S–305), as *Potamides* (*Liocerithium*) *sculptus,* with excellent toto-fig. 6, Pl. 17, exceedingly common and widespread in the Gulf.
Pt. Lobos on Espiritu Santo Island; as "mid-tide minute spired snails"; very abundant at Amortajada Bay; Marcial Point, high up; Concepcion Bay (no exact record other than place name; presumably sandy shore) at Angeles Bay; as "small turret-snails" at Puerto Refugio; Port San Carlos in Sonora.

FAMILY VERMETIDAE (p. 776 G. and G. 1931.)

REFERENCES

§ S–376 *Aletes squamigerus* Carpenter 1856.
p. 777 G. and G. 1931, Monterey to northern Peru and the Galápagos.
p. 44 Ricketts and Calvin 1939 (§ Y–3) with good toto-photograph in Pl. VIII. Exceedingly common in southern California.

East of La Paz, on coral rocks with *Vermicularia eburnea*. Reef at Marcial Point. Coronado Island.

§ S–377 *Vermetus contortus indentatus* Carpenter.
Mistaken for a serpulid worm, Pulmo Reef, and dispatched as such to the polychaet specialist who sent it to Drs. Hertlein and Hanna, to whom we are grateful for the above identification.

§ S–378 *Vermicularia eburnea* (Reeve) 1842.
p. 776 G. and G. 1931, San Pedro to Panama; South America.

Taken with *Aletes* at Pulmo Reef.

FAMILY LITTORINIDAE (p. 780 G. and G. 1931.)

§ S–379 *Littorina conspersa* Philippi.
p. 124 P. and L. 1932, rocks near high tide line, Cape San Lucas to Central America.

A white *Littorina* very common high on the rocks at Cape San Lucas, and taken less commonly at Pt. Lobos on Espiritu Santo Island.

§ S–380 *L. philippi* Carpenter.
p. 124 P. and L. 1932, Cape San Lucas to Central America.

Taken only once, high on the rocks at Amortajada Bay.

FAMILY ARCHITECTONIDAE (SOLARIIDAE)
p. 785 G. and G. 1931.)

§ S–381 *Heliacus radiatus* Menke.
p. 125 P. and L. 1932, "a few living on tide flats. La Paz."

Taken only once, at Coronado Island.

Family Hipponicidae (p. 788 G. and G. 1931.)

REFERENCES

§ S–382 *Hipponix antiquatus* (Linné). PL. 35 FIG. 1
Ancient Hoof-shell.
p. 788 G. and G. 1931, Crescent City, California, to Panama and Galápagos.
p. 126 P. and L. 1932, "under rocks between tides, common. Tres Marias."
Common limpets on rocks at foot of mangroves at Gabriel Bay on Espiritu Santo Island.

Family Crepidulidae (p. 789 G. and G. 1931.)

§ S–383 *Crepidula incurva* Broderip.
p. 125 P. and L. 1932, "on other shells in the same situation as *C. adunca* of the California coast." Guaymas to Central America.
Common (apparently on rocks) at Port San Carlos, Sonora.

§ S–384 *C. onyx* Sowerby 1824. PL. 35 FIG. 3
Onyx Slipper-shell.
p. 790 G. and G. 1931, Monterey to Panama.
p. 126 P. and L. 1932, "on dead shells on mud-flats," La Paz to Panama.
p. 213, with toto-fig. 98, Ricketts and Calvin 1939 (§ Y–3), common on other shells, especially on cones, in southern California.
East of La Paz; Puerto Escondido (distorted to fit the shells on which it lives); on *Phyllonotus bicolor* at Concepcion Bay; as "sand flat limpets" at Angeles Bay (taken presumably on rocks embedded in sand); also on rocks on the bouldery beach at Angeles Bay.

§ S–385 *C. squama* Broderip.
p. 126 P. and L. 1932, "under rocks between tides." Tres Marias; San Juan del Sur; Montijo Bay.
Pulmo Reef. San Carlos Bay. On rocks.

Family Calyptracidae (p. 793 G. and G. 1931.)

§ S–386 *Crucibulum imbricatum* (Sowerby) 1824. PL. 35 FIG. 6
Imbricated Cup and Saucer Limpet.
p. 793 G. and G. 1931. Gulf to Callao, Peru, and the Galápagos.
p. 125 P. and L. 1932 "on stones or other shells on mudflats." Guaymas and La Paz to Panama.

REFERENCES

Shell only at Pt. Lobos, Espiritu Santo Island; as a giant limpet at Puerto Escondido; removed from shell of giant conch, Puerto Escondido; Concepcion Bay on *Pinna*.

§ S–387 *C. spinosum* (Sowerby) 1824. "Cup and Saucer Limpet."
p. 793 G. and G. 1931, Trinidad, California, to Tome, Chile, and the Galápagos.
p. 125 P. and L. 1932, "on stones and dead bivalves on mud-flats," Guaymas and La Paz to Central America.

As small and highly ornamented limpets at Puerto Escondido; very common on rocks imbedded in sand at Angeles Bay; from *Pinna* (and in one case) from *Chione* at Gabriel Bay on Espiritu Santo Island; from rocks at base of mangroves at Gabriel Bay, Espiritu Santo Island, in several shapes and sizes, very common. Very spiny and ornamented, but the spines tend to be obsolete in large, old specimens.

FAMILY NATICIDAE (p. 796 G. and G. 1931.)

§ S–388 *Natica chemnitzii* Pfeiffer. PL. 35 FIG. 5
Variegated Moon-shell.
p. 16 Boone 1928 (§ S–4), as *N. macrochiensis* Gmelin, *chemnitzii* Pfr. Two specimens from Concepcion Bay.

A "variegated *Polinices*-like form" from the muddy sand flats at El Mogote; again noted in our collecting reports in exactly the same words from the sand flats of San Lucas Cove; and as a small "*Polinices*" from Estero de la Luna in Sonora.

§ S–389 *Polinices bifasciatus* (Gray) 1834.
p. 800 G. and G. 1931, Acapulco, Guaymas, and Cape San Lucas.

As one of the brown *Polinices,* the other being *reclusianus* in several examples, at El Mogote on the sandy mud flats.

§ S–390 *P. reclusianus* (Deshayes) 1839. PL. 35 FIG. 4
Recluz's Moon-shell.
p. 800 G. and G. 1931, good text figs., 13a, b, c, Crescent City, California, to Tres Marias Islands. Chile?
p. 126 P. and L. 1932, as *P. reclusiana* Petit var. "Sand bars between tides. Mazatlan; La Paz; Guaymas."
p. 183 Ricketts and Calvin (§ Y–3) as *reclusianus,* stated to be more common in, and almost restricted to the southern part of the area treated (Alaska to northern Lower California).

REFERENCES

As one of the brown *Polinices* on the sandy mud-flats at El Mogote, the other being *P. bifasciatus*. Common on the sand flats at Angeles Bay. Typically fawn-colored.

§ S–391 *P. uber* (Valenciennes) 1833.
p. 799 G. and G. 1931, San Martin Island, Lower California to Callao, Peru, and the Galápagos.
p. 127 P. and L. 1932, "beach shells in good condition," southern Mexico and Central America.

A small snail taken east of La Paz on flats with drowned coral heads; as a white *Polinices* on the sandy mud-flats of El Mogote.

FAMILY NERITIDAE
(not treated in G. and G.)

§ S–392 *Nerita bernhardi* Recluz.
p. 127 P. and L. 1932, "under rocks between tides." La Paz and Cape San Lucas to Central America.

Several specimens at Puerto Escondido. As a common half-tide shell at San Carlos Bay, presumably on rocky shore. As a small *Tegula*-like form at Port San Carlos, Sonora.
A "*Tegula*-like shell" from rocks among roots of mangroves in Gabriel Bay on Espiritu Santo Island.

§ S–393 *Nerita ornata* Sowerby.
p. 127 P. and L. 1932, "This is the more southern form of the large *N. scabricostata* . . ." Central America to Panama.

Common at Cape San Lucas. Very similar to N. scabricostata, below, which is common in the north.

§ S–394 *Nerita scabricostata* Lamarck. PL. 35 FIG. 2
Black and white Whorl-shell.
p. 127 P. and L. 1932, "on wave-beaten rocks near high tide line. Tres Marias; Mazatlan; La Paz; Cape San Lucas."

Exceedingly common at the south end of Tiburon Island; Gabriel Bay on Espiritu Santo Island, ecological niche uncertain. Superficially very similar to the above.

§ S–395 *Neretina picta* Sowerby.
p. 127 P. and L. 1932, "under rocks between tides," Guaymas and La Paz to Central America.

Taken only once, at Puerto Escondido.

SUPERFAMILY DOCOGLOSSA

FAMILY ACMAEIDAE (p. 809 G. and G. 1931.)

REFERENCES

§ S–396 *Acmaea atrata* Carpenter. PL. 34 FIG. 5

Coolie-hat Limpet.

p. 129 "on rocks between tides. Cape San Lucas."

A very common large limpet at Cape San Lucas; also at Pulmo Reef (with two other *Acmaea*); at Puerto Refugio as a high tide pool limpet; at the south end of Tiburon Island. At Cape San Lucas; the specimen photographed bore the barnacle *Chthamalus anisopoma* (§ N–19).

§ S–397 *A. dalliana* Pilsbry. Dall's Limpet. PL. 34 FIG. 6

A high littoral limpet at San Francisquito Bay; taken also at Angeles Bay; very common at Puerto Refugio.

San Francisquito specimens were encrusted with barnacles. The specimen photographed had *Tetraclita squamoso stalactifera f. confinis,* and others had *Chthamalus anisopoma* (§ N–19).

§ S–398 *A. discors* Philippi. Eroded Limpet. PL. 34 FIG. 3

p. 129 P. and L. 1932, "on rocks near low tide line, usually badly covered with coralline growth. Tres Marias; Mazatlan; Cape San Lucas."

Common at Cape San Lucas; with two other *Acmaea* on Pulmo Reef, large specimens are noted as having the upper surface heavily masked and corroded.

§ S–399 *A. fascicularis* Menke.

p. 129 P. and L. 1932, "under rocks between tides. Tres Marias; Mazatlan; Cape San Lucas."

Common at Cape San Lucas with other *Acmaea;* at Pulmo reef also with others.

§ S–400 *A. mesoleuca* Menke.

p. 129 P. and L. 1932, "under rocks between tides, not rare." La Paz and Cape San Lucas to Central America.

Very common at Puerto Refugio; at Port San Carlos, Sonora, some rather large, on smooth stones fairly high up.

§ S–401 *A. pediculus* Philippi.

p. 129 P. and L. 1932, "quite rare. Mazatlan; Manzanillo."

A single specimen was taken at the south end of Tiburon Island.

§ S–401a *Acmaea strigatella* Carpenter.

p. 129, P. and L. 1932, "rare Cape San Lucas."

Common on rocks about the foot of mangrove roots, Gabriel Bay on Espiritu Santo Island.

REFERENCES

§ S–402 *Acmaea* sp., too young for identifying, taken in very great quantity in the highest tide pools at Puerto Refugio. These uppermost pools, with their coralline algae, serve as nurseries for many of the species which are found as adults farther down. *Atrata, dalliana,* and *mesoleuca* occurred at this point in the lower reaches. It seems likely, however, that the majority of these, at least, are juvenile *A. atrata,* since a second batch sent to the U. S. National Museum, and picked at random, except that larger specimens were chosen, proved to be *atrata.*

§ S–403 *Lottia gigantea* Gray 1834.
p. 171 Dall 1921 (§ S–14) Crescent City to Guadalupe and Cedros Island; p. 159 Oldroyd 1927 (Vol. II, Pt. III) (§ S–23); p. 138 Keen 1937 (§ S–19) as *L. gigantea* Sowerby 1834. For some reason, this common southern California species seems not to be treated in Grant and Cale 1931. Ricketts and Calvin 1939 (§ Y–3), p. 20, with good photograph as Pl. 1.

Taken abundantly high on the shore at Amortajada Bay but the United States National Museum regards the identification as not entirely certain.

The specimens were strikingly smaller than, and otherwise bore little resemblance to, California *Lottia.*

Superfamily Rhipidoglossa

Family Turbinidae (p. 815 G. and G. 1931.)

§ S–404 *Callopoma fluctuosum* (Wood) 1828.
COLOR PL. 6 FIG. 1 also PL. 21 FIG. 2
p. 816 G. and G. 1931 as *Turbo (C.) fluctuosus,* Cedros and Gulf to Peru.
p. 127 P. and L. 1932 as *Turbo fluctuosus* "among stones between tides, common." Guaymas and La Paz to Acapulco.

Occasional on the drowned coral flats east of La Paz. Recorded also from the sand flats of San Lucas Cove, but this may have been an error of record or typography. As a large turban at Puerto Refugio on the rocks. Exceedingly abundant at the south end of Tiburon Island, in various sizes, including some very large specimens. Very much larger and more abundant in the north, but this may have been a seasonal factor incident to breeding, specimens having migrated inshore at this time.

FAMILY TROCHIDAE (p. 823 G. and G. 1931.)

REFERENCES

§ S–405 *Tegula (Omphalius) impressa* (Jonas) 1848.
p. 180, Oldroyd 1927 (§ S–23) Vol. II, Pt. III. San Diego to Tres Marias.

Taken with *Tegula* sp. (§ S–408) at Marcial Point: one of these shells only was living, the other had a hermit tenant. Angeles Bay, rocky shore. High tide pools at Puerto Refugio.

§ S–406 *Tegula (Omphalius) mariana* Dall.
p. 32 Lowe 1935 (§ S–21) "a few good living specimens taken under rocks at low tide." Punta Peñasco, Sonora.

With *Callopoma fluctuosum* on the coral flats east of La Paz. At Puerto Escondido, variable in appearance. South end of Tiburon Island. Nowhere common, we took no more than one or two each at any point.

§ S–407 *T. rugosa* A. Adams. PL. 37 FIG. 3
Variegated Turban.
p. 32 Lowe 1935 (§ S–21) "extra large specimens taken on upper side of rocks." Punta Peñasco, Sonora.

A large turban fairly common high up at San Carlos Bay in Lower California; and again very abundant on the gravelly and rocky shore at Port San Carlos, Sonora. Specimens from the latter station were sometimes encrusted with the barnacles *Chthamalus anisopoma* (§ N–19).

§ S–408 *Tegula* sp.

One only taken at Marcial Point, probably on rocks fairly high up. See § S–405.

SUPERFAMILY ZYGOBRANCHIA

FAMILY FISSURELLIDAE (p. 847, G. and G. 1931.)

§ S–409 *Diadora alta* C. B. Adams.
p. 128 P. and L. 1932, "under rocks at very low tide. Guaymas; La Paz; Mazatlan."

Pulmo Reef. South end of Tiburon Island.

§ S–410 *D. inequalis* (Sowerby) 1835. PL. 34 FIG. 4
Asymmetrical keyhole Limpet.
p. 850 G. and G. 1931, Santa Barbara, California, to Ecuador and the Galápagos.
p. 128 P. and L. 1932, "under rocks between tides, not rare." Guaymas, etc. to Panama.

REFERENCES

Common at Pt. Lobos on Espiritu Santo Island. High up on shore at Amortajada Bay. Puerto Escondido; Coronado Island, several; Puerto Refugio; south end of Tiburon Island; Port San Carlos, Sonora; Gabriel Bay on Espiritu Santo Island. The commonest keyhole limpet in the Gulf. Note the serpulid worm tubes (§ J–72 or 76) on the specimen photographed.

§ S–411 *Fissurella rugosa* Sowerby.

p. 128 P. and L. 1932, "on rocks between tides, often imbedded among algae." Cape San Lucas to Central America.

Pulmo Reef; Coronado Island; Puerto Refugio; south end of Tiburon Island; usually with *Diadora inequalis.*

SUPERFAMILY PTENOGLOSSA

FAMILY EPITONIIDAE. Usually small snails. None taken.

SUPERFAMILY GYMNOGLOSSA

Small snails, in the FAMILIES MELANELLIDAE and PYRAMIDELLIDAE. None taken.

§ S–412 SUMMARY: *Part I*

About 90 species of shelled snails, exclusive of the Tectibranchs, have been determined as a result of our short survey, but this must not be regarded as an index of the richness of the Gulf Gastropod fauna. A conchologist collecting over the area we covered, undoubtedly would report twice as many species, even without dredging. In the first place, we paid no especial attention to this group. We took no dead shells, and we restricted our activities entirely to the intertidal zone. But most important, we took no very small or minute specimens, thus automatically eliminating the several families in which the greatest numbers of species could be expected. However, it seems likely that we procured representatives of most of the common and obvious larger shore animals in the group.

From our findings it would appear that, although the Gulf may well be among the richest collecting grounds for shells in the whole world, few Gastropods assume here the importance of *Littorina, Tegula, Thais,* and *Acanthina* along the California coast. There are several perfectly huge conches, murices, etc., and many other individuals large by California standards, and a considerable variety of smaller shells. But the tendency is, on the whole, toward great numbers of species with few individuals per species, although this is less true of the northern half of the region.

Divergences in the reporting and reviewing of this section, as compared with the standards we set for other groups, probably should be mentioned. As with the Pelecypods, temporal limitations prevented a complete investigation. We attempted merely to provide references for geographic range, synonymy, description, and, if possible, good illustrative material, rather than to search out and cite each mention of the given species in our area. Unfortunately we were not always able to do even that. If the citations prove inadequate or poorly chosen, this should be attributed partly to the impossibility of our consulting important literature without the expenditure of considerable time and effort in chasing around the country. But after all, the aim throughout this project has been to provide a source book rather than a handbook, and merely to paint the sketchiest of pictures, or at the least to make materials for such a sketch available.

Due also to nomenclatural standards in conchology somewhat at variance with those applying to other groups, some of the information ordinarily symbolized in the citing of animals may not be available. Thus, in other groups the name of the original describer is parenthesized if the species is placed in a genus other than that to which it was attributed in the original description. Most of the authorities consulted in this group fail to conform to this custom, Grant and Gale excepted, and we have no choice but to follow them unless we trace back each species to its source.

Interestingly, there are fine papers easily available, usually with excellent illustrations, on the small, obscure, or rare species, but on the large, obvious, and showy forms, the very ones in which the traveler is most likely to be interested, there is little accessible information. In choosing, therefore, the commonest species for illustration, we may be rendering a real service to the traveler, student, and general zoologist.

§ S–413 *Part II*

The geographical analysis of our catch indicates the tropical affiliations of the Gulf fauna in this group. Of these

REFERENCES

90 common species, 11 could not be found by us readily in the literature. On certain others, we could find no direct statement of distribution, but assumed that their non-appearance in the very carefully worked-out California manuals and in available lists of the northern Lower California region, meant that they failed to range north of central Lower California. If, similarly, they were missing in the Strong and Hertlein, and in the Pilsbry and Lowe Panama lists, we assumed them to be not common at that point. On this basis, and utilizing what positive data could be had from Grant and Gale, etc., we rate the 79 appearing in lists accessible to us at the time we wrote up this group as follows:

Restricted to the Gulf	17
Ranging from the Gulf southward	42
Ranging from the Gulf northward	2
Ranging both north and south	18

Of the 20 which range north of the Gulf, 1 is recorded simply as from Lower California; 5 fail to range north of the Cedros Island complex; 1 extends a bit farther to San Martin Island. Not including a doubtful report, 4 extend to Southern California; 4 are reported from Monterey Bay; 1 from San Francisco; and 4 from the Crescent City-Trinidad area; none of these 9, however, is common north of southern California, and with one or two exceptions, all are reported from a single northern record each.

The shelled snails emphasize notably the faunal differences between the northern and southern part of the Gulf. Of the 90, the great majority, 52, were taken south of a line drawn from Santa Rosalia to Guaymas, or were markedly commoner south of this line. 15 were found in the north only, or were much larger and more common in the north. Only 23 occurred indiscriminately in the north and south. In actual field work, the collector at Angeles Bay or Puerto Refugio will note a difference even more striking, in comparing his picture with that got at Pulmo Reef or at Espiritu Santo Island. In the north the number of snails is distinctly limited, but there are great numbers of individuals, while the south, with its multitudes of species, has probably no disproportionate bulk of snails, because each species will be seen in only one or two examples.

From the list of 90, 23 can be considered common: the pulmonate limpet *Siphonaria aequilirata, Terebra variegata* (at Angeles Bay only, but exceedingly common there), *Oliva venulata, Engina ferruginosa, Nassarius tegula, Nitidella guttata* (the commonest small Gastropod), *Acanthina lugubris, Purpura patula, Phyllonotus bicolor, Phyllonotus nigritus, Thais centriquadrata, T. tuberculata, Cypraea annettae, Strombus gracilior, Cerithium maculosum* (exceedingly abundant at Port San Carlos, Sonora), *Littorina conspersa* (at Cape San Lucas only), *Crepidula onyx, Crucibulum spinosum, Nerita scabricostata,* one or two of the *Acmaea* (not surely differentiated in our field notes), *Callopoma fluctuosum, Tegula rugosa,* and *Diadora inequalis.* 8 of these range both north and south; 8 south only. 2 are restricted to the Gulf, and for the remainder there is inadequate information.

§ S–414 *Part III*

Most of them can be regarded therefore as truly tropical species. None, so far as we have been able to determine, has been reported outside the Panamic Province as here defined, except for the 9 species mentioned above as ranging into central or northern California. However, in comparing shelled Gastropods (and this applies to the Pelecypods also) with other groups, it must be borne in mind that conchologists and mollusk specialists have subjected shells to the most minute and comprehensive examinations and comparisons. Species differentials are far more subtle than in any other group of invertebrates, not even excepting the crabs. It seems reasonable to believe that, if similarly focused energy were devoted to other groups, it would be found that there, also, minute but constant differentials separate Panamic species from their almost identical West Indian, Indo-Pacific, or cosmopolitan relatives.

There are three well marked stages in the history of a given group. At first the various species along their native shores are discriminated and named by the naturalists of the various countries, who rarely specialize on any one group, but devote their energies to regional natural histories. Later, the first generation of specialists in the various groups will evaluate the previous descriptions and examine the type specimens, generally with the result that

REFERENCES

the regional species are merged into one—the original describers having had little opportunity for comparing materials outside their own country, though the same waters may wash the shores of an opposite country where similar animals were being described under different names. The second generation of specialists subsequently will review these groups, with better methods and more delicate criteria, discovering that actually there *are* minute but persistent differentials between these local races, at which time the names bestowed by the original describers will be revived as valid species.

In connection with the geographic distribution of shelled mollusks, it should be emphasized also that tenuous northward extensions of range could be expected for many Panamic species, based on the intense and careful collecting activities of the southern California conchologists. Every beach is combed over for rare species, every stray is eagerly pounced upon, and an imbalance is imposed on the literature by the reporting at Los Angeles of species rare to that area which are commoner to the south along shores less frequently visited. If quantitative results could be got and evaluated, the southern California area might be expected to represent for many species the extremely flat part of a frequency curve with a peak south of Cedros. It will be granted also that, *to some extent,* the presence or absence of a species in the area adjacent to its range is an index as to whether or not a good collector has gone over the region carefully and reported his results. There is no way out of this dilemma, but it would seem the part of clear thinking to take such facts into consideration.

SEA-HARES, NUDIBRANCHS, AND ONCHIIDS

(Naked Gastropods)

§ T–1 Bergh, R. 1894. "Die Opisthobranchien." *Albatross* 1891. Bull. Mus. Comp. Zool., Vol. 25 (10): 125-233, Pls. 1-12.

The following Panamic forms, usually from shallow water, are considered: *Glaucilla marginata* Bergh 1864 (p. 132, Cocos); *Pleurophyllidia californica* Cooper (p. 154, Panama, etc.); *Geitodoris immunda* gen. et sp. nov. (p. 167, Panama, with detail figures); *Chromodoris californiensis* Bergh 1879 (p. 181, region of Magdalena

REFERENCES

Bay): *C. agassizii* sp. nov. (p. 182, Panama, with detail figures): *Tridachia* (?) *diomedea* sp. nov. (p. 194, region of Magdalena Bay, with detail figures): *Pleurobranchus plumula* Mtg. (p. 197, region of La Paz, with detail figures); *Navarchus aenigmaticus* sp. nov. (p. 217, Panama, with detail figures and two fair toto-drawings). No bibliography.

§ T–2 Cockerell, T. D. A. and Sir C. Eliot 1905. "Notes on a collection of Californian nudibranchs." Journ. Malacol., Vol. 12: 31-63, Pls. 7 and 8.

San Pedro and La Jolla. Temperate genera mostly represented with a sprinkling of such tropical forms as *Chromodoris* and *Doridopsis*. Notes on natural history and color. Color sketches. No bibliography. 17 species, 5 of them new, some of which will have proven to be overlap forms or Panamic forms at the northerly limits of their ranges.

§ T–3 MacFarland, F. M. 1918. "The Dolabellinae." (*Albatross* 1899-1900). Mem. Mus. Comp. Zool., Vol. 35 (5): 299-348, 10 pls.

A monographic account of the chief subject, the extra-limital *Dolabella agassizi* sp. nov. from Easter Island, with fine color plate; incidental citations on p. 305-6 of *D. guayaquilensis* Petit in Sowerby 1870 and *D. californica* Stearns 1878, the latter from Mulege in the Gulf. Comprehensive bibliography.

§ T–4 .... 1924. "Opisthobranchiate Mollusca." Proc. Calif. Acad. Sci. (4) Vol. 13 (25): 389-420, Pls. 10-12.

From the 1921 Gulf Expedition. The sea-hares *Aglaja bakeri* sp. nov. and *Tethys parvula* (Guilding Ms. Mörch), and the elysid nudibranch *Tridachiella diomedea* (Bergh) 1894, all from San Marcos Island in the Gulf, and all with good toto-illustrations.

§ T–5 .... 1929. "*Drepania,* a genus of nudibranchiate mollusks new to California." Proc. Calif. Acad. Sci. (4) Vol. 18: 485-96, Pl. 35.

Complete description, with good toto-illustrations of a phanerobranchiate dorid from San Diego, length of living animal 16 mm. Probably representing an overlap or Panamic form, since there has been no report of it from the north.

§ T–6 O'Donoghue, C. H. 1926. "A List of Nudibranchiate Mollusca recorded from the Pacific coast of North America." Trans. Roy. Can. Inst., Vol. 15: 199-247.

Including 7 only species mentioned from stations below the U. S.-Mexican boundary: *Archidoris britannica* (equals *tuberculata*), p. 207; *Geitodoris complanata* (equals *immunda*), p. 208; *Glossodoris aegialia, agassizii,* and *californiensis,* p. 211; *Armina* (equals *Pleurophyllidia*) *californica,* p. 222; and *Coryphella californica,* p. 229. *Tridachiella*

REFERENCES

diomedea (Bergh) 1894 which we found commonly in the Gulf seems to be lacking. Comprehensive bibliography which, however, lacks MacFarland 1923 (§ T–9) and 1924 (§ T–4).

§ T–7 .... 1927. "Notes on a collection of nudibranchs from Laguna Beach." Journ. Ent. Zool., Vol. 19: 77-119 including 3 Pls.

Descriptions of 28 shore species of which 3 are new. Of the total, 7 have been reported only from southern California and probably represent overlap or Panamic species. 8 more extend north only to Monterey Bay. Bibliography of 33 titles. Detail figures only.

§ T–8 Stearns, R. E. C. 1878. "Description of a new species of *Dolabella* from the Gulf of California, with remarks on other rare or little-known species from the same region." Proc. Acad. Nat. Sci. Phil., Vol. 30: 395-401, Pl. 7.

D. californica, a large, dark-brown, warty sea-hare from Mulege. A *Murex,* a *Macron,* a *Cypraea,* and an extended description of what the author assumed to be *Onchidella carpenteri,* but which actually proved to be a new species, to be named *binneyi* (Stearns) 1893 § S–31.

§ T–9 There are other papers, as follows, concerned mostly with extra-limital and Californian species, some of which may turn out to be overlap forms, or which may appear in the Panamic fauna. Several of these, in the Semper series especially, were not accessible:

Bergh, R. 1878, (Heft 14) 1904, and 1905, (Bd. 9, Teil 6, Lief. 1 and 2) in Semper's Reis. Arch. Phil. (cites several species from the Gulf or which are known now to occur in the Gulf).

Cockerell, T. D. A., in Nautilus: Vol. 15: 90-91, 1901; Vol. 16: 19-20, 1902; Vol. 16: 117, 1903; Vol. 18: 131-32, 1905; Vol. 21: 106-7, 1908. In Journ. Malacol., Vol. 8: 121-2, 1901.

MacFarland, F. M. In Zool. Jahrb. Supp., Vol. 15: 515-36, 1912, with color lithograph. In Nautilus, Vol. 39: 1-27, 1923, Californian Acanthodoridae.

Mazzarelli e Zuccardi. In Boll. Soc. Nat. Napoli (1) Vol. 3: 47-54, 1889, the extra-limital *Aplysia chierchiana* sp. nov. (Peru), etc.

The apropos volumes of Pilsbry's *Manual* were inaccessible to us while this section of the work was in progress.

§ T–10 See also opisthobranch items in the general mollusk citations, especially: Bartsch and Rehder 1938 (§ S–3) who describe on p. 2, and illustrate on Pl. 5 *Aplysia cedroensis*

REFERENCES

sp. nov.; Rochebrune 1895a (§ S–30) who describes *Pleurobranchus digueti* and *Doris umbrella* from the La Paz area; Stearns 1893 (§ S–31) who lists *Dolabella* from the Gulf and cites his original 1878 description; and less important items as in § S–40.

LIST OF SPECIES TAKEN:

ORDER OPISTHOBRANCHIATA

SUB-ORDER TECTIBRANCHIATA

TRIBE APLYSOIDEA (Sea-hares)

FAMILY APLYSIIDAE

§ T–11 *Tethys* sp. probably *californica* (Cooper) 1863.
MacGinitie, G. E. 1934. "The egg-laying activities of the sea-hare, *Tethys californicus* (Cooper)." Biol. Bull. Vol. 67 (2): 300-303, stresses the enormous fecundity of this species.
MacGinitie 1935 (§ Y–26) fig. 20, p. 738. A sea-hare, formerly known as *Aplysia,* apparently the same animal we find in north temperate waters, but representing the small or southern California variety, of this sometimes very large aplysid, specimens of which at Elkhorn Slough near Monterey, may reach a weight of ten or fifteen pounds. The geographic range is from Bodega Bay, just north of San Francisco, (Hanna 1939, p. 34, Nautilus, Vol. 53 [1]), and commonly at Monterey Bay, to the Gulf.

Several specimens each time were taken at two widely separated points, on the sandy mud-flats at El Mogote, and in the rocky littoral at Puerto Refugio on Angel de la Guardia Island. The animals were specifically noted as greenish in the Puerto Refugio collecting report.

§ T–12 *Notarchus (Aclesia)* sp.

Large, white, with ornamental cerata resembling tritonid nudibranch more than a sea-hare. Taken at Pt. Lobos on Espiritu Santo Island, and on the drowned coral flats east of La Paz.

§ T–13 *Dolabella* sp. probably *californica* Stearns 1878.
Stearns 1878 (§ T–8) p. 395, figs. 1 and 2, Pl. 7; shells only. Mulege Bay, Gulf.
MacFarland 1918 (§ T–3) p. 306; Mulege.

REFERENCES

A large sea-hare which extruded the usual purple ink when captured. Very solid and almost cylindrical with suddenly truncated end. We referred to it in our notes as the "blunt-ended sea-hare." An eight-inch specimen which, when hardened, could barely be thrust into a two-quart jar, was captured at Puerto Escondido, and a similar animal, but notably green spotted, was taken at Gabriel Bay on Espiritu Santo Island.

TRIBE PLEUROBRANCHOIDEA

FAMILY PLEUROBRANCHIDAE

§ T–14 *Berthella plumula* (Montagu) 1803.
Bergh 1894 (§ T–1), p. 197, 10 fathoms in the La Paz region, as *Pleurobranchus plumula.* Pilsbry (§ T–9) p. 193. Tryon's *Manual of Conchology,* vol. 16, as *Pleurobranchus plumula.* North Sea, Atlantic coast of Europe, Mediterranean.

A specimen captured at Puerto Refugio is thought by Dr. MacFarland to belong undoubtedly here. Also see § T–16.

§ T–15 *Pleurobranchus digueti* Rochebrune 1895 cf.
Rochebrune 1895a (§ S–30), p. 239. Presumably from the La Paz area.

A small, red, naked tectibranch taken at Cape San Lucas, below the 0.0′ tide level and again at Pt. Lobos on Espiritu Santo Island, as a larger red tectibranch, in both cases on rocks. Also see § T–16.

§ T–16 In addition to the above, Pleurobranchids belonging probably to one or more species of the genus *Pleurobranchus,* were taken at Pt. Lobos, where the specimens were smaller than the above *P. digueti,* at Puerto Refugio where they were larger than the *B. plumula* reported above, at Angeles Bay, and at Port San Carlos, Sonora, fairly abundant at all four places. Collectively described in our field notes as "a sea-hare smaller than *Tethys,* foot flat, red or reddish with liver showing through very plainly on the dorsal surface." The one taken at Pt. Lobos on Espiritu Santo Island was very common, the two tentacles were obvious, and the size ranged from ½″ to 1″ or more in length. Specimens from Puerto Refugio were smaller, rounder, and more humped up, and no tentacles were noted, possibly a result of the killing process. The Sonora specimens were reported as orange-red and very abundant.

REFERENCES

In addition to these, a small white form was found at El Mogote on the sandy mud-flats, and a similar type at Puerto Escondido. It would seem that two or more undetermined species are represented here, probably differing from the *plumula* and *digueti* considered above, but no exact data can be found. Dr. MacFarland emphasizes the difficulties of dealing with the Pleurobranchidae and remarks that "much will have to be known about their internal structure before they can be straightened out satisfactorily."

SUB-ORDER NUDIBRANCHIATA

TRIBE DORIDOIDEA (Holohepatica)

CRYPTOBRANCHIATE DORIDS

§ T–17 A small white *Cadlina*-like dorid, undetermined, was taken in the rocky littoral at Pt. Lobos on Espiritu Santo Island.

§ T–18 A large seal-brown nudibranch, up to 2″ or 2½″ long was fairly abundant at Pt. Lobos, where several specimens were taken, and again at Coronado Island. Reminiscent of *Anisodoris* or of a large and handsome *Diaulula sandiegensis,* individuals of this species typically bore a number of contrasting black dots, which with the crispy white crepe gills, formed a pleasant contrast to the soft brown body tone.

Dr. MacFarland, in a personal communication, suggests that this may be what Rochebrune 1895a had in mind in writing up his inadequate description of *Doris umbrella* (§ S–30).

PHANEROBRANCHIATE DORIDS (a group which includes such forms as *Triopha* and *Hopkinsia* of the North Pacific).

§ T–19 *Aegires* sp.

Genus on p. 213, O'Donoghue, 1926 (§ T–6).

A single specimen was taken at Puerto Refugio, littoral rocky, noted on the collecting report for that day as "small, elongate, dotted."

TRIBE AEOLIDOIDEA

(or Cladohepatica, including *Dendronotus, Pleurophyllidia, Hermissenda,* etc.)

REFERENCES

§ T–20 *Chioraera Leonina* Gould.
O'Donoghue 1926 (§ T–6) p. 226, Alaska to Santa Barbara. See also Guberlet, J. E. 1928. "Observations on the spawning habits of *Melibe.*" Publications of the Puget Sound Biological Station, Vol. 6: 262-70.

A pelagic form, identical with the free-swimming nudibranch common in Puget Sound and formerly called *Melibe.* Pelagic, taken at night by net at anchorage, Puerto Refugio, where by the light of a lamp hung over the side, it was seen swimming past the boat with the characteristically twisting movement, just as it does in more northern waters, and autotomizing its appendages just as quickly when captured.

TRIBE ELYSIOIDEA

(Section Ascoglossa in the alternate classification, wherein all the other tribes belong in the Section Sacoglossa)

§ T–21 *Tridachiella diomedea (Bergh)* 1894.
MacFarland 1924 (§ T–4) p. 406, with excellent toto-illustrations Pl. 10; Gulf only.

One of the most spectacular and brilliantly colored small animals captured on the trip. MacFarland 1924 says of it, p. 405, "strikingly similar at first glance to a polyclad worm" and earlier in the same paper, speaking of the genus *Tridachia* from which his new genus is split, he notes the "extraordinary development of the parapodia, which possess very long and undulating margins . . . from which fancied resemblance to a lettuce leaf the generic name arises."

Possibly the most common nudibranch we encountered. Noted as a "green nudibranch-like form with fimbriated margins," at Pt. Lobos on Espiritu Santo Island, on the drowned coral flats east of La Paz, at El Mogote, in the northern part of the Gulf at Tiburon Island, and from an unspecified locality.

ORDER PULMONATA

FAMILY ONCHIIDAE

§ T–22 We took a naked marine pulmonate as one of the commonest rocky shore animals in the upper part of the Gulf, specimens occurring very abundantly at San Carlos Bay, at Angeles Bay, at Puerto Refugio, and elsewhere, but

REFERENCES

solely in the north. This was originally assumed by Dr. Berry to be *Onchidella binneyi* (sometimes *Onchidiella*) [Stearns 1893 (§ S–31, with good toto-fig. 1-2, Pl. 50) identified with *O. carpenteri* (Binney) 1860 by Stearns 1878 (§ T–8)] since it came from the same region, San Francisquito, Las Animas, and Angeles Bay, and since it fitted the description and habitat, "under stones at low tide, overlapping, very abundant." Dr. Berry, however suggested that we consult Dr. MacFarland in the matter, who turned our material over to Dr. Hanna, to whom we are grateful for the following determination:

Onchidium lesliei Stearns.
per description in Stearns 1893a (§ S–32) p. 383, with good toto-illustration as fig. 2 and 3, on Pl. 51; from the Galápagos Islands.

§ T–23 There is an important consideration of these curious semi-terrestrial pulmonates, concerned, however, solely with Chilean and Magellanic species, with brilliant lithographic illustrations, in *Fauna Chilensis,* as follows: von Wissell, Dr. Kurt. 1898. "Beiträge zur Anatomie der Gattung Onchidiella." Zool. Jahrb. Abt. f. syst., Supp. Vol. 4: 583-640, Pls. 34-36.
Of the Onchiids, common in Arctic and Antarctic waters, two are known from the Galápagos (brought there presumably by the Antarctic Humboldt current); one, *Onchidella steindachneri,* is abundant; and the other, *Onchidium lesliei*—the species we found so common in the Gulf—is rare. The only species reported heretofore as abundant in the Gulf has been *Onchidella binneyi;* apparently, therefore, two species occur commonly in the Gulf, *O. binneyi,* and *O. lesliei* which seems to have migrated upward from Galápagos. Our procuring of only the latter must be attributed to chance, or to the collecting of selected specimens.

§ T–24 SUMMARY

12 or more species of naked Gastropods seem to have been taken during the course of our short survey, all along shore in the Gulf. There were 6 or more sea-hares and similar forms, 5 nudibranchs, a total of 11 opisthobranchs, and 1 pulmonate. The commonest nudibranch was the

REFERENCES

brilliant and fimbriated *Tridachiella,* in distribution limited to the southern part of the Gulf, and simulating closely, both in appearance and locomotion, a Cotylean flatworm. The commonest tectibranchs were of an undetermined species of *Pleurobranchus.* The pulmonates, originally described from Galápagos, were exceedingly common in the northern part of the Gulf. Of the previously known tectibranchs, 2 are restricted to the Gulf, one ranges to the north as far as Monterey Bay, the fourth is more or less cosmopolitan. Of the described nudibranchs, two (if § T–18 is substantiated as Rochebrunes *umbrella)* are restricted to the Gulf, and the third has a most spectacular geographic range. If *Chioraera leonina* includes *C. dalli,* as seems to be accepted, the range is from the Aleutian Islands, clear on down to the Gulf. In any case, the range extends to British Columbia, where this curious pelagic form is very common in eel grass beds.

CLASS AMPHINEURA

Order Opisthobranchiata

Sub-Order Tectibranchiata

Shelled Tectibranchs

§ T–101 Baker, Fred and G. Dallas Hanna 1927. "Marine Mollusks of the order Opisthobranchiata." Proc. Calif. Acad. Sci. (4) Vol. 16 (5): 123-135, Pl. 4.

Shelled Tectibranchs only. 12 bubble-shells, etc., in the genera *Acteocina, Atys, Bullaria, Cylichnella, Haminoea,* and *Retusa,* 7 of which are new. The new species and Dall's *Retusa pasiana* are illustrated by toto-figures of the shell.

§ T–102 Two additional short articles on *Haminoea virescens* will be found in Nautilus Vol. 46 (4) and Vol. 47 (1).
See also bubble-shell items in Dall 1919a (§ S–317a) and in Ricketts and Calvin 1939 (§ Y–3), who on p. 295 cite references for *Bullaria* and *Haminoea* (as *Haminaea*).

§ T–103 ACKNOWLEDGMENT: For determining the *Bulla* we have to thank the Department of Mollusks at the United States National Museum; for the Haminoea, Dr. Hanna at the California Academy of Sciences.

LIST OF SPECIES TAKEN:

TRIBE BULLOIDEA

REFERENCES

§ T–104 *Bulla gouldiana* Pilsbry, also as *Bullaria gouldiana* (Pilsbry) 1893.
Grant and Gale 1931 (§ S–16) p. 457, as *Bullus gouldianus.* Range: Santa Barbara to Mazatlan. Baker and Hanna 1927 (§ T–101), p. 127, at 13 stations in the Gulf, north and south. Ricketts and Calvin 1939 (§ Y–3), p. 207, with good toto-illustration, fig. 4, Pl. 40, as *Bullaria Gouldii.*

A few only specimens of these, noted in our collecting report as "gray Bullaria(?)" were taken on the sandy mud-flats at El Mogote.

§ T–105 *Haminoea strongi* Baker and Hanna 1927.
Baker and Hanna 1927 (§ T–101) p. 130, with illustration of shell as fig. 2, Pl. 4. Found very abundantly in the Gulf intertidal at 14 stations.

A small bubble-shell very abundant in the tide pools at San Carlos Bay in Lower California and at Puerto Refugio. Very small juveniles were a feature of the uppermost tide pools at Puerto Refugio also.

§ T–106 Undetermined bubble-shells of one or more additional species were taken also at Puerto Escondido and at San Carlos Bay with the above *H. strongi.*

SUBCLASS POLYPLACOPHORA (CHITONS)

INTRODUCTION

§ U–1 Among the most varied, abundant, and noticeable of the sessile animals of rocky and bouldery shores along the entire Pacific North American coasts are the chitons, popularly known as "sea cradles," but definite information on these animals is hard to obtain. There seems to be at present only one specialist, Dr. S. Stillman Berry, of Redlands, California, actively and seriously engaged in problems of this sort in the west, possibly in the entire United States. However, in the Rev. Elwood B. Hunter of Pacific Grove there is promise that the field will be augmented.

The literature is equally difficult. There is no applicable monograph, except for the obsolete and costly *Manual of Conchology* volumes. Ecologists and collectors must de-

REFERENCES

pend on the apropos sections of Pratt's *Manual,* which is almost entirely inadequate for Pacific coast material; and on the necessarily incomplete information in Keep-Baily (which devotes only 8 pp. to the description of spp. and which lacks a bibliography); in Johnson and Snook; in Ricketts and Calvin; and in Oldroyd; in addition to the separates listed below (where the sparse Panamic literature specific to chitons is reviewed) and in the Ricketts and Calvin bibliography. There is no book or paper wherein the average zoologist can determine even the commonest Panamic species, without doing work that puts him on the way to becoming a specialist in the group.

§ U–2 See also chiton references in the general mollusk papers cited in the "S" sections. There are but few papers devoted specifically to chitons, this group having been treated (especially in the earlier literature) with the shells or in connection with the Gastropods. Of the more recent Panamic general mollusk papers, chitons are mentioned importantly in only a few. Bartsch 1938 (§ S–3) lists *Nuttalina fluxa* Carpenter, from Cedros Island (p. 15), and two *Chiton (Radsia)* species from Galápagos, p. 17.

Boone 1928 (§ S–4) p. 17, mentions *Ischnochiton conspicuus* Carpenter, from Concepcion Bay in the Gulf. Boone 1933 (§ S–4a) lists 3 species of *Chiton* from the Galápagos, p. 200-202.

Among the deep-water forms, a few shallow-water chitons are mentioned by Dall 1908 (§ S–11): pp. 355-6, *Callistochiton periconis* n. sp., *Chiton* sp., and *Ischnochiton ophioderma* n. sp. He also lists on p. 436 4 spp. including *Chiton stokesii* Broderip, and *Ischnochiton pectinulatus* Carpenter from Panama Bay shore—the former being found again on shore at Cocos (p. 437).

Lowe, 1935 (§ S–21) lists 10 species, 4 of them undetermined, from a rocky point high up in the Gulf on the Sonora shore: *Chiton virgulatus, Ischnochiton acrior* Carpenter, and *I. clathratus* are mentioned as common under rocks; *Ischnochiton limaciformis, Callistochiton infortunatus,* and *Acanthochites diegensis* Pilsbry are reported as less common, the first and last on the reef—*infortunatus,* under rocks.

REFERENCES

Pilsbry and Lowe 1932 (§ S–26) have the largest recent annotated list from the region (p. 129-30), 19 species being reported; only *Chiton stokesi* and *C. albolineatus* are reported as common, and both of these from Mazatlan to Panama only; *Nuttallina mexicana* Pilsbry, *Chaetopleura raripustulosa, Callistochiton infortunatus* and *pulchellus* are reported from the Guaymas region; *Ischnochiton limaciformis, Chiton virgulatus,* and *Acanthochitona exquisita* are reported from La Paz or Cape San Lucas. Shorter references also will be found in Dall 1902 (§ S–10), 1909a (§ S–13), and 1921 (§ S–14), and in Tomlin 1927-8 (§ S–40).

§ U–3 Bergenhayn, J. R. M. 1937. "Polyplacophoren von den Galápagos-Inseln."
No. 14 in: "The Norwegian Zoological Expedition to the Galápagos Islands, 1925, conducted by Alf Wollebaeck." Medd. Zool. Mus. Oslo, N:r 27: 313-324, 2 pls.

Four species only were taken. The two *Chiton* (*Radsia*) species had been previously known from this locality, but *Callistochiton gabbi* Pilsbry had been recorded formerly from the Gulf only, and *Callistochiton shuttleworthianus* Pilsbry only from Key West, Florida. Detail figures only. In a personal communication, Dr. Berry notes *Shuttleworthianus* as an exceedingly doubtful Pacific record.

§ U–4 Berry, S. Stillman. 1919. "A new *Lepidozona* from Southern California." No. 6 in: "Notes on West American Chitons, II." Proc. Calif. Acad. Sci. (4) Vol. 9 (1): 18-21, Pl. 8.

Ischnochiton (*Lepidozona*) *asthenes* n. sp., from 70 specimens taken under stones at low tide in Southern California, probably representing an overlap or Panamic form. Excellent toto-illustration.

§ U–5 .... 1922. "Fossil chitons of western North America." Proc. Calif. Acad. Sci. (4) Vol. 11 (18): 399-526, 16 pls. 16 text figs.

Since there is no comprehensive account of living Pacific North American chitons, this monograph of fossil species (all but 7 of which still occur in the region) must be regarded as most nearly satisfying the requirements for a key paper. 28 living species are represented of which 22 (plus 2?) occur in southern California, 14 (plus 1?) in Lower California, and 3 are doubtfully Panamic (*Acanthochitona avicula, Ischnochiton conspicuus,* and *I. acrior*). No toto-illustrations. Good bibliography, but entirely paleontological.

REFERENCES

§ U–6 .... 1925. "The species of *Basiliochiton.*" Proc. Acad. Nat. Sci. Phil., Vol. 77: 23-29, one plate.

B. lobium n. sp., described, with good toto-figure, from a single tide pool specimen taken at La Jolla, is probably an overlap or Panamic form. (However, Berry in a personal communication regards this as possibly northern.)

§ U–7 .... 1925a. "New or little known southern Californian *Lepidozonas.*" Proc. Malacol. Soc., Vol. 16 (5): 228-31, 1 pl.

Two new species of *Ischnochiton* occurring in this subgenus, one from deep water only. Good toto-figures. The littoral species especially represents probably a Panamic or overlap form. (Berry now regards these also as possibly northern.)

§ U–8 .... 1926. "Fossil chitons from the Pleistocene of San Quentin Bay, Lower California." Amer. Journ. Sci., Vol. 12: 455-56.

Among others previously recorded (§ U–5) both living and fossil, "six valves of" [*Chaetopleura lanuginosa* (Carpenter)], "this characteristic Lower California chiton are reported" as fossil.

§ U–9 .... 1931. "A re-description, under a new name, of a well-known Californian chiton." Proc. Malacol. Soc., Vol. 19 (V): 255-8, 1 pl.

Detail illustrations for differentiating the true *Ischnochiton clathratus,* a large Panamic chiton, from the smaller common southern California species which was heretofore known by that name but which is herewith described, with a good toto-figure, as *californiensis.* Bibliography includes Carpenter 1864 and 1866, Reeve 1847, and Sowerby 1840, all of which carry information on Pacific chitons.

§ U–10 Dall, W. H. 1919. "Descriptions of new species of chitons from the Pacific coast of America." Proc. U.S.N.M., Vol. 55: 499-516, no illustrations.

8 new species are from Magdalena Bay and the Gulf, one from Panama, one from Galápagos. Many others are from southern California, mostly littoral.

§ U–11 Pilsbry, H. A. 1892-93. "Polyplacophora." Vol. 14, 1892 and part of 15, 1893, First Series, *Manual of Conchology,* Acad. Nat. Sci. Phil. as follows: Vol. 14, xxxiv plus 350 pp., 68 pls., Vol. 15, 132 pp., 17 pls.

Morphology and classification of the chitons, with analytical keys and descriptions, synonymy, and geographic distribution of all species known to that date. Good plates often with toto-illustrations. Résumé of the literature but otherwise no bibliography except as in the

REFERENCES

synonymy citations. The generic index to both volumes is on pp. 122-133 of Vol. 15.

§ U–11a In a personal communication, Dr. Berry mentions two additional short papers on Pacific coast chitons which should be cited. We have been unable to consult them:

Baker, F. 1937. "Notes on *Ischnochiton ophioderma* and *Milneria kelseyi*." Nautilus, Vol. 50 (3): 86.

Pilsbry, H. A. 1893. "*Acanthochites exquisitus*." Nautilus, Vol. 7 (8): 95-96 (correction of original locality to Las Animas Bay).

§ U–12 ACKNOWLEDGMENT: For determining the following species, for assistance with, and gifts of literature, for co-operation in the matter of photographs mostly of specimens in his personal collection, for additional assistance too numerous to specify, but most of all for checking this portion of the phyletic catalogue, we have to thank Dr. S. Stillman Berry of Redlands, California. We are also indebted to the Rev. Elwood B. Hunter of Pacific Grove for suggestions, for help with the literature, and for his preliminary reading of this section.

LIST OF SPECIES TAKEN

§ U–13 *Acanthochitona arragonites* (Carpenter) 1857.

FAMILY ACANTHOCHITONIDAE (Acanthothidae)

Pilsbry 1893 (§ U–11) p. 25; Mazatlan, on shell of *Spondylus,* presumably littoral.

"A single specimen with the other *Acanthochitona* from Gabriel Bay (Espiritu Santo Island) is the first specimen which seems reasonably referable to Carpenter's species in the Mazatlan Catalogue that I have ever seen or heard of. Carpenter's specimen was excessively juvenile, but this fits his description quite closely in the more essential details. For my work this find is a real treasure. The very wide exposed portions of the valves differentiate it readily from *A. exquisita*." Personal communication from Dr. Berry.

§ U–14 *Acanthochitona exquisita* (Pilsbry) 1893. PL. 27 FIG. 1 Bristly chiton.

FAMILY ACANTHOCHITONIDAE (Acanthochitidae)

Pilsbry 1893 (§ U–11) p. 23, n. sp., but lacking toto-illustration; La Paz.

REFERENCES

Pilsbry and Lowe 1932 (§ S–26) p. 130, report it from La Paz, under stones between tides in sandy mud.

The second commonest chiton encountered, a fuzzy or furry species, with plates characteristically almost hidden by the mantle bristles, which, in carefully dried specimens may be brilliantly iridescent.

Taken at Marcial Pt.; Puerto Escondido; Coronado Island; a few each. Abundant at San Carlos Bay where the juveniles reminded us of *Nuttallina*. Abundant at Angeles Bay. Very common at Puerto Refugio under rock with *I. clathratus*. A few at Gabriel Bay on Espiritu Santo Island. The photograph is from a dried specimen 27 mm. long, from San Carlos Bay, in the collection of Dr. Berry.

§ U–15 *Callistochiton infortunatus* Pilsbry 1893.

FAMILY ISCHNOCHITONIDAE PL. 27 FIG. 2

p. 266, Pilsbry 1893 (§ U–11) detail figure only; Ecuador, La Paz, west Mexico.

Lowe 1935 (§ S–21) p. 32; "not common under rocks." Punta Peñasco.

Pilsbry and Lowe 1932 (§ S–26) p. 129, Guaymas and Taboga Island (Panama) under stones between tides. Hunter, in a personal communication reports "one specimen taken ½ mile west of Miramar (Guaymas) in fairly muddy sand, under rocks at low tide level."

A small and deeply sculptured species taken only once, at Port San Carlos in Sonora. The specimen photographed measured 13.8 mm.

§ U–16 *Chaetopleura* aff. *lurida* (Sowerby) 1832. Hairy chiton.

FAMILY ISCHNOCHITONIDAE PL. 27 FIG. 3

Pilsbry 1892 (§ U–11) p. 33, toto-fig. 54, Pl. 12, Santa Elena, west Colombia. Several varieties, Cape San Lucas to Panama.

Pilsbry and Lowe 1932 (§ S–26) p. 129, under stones between tides at six points between Panama and Acapulco.

A small and intricately sculptured species in which the girdle is covered with sparse but long hairs. Referred to in our collecting notes as a "small, hairy chiton" or as a *Mopalia*-like type." Taken only twice, at Pt. Lobos (Espiritu Santo Island) and at Angeles Bay. The specimen photographed was from Pt. Lobos, 18 mm. long in its curled state.

§ U–17 *Chiton virgulatus* Sowerby 1840. Common Gulf chiton.

COLOR PL. 7 FIG. 2 also PL. 27 FIG. 4

FAMILY CHITONIDAE

Pilsbry 1892 (§ U–11) p. 166, good toto-fig. 54, Pl. 32. Magdalena Bay and Gulf of California. Lowe 1935 (§ S–21)

REFERENCES

p. 32, abundant under rocks Punta Peñasco. Pilsbry and Lowe 1932 (§ S–26) p. 130. La Paz, under stones between tides.

The commonest chiton in the gulf, very handsome, brilliant and clear-cut, brown to slate-colored, up to 3″ long (but Hunter has one specimen 90 mm. long, from Miramar, Sonora, high up, and 5 individuals over 3″). Strongly photonegative, under well-buried stones only, but occurring most abundantly in the upper half of the intertidal. At one or two stations it was one of the uppermost animals along the shore. Hunter reports in a personal communication, abundant material both from rocks in black muddy sand and from a wave washed rock sunk into sparse coarse sand fronting the open Gulf; both in the Guaymas area and both high up above mean tide level, indicating considerable adaptability.

A few each at Marcial Pt. and at Puerto Escondido, but occurring by the hundreds at San Carlos Bay and Angeles Bay. Abundant also at Puerto Refugio and at Tiburon Island. Vividly kelp-green colored specimens at San Carlos Bay probably belong here as a color-phase, but may be a new variety or species. The photographed specimen was 78.1 mm. long, from San Carlos Bay.

§ U–18 *Ischnochiton (Lepidozona) clathratus* (Reeve) 1847.
FAMILY ISCHNOCHITONIDAE PL. 26 FIG. 5

Berry 1931 (§ U–9) p. 257, differentiated from the similar *californiensis*. Pilsbry 1892 (§ U–11) p. 128, toto-fig. 34, Pl. 26. Monterey and San Diego to La Paz, but the northern records apply to Berry's *californiensis*. Lowe 1935 (§ S–21), p. 32, plentiful under rocks at Punta Peñasco.

Fairly large, deeply etched, with rough felty mantle, recalling *I. cooperi* of Monterey.

Taken at Puerto Escondido. A common under-rock chiton at Puerto Refugio. A large *Ischnochiton*, probably of this species, was taken also at Angeles Bay.

§ U–19 *Ischnochiton (Radsiella) tridentatus* Pilsbry 1892.
Smooth Gulf chiton. PL. 27 FIG. 5
FAMILY ISCHNOCHITONIDAE

Pilsbry 1892 (§ U–11) p. 140, n. sp., toto-fig. 35, Pl. 16, La Paz and Gulf of California. Hunter, in a personal communication reports this as the most numerous chiton in the Guaymas area. "Found especially just west of the

REFERENCES

concrete boat pier near the hotel west of Miramar Beach, and under rocks along the bay just east of the point west of Miramar which marks the area on the open gulf front. The bottom where these were found seemed to be almost bare of sand or mud sometimes, and never were they found in a spot with a very deep bottom."

Possibly the third most common chiton in the Gulf, superficially a nondescript species, usually small, with hardly any evident sculpture other than microscopic granulations. A large specimen was taken at Marcial Pt. The smallest were taken, high up in the tide pools at Puerto Refugio, exposed to sun and surf, with *Nuttallina.* Specimens were taken also at Puerto Escondido, Concepcion Bay, San Carlos Bay, San Francisquito Bay, and on Tiburon Island. The specimen photographed was 19.5 mm. long, from Puerto Refugio.

§ U–20 *Ischnochiton (Stenoplax) limaciformis* (Sowerby) 1832.

FAMILY ISCHNOCHITONIDAE PL. 27 FIG. 6

Pilsbry 1892 (§ U–11) p. 57 with 4 toto-illustrations on Pl. 16. West Indies, Central America, Peru. Pilsbry and Lowe 1932 (§ S–26) p. 129; Taboga (Panama), Nicaragua, Mazatlan, La Paz, and Cape San Lucas; under stones at extreme low tide; rare. Hunter reports taking about 16 specimens ranging from 13 to 38.3 mm. in length, and from 5.2 to 14.3 mm. in width, along the beach west of Miramar at about 0′ tide level under rocks on a fine sand bottom.

A characteristically elongate species. Taken at Pt. Lobos on Espiritu Santo Island, and at Tiburon Island. Appearing in Dr. Berry's correspondence as "one of the smaller of your long narrow chitons." The specimen photographed was 35.2 mm. long, from Tiburon Island.

§ U–21 *Ischnochiton (Stenoplax) conspicuus* (Carpenter) in Dall 1897.

FAMILY ISCHNOCHITONIDAE

Pilsbry 1892 (§ U–11) p. 63, toto-figs. 91-93, Pl. 15 in Section *Stenoradsia,* Santa Barbara to Magdalena Bay.
There is a good toto-illustration as Pl. XIX in Ricketts and Calvin 1939 (§ Y–3). Berry 1922 (§ U–5) p. 465; range recorded as Santa Barbara to Magdalena Bay between tides. Boone 1928 (§ S–4) p. 17, records it from Concepcion Bay in the Gulf.

A single specimen, apparently referable to this very large species which is one of the most characteristic under-rock animals at La Jolla and Ensenada, was taken at San Carlos Bay.

REFERENCES

§ U–22 *Nuttallina* sp. cf. *allantophora* Dall 1919.
FAMILY LEPIDOCHITONIDAE (Callochitonidae)

PL. 26 FIG. 6

Dall 1919 (§ U–10) p. 502, from Las Animas Bay in the Gulf. No figure. Lowe 1935 (§ S–21) p. 32, reports a very small species of *Nuttallina* on the outer rocks at Punta Peñasco.

A typical *Nuttallina*, small, eroded, with red-brown mantle, taken in great quantity at Puerto Refugio, high up, on rocks exposed to sun and surf. The most positively phototropic chiton occurring in the Gulf. The photographed specimen was 17.3 mm. long as curled.

§ U–23 A few specimens remain not yet determined, such as the minute *Mopalia*-like form, and a nondescript brown chiton, both from Angeles Bay rocks, and small chitons fairly abundant and exposed in rock niches at Marcial Reef.

§ U–24 In a personal communication Hunter reports taking two additional species in the Guaymas area. "A small species of *Nuttallina* was taken in open tide pools on a rocky shelf on the open Gulf front. The pools were only 3 or 4" deep, and the specimens were out in open sunlight, with coralline algae surrounding them, but were seldom found on it or hiding directly in it. One specimen was found on a *Pinna* taken in a crevice on this shelf." He also, like most other careful collectors in this region (we cannot understand why we failed to find it!) records *Ischnochiton (Stenoradsia) acrior* Carpenter. "Eight specimens were taken at mid-tide level on rocks with a fine sand and mud bottom, west of Miramar. The anterior valve is concave in outline as seen from the lateral view which suggests *I. conspicuus.* The girdle is also similar to this species, having a velvety appearance. However the valves are not as heavy and have a finer and deeper sculpture. . . . The color is olive-green with pink where the valves have been eroded."

§ U–25 SUMMARY

From the above it will be seen that chitons are characteristic and abundant constituents of the Gulf fauna, especially in the upper part. In the southern part, from Cape San Lucas up to Puerto Escondido, they were notably scarce, although suitable ecological niches were present. At San Carlos Bay, *Chiton virgulatus* was possibly

REFERENCES

the commonest invertebrate, and this species, the predominant chiton in the Gulf, was abundant at all the suitable stations in the north, and occurred sparsely even in the south. The exquisite *Acanthochitona* was abundant also in the north, and one or two specimens were recorded from southern stations.

In this group, the Gulf fauna is strongly unique, 5 of the 10 species, including *Acanthochitona exquisita* being entirely restricted to Gulf waters (including Mazatlan) and one additional species, *Chiton virgulatus,* extending only to Magdalena Bay. The 3 most common forms are confined to the region. Only one, *I. conspicuus,* extends to the north (Santa Barbara). The other 3 range south to Colombia, Ecuador, and Peru, one being reported also from the West Indies. None of the 10, with the exception of this last, occur anywhere outside Panamic waters, and only one is known from north of Cedros Island.

SUBCLASS APLACOPHORA

§ U–26 Comprising the entirely deep-water Solenogastres, wormlike mollusks, none of which was taken. There is a monograph of the Panamic species based on material dredged by the *Albatross* as follows:

§ U–27 Heath, Harold, 1911. "The Solenogastres." *Albatross* 1899-1900. Mem. Mus. Comp. Zool., Vol. 45 (1): 180 pp. 40 pls.

Descriptions, with excellent toto-illustrations, of 31 species, 21 to 2228 fathoms, but mostly fairly deep, and mostly from north of the area we consider. One only species was Panamic, off the Revillagigedo Islands, but in 460 fathoms. The most southerly of the others (on the North American coast) was southern California.

CLASS CEPHALOPODA

Octopi and Squids, etc.

§ U–101 Berry, S. Stillman 1911. "A note on the genus *Lolliguncula.*" Proc. Acad. Nat. Sci. 1911: 100-105, Pl. 6.

Lolliguncula (?) *panamensis* n. sp. described, with good toto-illustration, from Panama and Ecuador.

REFERENCES

§ U–102 .... 1912. "A review of the cephalopods of western North America." Bull. Bur. Fish. Vol. 30: 269-336, Pls. 32-56, 18 text figs.

Although concerned only with species occurring in the Bering Sea to Coronados Islands range, several Panamic and overlap forms are treated, and, since any other monographic treatment of Pacific coast cephalopods is lacking, this must serve as the key account. Fine bibliography, complete to its date. Excellent toto- and detail figures. Should be used with the following:

§ U–103 .... 1913. "Notes on some West American cephalopods." Proc. Acad. Nat. Sci. Phil., Febr. 1913; 72-77, 2 text figs.

Addenda and errata applicable to the above. The Californian and Central American squid previously identified with *M. hoylei,* p. 305, becomes a new species, *heteropsis.* Bibliography includes Berry's 1912 paper on Laguna Beach cephalopods.

§ U–104 .... 1929. "*Loliolopsis chiroctes,* a new genus and species of squid from the Gulf of California." Trans. San Diego Soc. Nat. Hist., Vol. 5 (18): 263-282, 2 pls. 9 text figs.

Described, with excellent toto-figures, from a series obtained at Puerto Escondido.

§ U–105 Hoyle, W. B. 1904. "Reports on the Cephalopoda." *Albatross* 1891 and 1899-1900. Bull. Mus. Comp. Zool., Vol. 43 (1): 1-71, 12 pls.

About 30 species of which 6 are described as new, and about 13, with one new species, respectively, from the first and second trips; several from the region of Acapulco. Mostly deep, but 2 species of *Argonauta,* and one each of *Tremoctopus, Loligo* (*diomedeae* n. sp.), *Symplectoteuthis, Onychoteuthis, Abraliopsis, Cranchia,* and *Taonius* were at or near the surface in or about the Panamic area. Consideration of luminous organs. There are many good toto-illustrations, some colored. Good bibliography.

§ U–106 Jatta, Giuseppe. 1889. "Elenco dei cefalopodi della *Vettor Pisani.*" Boll. Soc. Nat. Napoli (1) Vol. 3: 63-67.

Four species of *Octopus* from Payta and from Panama, including *O. chierchiae* n. sp., *Loligo brasiliensis* Blainville from Payta and Panama, and a pelagic *Taonius* supposedly from 100 meters.

§ U–106a .... 1899. "Sopra alcuni cefalopodi della *Vettor Pisani.*" Boll. Soc. Nat. Napoli (1) Vol. 12: 17-32, Pl. 1.

Not consulted. Berry reports: "*Octopus chierchiae* is here figured and a more complete report made on all the *Vettor Pisani* material."

REFERENCES

§ U–107 Perrier, E. and A. T. de Rochebrune 1894. "Sur un Octopus nouveau de la Basse Californie, habitant les coquilles des mollusques bivalves." Comp. Rend. Acad. Sci. Paris, 118: 770-773.

6 examples of *O. digueti* n. sp., were taken at La Paz, four living in the shells of *Cytherea* and *Pecten*. Also see § 109a.

§ U–108 Robson, G. C. 1929. "A monograph of the recent Cephalopoda. Part I. Octopodinae." British Museum, 236 pp., 7 pls., 88 text figs.

§ U–109 .... Part II. 1932. The Octopoda (excluding the Octopodinae). British Museum, 359 pp., 6 pls., 79 text figs.

Articles on natural history, structure, distribution, etc., with comprehensive bibliography, complete for each part. P. 37, Part I "The Magellanic region and the North Pacific seem to be foci of local differentiation, giving rise to fairly distinctive groups." Panamic species on p. 79, 84, 150, 152, and 197, all littoral, in Part I. Hoyle's reference to a Nicaragua specimen as *O. tehuelchus* Orbigny, p. 147, is regarded as questionable. In Part II, 12 Panamic forms are treated, but most are deep or abyssal. However, three species of *Argonauta*, p. 181, 197, and 198 are reported as pelagic and at the surface. The cosmopolitan *Ocythoë tuberculata* is reported on p. 201 from Catalina Island, and may occur southward into the more strictly Panamic range. The p. 309 record of *Graneledone verrucosa* (Verrill) from Panama is regarded as doubtful.

§ U–109a Rochebrune, A. T. de 1896. "Étude sur une forme nouvelle du genre *Octopus*." Arch. Mus. Paris, Vol. 8: 75-86, pls.

Not consulted. *O. digueti*.

§ U–110 Thomsen (née Nielsen), Ellen 1931-32. "Remarks on the distribution and systematic position of *Graneledone verrucosa* (Verrill)." Vid. Medd fra Dansk naturh. Foren., Vol. 92: 293-299, 1 pl.

Mention of 3 male *Eledone verrucosa* from deep water, Gulf of Panama. Berry, however, in a personal communication, questions any Pacific records of this species, and refers to his note in the Proc. Acad. Nat. Sci. Phil., for 1917, pp. 2-4.

§ U–110a Verrill, A. E. 1883. "Descriptions of two species of *Octopus* from California." Bull. Mus. Comp. Zool., Vol. 11 (6): 117-124, Pls. 4-6.

Including *Octopus bimaculatus*.

REFERENCES

§ U–111 See also appropriate items in the general mollusk papers listed under "S" References, especially Dall 1871, (§ S–41) who describes a squid from the Gulf; Boone 1928 (§ S–4) who lists an *Argonauta, Polypus americanus* and *bimaculatus, Loligo diomedeae,* and another squid, all but the first and last from the Gulf. Less important cephalopod items will be found in § S–4a, S–10, S–11, and S–22.

§ U–112 ACKNOWLEDGMENT: For the following identifications, for the securing of the *Octopus* sp. photographs, and for checking over this portion of the phyletic catalogue, we are grateful to Dr. S. Stillman Berry of Redlands, California.

LIST OF SPECIES TAKEN:

§ U–113 Two juvenile loliginids, as yet undetermined, and possibly indeterminable because of their immaturity, were taken as minute pelagic squid at San Lucas Cove, south of Santa Rosalia, attracted to the light hung overside at our night anchorage.

§ U–114 *Lolliguncula* (?) *panamensis* Berry 1911.
Berry 1911 (§ U–101) p. 100, etc. good toto-figures, Panama and Ecuador.

A squat, chunky squid, females only of which were taken aboard the Japanese shrimp dredger off Guaymas. At first not distinguished from the more abundant *Loligo opalescens.* Berry writes: ". . . the first material of the species which has come to hand since my description of the species in 1911. . . . As no males have been seen, the generic position of *panamensis* cannot be taken as fully established."

§ U–115 *Loligo opalescens* Berry 1911.
Berry 1912 (§ U–102) p. 294, with good toto-illustrations on Pl. 44-45, recorded from Puget Sound to San Diego.

Taken in some quantity, with the above *Lolliguncula panamensis* aboard the Japanese shrimp dredger south of Guaymas. This constitutes the southernmost record of this common Californian form.

§ U–116 *Octopus* sp. PL. 25 FIGS. 2 AND 3
Dr. Berry has been unable to determine this species, the only octopus collected other than the several representatives of *bimaculatus,* below. A large smooth species taken in a cavern under a pile of rocks on the sand flats of Angeles Bay.

REFERENCES

§ U–117 *Octopus bimaculatus* Verrill 1883.

Berry 1912 (§ U–102) p. 278, good toto-illustrations on Pl. 34; range: San Pedro to Panama. Ricketts and Calvin 1939 (§ Y–3) also has an illustration as fig. 5, Pl. 21 which unfortunately shows only one of the two spots, and that not very clearly.

All the octopi we collected except that listed above turned out to be examples of this species, common in southern California and northern Lower California. There was also an octopus seen but not captured at Pulmo Reef, which could not therefore be determined. Preserved specimens, however, from the following localities were checked as *bimaculatus*, the characteristically two-spotted species; La Paz, San Carlos Bay, Angeles Bay (2 specimens taken on the rocky flats), Puerto Refugio.

§ U–118 SUMMARY

In checking over our results, it becomes apparent to us that we failed to get a representative cephalopod picture of the region. In an area presumably ideally adapted to these animals (to octopi especially for which the Gulf is noted) and where even the most cursory collecting, such as that of the Bingham Expedition, turned out 5 species, we report only 2 octopi and 2 or 3 squid. Why this should be is hard to say. We covered the region fairly carefully, considering the time limitations, and devoted a fair pro ratum of our attention to this group, capturing, with one exception, every cephalopod encountered. Of our few certainly identified species, the octopus is definitely Panamic, reaching in southern California its northerly limit. One squid is northern (occurring to Puget Sound), the other ranges from Guaymas south to Ecuador, both species, strangely enough, having been found at the extremes of their respective ranges, the one at its northernmost limit, the other being by these reports extended many hundred miles to the south.

Phylum Chordata

Subphylum Hemichordata—Enteropneusta

REFERENCES

§ V–1 Horst, C. J. van der 1927-1939. "Hemichordata." Bronns Klass. u. Ord. d. Tier-Reichs. Bd. 4, Abt. 4, Buch 2, Teil 2: 737 pp. 733 figs. Leipzig, 1939.

Cited for its value as a monograph and as a general orientation account. Dr. Bullock writes: ". . . finally completed, this generous monograph makes survey of any Enteropneust problem easy, includes synopses of every known species. Points out priority of *Saccoglossus* over *Dolichoglossus*," p. 406.

Schizocardium peruvianum Spengel 1893 with toto-figure, p. 692, from Pisco, Peru. *Ptychodera flava* Eschscholtz 1825 with good toto-figures, p. 720. Lieferung 6, 1939, with 137 pp. covers the species descriptions.

§ V–2 . . . . 1930. "Observations on some Enteropneusta." No. 51: Papers from Dr. Th. Mortensen's Pacific Expedition, 1914-16. Vidensk. Medd. fra Dansk naturh. Foren., Vol. 87: 135-200, 64 text figs.

Dolichoglossus pusillus (now *Saccoglossus pusillus*) formerly a MSS. species of Ritter, is finally described after many years, from San Diego, San Pedro, Monterey, and Bolinas Bay, with good toto-fig. 22. *Glossobalanus* sp. is described from one poorly preserved specimen taken at Taboguilla, Panama, p. 173. Good bibliography.

§ V–3 Ritter, Wm. E. 1894. "On a new *Balanoglossus* larva from the coast of California, and its possession of an endostyle." Zool. Anz., Vol. 17: 24-30, 2 text figs.

Tornaria larvae taken in the plankton tow at Avalon, southern California, in August, described from a few specimens. May prove to be an overlap form as defined herein.

§ V–4 . . . . and B. M. Davis 1904. "Studies on the ecology, morphology, and speciology of the young of some Enteropneusta of western North America." Univ. Calif. Publ. Zool., Vol. 1 (5): 171-210, Pls. 17-20.

An account chiefly of two tornaria larvae, *ritteri* of Spengel, and *hubbardi* sp. nov., in the plankton of San Diego, with some information on adult *D.* (now *Saccoglossus*) *pusillus* including mention of the absence of the tornaria stage in this form.

§ V–5 Spengel, J. W. 1893. "Die Enteropneusten. . . ." Fauna u. Flora des Golfes von Neapel, Vol. 18: 758 pp. 27 pls.

On p. 216-22 there is mention of *Schizocardium peruvianum*, "a species created to accommodate a single fragment sent to Prof. Spengel fifty-plus years ago, from Peru. This is the only representative

REFERENCES

of the Enteropneusta known from the entire west coast of South America. It has not been reported since, but I have one specimen, taken by Dr. Schmitt (U.S.N.M.) on a Hancock expedition recently." (Communication from Dr. Bullock.) Probably south of our area, but listed here for the sake of completeness, and because so little is known of this group in the entire area that that little can be readily abstracted.

§ V–6 Stiasny, Gustav 1922. "Die Tornaria-Sammlung von Dr. Th. Mortensen." Vidensk. Medd. fra Dansk Naturh. Foren., Vol. 73: 123-138, 9 text figs.

Tornaria tabogae, nov. sp., p. 130, from Taboga, Panama, text fig. 7, one specimen, pelagic. Good bibliography.

§ V–7 Trewavas, Ethelwynn 1931. "Eteropneusta." British Museum (Nat. Hist.), Great Barrier Reef Expedition 1928-29, Vol. 4 (2): 39-67, 18 text figs.

Seven specimens of *Ptychodera flava* are reported from the Galápagos, p. 44, collected by Dr. Crossland on the St. George Expedition. There is also a good toto-drawing of an incomplete specimen from the Great Barrier Reef, as text fig. 1.
A find of *Ptychodera flava* Eschscholtz is mentioned from Galápagos.

§ V–8 See also the report of a great Balanoglossus, from the La Paz area, still undescribed, in Gravier's 1905 Bull. Mus. Hist. Nat. Paris, and Bull. Soc. Phil. Paris articles on its commensals (§ J–8) with a toto-figure of the posterior portion to show the host with its polynoid and *Lysiosquilla* guests.

§ V–9 Due to the co-operation of Dr. Bullock, we are able to present a fairly comprehensive abstract of what has been done to date with this group in the area under consideration. In addition to the items mentioned in the above citations, Bullock remarks: "To this list I can add only the following records of material that has come to my attention but which has not been identified—a laborious process involving the preparation of serial histologic sections:
"1. *Saccoglossus pusillus* (Ritter) occurs in the Estero de Punta Banda at Ensenada, as would be expected from its comparative abundance in the mud-flats of southern California.
"2. A species of *Balanoglossus* proper has been taken by Glassell from the coast of Sonora. It may be the same as Gravier's species.
"3. Two or three records, unknown even as to genus are

REFERENCES

in the Hancock collections from various points on the coast of Lower California, and to this group belong your finds."

§ V–10 ACKNOWLEDGMENT: For most of the information supplied herein, for help with the literature, for determining (so far as could be done) the fragmentary material we collected, for the illustrations herein reproduced, and for checking this portion of the phyletic catalogue, we have to thank Dr. T. H. Bullock, of the Osborn Zoological Laboratory at Yale University.

LIST OF SPECIES TAKEN:

§ V–11 *Balanoglossus* sp. PL. 36 FIG. 2

Posterior fragments only "of the last few inches of an animal that certainly exceeded 30 inches in length." The animals were very abundant as evidenced by castings seen frequently on estuary sandy mud-flats at many points in the Gulf, as at El Mogote, at Angeles Bay, etc. The fragments were collected at San Lucas Cove, Baja California, and at Estero de la Luna, Sonora. At the former point, they occurred fairly high in the intertidal, and abundantly over a sand flat area many acres in extent. Although among the most common and characteristic of all estuary inhabitants, their soft bodies which break at the slightest strain, and their great size and consequent depth in the substratum, make it almost impossible for one to dig them out, and we were unsuccessful in procuring a single whole animal, or even any of the anterior, submerged portions. Probably this is the species mentioned and illustrated, but not described by Gravier 1905 (§ J–8).

§ V–12 *Ptychodera flava* Eschscholtz, cf. PL. 36 FIG. 1
Trewavas 1931 (§ V–7). Horst 1927-39 (§ V–1) p. 720.

This seems to be the form taken at least once under rock in the littoral at Pt. Lobos, Espiritu Santo Island, in a substratum of sandy mud on the bouldery shore. Bullock remarks of *flava* "the first described Enteropneust and the best known Indo-Pacific species, abundant throughout the tropics; has been reported from the Galápagos, and the Hancock expeditions have brought back a number of specimens from these islands. I expect you will find them on the mainland—near the surface of coarse, usually coral sand." The specimen noted was slightly slimy, brown, about an inch long, thick cucumber-shaped. Reminded one of a thick-skinned echiuroid, or of a completely contracted edwardsiidian anemone.

§ V–13 A portion of a *Balanoglossus*-like form was taken also in the sandy substratum under boulders at Pt. Lobos on Espiritu Santo Island. This has not yet been determined.

Subphylum Urochordata—(The Tunicates)

REFERENCES

§ V–101 Ritter, W. E. 1907. "The ascidians collected by the U. S. Fisheries steamer *Albatross* on the coast of California during the summer of 1904." Univ. Calif. Publ. Zool., Vol. 4 (1): 1-52, Pls. 1-3.

14 species, 12 of them new. Entirely dredged material from 33 fathoms to very deep water, from Monterey Bay and southern Californian stations, 10 of the species occurring in the south and 6 in depths of less than 100 fathoms. We found one species on shore in the Gulf. Bibliography of 18 citations. Well illustrated in toto-drawings of most of the species.

§ V–102 Ritter, W. E. 1909. "*Halocynthia johnsoni* n. sp. A comprehensive inquiry as to the extent of law and order that prevails in a single animal species." Univ. Calif. Publ. Zool., Vol. 6 (4): 65-114, Pls. 7-14.

From southern California, probably representing an overlap form. Good toto-figures.

§ V–103 .... 1913. "The simple ascidians from the northeastern Pacific in the collection of the U. S. National Museum." Proc. U.S.N.M., Vol. 45: 427-505, Pls. 33-36.

Only one may represent a Panamic or overlap form, *Phallusia vermiformis* n. sp., p. 496, from 30 fathoms off southern California, although a northern species, *Styela hemicaespitosa,* p. 471, falls within our overlap limits, many specimens having been taken so far south as 28°, at 44 fathoms. Good bibliography.

§ V–104 Ritter, W. E. and Ruth A. Forsyth 1917. "Ascidians of the littoral zone of southern California." Univ. Calif. Publ. Zool., Vol. 16 (24): 439-512, Pls. 38-46.

24 species, 21 of them compound, 8 simple. 21 are new species or varieties. All littoral. 11 are illustrated by toto-figures. Bibliography.

§ V–105 Van Name, W. G. 1930. "The Ascidians of Porto Rico and the Virgin Islands." Sci. Survey of Porto Rico and the Virgin Islands, N. Y. Acad. Sci., Vol. 10 (4): 403-535, Pls. 5-8, 73 text figs.

Supplementing the 1921 West Indies monograph. An excellent orientation account, especially for tropical species. General characteristics, anatomy, collecting and preserving directions, natural history, bibliography. Diagnosis of families and genera. Description of one tropicopolitan species, *Cystodytes dellechiajae,* which we found abundant in the Gulf.

REFERENCES

§ V–106 .... 1931. "New North and South American ascidians." Bull. Am. Mus. Nat. Hist., Vol. 61 (6): 207-225, 8 text figs.

Stolonica zorritensis n. sp. p. 218 with small toto-drawing as fig. 6c, and *Pyura bradleyi* n. sp. p. 221, toto-fig. 7e, from Zorritos, in northern Peru, presumably along shore. Description also of *Clavelina huntsmani* n. sp. from Monterey and British Columbia, to which one of our Gulf species is closely related.

§ V–107 There are also references to entirely pelagic tunicates, mostly of the Southern California area, but which can be expected throughout Panamic and possibly even throughout the tropic Pacific waters, as follows:

Essenberg 1926. "Copelata from the San Diego region. . . ." Univ. Calif. Publ. Zool., Vol. 28 (22): 399-521, 170 text figs.

Johnson, Myrtle E. 1910. "A quantitative study of the development of the *Salpa* chain in *S. fusiformis-runcinata*." Univ. Calif. Publ. Zool., Vol. 6 (7): 145-176.

Ritter and Byxbee 1905. "The pelagic Tunicata." *Albatross*, 1899-1900. Mem. Mus. Comp. Zool., Vol. 26 (5): 195-214, Pls. 1-2.

§ V–108 ACKNOWLEDGMENT: For determining our Gulf specimens, for checking over this portion of the phyletic catalogue, and for much co-operation and assistance, both in this connection and previously, we have to thank Dr. Willard G. Van Name, American Museum of Natural History, New York City.

LIST OF SPECIES TAKEN:

Compound Tunicates

FAMILY SYNOICIDAE

p. 422 Van Name 1930 (§ V–105)

§ V–109 *Polyclinum* sp. probably new species.

Genus on p. 422 in the above mentioned Van Name paper.

"Forms large rounded colonies of soft consistency, not raised on a pedicel." Van Name, personal communication. Large colonies at Gabriel Bay on Espiritu Santo Island, noted as "mahogany colored in life." Small, circular, brown colony taken at El Mogote.

REFERENCES

§ V–110 *Amaroucium californicum* Ritter and Forsyth 1917. Ritter and Forsyth 1917 (§ V–104) p. 483 with detail figure only. Puget Sound surely (and possibly Shumagin Islands, Alaska) to San Diego.

"Very common and very variable in size, shape, and color. Many of the Gulf of California specimens are rather dark-colored (the rest smoky gray), and have conspicuously red zooids." (Personal communication from Dr. Van Name.)

The commonest compound tunicate along the cold, surf-swept California coast, and the commonest again here in the Gulf under utterly different physical conditions. Pulmo Reef (poor specimen, identification doubtful); Pt. Lobos on Espiritu Santo Island; La Paz area; Concepcion Bay (very abundant encrusting on the shells of *Pinna*); Angeles Bay rocky shore; Puerto Refugio; south end of Tiburon Island.

FAMILY DIDEMNIDAE, p. 428 Van Name 1930 (§ V–105)

§ V–111 *Didemnum carnulentum* Ritter and Forsyth 1917. Ritter and Forsyth 1917 (§ V–104) p. 470, with toto-illustration of colony as fig. 11, Pl. 39; southern California.

"Forms thin incrusting colonies white in preservation, often reddish or yellow in life, of a chalky consistency due to minute stellate spicules. Surface varies from quite rough to rather smooth." (Personal communication from Dr. Van Name.)

Another very abundant southern California species which is also one of the abundant Gulf forms. The commoner of the only two tunicates found at Cape San Lucas, La Paz, Coronado Island (mistaken for a white sponge). Concepcion Bay, on shells submerged a foot or more at low tide, very common. Angeles Bay, rocky shore. San Francisquito Bay (mistaken for a white sponge). Puerto Refugio on Angel de la Guardia. Flesh-colored. Probably the second commonest tunicate we found in the Gulf.

§ V–112 *Trididemnum opacum* Ritter 1907. (Equals *T. della vallei* Ritter and Forsyth 1917, p. 472 (§ V–104) detail figures only, from Southern California.) Ritter 1907 (§ V–101), p. 42, one colony from 33 fathoms off San Nicholas Island, southern California, small toto-illustration as fig. 40, Pl. 3.

A poorly preserved specimen from Cape San Lucas was referred to this species, "related to last species but with 3 instead of 4 rows of stigmata." (Personal communication from Dr. Van Name.)

FAMILY POLYCITORIDAE, Van Name 1930
(§ V–105) p. 443

REFERENCES

§ V–113 *Eudistoma* sp.
Probably new species or subspecies. Genus diagnosis on p. 448 of Ritter and Forsyth 1917 (§ V–104), and on p. 445 Van Name 1930 (§ V–105).

"Forms rather soft, fleshy incrusting colonies of a gelatinous consistency, dark-brown or purplish in preservation." Related to *E. diaphanes* Ritter and Forsyth (§ V–104), p. 469, no illustration, San Francisco to San Diego. "Anterior parts of zooids with much brown or purplish pigment." (Personal communication from Dr. Van Name.) Another exceedingly common encrusting Gulf tunicate. Taken at Pt. Lobos on Espiritu Santo Island; on the reef at Marcial Point; at Puerto Escondido; at Concepcion Bay; Angeles Bay; Puerto Refugio; and at the south end of Tiburon Island.

§ V–114 *Cystodytes dellechiajei* (Della Valle) 1877.
Van Name 1930 (§ V–105) p. 452, from the West Indies and (Della Valle) from Naples, probably more or less tropicopolitan.

"Forms black or dark purplish-brown incrusting colonies with rather smooth soft surface. When cut open the lower ends of the zooids are seen to be inclosed in white calcareous capsules composed of minute disk-shaped spicules, which however are often much broken." (Personal communication from Dr. Van Name.)

Noted in our collecting records as "smooth, black, encrusting," and "purple encrusting . . ." Taken east of La Paz, at Puerto Escondido and at Coronado Island (where it was mistaken for a sponge), all in the southern part of the Gulf.

§ V–115 *Clavelina* sp. Semi-compound tunicate. PL. 32 FIG. 4
Related to *C. huntsmani* Van Name 1931 (§ V–106) but representing probably a new species. Taken in the La Paz area only, where it was one of the abundant and characteristic forms found in the interstices of old coral heads.

FAMILY BOTRYLLIDAE, Van Name 1930
(§ V–105) p. 474.

§ V–116 *Botrylloides diegensis* Ritter and Forsyth 1917.
Ritter and Forsyth 1917 (§ V–104) p. 462, with good totofigure 49 on Pl. 43, from San Diego.

REFERENCES

"Very common species forming thin soft incrusting colonies often of large extent. Preserved colonies are some shade of purple, in life its color varies, and the anterior ends of the zooids are often marked with bright colored (yellow, orange, or greenish) pigment so the colonies are very handsome and conspicuous." (Personal communication from Dr. Van Name.)

Noticeable especially in the La Paz area where the handsomely-colored colonies encrust concavities in the dead coral heads. Taken also at Gabriel Bay on Espiritu Santo Island; and in the north, at San Francisquito Bay where the colony was very light colored.

Simple Tunicates

Family Pyuridae, Van Name 1930 (§ V–105) p. 495.

§ V–117 *Pyura* sp. probably new species.

The genus is diagnosed on p. 495 of the above 1930 paper, and the similar or synonymous genus *Halocynthia* on p. 442, Ritter and Forsyth, 1917 (§ V–104).

An "elongate, simple ascidian, entirely red or deep pink." Noted as a "red, sea-squirt" and taken at Pt. Lobos on Espiritu Santo Island; at Coronado Island; and at Puerto Escondido.

Family Ascidiidae. Van Name 1930 (§ V–105) p. 459.

§ V–118 *Ascidia* sp. Probably new species or subspecies.

Genus diagnosed on p. 442, Ritter and Forsyth 1917 and on p. 463, Van Name, 1930.

"Large, (40-50 mm. long) simple ascidian, watery grayish, semi-transparent. Allied to *A. sydneiensis* Stimpson, 1855 and perhaps best considered as an undescribed subspecies of it." (Personal communication from Dr. Van Name.)

Noted as a large sea-squirt the color of water, attached with *Eudistoma* and *Amaroucium,* to the shells of living *Pinna* at Concepcion Bay. Found also at Puerto Escondido and at Puerto Refugio.

§ V–119 SUMMARY

10 species were taken, of which two only were simple tunicates and 8 were compound. Half of the total represent new species or subspecies. Geographically, there are one tropicopolitan or near tropicopolitan, and 4 Californian species. Of the new species, at least two are closely related to (possibly subspecies of) northern Pacific forms. The tunicates therefore seem to represent one of the two

groups (sponges comprising the other) in which the Californian and northern Pacific affiliations are marked. Three of the four commonest Gulf tunicates are Californian forms. The commonest, *A. californicum,* is also the commonest tunicate on Californian shores, extends at least to Puget Sound, and possibly to western Alaska. The second, *Didemnum carnulentum* is common only in the San Diego area (La Jolla). The third, *Eudistoma* sp., is most nearly related to *E. diaphanes* which is common from San Francisco to San Diego. And the fourth, *Botrylloides diegensis,* is very abundant in the San Diego area, but not reported to the north.

As with the sponges, the southern California collector would feel very much at home in the Gulf, the tropical forms representing a minority in the total number of species, very much a minority in total bulk, and being pretty well confined to the southern part. 3 of the new species, *Polyclinum* sp., *Pyura* sp., and *Clavelina* sp., seem to be confined to that portion of the Gulf which is south of Puerto Escondido. Interesting also in connection with the sponges is the fact that at least 2 species of tunicates, *Didemnum* (§ V–111) and *Cystodytes* (§ V–114) so resemble white and black sponges respectively that they were sorted and labeled as such in our preliminary allocations.

Subphylum Cephalochordata (Leptocardia)

The Lancelets

§ V–201 Hubbs, Carl L. 1922. "A list of the lancelets of the world . . ." Occ. Papers, Mus. Zool., Univ. Michigan, No. 105: 16 pp.

A revision, with keys to the families and genera. 2 species are recorded from the Pacific American coasts, including (p. 13) *Branchiostoma elongatum* from Galápagos and Chile. No bibliography.

§ V–202 See also lancelet references in citations applying to the fishes, pp. 570-8, especially in Meek and Hildebrand 1923 (§ W–9).

§ V–203 *Branchiostoma californiense* Andrews 1893.
Meek and Hildebrand 1923 (§ W–9), taken abundantly in sand dredger at Chame Point, Panama, p. 28.

REFERENCES

Hubbs, 1922 (§ V–201), p. 11, description, Monterey to San Luis Gonzales, Gulf.

The only species occurring in the region. Determined by Dr. L. P. Schultz, Curator of Fishes, United States National Museum. Taken at two only points in the Gulf, and abundantly at neither, whereas we expected to find this sand flat animal widespread and common as it has been in southern California in the past. A few specimens were taken on the outside of a sand bar at San Lucas Cove (south of Santa Rosalia), and again on the sand flats at Angeles Bay. The tide in both cases was inadequate for collecting of this sort, however, and it may well be that we took merely the high-up strays of large colonies deeper down.

Subphylum Craniata (Vertebrata)

CLASS PISCES (FISHES)

§ W–1 Barnhart, P. S. 1936. Marine Fishes of Southern California. pp. 209, 290 figs. Univ. Calif. Press.

A popular account, profusely illustrated by toto-drawings, of 372 spp. known from southern California, with ranges. Many extend into the Gulf. Bibliography cites the following works apropos to the Panamic region, but which we do not list: Garmann 1899 (*Albatross* 1891, etc., Deep Sea Fishes), Hubbs 1918c (Flight of the California Flying Fish), Townsend and Nichols 1925, *Albatross* Gulf Deep Sea Fishes; Wales 1932 (Report on two collections of Lower California Marine Fishes), and the necessary general orientation accounts: Jordan 1905; Jordan and Evermann 1896-1900; Jordan, Evermann, and Clark 1928; Starks 1921; and Walford 1931a.

§ W–1a Bean, Barton A. and Alfred C. Weed 1910. "A review of the venomous toadfishes." Proc. U.S.N.M., Vol. 38: 511-526, 8 text figs., 4 pls.

Two are shallow water Panamic forms, p. 512: "To this account we may add that in the specimen of *Thalassophryne reticulata* which we very carefully examined, and in which the skin over the point of the opercular spine had not been ruptured, the poison sac lay along the whole outer surface of the spine and not merely at its base. The sac is so placed that any pressure tending to cause the spine to pierce the skin would produce a corresponding pressure on the contents of the sac and cause the poison to flow into the wound with considerable force. In one specimen the pressure that exposed the point of the spine in a fish that had been in alcohol nearly thirty years caused the contents of the sac to be ejected to a distance of 2 or 3 feet. . . . In 1865 a letter from Captain Dow to Doctor Günther was read before the Zoological Society of London, in which he described the poison as producing fever similar to the effects of the sting of a

REFERENCES

scorpion. He adds that serious effects from the poison are very rarely known." In several species the opercular spine was found to be hollow and highly specialized for injecting the contents of the specialized poison duct.

§ W–2 Beebe, Wm. and John Tee Van 1938. "Seven new marine fishes from Lower California." Zoologica, Vol. 23: 299-312. 3 pls., 5 text figs.

One *Mobula* (eagle ray), one "eel," one flatfish, one Scorpaenid, one *Ammodytes,* and two blennies.

§ W–3 Bolin, Rolf L. 1939. "A review of the Myctophid fishes of the Pacific coast of the United States and of Lower California." Stanford Ichthyological Bulletin, Vol. 1 (4): 89-156, 27 text figs. Photolith.

Monograph of the lantern fish within the region involved, which includes the Gulf of California. Descriptions and toto-drawings of 20 species of these pelagic and bathypelagic forms, 3 of them new. 7 are Panamic and 6 are probably overlap.

§ W–4 Breder, C. M., Jr. 1928. "Elasmobranchii from Panama to Lower California." Bull. Bingham Ocean. Coll., Vol. 2 (1): 1-13, 12 figs.

19 species, 2 of them new. 21 toto-photos or drawings, of 8 species. Shore to 30 fathoms, but mostly seined from shore. No bibliography.

§ W–5 .... 1928a. "Nematognathi, Apodes, Isospondyli, Synentognathi, and Thoracostraci from Panama to Lower California." Bull. Bingham Ocean. Coll., Vol. 2 (2): 1-25, 10 text figs.

37 Pacific species, 2 of which are new. 2 new genera are described. 7 toto or near toto-drawings of 4 species. Shore to 620 fathoms, but mostly at or close to surface. Short bibliography.

§ W–6 .... 1936. "Heterosomata to Pediculati from Panama to Lower California." Bull. Bingham Ocean. Coll., Vol. 2 (3): 1-56, 19 text figs.

172 Pacific species, 8 (including one subspecies) of which are new. One new genus is described. Mostly seined on or near shore. 20 toto-drawings of 16 species. Addendum includes three species omitted from the previous Article 2. Bibliography of 16 titles.

§ W–6a Kendall, Wm. C. and Lewis Radcliffe 1912. "The shore fishes." *Albatross* 1904-05. Mem. Mus. Comp. Zool., Vol. 35 (3): 77-172, 8 pls.

Within the Panamic area, collections were made at Acapulco, Panama Bay, and the Galápagos. Of the total 227 species reported upon, 163 were from one of the above three stations, or from the open ocean

REFERENCES

between these points and Polynesia. Excellent toto-illustrations of the new species.

§ W-7 Parr, A. E. 1931. "Deepsea fishes from off the western coast of North and Central America." Bull. Bingham Ocean. Coll., Vol. 2 (4): 1-53, 18 figs.

Bathypelagic material from intermediate and great depths in the region involved; hence merely of subsidiary interest here.

§ W-8 Herre, A. W. C. T. 1924. "Poisonous and worthless fishes. An account of the Philippine Plectognaths." Phil. Journ. Sci., Vol. 25 (4): 415-510, 2 color plates.

A general account of the Indo-Pacific poisonous fishes of the FAMILIES BALISTIDAE, TETRAODONTIDAE, and DIODONTIDAE, similar or almost identical species of which occur in the Panamic area.

§ W-9 Meek, S. E. and S. F. Hildebrand 1923, 1925, and 1928. "The Marine Fishes of Panama." Pts. 1, 2, and 3. Field Mus. Nat. Hist., Zool. Ser., Vol. 15: 1045 pp., 102 pls.

Including the lancelets. 450 species are listed from the Pacific, with synonymy, geographic and often bathymetric distribution, and ecological data, usually with descriptions, and often with toto-illustrations. 403 of these are actually recorded from Panama Bay. 72 species occur on both sides of the Isthmus. All are shore forms, in water up to 50 fathoms. Index for all three parts starting p. 1019. Key paper for fishes, p. 9: "The ichthyological fauna of the Pacific coast of Panama appears to range from the Gulf of California to Ecuador."

§ W-10 Osborn, A. C. and J. T. Nichols 1916. "Shore fishes collected by the *Albatross* Expedition in Lower California with descriptions of new species." Bull. Am. Mus. Nat. Hist., Vol. 35 (16): 139-181, 15 text figs.

185 spp. and varieties, of which 14 are described as new. Shore to 13 fathoms. Many stations from Guadalupe and Cedros Island to Cape San Lucas and far up in the Gulf. No bibliography.

§ W-11 Pellegrin, Dr. J. 1908. "Sur un grand poisson percoide peu connu du Golfe de Californie (Epinephelus rosaceus Streets)." Bull. Mus. Hist. Nat. Paris, Vol. 14: 349-352.

Collected by the indefatigable L. Diguet. Very large (1½ M.) called "Cabrilla" by the inhabitants.

§ W-12 Reid, Earl R. 1940. "A new genus and species of pearl fish, FAMILY CARAPIDAE, from off Gorgona Island, Colombia." A. Hancock Pac. Exp., Vol. 9 (2): 47-50, Pl. 6.

Encheliophiops hancocki, n. sp., known from a single specimen 74.8 mm. long, taken from shore in the coral *Pocillopora.*

REFERENCES

§ W–13 Schmitt, W. L. and L. P. Schultz 1940. "List of the fishes taken on the Presidential Cruise of 1938." Smith. Misc. Coll., Vol. 98 (25): 10 pp.

Annotated catalogue of 43 Pacific species, plus a few Caribbean forms.

§ W–14 Terron, C. C. 1932. "Lista de los peces de las costas de la Baja California." An. Inst. Biol., Vol. 3: 75-80.

By families. Scientific and common Spanish names, localities.

§ W–15 Additional apropos papers will be found in the Bull. U.S. Nat. Mus., Vol. 7 (Streets, 1877, p. 43, etc.): in Zoologica, Vol. 5 (Nichols, 1924, p. 62-65), Vol. 25 (Tee-Van, 1940, p. 53-64): and in A. Hancock Pac. Exp., Vol. 2 (Myers and Reed, 1936, p. 7-10): ditto (Ginsburg, 1938, 109-121): Vol. 9 (Seale, 1940, p. 1-46): ditto (Herald, in press, 194—, p. 51-67): ditto (Myers and Wade, in press, 194—, p. —); Copeia 1938 (No. 3, Brock, p. 128-131).

As with shells, no attempt has been made to compile a comprehensive bibliography of the fish of the Panamic area, even to the degree that this has been attempted in other groups. Ichthyology is a highly specialized and a highly organized subject, and the construction of such a bibliography, however great the need, must await the services of a specialist.

§ W–16 The Gulf is famed for its fish resources, important alike for commerce, sport, and science. However, due to personnel and to the deliberately limited objectives of the trip, no attempt was made to procure representatives of this group—a task in itself, and one to which more focused energy could be devoted profitably. But certain species which were obviously characteristic turned up so persistently along shore, others were got so readily at night by light hung over the side, and opportunities for collecting representative bottom fish were so tempting on the Japanese shrimp boats, that we assembled a small collection. Consisting of some 46 lots with possibly nearly as many species, this material has been forwarded to the United States National Museum, but the determinations will hardly be available by the time this goes to press. In the meanwhile, however, Dr. Rolf Bolin of the Hopkins Marine Station has been good enough to identify several of the more important or spectacular species per the following list:

REFERENCES

§ W–16a ACKNOWLEDGMENT: For checking this portion of the phyletic catalogue, for determining the ten species mentioned below, and for help with and access to the literature, we are indebted to Dr. Rolf Bolin of the Hopkins Marine Station at Pacific Grove.

LIST OF SPECIES TAKEN:

§ W–17 Leptocephalus larva of *Albula vulpes* (Linnaeus) 1758. PL. 40 FIG. 2

p. 178 Meek and Hildebrand 1923 (§ W–9). p. 15 Barnhart 1936 (§ W–1) toto-fig. 42, tropicopolitan, on the Pacific coast sometimes occurring as far north as Monterey, p. 4, Breder 1928a (§ W–5). Bolin notes that a stray specimen was reported recently from Sausalito by Myers, p. 83-85, Copeia No. 2 of July 31, 1936.

Taken very abundantly at Concepcion Bay, as a transparent ribbonfish attracted to the boat by a light hung over the side at night.

§ W–18 *Balistes polylepis* Steindachner 1876. PL. 31 FIG. 2

Trigger Fish. Puerco.

p. 790 Meek and Hildebrand 1928 (§ W–9), Lower California to Peru. p. 94 Barnhart 1936 (§ W–1) toto-fig. 284, considers the range as Lower California to San Pedro and Catalina Island. Not listed in Breder 1936 (§ W–6).

The photograph is from a 12″ specimen taken in shallow water in the La Paz area. Presumably the same form was taken at Concepcion Bay, and on the Japanese shrimp trawler outside Guaymas. Thought to be common and widely distributed along shore in the Gulf.

§ W–19 *Cypselurus californicus* (Cooper). PL. 31 FIG. 1

California Flying Fish. Not listed in Meek and Hildebrand (§ W–9). p. 24 Barnhart 1936 (§ W–1) toto-fig. 80, Cape San Lucas to Pt. Conception. Not listed in Breder 1928a (§ W–5).

Very common off the west coast of Lower California especially in the stretch from Magdalena Bay to Cedros Island, but not seen within the Gulf. The photographed specimen was about 11″ in length.

§ W–20 *Diodon* sp. (*hystix* or *holocanthus,* both Linnaeus 1758). PL. 30 FIG. 1

Puffer. Porcupine fish. "Sapos" in Spanish.

p. 827 and 829 Meek and Hildebrand 1928 (§ W–9), tropicopolitan. Not listed in Barnhart 1938 (§ W–1). *D. holocanthus* is listed as abundant from Cape San Lucas and Panama, in Breder 1936 (§ W–6), p. 52.

REFERENCES

The specimen photographed was small, about 4″ long, taken by hook and line in shallow water at Pt. Lobos, Espiritu Santo Island, where the boat was at anchor. *Diodons* of one species or another (not yet differentiated) were taken also at Pulmo Reef, and in the La Paz area. It would appear that these famous and fantastic tropical forms are restricted to the southern third of the Gulf.

§ W–21 *Encheliophiops hancocki* Reid 1940.

Reid 1940 (§ W–11) p. 47, toto-figure on Pl. 6, described from a single specimen 74.8 mm. in length, taken in coral along shore, Gorgona Island, Colombia.

A single specimen of this elongate Carapid, belonging to one of the genera which has replaced the well-known name, *Fierasfer,* traditional commensals of holothurians and of pearl oysters, was taken at Pt. Lobos on Espiritu Santo Island in the common shore cucumber *Holothuria lubrica* (§ L–25). At first thought to represent a new species, Dr. Bolin finally has assigned this to *hancocki* after communicating with Dr. Reid who made a re-examination of the type specimen.

§ W–22 *Fodiator acutus* (Cuvier and Valenciennes).

PL. 30 FIG. 2

Acute Nosed Flying Fish. "Volador" in Spanish.

p. 243, Meek and Hildebrand 1923a (§ W–9), "known from both shores of tropical America." p. 15 Breder 1928 (§ W–5), 48 specimens from Panama and in the Gulf as far north as San Felipe Bay, 37 to 162 mm. total length. Mentioned on p. 23, with toto-figure 79, Barnhart 1938 (§ W–1), as occurring occasionally as far north as San Pedro.

The specimen photographed, about 6″ long, flew aboard, or was caught by net overside, at night while the ship was at anchor with decks lighted, in Concepcion Bay. In parts of the Gulf, these small, fast-flying, sharp-nosed projectiles constitute such a menace that decks of anchored boats must be kept dark. Juveniles of presumably the same species were also taken near a light hung over the side at the same place.

§ W–23 *Hippocampus ingens* Girard 1858.

Giant Sea Horse. Caballo del Mar.

p. 256, Meek and Hildebrand 1923 (§ W–9), California southward to Panama. p. 23 Breder 1928a (§ W–5), color in life predominantly orange. p. 34, toto-figure 117, Barn-

REFERENCES

hart 1938 (§ W–1), normal range Magdalena Bay southward, but type locality is reported as San Diego.

Several of these were taken in the hauls of the Japanese shrimp trawlers south of Guaymas. The fresh material is brilliantly colored with specklings of chromatophores that look like delicate India-ink markings. The males are much prized in the curio-stores of Guaymas, La Paz, etc., because of the brood pouches which are facetiously considered to be closed by zippers.

§ W–24 *Hypsoblennius gentilis* (Girard) 1854.

Not mentioned in Meek and Hildebrand (§ W–9) or in the Bingham Reports (§ W–5 and 6). p. 87, toto-figure 262, Barnhart 1938 (§ W–1) Gulf of California to Monterey.

A similar species, known in the north as the tide pool blenny, has been described as a "small, omnivorous, and pugnacious fish." Several specimens were taken at Concepcion Bay: one, a minute form brought on board with sponges and "Hacha" shells, (*Pinna* sp.) viciously attacked the molesting finger of one of us, achieving results with minute needle-like teeth quite out of proportion to its size—being known hereafter to all involved as "the-fish-that-bit-Carol." The specimen photographed was about 4″ long. Dr. Bolin determined this only provisionally.

§ W–25 *Myrichthys tigrinus* Girard 1859. PL. 40 FIG. 1

p. 149, Meek and Hildebrand 1923 (§ W–9) Oregon (?) to Panama. Breder 1928a (§ W–5) lists 10 specimens from the Gulf. Not mentioned in Barnhart 1938 (§ W–1).

At Concepcion Bay, where several specimens of this agile and highly colored eel appeared in the circle illuminated by a light hung overside, they were taken at first for the true sea serpents known to be common farther south.

§ W–26 *Sphaeroides annulatus* (Jenyns) 1842. *Botete*

p. 816, toto-photograph on Pl. 78 Meek and Hildebrand 1928 (§ W–9) from California to Peru and the Galápagos. As *Cheilichthys annulatus,* p. 49 Breder 1936 (§ W–6), 21 specimens from Panama and from the Gulf. Not listed in Barnhart 1938 (§ W–1).

This is one of the forms first made known to science by Darwin's "Beagle" collections. One of the commonest of all shore fish in the Gulf, having been seen at practically every suitable station. The large (15″) specimen illustrated was taken in the La Paz area where the liver is regarded as poisonous. Additional specimens were collected from Puerto Escondido and at San Gabriel Bay on Espiritu

REFERENCES

Santo Island. In Agiabampo Estuary, numerous individuals could be seen on eel grass flats in low tide depths of from a few inches to a few feet; as we waded along collecting they would lie inert until we almost touched them, at which they clear out at high speed, muddying up the water with their activity. Clavigero (§ Y–12, Lake and Gray translation) records that in 1706, some of Father Ugarte's soldiers died from eating these livers (p. 213).

§ W–27 In addition to small specimens of some 30 other species which were preserved biologically, we brought aboard many large individuals of these and other forms which were discarded without adequate record, and several food fish not surely determinable. The following, however, were identified with fair certainty:

Gyropleurodus francisci (Girard).

The horned shark; p. 7 Barnhart 1938 (§ W–1), toto fig. 5; taken at San Francisquito.

Sphyrna zygaena (Linn).

The hammer-head shark, called "Cornuda" locally. Barnhart, p. 9, toto-fig. 19. Taken at Estero de la Luna, on Japanese shrimp boats at Guaymas. Carcasses at Cape San Lucas.

Cynias lunulatus (Jordan and Gilbert) probably.

A smooth hound, p. 8, Barnhart 1938 (§ W–1), toto-fig. 13. Taken at San Francisquito Bay.

Manta birostris (Walbaum).

The Manta ray, p. 14 toto-fig. 40, Barnhart, 1938 (§ W–1). Many times harpooned off Pulmo, off Espiritu Santo Island (Pt. Lobos), River Mayo (near Estero de la Luna).

Netuma platypogon (Günther) probably.

A catfish, p. 146, Osborn and Nichols, 1916 (§ W–10). Abundant by hook and line outside Guaymas Harbor.

Sphyraena argentea Girard possibly.

A barracuda-like form, possibly the California barracuda, p. 36, toto-fig. 123, Barnhart, 1938 (§ W–1). Four or five specimens harpooned at Tiburon Island.

Katsuwonus pelamis (Linn.) 1758.

Skipjack, called Barrilete. Barnhart, 1938 (§ W–1), toto-fig. 126, as K. vagans (Lesson) 1828. Taken south of Magdalena Bay; off Pulmo; south of Puerto Escondido.

Sarda chiliensis (Girard).

Bonito, called the same in Spanish, Barnhart (§ W–1), toto-fig. 128, 4 or 5 caught north of Magdalena Bay.

REFERENCES

Neothunnus macropterus (Schlegel).

Yellowfin. Barnhart (§ W–1), p. 37, toto-fig. 130. 2 off Espiritu Santo Island.

Germo alalunga (Gmelin).

Albacore, or rather a cross (?) between Albacore and Yellowfin which the fishermen say is fairly common. Barnhart (§ W–1), toto-fig. 131. Taken north of Loreto.

Scomberomorus sierra Jordan & Starks.

Sierra, called the same in Spanish. Barnhart (§ W–1), toto-fig. 132. Finest food fish we took, taken south of Puerto Escondido, north of Loreto, south of Tiburon Island.

Coryphaena equisetis Linn.

Dolphin, called Gallo. Barnhart (§ W–1), p. 39, toto-fig. 141. 2 taken off Cabo Falso; the second finest food fish taken on the whole trip.

GENERAL REFERENCES

I. Apropos books on natural history, oceanography, procedure:

REFERENCES

§ Y–1 Fowler, G. H. and Allen, E. J. 1928. *Science of the Sea. An elementary handbook of practical oceanography for travellers, sailors, and yachtsmen.* Prepared by the Challenger Society. Oxford, Clarendon Press. 502 pp., 220 figs., 11 charts.

Cited specifically for its zoogeographical considerations on p. 290, 292, and 294 with Chart XI, which consider also the area we treat, and for the chapter on tropical collecting. In addition there is much practical information on dredging and on general collecting, with data on equipment and procedure. Chapter XI considers methods of preserving, labeling, and storing marine organisms, important items about which there is little information available. The formalin remarks on pp. 388 ff. should be regarded, however, with caution; it becomes increasingly apparent that formalin is to be preferred over alcohol as a permanent preservative only for a few groups, fish and jellyfish notably, but that in many groups its effects are disastrous in the long run. Worst of all, its constant use almost invariably results sooner or later in an allergic condition on the part of the user.

§ Y–2 Johnson, Myrtle E. and Snook, H. J. 1935. *Seashore Animals of the Pacific Coast.* Macmillan Co., N. Y. 659 pp., 700 text figs., 12 color plates.

Although concerned primarily with animals north of Mexico, a great many overlap and some true Panamic forms are treated. The San Diego residence of one of the authors promises that southern California animals, many of which range southwardly, will have been given especial consideration. 15-page bibliography. Glossary.

§ Y–3 Ricketts, E. F. and Calvin, J. 1939. *Between Pacific Tides.* Stanford University Press. 320 pp., plus 91 pages of illustrations.

An account of the habits and habitats of some five hundred of the common and obvious intertidal forms between Sitka, Alaska, and northern Mexico, with emphasis on ideas and principles rather than on descriptions of the animals. Bibliography of four hundred titles.

§ Y–4 Rioja, Dr. Enrique. 1931. *El Mar Acuaria del Mundo.* Editorial Séneca, México. 405 pp., 56 text figs.

Cited here chiefly for the benefit of Mexican readers. An interesting and well written popular account, illustrated with drawings by the author. The Spanish equivalent of Russell and Yonge, *The Seas.* The work of an eminent Spanish marine biologist and polychaet specialist,

REFERENCES

known also for his popular scientific writings, now professor of invertebrate zoology at the Institute of Biology in Mexico City.

§ Y–5 Schenck, Edward T. and J. H. McMasters. 1936. *Procedure in Taxonomy.* Stanford University Press. 72 pp.

Because so much of marine biology is conditioned by taxonomic discipline, some knowledge of the sub-science which concerns itself with the classifying and naming of living organisms is desirable. A statement of the problems that have made of this subject a veritable Augean stables, and of their ameliorations and possible solutions. Although intended primarily for paleontologists, these generalizations and suggestions are applicable also to biology in general.

Significant remarks on nomenclature incidentally, will be found on p. 81 ff. of Grant and Gale, 1931 (§ S–16).

§ Y–6 See also pp. 302-306 of Ricketts and Calvin 1939 (§ Y–3) wherein several apropos items are cited which need not be repeated here.

II. Bibliography of the Gulf and of other Panamic areas: Books

§ Y–7 Bancroft, G. 1932. *The Flight of the Least Petrel.* G. P. Putnam's Sons, N. Y. and London. 396 pp., 46 illustrations. Map.

The log of an ornithologist's encirclement of Lower California by small boat. Detailed accounts of the shore line and of anchorages. Much of the work was done in the Gulf, the east coast of Lower California having been explored clear up into the mouth of the Colorado River.

§ Y–8 Beebe, William. 1924. *Galápagos, World's End.* G. P. Putnam's Sons, N. Y. 443 pp., 83 photographs and 24 colored illustrations.

The popular log of a biological expedition. Considerations of Panamic marine invertebrates will be found on pp. 39, 64 ff., 129 ff., 154, and 328. Comprehensive data on the marine iguana. Galápagos bibliography.

§ Y–9 .... 1938. *Zaca Venture.* Harcourt, Brace and Co., N. Y. 308 pp., 24 illustrations.

A popular and quite satisfying account of the Templeton Crocker Expedition to Cedros Island, Turtle Bay, Magdalena Bay, Cape San Lucas, Santa Inez Bay, and other stations chiefly in the Panamic area. Important considerations of the marine invertebrates which formed a large part of the collections. The significant observations at Santa Inez and Arena Bank were at points well within the Gulf. The unconventional attitude which may have been a factor in the number of new species sought and taken is reflected in the p. 139 remarks: ". . . utter inadequacy of present human scientific descriptions . . . our conventional diagnoses exclude not only the changes

REFERENCES

due to the emotions of fish, but a further fifty per cent must be deducted because of our laziness and diurnal habits."

§ Y–10 Coolidge, Dane and Mary R. 1939. *The Last of the Seris.* E. P. Dutton and Co., N. Y. 264 pp. Illustrations.

Some incidental natural history and geographic information, in an account based on an extended stay with the Seri Indians at Kino Bay in Sonora.

§ Y–11 Hakluyt, Richard, and Goldsmid, Edmund 1890. *The Voyages of the English Nation to America.* E. and G. Goldsmid, Edinburgh.

In Vol. III, p. 317-377, under the title "The First and Second Discovery of the Gulfe of California. . ." there will be found interesting items on the Isla de Perlas, on great tides (presumably in the Colorado River entrance), and on the Indians, with some incidental natural history and much geographic information which it is possible occasionally to correlate with present place names. A very readable account, replete with such ringing sentences as these of Ulloa, (p. 320), "The Countrey is plaine, but farre within the land they saw great and small hills extending themselves a great way, and being very faire and pleasant to behold." (p. 324), "and that so great a countrey, that I suppose if it should so continue further inwarde, there is country ynough for many yeeres to conquer . . . and on our left hande we descried a plaine countrey, and saw in the night certaine fires."

§ Y–12 Lake, Sara E. and Gray, A. A., translators and editors. 1937. *The History of [Lower] California,* by Don Francisco Javier Clavigero, S.J. Stanford University Press. pp. 413 plus xxvii. Portrait and 2 maps.

Clavigero (1731-1787) completely translated for the first time into English, emerges as a disciplined, competent, and significant writer. For factual geographic information, this account is pertinent even today.

§ Y–13 Lamb, Dana, and June Cleveland 1938. *Enchanted Vagabonds.* Harper and Bros., N. Y., pp. 415, 48 illustrations from photographs.

The record of a canoe trip covering several years, along the west coast of Mexico and Central America. The first 11 chapters of this romantically entitled account deal with Lower California and the Gulf. There is some incidental information on the more obvious animals. If objective truths are recorded here—and there seem to be no factual reasons for discrediting these "believe it or not" episodes—these people must know this coast very intimately, and know it from bitter experience, one moment of which is more effective than hours of arm-chair study. More detailed statements of procedure, methods, and equipment would have been welcome, whereupon the results of their grief, work, and experience might achieve supra-personal value.

REFERENCES

§ Y–14 North, A. W. 1910. *Camp and Camino in Lower California.* Baker and Taylor Co., N. Y. pp. 346, 26 pls., 1 map.

A popular account of wandering, hunting, and exploring the length of the peninsula. Much of this country seems to have been less sparsely populated during these days than in 1923 when one of us started investigating the marine fauna of its northwestern coast. Very naïve information on poisonous animals (p. 133). Mirage in the Gulf country; the nostalgic call of pigeons. The visits of some of the early biological explorers, Professor Gabb being mentioned by name, had already become legends. There seems still to have been a curious liaison, then, between the days of the padres and our time. North reports seeing an Indian woman reputedly more than 100 years old, worshipping as she had learned from the padres, within the tumbled walls of a mission long since abandoned.

§ Y–15 U. S. Navy, Hydrographic Office. 1937. "Sailing Directions for the west coasts of Mexico and Central America." H. O. No. 84, 430 pp., text figs.

Includes two maps, one of which comprises a directory of marine charts. Detailed and accurate information on the topography of the shore. The information on lights and on port usages, but especially on piers, buoys, and harbor works, is slightly out of date for the parts of Mexico we visited.

III. Bibliography of the Gulf and of Panamic areas: papers and articles.

§ Y–16 Agassiz, Alexander 1892. "General sketch of the expedition of the *Albatross* from February to May, 1891." Bull. Mus. Comp. Zool., Vol. 23: 89 pp., 22 pls.

Concerned chiefly with dredging and oceanography within the Panamic area, but usually at great depths.

§ Y–17 Beebe, Wm. 1923. "Williams Galápagos Expedition." Zoologica, Vol. 5 (1): 2-20.

Narrative which was expanded into the book § Y–8. Scant information on marine invertebrates.

§ Y–18 .... 1937. "The Templeton Crocker Expedition . . ." Zoologica, Vol. 22: 33-46, pl. and text fig.

Itinerary, lists of stations, nets, and dredges. See § Y–9, *Zaca Venture.*

§ Y–19 .... and John Tee Van 1938. "Eastern Pacific Expeditions of the New York Zoological Society . . ." Zoologica, Vol. 23: 287-298, 2 text figs.

Ensenada to Panama, not in the Gulf. First station was Cedros Island. Itinerary, maps, list of stations, nets, and dredges.

REFERENCES

§ Y-20 California Automobile Association. Log of Lower California and the Gulf of California. Outing Bureau of the Automobile Club of Southern California, Los Angeles.

Revised from time to time. The most accurate and practical information we have found on this little known area. Mileages, road conditions, supplies, and short descriptions of the country. Yachting information, fishing conditions, description of ports and anchorages.

§ Y-21 Crossland, C. 1927. "Expedition to the South Pacific on the S. Y. *St. George,* Marine Ecology . . ." Trans. Roy. Soc. Edinburgh, Vol. 55 (3): 331-554, 1 pl., 11 text figs.

The section on the tropical Pacific coast of America mentions water temperatures, difference between Panama and Galápagos in fauna and in physical conditions, regards the littoral fauna of the Panama west coast mainland as the richest the author has encountered. Information on corals and worms, pearl oysters. Description of fauna, flora, and environmental conditions at Panama, at the Pearl Islands, and at Galápagos. Much of the article, however, and all the illustrations deal with the coral reefs of the South Seas.

§ Y-22 Diguet 1895. "Note sur une exploration de la Basse-Californie par M. Diguet, chargé d'une mission par le muséum." Bull. Mus. Hist. Nat. Paris, Vol. 1: 28-30.

Record of a six-months trip. Santa Rosalia, Mulege, Loreto and La Paz. The most important collections came from the La Paz area, especially from pearl divers working around Espiritu Santo Island. There is evidence from the literature that M. Diguet was indefatigable, that he collected widely, intensively, and aggressively, procuring most of the common shore animals, including several which, because of difficulty of capture despite their abundance have not been taken since.

§ Y-23 Eisen, Gustav. 1895. "Explorations in the Cape Region of Baja California in 1894, with references to former expeditions of the California Academy of Sciences." Proc. Calif. Acad. Sci. (2) Vol. 5: 733-775, Pls. 72-75.

A consideration chiefly of the land and of land life, with some general information, p. 734: "Indeed when the Academy began these explorations, there remained no other country within our reach which was less known, more misjudged, less understood, and about which more conjectures were made and less real facts known." (A situation which is almost equally true today!)

§ Y-24 Hanna, G. Dallas 1925. "Expedition to Guadalupe Island, Mexico. General Report." Proc. Calif. Acad. Sci. (4) Vol. 12 (6): 55-72, Map. Pls. 15-19.

Log of six weeks' trip to Guadalupe Island, Cedros Island, Magdalena Bay, and other points on the Lower California west coast. Although the emphasis was on vertebrates, especially on elephant seals, abalones and other mollusks are mentioned; there is a description of the marine gardens at Cedros, etc.

§ Y–25 .... "Expedition to the Revillagigedo Islands, Mexico, in 1925. General Report." Proc. Calif. Acad. Sci. (4) Vol. 15 (1): 1-113, 7 text figs. 10 pls.

A general collecting expedition extending over a period of about 10 weeks, with stops at Guadalupe and Cedros Island, and at several points on the Pacific coast of Lower California. Although emphasis was on fossils and vertebrates, invertebrates were collected by Dr. Hanna and Mr. Jordan. Clarion Island is surrounded by a true coral reef. Cape San Lucas is mentioned on pp. 79-80.

§ Y–26 MacGinitie, G. E. 1935. "Ecological aspects of a California marine estuary." American Midland Naturalist, Vol. 16 (6): 629-765, 21 text figs.

Apropos here only because it has been cited several times in our text, and because it records occasional observations on the estuary at Corona del Mar in southern California where a number of Panamic and overlap species occur. There is an excellent photograph of *Tethys* on p. 738, as fig. 20. Practical collecting and general procedure information, in addition to the reports of a survey of several years which turned out a total list of 207 species, 22 of them new.

§ Y–27 Mortensen, Th. 1918. "Observations on protective adaptations and habits, mainly in marine animals." Vidensk. Medd. fra Dansk naturh. Foren., Vol. 69: 57-96, 1 pl., 19 text figs.

Within the Panamic area, instances of presumed protective mimicry among marine invertebrates are described (p. 64-5) for communities of several species of fish which resemble floating wood chips, and (p. 66) drops of water, (p. 68) the shelled snail *Ovulum* occurring on gorgonians, (p. 69) *Cypraea* on corals, (p. 72) masked tunicates, (p. 77) shrimps, (p. 82) the relations between ophiurids and gorgonians.

§ Y–28 Nelson, Edward W. 1911. "A land of drought and desert, Lower California." National Geographic Magazine, Vol. 22: 443-474, many illustrations.

Of the popular commentaries cited in the below-mentioned bibliography (including Ellison 1905, North 1908, etc.), we consider only his own article worth listing. The others are chiefly literary, historical, or timely. This is a good, sober account, a condensed source book, with excellent illustrations, chiefly of the flora.

§ Y–29 .... 1921. "Lower California and its natural resources." Mem. Nat. Acad. Sci., Vol. 16 (1): 1-194, 34 pls., 1 map.

REFERENCES

An account of the Biological Survey's 1905-6 expeditions by land into Lower California, and a summary of what had been published up to 1919 on the region, together with a 25 page annotated bibliography of some 454 titles, consisting of general, geographical, geological, botanical, and vertebrate zoology items, with a few conchological citations.

§ Y–29a Schenck, Hubert G. and A. Myra Keen 1936. "Marine Molluscan Provinces of Western North America." Proc. Amer. Phil. Soc., Vol. 76 (6): 921-938, 6 text figs., and ...... 1937. "An index method for comparing molluscan faunules." Vol. 77 (2): 161-182, 4 text figs.

Statement of a biometric method for differentiating zoogeographical provinces, using the total ranges of 1948 species of shelled mollusks listed by Dall 1921 for the Pacific coast north of San Diego. On the basis of this total evidence, which derives from shallow and dredged, common and rare species, the following provinces are adduced: Arctic, from 72° to 58° N. (Icy Point, above Sitka); overlapping an Aleutian from 62° N. (provisional) to 42° N. (Cape Mendocino); which in turn overlaps the California Province from Cape Flattery (48° N.) to Cape San Lucas (23° N.). "If, however, shore collections or shallow dredgings alone were considered, a fairly sharp line of division would appear in the vicinity of Point Conception. We have little information on tropical or sub-tropical West American mollusks the ranges of which do not extend north to San Diego; hence, we can only tentatively indicate Cape San Lucas as the boundary. It may well be, particularly in view of the demonstrated areal boundaries to the north, that the Californian and Panamic Provinces are separated by an overlap area, probably along the outer coast of Lower California. The entire Gulf of California falls in the Panamic province" (p. 931, 1936). In the 1937 paper they note that the littoral districts conform only in a general way to the provinces above specified (p. 166), and remark the critical nature of Pt. Conception as a barrier for shore species.

§ Y–30 Schmitt, W. L. 1939. "Decapod and other crustacea collected on the Presidential Cruise of 1938 (with an introduction and station data)." Smith. Misc. Coll., Vol. 98 (6): 29 pp., 3 pls.

Narrative, with some illustrations. Magdalena Bay and several other stations in the Panamic area were visited.

§ Y–31 Slevin, J. R. 1923. "Expedition of the California Academy of Sciences to the Gulf of California in 1921. General Account." Proc. Calif. Acad. Sci. (4) 12 (6): 55-72, map.

Almost three months were spent in a survey entirely within the Gulf from Ceralbo Island to Georges Island at 31° N. latitude. 8 scientists were aboard; insects, vertebrates, fossils, and plants were specifically sought, but mollusks were collected among other invertebrates.

REFERENCES

§ Y–32 Thorade, H. 1909. "Uber die Kalifornische Meeresströmung." Inaugural-Dissertation. . . . Universität zu Göttingen. pp. 31, 3 folded color plates.

Tables and charts to show the water temperatures and oceanic currents by the month for the Pacific coast from Vancouver to San Blas. In the southern part of the Gulf, as at Pulmo Reef, the water temperatures are consistently high, as high as at Mazatlan. Farther up in the Gulf, the temperatures are lower for the year and especially during the winter, but achieve a high maximum, 29° C. at 25° N. in August. Opposite, on the outer coast, at Cedros, the temperature is 19°, and 26° at Cape San Lucas. In May, per the chart reproduced herewith, in the few miles represented by the stretch of coastline from just north of Cape San Lucas on the outside, to Pulmo Reef just within the Gulf, the isotherms run from 18 to 24°. No temperatures are reported above Guaymas.

§ Y–33 Townsend, C. H. 1891. "Report upon the pearl fishery of the Gulf of California." Bull. U. S. Fish Comm., Vol. 9 for 1889: 91-94. Illustrations.

At that time work by a large concessionaire had been going on for 15 years, from May to October. Even then the years were "spotty." Consideration of the pearl fisheries elsewhere, at Acapulco and from Magdalena Bay northward.

§ Y–34 .... 1916. "Voyage of the *Albatross* to the Gulf of California in 1911." Bull. Amer. Mus. Nat. Hist., Vol. 25: 399-476, 45 text figs., map.

Two months' cruise from San Francisco to Angel de la Guardia and return, with 8 scientists and collectors aboard. Chief attention was paid to plants and to vertebrates, especially to the fish, 185 shore species having been obtained with up to 20 men working the large beach seines. The collection of echinoderms was large, and other invertebrates were obtained. There were 27 dredge hauls from 284 to 1760 fathoms, and 30 shore stations and anchorages were occupied. The mean water temperature in the Gulf was 68°, the air 71° from March 26 to April 20. Narrative with collecting reports. Photographs of towns and anchorages. Detailed accounts of the commercial fisheries, for pearl oysters, spiny lobsters, turtles, oysters, abalones, and food fishes. Tables of temperatures, dredging stations, etc. Surface temperatures as low as 57° F. were recorded in the northern part of the Gulf.

§ Y–35 See also various popular articles on Lower California, especially on shell collecting, in Orcutt's "West American Scientist" a magazine published from 1913 to 1921 in San Diego. There are a number of additional general reports of scientific expeditions to the Galápagos (which has been a favorite site for zoological research for a hundred years), and several popular reports.

GLOSSARY

OF TERMS AS USED IN THIS WORK

ABORAL. The upper surface of a starfish, brittle-star, or sea-urchin, as opposed to the under or oral surface whereon the mouth is situated.

ALGAE. Simple plants, often unicellular; the higher forms include the seaweeds.

AMBULACRAL GROOVE. A furrow bisecting the underside of the rays of starfish through which the tube feet are protruded.

AMPHIPOD. Literally, "paired-legs." Minute shrimp-like crustaceans, laterally compressed; the beach hoppers, sand fleas, skeleton shrimps, etc.

ANASTOMOSING. Dictionary definition: "Union or intercommunication of any system or network of lines, branches, streams, or the like."

ASSOCIATION. An assemblage of animals having ecologically similar requirements.

ATOKOUS. The sexually immature stage of certain polychaet worms.

AUTONOMY. Reflex, or seemingly voluntary, separation of a part or a limb from the body, followed by regeneration.

BUNODID ANEMONE. One of a family of sea-anemones characterized by a bumpy or warty body wall.

CALCAREOUS. Containing deposits of calcium carbonate; calcification.

CERATA. Dorsal projections which take the place of gills.

COMMENSAL. An organism living in, with, or on another, generally partaking of the same food.

COSINE WAVE. A wave graphically represented by a curving line, the peaks and troughs of which are equal and complementary.

CTENOPHORE. A type of jellyfish characterized by the possession of meridional rows of vibrating plates which propel and orient the animal.

DACTYL. Term applied to the last joint of a crustacean leg.

DEHISCENCE. A bursting discharge, usually of eggs or sperm.

DROWNED CORAL FLAT. A flat containing coral, some heads of which have been suffocated by sand.

ECHIUROID. A worm-like animal related to the sipunculids, in which the body is variably sac-like, usually with thin skin, and having often a spoon-shaped proboscis.

ECOLOGY. The study of the mutual relations between an organism and its physical and sociological environment.

ELYTRA. Shield-like scales of certain worms.

ENDEMIC. Dictionary example: "An *endemic disease* is one which is constantly present to a greater or less degree in any place, as distinguished from an *epidemic disease*, which prevails widely at some time, or periodically. . . ."

EPITOKOUS. Sexually mature stage in polychaet worms, characterized by changes of the posterior end which enable normally crawling worms to be free-swimming.

ETIOLOGY. Dictionary definition: "1. The science, doctrine, or demonstration of causes, especially the investigation of the causes of any disease. 2. The assignment of a cause or reason; as, the *etiology* of a historical custom."

FLORIATE. Flower-like.

GASTROPOD. Literally, "stomach-foot." Belonging to a group of animals comprising the snails, slugs, sea-hares, etc.

GYMNOBLAST. Belonging to a group (of hydroids) in which the polyps lack the skeletal cups of other hydroids into which the soft parts can be withdrawn.

HOLOTHURIAN. Sea-cucumber. One of a group of echinoderms, or spiny-skinned animals, some varieties of which, under the commercial name *bêche-de-mer* or *trepang*, are used by the Chinese for food.

HYDROID. A small, plant-like, usually colonial animal.

INTERTIDAL. See *Littoral*.

INTROVERT. A closed tubular pocket capable of being unrolled and extended inside out.

ISOPOD. Literally, "same legs." Usually small crustaceans in which all the legs are similar, comprising the pill-bugs, sow-bugs, and many marine forms.

ISOTHERM. A line joining or marking equal temperatures.

LITTORAL. Region of the shore bounded by its highest normal submergence at high tide and most extreme emergence at low tide. Intertidal.

MUTATION. In the life history of a species, the sudden appearance of a new trait that breeds true and becomes eventually one of the characters of the species or of the new species thus formed.

MYSIDS. Usually minute crustacea, called "opossum shrimps" because of their possession of marsupial plates within which the young develop.

NUDIBRANCH. Literally, "naked gill." One of a group of shell-less gastropods, often brilliantly colored and of delicately beautiful form.

OPHIURAN. Brittle-star or serpent-star. Members of one of the five classes of echinoderms or spiny-skinned animals.

PAPILLA. Small elevation; in holothurians, modified tube feet not used for locomotion.

PELAGIC. Free-floating at or near the surface of the sea.

PLANKTON. Generally microscopic plant and animal life floating or weakly swimming in the upper layer of a body of water.

POLYCHAETS. Usually elongate worms characterized by the possession of abundant chaetae or bristles.

POLYCLADS. Flatworms in which the intestinal tract has extensive ramifications.

POLYP. An invertebrate having a hollow cylindrical body, closed and attached at one end and opening at the other by a central mouth surrounded by tentacles. May be an individual (as an anemone) or a member of a colony (as a coral polyp).

PORCELLANIDS. Crabs of the family Porcellanidae, often called porcelain crabs because of the carapace texture of typical examples.

QUATERNARY, OR RECENT. The latest of the epochs into which geologists divide the history of the earth. Late Quaternary includes the present time.

RESPIRATORY TREE. The respiratory organ of holothurians; so named because it resembles a tree inside out. Fresh water is taken in at what corresponds to the trunk and penetrates to the delicate branches, which provide great absorption area in proportion to the volume.

SCALAR. Mathematical term. An abstract quantity having magnitude but not direction, such as volume, mass, weight, time, electrical charge, and always indicated by a real number.

SERPULID. A polychaet worm which builds a calcareous tube, usually coiled.

SESSILE. Attached, therefore not moving.

SIPHONOPHORE. A type of jellyfish. The Portuguese man-o'-war and other spectacular forms belong to this group.

SIPUNCULIDS. Worm-like animals characterized (among other things) by the possession of an introvert, and of rough, cuticle-like skin. Capable of great expansion; contracted, some of them merit the name peanut worm.

SYNDROME. A group of signs and symptoms occurring together and characterizing a disease.

SYNONYMY. The various names used to designate a given species or group.

TAXONOMY. A sub-science of biology concerned with the classification of animals according to natural relationships and with the rules governing the system of nomenclature.

TECTIBRANCHS. A group of sometimes shell-less gastropods to which belong the sea-hares and bubble-shells.

TELEOLOGY. The assumption of predetermined design, purpose, or ends in Nature by which an explanation of phenomena is postulated.

TENSOR. A mathematical term for the stretching factor which is necessary to change one vector, or force, into another vector having a different amount of force and direction. (Thus, if one imagines a given force *A* traveling south at 40 miles an hour, and another force *B* traveling southeast at 60 miles an hour, mathematically to translate force *A* into force *B*, the factor which changes one into the other must have not only force and direction, but stretching power, to pull *A* equal to *B*, and that factor is called the *tensor*.) Tensor is the quantity necessary in Einsteinian physics to translate vectors from one set of co-ordinates (frame of reference) to another.

TEREBELLID WORM. A polychaet worm which builds a sandy or pebbly tube, cemented usually to the underside of rocks by its own mucus.

THIGMOTROPISM. An innate tendency to seek enclosing contact with a solid or rigid surface, as in a burrow.

TROPISM. Innate involuntary movement of an organism or any of its parts toward (positive) or away from (negative) a stimulus.

TURBELLARIAN WORMS. The large group of flatworms to which the polyclads belong.

UBIQUITOUS. Occurring everywhere (though not necessarily abundantly) in the total area under consideration.

VECTOR. A mathematical term for an abstract quantity such as velocity, acceleration, or force, having *both* magnitude and direction. It may also have position in space, but this is not necessary. A vector is symbolized or represented by an arrow.

XEROPHYTIC. Plants structurally adapted to withstand drought.

ZOOID. Individual member of a colony or compound organism, having more or less independent life of its own.

INDEX